T0180995

COMPUTATIONAL
PHYSICS
Fortran Version

COMPUTATIONAL PHYSICS
Fortran Version

Steven E. Koonin
Professor of Theoretical Physics
California Institute of Technology

Dawn C. Meredith
Assistant Professor of Physics
University of New Hampshire

Advanced Book Program

CRC Press
Taylor & Francis Group
Boca Raton London New York

CRC Press is an imprint of the
Taylor & Francis Group, an **informa** business

First published 1990 by Westview Press

Published 2018 by CRC Press
Taylor & Francis Group
6000 Broken Sound Parkway NW, Suite 300
Boca Raton, FL 33487-2742

ISBN 13: 978-0-201-38623-3 (pbk)
ISBN 13: 978-0-201-12779-9 (hbk)

Visit the Taylor & Francis Web site at
http://www.taylorandfrancis.com

and the CRC Press Web site at
http://www.crcpress.com

This book was typeset by Elizabeth K. Wood using the TEX text-processing system running on a Digital Equipment Corporation VAXStation 2000 computer.

Following is a list of trademarks used in the book:
VAX, VAXstation, VMS, VT are trademarks of Digital Equipment Corporation.
also UIS, HCUIS***
CA-DISSPLA is a trademark of Computer Associates International, Inc.
UNIX is a trademark of AT&T.
IBM-PC is a trademark of International Business Machines.
Macintosh is a trademark of Apple Computer, Inc.
Tektronix 4010 is a trademark of Tektronix Inc.
GO-235 is a trademark of GraphOn Corporation.
HDS3200 is a trademark of Human Designed Systems.
TeX is a trademark of the American Mathematical Society.
PST is a trademark of Prime Computers, Inc.
Kermit is a trademark of Walt Disney Productions.
SUN is a registered trademark of SUN Microsystems,Inc.

Library of Congress Cataloging-in-Publication Data

Koonin, Steven E.
  Computational physics (FORTRAN version) / Steven E. Koonin, Dawn Meredith.
    p. cm.
  Includes index.
  1. Mathematical physics—Data processing. 2. Physics—Computer programs.
3. FORTRAN (computer program language) 4. Numerical analysis. I. Meredith, Dawn.
II. Title.
QC20.7.E4K66 1990   530. 1′5′02855133—dc19   90-129

Preface

Computation is an integral part of modern science and the ability to exploit effectively the power offered by computers is therefore essential to a working physicist. The proper application of a computer to modeling physical systems is far more than blind "number crunching," and the successful computational physicist draws on a balanced mix of analytically soluble examples, physical intuition, and numerical work to solve problems that are otherwise intractable.

Unfortunately, the ability "to compute" is seldom cultivated by the standard university-level physics curriculum, as it requires an integration of three disciplines (physics, numerical analysis, and computer programming) covered in disjoint courses. Few physics students finish their undergraduate education knowing how to compute; those that do usually learn a limited set of techniques in the course of independent work, such as a research project or a senior thesis.

The material in this book is aimed at refining computational skills in advanced undergraduate or beginning graduate students by providing direct experience in using a computer to model physical systems. Its scope includes the minimum set of numerical techniques needed to "do physics" on a computer. Each of these is developed in the text, often heuristically, and is then applied to solve non-trivial problems in classical, quantum, and statistical physics. These latter have been chosen to enrich or extend the standard undergraduate physics curriculum, and so have considerable intrinsic interest, quite independent of the computational principles they illustrate.

This book should not be thought of as setting out a rigid or definitive curriculum. I have restricted its scope to calculations that satisfy simultaneously the criteria of illustrating a widely applicable numerical technique, of being tractable on a microcomputer, and of having some particular physics interest. Several important numerical techniques have therefore been omitted, spline interpolation and the Fast Fourier Transform among them. *Computational Physics* is perhaps best thought of as establishing an environment offering opportunities for further exploration. There are many possible extensions and embellishments of the material presented; using one's imagination along these lines is one of the more rewarding parts of working through the book.

Computational Physics is primarily a physics text. For maximum benefit, the student should have taken, or be taking, undergraduate courses in classical mechanics, quantum mechanics, statistical mechanics, and advanced calculus or the mathematical methods of physics. This is *not* a text on numerical analysis, as there has been no attempt at rigor or completeness in any of the expositions of numerical techniques. However, a prior course in that subject is probably not essential; the discussions of numerical techniques should be accessible to a student with the physics background outlined above, perhaps with some reference to any one of the excellent texts on numerical analysis (for example, [Ac70], [Bu81], or [Sh84]). This is also *not* a text on computer programming. Although I have tried to follow the principles of good programming throughout (see Appendix B), there has been no attempt to teach programming *per se*. Indeed, techniques for organizing and writing code are somewhat peripheral to the main goals of the book. Some familiarity with programming, at least to the extent of a one-semester introductory course in any of the standard high-level languages (BASIC, FORTRAN, PASCAL, C), is therefore essential.

The choice of language invariably invokes strong feelings among scientists who use computers. Any language is, after all, only a means of expressing the concepts underlying a program. The contents of this book are therefore relevant no matter what language one works in. However, *some* language had to be chosen to implement the programs, and I have selected the Microsoft dialect of BASIC standard on the IBM PC/XT/AT computers for this purpose. The BASIC language has many well-known deficiencies, foremost among them being a lack of local subroutine variables and an awkwardness in expressing structured code. Nevertheless, I believe that these are more than balanced by the simplicity of the language and the widespread fluency in it, BASIC's almost universal availability on the microcomputers most likely to be used with this book, the existence of both BASIC interpreters convenient for writing and debugging programs and of compilers for producing rapidly executing finished programs, and the powerful graphics and I/O statements in this language. I expect that readers familiar with some other high-level language can learn enough BASIC "on the fly" to be able to use this book. A synopsis of the language is contained in Appendix A to help in this regard, and further information can be found in readily available manuals. The reader may, of course, elect to write the programs suggested in the text in any convenient language.

This book arose out of the Advanced Computational Physics Laboratory taught to third- and fourth-year undergraduate Physics majors at Caltech during the Winter and Spring of 1984. The content and presentation have benefitted greatly from the many inspired suggestions of M.-C. Chu, V. Pönisch, R. Williams, and D. Meredith. Mrs. Meredith was also of great assistance in producing the final form of the manuscript and programs. I also wish to thank my wife, Laurie, for her extraordinary patience, understanding, and support during my two-year involvement in this project.

Steven E. Koonin
Pasadena
May, 1985

Preface to the FORTRAN Edition

At the request of the readers of the BASIC edition of *Computational Physics* we offer this FORTRAN version. Although we stand by our original choice of BASIC for the reasons cited in the preface to the BASIC edition it is clear that many of our readers strongly prefer FORTRAN, and so we gladly oblige. The text of the book is essentially unchanged, but all of the codes have been translated into standard FORTRAN-77 and will run (with some modification) on a variety of machines. Although the programs will run significantly faster on mainframe computers than on PC's, we have not increased the scope or complexity of the calculations. Results, therefore, are still produced in "real time", and the codes remain suitable for interactive use.

Another development since the BASIC edition is the publication of an excellent new text on numerical analysis (*Numerical Recipes* [Pr86]) which provides detailed discussions and code for state-of-the-art algorithms. We highly recommend it as a companion to this text.

The FORTRAN versions of the code were written with the help of T. Berke (who designed the menu), G. Buzzell, and J. Farley. We have also profited from the suggestions of many colleagues who have generously tested the new codes or pointed out errors in the BASIC edition. We gratefully acknowledge a grant from the University of New Hampshire DISCovery Computer Aided Instruction Program which provided the initial impetus for the translation. Lastly, our thanks go to E. Wood for typesetting the text in TEX.

Steven E. Koonin
Pasadena, CA
August, 1989

Dawn C. Meredith
Durham, NH

How to Use This Book

This book is organized into chapters, each containing a text section, an example, and a project. Each text section is a brief discussion of one or several related numerical techniques, often illustrated with simple mathematical examples. Throughout the text are a number of exercises, in which the student's understanding of the material is solidified or extended by an analytical derivation or through the writing and running of a simple program. These exercises are indicated by the symbol ■.

Also located throughout the text are tables of numerical errors. Note that the values listed in these tables were taken from runs using BASIC code on an IBM-PC. When the machine precision dominates the error, your values obtained with FORTRAN code may differ.

The example and project in each chapter are applications of the numerical techniques to particular physical problems. Each includes a brief exposition of the physics, followed by a discussion of how the numerical techniques are to be applied. The examples and projects differ only in that the student is expected to use (and perhaps modify) the program that is given for the former in Appendix B, while the book provides guidance in writing programs to treat the latter through a series of steps, also indicated by the symbol ■. However, programs for the projects have also been included in Appendix C; these can serve as models for the student's own program or as a means of investigating the physics without having to write a major program "from scratch". A number of suggested studies accompany each example and project; these guide the student in exploiting the programs and understanding the physical principles and numerical techniques involved.

The programs for both the examples and projects are available either over the Internet network or on diskettes. Appendix E describes how to obtain files over the network; alternatively, there is an order form in the back of the book for IBM-PC or Macintosh formatted diskettes. The codes are suitable for running on any computer that has a FORTRAN-77 standard compiler. (The IBM BASIC versions of the code are also available over the network.) Detailed instructions for revising and running the codes are given in Appendix A.

A "laboratory" format has proved to be one effective mode of presenting this material in a university setting. Students are quite able to work through the text on their own, with the instructor being available for consultation and to monitor progress through brief personal interviews on each chapter. Three chapters in ten weeks (60 hours) of instruction has proved to be a reasonable pace, with students typically writing two of the projects during this time, and using the "canned" codes to work through the physics of the remaining project and the examples. The eight chapters in this book should therefore be more than sufficient for a one-semester course. Alternatively, this book can be used to provide supplementary material for the usual courses in classical, quantum, and statistical mechanics. Many of the examples and projects are vivid illustrations of basic concepts in these subjects and are therefore suitable for classroom demonstrations or independent study.

Contents

The problem with computers is that they only give answers
—attributed to P. Picasso

Chapter 1

Basic
Mathematical
Operations

Three numerical operations — differentiation, quadrature, and the finding of roots — are central to most computer modeling of physical systems. Suppose that we have the ability to calculate the value of a function, $f(x)$, at any value of the independent variable x. In differentiation, we seek one of the derivatives of f at a given value of x. Quadrature, roughly the inverse of differentiation, requires us to calculate the definite integral of f between two specified limits (we reserve the term "integration" for the process of solving ordinary differential equations, as discussed in Chapter 2), while in root finding we seek the values of x (there may be several) at which f vanishes.

If f is known analytically, it is almost always possible, with enough fortitude, to derive explicit formulas for the derivatives of f, and it is often possible to do so for its definite integral as well. However, it is often the case that an analytical method cannot be used, even though we can evaluate $f(x)$ itself. This might be either because some very complicated numerical procedure is required to evaluate f and we have no suitable analytical formula upon which to apply the rules of differentiation and quadrature, or, even worse, because the way we can generate f provides us with its values at only a set of discrete abscissae. In these situations, we must employ approximate formulas expressing the derivatives and integral in terms of the values of f we can compute. Moreover, the roots of all but the simplest functions cannot be found analytically, and numerical methods are therefore essential.

This chapter deals with the computer realization of these three basic operations. The central technique is to approximate f by a simple function (such as first- or second-degree polynomial) upon which these

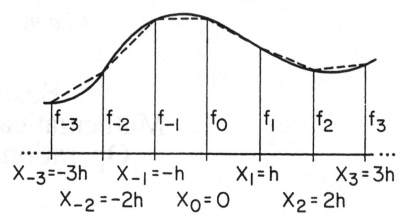

Figure 1.1   Values of f on an equally-spaced lattice. Dashed lines show the linear interpolation.

operations can be performed easily. We will derive only the simplest and most commonly used formulas; fuller treatments can be found in many textbooks on numerical analysis.

1.1 Numerical differentiation

Let us suppose that we are interested in the derivative at $x = 0$, $f'(0)$. (The formulas we will derive can be generalized simply to arbitrary x by translation.) Let us also suppose that we know f on an equally-spaced lattice of x values,

$$f_n = f(x_n) \; ; \; x_n = nh \; (n = 0, \pm 1, \pm 2, \ldots) \,,$$

and that our goal is to compute an approximate value of $f'(0)$ in terms of the f_n (see Figure 1.1).

We begin by using a Taylor series to expand f in the neighborhood of $x = 0$:

$$f(x) = f_0 + x f' + \frac{x^2}{2!} f'' + \frac{x^3}{3!} f''' + \ldots \,, \tag{1.1}$$

where all derivatives are evaluated at $x = 0$. It is then simple to verify that

$$f_{\pm 1} \equiv f(x = \pm h) = f_0 \pm h f' + \frac{h^2}{2} f'' \pm \frac{h^3}{6} f''' + \mathcal{O}(h^4) \,, \tag{1.2a}$$

$$f_{\pm 2} \equiv f(x = \pm 2h) = f_0 \pm 2h f' + 2h^2 f'' \pm \frac{4h^3}{3} f''' + \mathcal{O}(h^4) \,, \tag{1.2b}$$

where $\mathcal{O}(h^4)$ means terms of order h^4 or higher. To estimate the size of such terms, we can assume that f and its derivatives are all of the same order of magnitude, as is the case for many functions of physical relevance.

Upon subtracting f_{-1} from f_1 as given by (1.2a), we find, after a slight rearrangement,

$$f' = \frac{f_1 - f_{-1}}{2h} - \frac{h^2}{6} f''' + \mathcal{O}(h^4) . \tag{1.3a}$$

The term involving f''' vanishes as h becomes small and is the dominant error associated with the finite difference approximation that retains only the first term:

$$f' \approx \frac{f_1 - f_{-1}}{2h} . \tag{1.3b}$$

This "3-point" formula would be exact if f were a second-degree polynomial in the 3-point interval $[-h, +h]$, because the third- and all higher-order derivatives would then vanish. Hence, the essence of Eq. (1.3b) is the assumption that a quadratic polynomial interpolation of f through the three points $x = \pm h, 0$ is valid.

Equation (1.3b) is a very natural result, reminiscent of the formulas used to define the derivative in elementary calculus. The error term (of order h^2) can, in principle, be made as small as is desired by using smaller and smaller values of h. Note also that the symmetric difference about $x = 0$ is used, as it is more accurate (by one order in h) than the forward or backward difference formulas:

$$f' \approx \frac{f_1 - f_0}{h} + \mathcal{O}(h) ; \tag{1.4a}$$

$$f' \approx \frac{f_0 - f_{-1}}{h} + \mathcal{O}(h) . \tag{1.4b}$$

These "2-point" formulas are based on the assumption that f is well approximated by a linear function over the intervals between $x = 0$ and $x = \pm h$.

As a concrete example, consider evaluating $f'(x = 1)$ when $f(x) = \sin x$. The exact answer is, of course, $\cos 1 = 0.540302$. The following FORTRAN program evaluates Eq. (1.3b) in this case for the value of h input:

```
C chap1a.for
      X=1.
      EXACT=COS(X)
```

```
10      PRINT *, 'ENTER VALUE OF H (.LE. 0 TO STOP)'
        READ *, H
        IF (H .LE. 0) STOP
        FPRIME=(SIN(X+H)-SIN(X-H))/(2*H)
        DIFF=EXACT-FPRIME
        PRINT 20,H,DIFF
20      FORMAT (' H=',E15.8,5X,'ERROR=',E15.8)
        GOTO 10
        END
```

(If you are a beginner in FORTRAN, note the way the value of H is requested from the keyboard, the fact that the code will stop if a non-positive value of H is entered, the natural way in which variable names are chosen and the mathematical formula (1.3b) is transcribed using the SIN function in the sixth line, the way in which the number of significant digits is specified when the result is to be output to the screen in line 20, and the jump in program control at the end of the program.)

Results generated with this program, as well as with similar ones evaluating the forward and backward difference formulas Eqs. (1.4a,b), are shown in Table 1.1. (All of the tables of errors presented in the text were generated from BASIC programs; the numbers may vary from those obtained from FORTRAN code, especially when numerical roundoff dominates.) Note that the result improves as we decrease h, but only up to a point, after which it becomes worse. This is because arithmetic in the computer is performed with only a limited precision (5–6 decimal digits for a single precision BASIC variable), so that when the difference in the numerator of the approximations is formed, it is subject to large "round-off" errors if h is small and f_1 and f_{-1} differ very little. For example, if $h = 10^{-6}$, then

$$f_1 = \sin(1.000001) = 0.841472 \; ; \quad f_{-1} = \sin(0.999999) = 0.841470 \; ,$$

so that $f_1 - f_{-1} = 0.000002$ to six significant digits. When substituted into (1.3b) we find $f' \approx 1.000000$, a very poor result. However, if we do the arithmetic with 10 significant digits, then

$$f_1 = 0.8414715251 \; ; \quad f_{-1} = 0.8414704445 \; ,$$

which gives a respectable $f' \approx 0.540300$ in Eq. (1.3b). In this sense, numerical differentiation is an intrinsically unstable process (no well-defined limit as $h \to 0$), and so must be carried out with caution.

Table 1.1   Error in evaluating $d\sin x/dx|_{x=1} = 0.540302$

h	Symmetric 3-point Eq. (1.3b)	Forward 2-point Eq. (1.4a)	Backward 2-point Eq. (1.4b)	Symmetric 5-point Eq. (1.5)
0.50000	0.022233	0.228254	−0.183789	0.001092
0.20000	0.003595	0.087461	−0.080272	0.000028
0.10000	0.000899	0.042938	−0.041139	0.000001
0.05000	0.000225	0.021258	−0.020808	0.000000
0.02000	0.000037	0.008453	−0.008380	0.000001
0.01000	0.000010	0.004224	−0.004204	0.000002
0.00500	0.000010	0.002108	−0.002088	0.000006
0.00200	−0.000014	0.000820	−0.000848	−0.000017
0.00100	−0.000014	0.000403	−0.000431	−0.000019
0.00050	0.000105	0.000403	−0.000193	0.000115
0.00020	−0.000163	−0.000014	−0.000312	−0.000188
0.00010	−0.000312	−0.000312	−0.000312	−0.000411
0.00005	0.000284	0.001476	−0.000908	0.000681
0.00002	0.000880	0.000880	0.000880	0.000873
0.00001	0.000880	0.003860	−0.002100	0.000880

It is possible to improve on the 3-point formula (1.3b) by relating f' to lattice points further removed from $x = 0$. For example, using Eqs. (1.2), it is easy to show that the "5-point" formula

$$f' \approx \frac{1}{12h}[f_{-2} - 8f_{-1} + 8f_1 - f_2] + \mathcal{O}(h^4) \qquad (1.5)$$

cancels all derivatives in the Taylor series through fourth order. Computing the derivative in this way assumes that f is well-approximated by a fourth-degree polynomial over the 5-point interval $[-2h, 2h]$. Although requiring more computation, this approximation is considerably more accurate, as can be seen from Table 1.1. In fact, an accuracy comparable to Eq. (1.3b) is obtained with a step some 10 times larger. This can be an important consideration when many values of f must be stored in the computer, as the greater accuracy allows a sparser tabulation and so saves storage space. However, because (1.5) requires more mathematical operations than does (1.3b) and there is considerable cancellation among the various terms (they have both positive and negative coefficients), precision problems show up at a larger value of h.

Formulas for higher derivatives can be constructed by taking appro-

Table 1.2   4- and 5-point difference formulas for derivatives

	4-point	5-point
hf'	$\pm\frac{1}{6}(-2f_{\mp1} - 3f_0 + 6f_{\pm1} - f_{\pm2})$	$\frac{1}{12}(f_{-2} - 8f_{-1} + 8f_1 - f_2)$
h^2f''	$f_{-1} - 2f_0 + f_1$	$\frac{1}{12}(-f_{-2} + 16f_{-1} - 30f_0 + 16f_1 - f_2)$
h^3f'''	$\pm(-f_{\mp1} + 3f_0 - 3f_{\pm1} + f_{\pm2})$	$\frac{1}{2}(-f_{-2} + 2f_{-1} - 2f_1 + f_2)$
$h^4f^{(iv)}$	$\cdots$	$f_{-2} - 4f_{-1} + 6f_0 - 4f_1 + f_2$

priate combinations of Eqs. (1.2). For example, it is easy to see that

$$f_1 - 2f_0 + f_{-1} = h^2 f'' + \mathcal{O}(h^4) , \qquad (1.6)$$

so that an approximation to the second derivative accurate to order h^2 is

$$f'' \approx \frac{f_1 - 2f_0 + f_{-1}}{h^2} . \qquad (1.7)$$

Difference formulas for the various derivatives of f that are accurate to a higher order in h can be derived straightforwardly. Table 1.2 is a summary of the 4- and 5-point expressions.

■ **Exercise 1.1** Using any function for which you can evaluate the derivatives analytically, investigate the accuracy of the formulas in Table 1.2 for various values of h.

1.2 Numerical quadrature

In quadrature, we are interested in calculating the definite integral of f between two limits, $a < b$. We can easily arrange for these values to be points of the lattice separated by an even number of lattice spacings; i.e.,

$$N = \frac{(b - a)}{h}$$

is an even integer. It is then sufficient for us to derive a formula for the integral from $-h$ to $+h$, since this formula can be composed many times:

$$\int_a^b f(x)\,dx = \int_a^{a+2h} f(x)\,dx + \int_{a+2h}^{a+4h} f(x)\,dx + \ldots + \int_{b-2h}^b f(x)\,dx .$$

$$(1.8)$$

The basic idea behind all of the quadrature formulas we will discuss (technically of the closed Newton-Cotes type) is to approximate f between $-h$ and $+h$ by a function that can be integrated exactly. For example, the simplest approximation can be had by considering the intervals $[-h, 0]$ and $[0, h]$ separately, and assuming that f is linear in each of these intervals (see Figure 1.1). The error made by this interpolation is of order $h^2 f''$, so that the approximate integral is

$$\int_{-h}^{h} f(x)\,dx = \frac{h}{2}(f_{-1} + 2f_0 + f_1) + \mathcal{O}(h^3)\,, \qquad (1.9)$$

which is the well-known trapezoidal rule.

A better approximation can be had by realizing that the Taylor series (1.1) can provide an improved interpolation of f. Using the difference formulas (1.3b) and (1.7) for f' and f'', respectively, for $|x| < h$ we can put

$$f(x) = f_0 + \frac{f_1 - f_{-1}}{2h}x + \frac{f_1 - 2f_0 + f_{-1}}{2h^2}x^2 + \mathcal{O}(x^3)\,, \qquad (1.10)$$

which can be integrated readily to give

$$\int_{-h}^{h} f(x)\,dx = \frac{h}{3}(f_1 + 4f_0 + f_{-1}) + \mathcal{O}(h^5)\,. \qquad (1.11)$$

This is Simpson's rule, which can be seen to be accurate to two orders higher than the trapezoidal rule (1.9). Note that the error is actually better than would be expected naively from (1.10) since the x^3 term gives no contribution to the integral. Composing this formula according to Eq. (1.8) gives

$$\int_{a}^{b} f(x)\,dx = \frac{h}{3}[f(a) + 4f(a + h) + 2f(a + 2h) + 4f(a + 3h) \qquad (1.12)$$
$$\dots + 4f(b - h) + f(b)]\,.$$

As an example, the following FORTRAN program calculates

$$\int_{0}^{1} e^x\,dx = e - 1 = 1.718282$$

using Simpson's rule for the value of $N = 1/h$ input. [Source code for the longer programs like this that are embedded in the text are contained on the *Computational Physics* diskette or available over Internet (see Appendix E); the shorter codes can be easily entered into the reader's computer from the keyboard.]

```
C chap1b.for
          FUNC(X)=EXP(X)                    !function to integrate
          EXACT=EXP(1.)-1.
30        PRINT *,'ENTER N EVEN (.LT. 2 TO STOP)'
          READ *, N
          IF (N .LT. 2) STOP
          IF (MOD(N,2) .NE. 0) N=N+1
          H=1./N
          SUM=FUNC(0.)                      !contribution from X=0
          FAC=2                             !factor for Simpson's rule
          DO 10 I=1,N-1                     !loop over lattice points
             IF (FAC .EQ. 2.) THEN         !factors alternate
                FAC=4
             ELSE
                FAC=2.
             END IF
             X=I*H                          !X at this point
             SUM=SUM+FAC*FUNC(X)            !contribution to the integral
10        CONTINUE
          SUM=SUM+FUNC(1.)                  !contribution from X=1
          XINT=SUM*H/3.
          DIFF=EXACT-XINT
          PRINT 20,N,DIFF
20        FORMAT (5X,'N=',I5,5X,'ERROR=',E15.8)
          GOTO 30                           !get another value of N
          END
```

Results are shown in Table 1.3 for various values of N, together with the values obtained using the trapezoidal rule. The improvement from the higher-order formula is evident. Note that the results are stable in the sense that a well-defined limit is obtained as N becomes very large and the mesh spacing h becomes small; round-off errors are unimportant because all values of f enter into the quadrature formula with the same sign, in contrast to what happens in numerical differentiation.

An important issue in quadrature is how small an h is necessary to compute the integral to a given accuracy. Although it is possible to derive rigorous error bounds for the formulas we have discussed, the simplest thing to do in practice is to run the computation again with a smaller h and observe the changes in the results.

Table 1.3   Errors in evaluating $\int_0^1 e^x \, dx = 1.718282$

N	h	Trapezoidal Eq. (1.9)	Simpson's Eq. (1.12)	Bode's Eq. (1.13b)
4	0.2500000	−0.008940	−0.000037	−0.000001
8	0.1250000	−0.002237	0.000002	0.000000
16	0.0625000	−0.000559	0.000000	0.000000
32	0.0312500	−0.000140	0.000000	0.000000
64	0.0156250	−0.000035	0.000000	0.000000
128	0.0078125	−0.000008	0.000000	0.000000

Higher-order quadrature formulas can be derived by retaining more terms in the Taylor expansion (1.10) used to interpolate f between the mesh points and, of course, using commensurately better finite-difference approximations for the derivatives. The generalizations of Simpson's rule using cubic and quartic polynomials to interpolate (Simpson's $\frac{3}{8}$ and Bode's rule, respectively) are:

$$\int_{x_0}^{x_3} f(x)\, dx = \frac{3h}{8}[f_0 + 3f_1 + 3f_2 + f_3] + \mathcal{O}(h^5) ; \qquad (1.13a)$$

$$\int_{x_0}^{x_4} f(x)\, dx = \frac{2h}{45}[7f_0 + 32f_1 + 12f_2 + 32f_3 + 7f_4] + \mathcal{O}(h^7) . \qquad (1.13b)$$

The results of applying Bode's rule are also given in Table 1.3, where the improvement is evident, although at the expense of a more involved computation. (Note that for this method to be applicable, N must be a multiple of 4.) Although one might think that formulas based on interpolation using polynomials of a very high degree would be even more suitable, this is not the case; such polynomials tend to oscillate violently and lead to an inaccurate interpolation. Moreover, the coefficients of the values of f at the various lattice points can have both positive and negative signs in higher-order formulas, making round-off error a potential problem. It is therefore usually safer to improve accuracy by using a low-order method and making h smaller rather than by resorting to a higher-order formula. Quadrature formulas accurate to a very high order can be derived if we give up the requirement of equally-spaced abscissae; these are discussed in Chapter 4.

■ **Exercise 1.2** Using any function whose definite integral you can compute analytically, investigate the accuracy of the various quadrature methods discussed above for different values of h.

Some care and common sense must be exercised in the application of the numerical quadrature formulas discussed above. For example, an integral in which the upper limit is very large is best handled by a change in variable. Thus, the Simpson's rule evaluation of

$$\int_1^b dx \, x^{-2} \, g(x)$$

with $g(x)$ constant at large x, would result in a (finite) sum converging very slowly as b becomes large for fixed h (and taking a very long time to compute!). However, changing variables to $t = x^{-1}$ gives

$$\int_{b^{-1}}^1 g(t^{-1}) \, dt \,,$$

which can then be evaluated by any of the formulas we have discussed.

Integrable singularities, which cause the naive formulas to give nonsense, can also be handled in a simple way. For example,

$$\int_0^1 dx (1 - x^2)^{-1/2} \, g(x)$$

has an integrable singularity at $x = 1$ (if g is regular there) and is a finite number. However, since $f(x = 1) = \infty$, the quadrature formulas discussed above give an infinite result. An accurate result can be obtained by changing variables to $t = (1 - x)^{1/2}$ to obtain

$$2 \int_0^1 dt \, (2 - t^2)^{-1/2} g(1 - t^2) \,,$$

which is then approximated with no trouble.

Integrable singularities can also be handled by deriving quadrature formulas especially adapted to them. Suppose we are interested in

$$\int_0^1 f(x) \, dx = \int_0^h f(x) \, dx + \int_h^1 f(x) \, dx \,,$$

where $f(x)$ behaves as $Cx^{-1/2}$ near $x = 0$, with C a constant. The integral from h to 1 is regular and can be handled easily, while the integral from 0 to h can be approximated as $2Ch^{1/2} = 2hf(h)$.

■ **Exercise 1.3**   Write a program to calculate

$$\int_0^1 t^{-2/3} (1 - t)^{-1/3} \, dt = 2\pi/3^{1/2}$$

using one of the quadrature formulas discussed above and investigate its accuracy for various values of h. (Hint: Split the range of integration into two parts and make a different change of variable in each integral to handle the singularities.)

1.3 Finding roots

The final elementary operation that is commonly required is to find a root of a function $f(x)$ that we can compute for arbitrary x. One sure-fire method, when the approximate location of a root (say at $x = x_0$) is known, is to guess a trial value of x guaranteed to be less than the root, and then to increase this trial value by small positive steps, backing up and halving the step size every time f changes sign. The values of x generated by this procedure evidently converge to x_0, so that the search can be terminated whenever the step size falls below the required tolerance. Thus, the following FORTRAN program finds the positive root of the function $f(x) = x^2 - 5$, $x_0 = 5^{1/2} = 2.236068$, to a tolerance of 10^{-6} using $x = 1$ as an initial guess and an initial step size of 0.5:

```
C chap1c.for
        FUNC(X)=X*X-5.            !function whose root is sought
        TOLX=1.E-06              !tolerance for the search
        X=1.                    !initial guess
        FOLD=FUNC(X)            !initial function
        DX=.5                   !initial step
        ITER=0                  !initialize count
10      CONTINUE
        ITER=ITER+1                !increment iteration count
        X=X+DX                     !step X
        PRINT *,ITER,X,SQRT(5.)-X  !output current values
        IF ((FOLD*FUNC(X)) .LT. 0) THEN
          X=X-DX                   !if sign change, back up
          DX=DX/2                  ! and halve the step
        END IF
        IF (ABS(DX) .GT. TOLX) GOTO 10
        STOP
        END
```

Results for the sequence of x values are shown in Table 1.4, evidently converging to the correct answer, although only after some 33 iterations. One must be careful when using this method, since if the initial step size

Table 1.4   Error in finding the positive root of $f(x) = x^2 - 5$

Iteration	Search	Newton Eq. (1.14)	Secant Eq. (1.15)
0	1.236076	1.236076	1.236076
1	0.736068	−0.763932	−1.430599
2	0.236068	−0.097265	0.378925
3	−0.263932	−0.002027	0.098137
4	−0.013932	−0.000001	−0.009308
5	0.111068	0.000000	0.000008
6	−0.013932	0.000000	0.000000
⋮	⋮	⋮	⋮
33	0.000001	0.000000	0.000000

is too large, it is possible to step over the root desired when f has several roots.

■ **Exercise 1.4** Run the code above for various tolerances, initial guesses, and initial step sizes. Note that sometimes you might find convergence to the negative root. What happens if you start with an initial guess of −3 with a step size of 6?

A more efficient algorithm, Newton-Raphson, is available if we can evaluate the derivative of f for arbitrary x. This method generates a sequence of values, x^i, converging to x_0 under the assumption that f is locally linear near x_0 (see Figure 1.2). That is,

$$x^{i+1} = x^i - \frac{f(x^i)}{f'(x^i)} . \tag{1.14}$$

The application of this method to finding $5^{1/2}$ is also shown in Table 1.4, where the rapid convergence (5 iterations) is evident. This is the algorithm usually used in the computer evaluation of square roots; a linearization of (1.14) about x_0 shows that the number of significant digits doubles with each iteration, as is evident in Table 1.4.

The secant method provides a happy compromise between the efficiency of Newton-Raphson and the bother of having to evaluate the derivative. If the derivative in Eq. (1.14) is approximated by the differ-

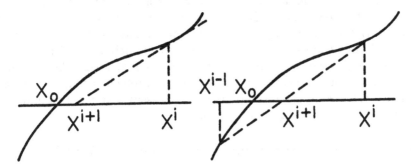

Figure 1.2 Geometrical bases of the Newton-Raphson (left) and secant (right) methods.

ence formula related to (1.4b),

$$f'(x^i) \approx \frac{f(x^i) - f(x^{i-1})}{x^i - x^{i-1}} \,,$$

we obtain the following 3-term recursion formula giving x^{i+1} in terms of x^i and x^{i-1} (see Figure 1.2):

$$x^{i+1} = x^i - f(x^i)\frac{(x^i - x^{i-1})}{f(x^i) - f(x^{i-1})} \,. \tag{1.15}$$

Any two approximate values of x_0 can be used for x^0 and x^1 to start the algorithm, which is terminated when the change in x from one iteration to the next is less than the required tolerance. The results of the secant method for our model problem, starting with values $x^0 = 0.5$ and $x^1 = 1.0$, are also shown in Table 1.4. Provided that the initial guesses are close to the true root, convergence to the exact answer is almost as rapid as that of the Newton-Raphson algorithm.

■ **Exercise 1.5** Write programs to solve for the positive root of $x^2 - 5$ using the Newton-Raphson and secant methods. Investigate the behavior of the latter with changes in the initial guesses for the root.

When the function is badly behaved near its root (e.g., there is an inflection point near x_0) or when there are several roots, the "automatic" Newton-Raphson and secant methods can fail to converge at all or converge to the wrong answer if the initial guess for the root is poor. Hence, a safe and conservative procedure is to use the search algorithm to locate x_0 approximately and then to use one of the automatic methods.

■ **Exercise 1.6**   The function $f(x) = \tanh x$ has a root at $x = 0$. Write a program to show that the Newton-Raphson method does not converge for an initial guess of $x \gtrsim 1$. Can you understand what's going wrong by considering a graph of $\tanh x$? From the explicit form of (1.14) for this problem, derive the critical value of the initial guess above which convergence will not occur. Try to solve the problem using the secant method. What happens for various initial guesses if you try to find the $x = 0$ root of $\tan x$ using either method?

1.4 Semiclassical quantization of molecular vibrations

As an example combining several basic mathematical operations, we consider the problem of describing a diatomic molecule such as O_2, which consists of two nuclei bound together by the electrons that orbit about them. Since the nuclei are much heavier than the electrons we can assume that the latter move fast enough to readjust instantaneously to the changing position of the nuclei (Born-Oppenheimer approximation). The problem is therefore reduced to one in which the motion of the two nuclei is governed by a potential, V, depending only upon r, the distance between them. The physical principles responsible for generating V will be discussed in detail in Project VIII, but on general grounds one can say that the potential is attractive at large distances (van der Waals interaction) and repulsive at short distances (Coulomb interaction of the nuclei and Pauli repulsion of the electrons). A commonly used form for V embodying these features is the Lennard-Jones or 6–12 potential,

$$V(r) = 4V_0 \left[\left(\frac{a}{r} \right)^{12} - \left(\frac{a}{r} \right)^{6} \right], \qquad (1.16)$$

which has the shape shown in the upper portion of Figure 1.3, the minimum occurring at $r_{\min} = 2^{1/6} a$ with a depth V_0. We will assume this form in most of the discussion below. A thorough treatment of the physics of diatomic molecules can be found in [He50] while the Born-Oppenheimer approximation is discussed in [Me68].

The great mass of the nuclei allows the problem to be simplified even further by decoupling the slow rotation of the nuclei from the more rapid changes in their separation. The former is well described by the quantum mechanical rotation of a rigid dumbbell, while the vibrational states of relative motion, with energies E_n, are described by the bound

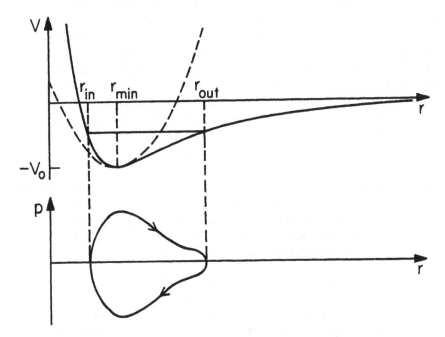

Figure 1.3 (Upper portion) The Lennard-Jones potential and the inner and outer turning points at a negative energy. The dashed line shows the parabolic approximation to the potential. (Lower portion) The corresponding trajectory in phase space.

state solutions, $\psi_n(r)$, of a one-dimensional Schroedinger equation,

$$\left[-\frac{\hbar^2}{2m}\frac{d^2}{dr^2} + V(r)\right]\psi_n = E_n\psi_n . \tag{1.17}$$

Here, m is the reduced mass of the two nuclei.

Our goal in this example is to find the energies E_n, given a particular potential. This can be done exactly by solving the differential eigenvalue equation (1.17); numerical methods for doing so will be discussed in Chapter 3. However, the great mass of the nuclei implies that their motion is nearly classical, so that approximate values of the vibrational energies E_n can be obtained by considering the classical motion of the nuclei in V and then applying "quantization rules" to determine the energies. These quantization rules, originally postulated by N. Bohr and Sommerfeld and Wilson, were the basis of the "old" quantum theory from which the modern formulation of quantum mechanics arose. However, they can also be obtained by considering the WKB approximation to the wave equation (1.17). (See [Me68] for details.)

Confined classical motion of the internuclear separation in the potential $V(r)$ can occur for energies $-V_0 < E < 0$. The distance between the nuclei oscillates periodically (but not necessarily harmonically) between inner and outer turning points, r_{in} and r_{out}, as shown in Figure 1.3. During these oscillations, energy is exchanged between the kinetic energy of relative motion and the potential energy such that the total energy,

$$E = \frac{p^2}{2m} + V(r) \,, \tag{1.18}$$

is a constant (p is the relative momentum of the nuclei). We can therefore think of the oscillations at any given energy as defining a closed trajectory in phase space (coordinates r and p) along which Eq. (1.18) is satisfied, as shown in the lower portion of Figure 1.3. An explicit equation for this trajectory can be obtained by solving (1.18) for p:

$$p(r) = \pm \left[2m\left(E - V(r)\right)\right]^{1/2} \,. \tag{1.19}$$

The classical motion described above occurs at *any* energy between $-V_0$ and 0. To quantize the motion, and hence obtain approximations to the eigenvalues E_n appearing in (1.17), we consider the dimensionless action at a given energy,

$$S(E) = \oint k(r)\, dr \,, \tag{1.20}$$

where $k(r) = \hbar^{-1} p(r)$ is the local de Broglie wave number and the integral is over one complete cycle of oscillation. This action is just the area (in units of $\hbar$) enclosed by the phase space trajectory. The quantization rules state that, at the allowed energies E_n, the action is a half-integral multiple of 2π. Thus, upon using (1.19) and recalling that the oscillation passes through each value of r twice (once with positive p and once with negative p), we have

$$S(E_n) = 2\left(\frac{2m}{\hbar^2}\right)^{1/2} \int_{r_{in}}^{r_{out}} \left[E_n - V(r)\right]^{1/2} dr = \left(n + \frac{1}{2}\right)2\pi \,, \tag{1.21}$$

where n is a nonnegative integer. At the limits of this integral, the turning points r_{in} and r_{out}, the integrand vanishes.

To specialize the quantization condition to the Lennard-Jones potential (1.16), we define the dimensionless quantities

$$\epsilon = \frac{E}{V_0} \ , \quad x = \frac{r}{a} \ , \quad \gamma = \left(\frac{2ma^2 V_0}{\hbar^2}\right)^{1/2} ,$$

so that (1.21) becomes

$$s(\epsilon_n) \equiv \frac{1}{2} S(\epsilon_n V_0) = \gamma \int_{x_{\text{in}}}^{x_{\text{out}}} \left[\epsilon_n - v(x)\right]^{1/2} dx = \left(n + \frac{1}{2}\right)\pi \ , \quad (1.22)$$

where

$$v(x) = 4\left(\frac{1}{x^{12}} - \frac{1}{x^6}\right)$$

is the scaled potential.

The quantity γ is a dimensionless measure of the quantum nature of the problem. In the classical limit ($\hbar$ small or m large), γ becomes large. By knowing the moment of inertia of the molecule (from the energies of its rotational motion) and the dissociation energy (energy required to separate the molecule into its two constituent atoms), it is possible to determine from observation the parameters a and V_0 and hence the quantity γ. For the H_2 molecule, $\gamma = 21.7$, while for the HD molecule, $\gamma = 24.8$ (only m, but not V_0, changes when one of the protons is replaced by a deuteron), and for the much heavier O_2 molecule made of two ^{16}O nuclei, $\gamma = 150$. These rather large values indicate that a semiclassical approximation is a valid description of the vibrational motion.

The FORTRAN program for Example 1, whose source code is contained in Appendix B and in the file EXMPL1.FOR, finds, for the value of γ input, the values of the ϵ_n for which Eq. (1.22) is satisfied. After all of the energies have been found, the corresponding phase space trajectories are drawn. (Before attempting to run this code on your computer system, you should review the material on the programs in "How to use this book" and in Appendix A.)

The following exercises are aimed at increasing your understanding of the physical principles and numerical methods demonstrated in this example.

■ **Exercise 1.7** One of the most important aspects of using a computer as a tool to do physics is knowing when to have confidence that the program is giving the correct answers. In this regard, an essential test is the detailed quantitative comparison of results with what is known in analytically soluble situations. Modify the code to use a parabolic potential (in subroutine POT, taking care to heed the instructions given there), for which the Bohr-Sommerfeld quantization gives the exact eigenvalues of the Schroedinger equation: a series of equally-spaced energies, with the lowest being one-half of the level spacing above the minimum of the potential. For several values of γ, compare the numerical results for this case with what you obtain by solving Eq. (1.22) analytically. Are the phase space trajectories what you expect?

■ **Exercise 1.8** Another important test of a working code is to compare its results with what is expected on the basis of physical intuition. Restore the code to use the Lennard-Jones potential and run it for $\gamma = 50$. Note that, as in the case of the purely parabolic potential discussed in the previous exercise, the first excited state is roughly three times as high above the bottom of the well as is the ground state and that the spacings between the few lowest states are roughly constant. This is because the Lennard-Jones potential is roughly parabolic about its minimum (see Figure 1.3). By calculating the second derivative of V at the minimum, find the "spring constant" and show that the frequency of small-amplitude motion is expected to be

$$\frac{\hbar\omega}{V_0} = \frac{6 \times 2^{5/6}}{\gamma} \approx \frac{10.691}{\gamma} . \tag{1.23}$$

Verify that this is consistent with the numerical results and explore this agreement for different values of γ. Can you understand why the higher energies are more densely spaced than the lower ones by comparing the Lennard-Jones potential with its parabolic approximation?

■ **Exercise 1.9** Invariance of results under changes in the numerical algorithms or their parameters can give additional confidence in a calculation. Change the tolerances for the turning point and energy searches or the number of Simpson's rule points (this can be done at run-time by choosing menu option 2) and observe the effects on the results. Note that because of the way in which the expected number of bound states is calculated (see the end of subroutine PARAM) this quantity can change if the energy tolerance is varied.

■ **Exercise 1.10** Replace the searches for the inner and outer turning points by the Newton-Raphson method or the secant method. (When ILEVEL$\neq$ 0, the turning points for ILEVEL − 1 are excellent starting values.) Replace the Simpson's rule quadrature for s by a higher-order formula [Eqs. (1.13a) or (1.13b)] and observe the improvement.

Table 1.5   Experimental vibrational energies of the H_2 molecule

n	E_n (eV)	n	E_n (eV)
0	−4.477	8	−1.151
1	−3.962	9	−0.867
2	−3.475	10	−0.615
3	−3.017	11	−0.400
4	−2.587	12	−0.225
5	−2.185	13	−0.094
6	−1.811	14	−0.017
7	−1.466		

■ **Exercise 1.11**   Plot the ϵ_n of the Lennard-Jones potential as functions of γ for γ running from 20 to 200 and interpret the results. (As with many other short calculations, you may find it more efficient simply to run the code and plot the results by hand as you go along, rather than trying to automate the plotting operation.)

■ **Exercise 1.12**   For the H_2 molecule, observations show that the depth of the potential is $V_0 = 4.747$ eV and the location of the potential minimum is $r_{min} = 0.74166$ Å. These two quantities, together with Eq. (1.23), imply a vibrational frequency of

$$\hbar\omega = 0.492V_0 = 2.339 \text{ eV} ,$$

more than four times larger than the experimentally observed energy difference between the ground and first vibrational state, 0.515 eV. The Lennard-Jones shape is therefore not a very good description of the potential of the H_2 molecule. Another defect is that it predicts 6 bound states, while 15 are known to exist. (See Table 1.5, whose entries are derived from the data quoted in [Wa67].) A better analytic form of the potential, with more parameters, is required to reproduce simultaneously the depth and location of the minimum, the frequency of small amplitude vibrations about it, and the total number of bound states. One such form is the Morse potential,

$$V(r) = V_0\left[(1 - e^{-\beta(r-r_{min})})^2 - 1\right] , \qquad (1.24)$$

which also can be solved analytically. The Morse potential has a minimum at the expected location and the parameter β can be adjusted to fit the curvature of the minimum to the observed excitation energy of the first

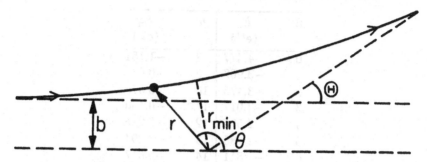

Figure I.1   Quantities involved in the scattering of a particle by a central potential.

vibrational state. Find the value of β appropriate for the H_2 molecule, modify the program above to use the Morse potential, and calculate the spectrum of vibrational states. Show that a much more reasonable number of levels is now obtained. Compare the energies with experiment and with those of the Lennard-Jones potential and interpret the latter differences.

Project I: Scattering by a central potential

In this project, we will investigate the classical scattering of a particle of mass m by a central potential, in particular the Lennard-Jones potential considered in Section 1.4 above. In a scattering event, the particle, with initial kinetic energy E and impact parameter b, approaches the potential from a large distance. It is deflected during its passage near the force center and eventually emerges with the same energy, but moving at an angle Θ with respect to its original direction. Since the potential depends upon only the distance of the particle from the force center, the angular momentum is conserved and the trajectory lies in a plane. The polar coordinates of the particle, (r, θ), are a convenient way to describe its motion, as shown in Figure I.1. (For details, see any textbook on classical mechanics, such as [Go80].)

Of basic interest is the deflection function, $\Theta(b)$, giving the final scattering angle, Θ, as a function of the impact parameter; this function also depends upon the incident energy. The differential cross section for scattering at an angle Θ, $d\sigma/d\Omega$, is an experimental observable that is related to the deflection function by

$$\frac{d\sigma}{d\Omega} = \frac{b}{\sin\Theta}\left|\frac{db}{d\Theta}\right|.\qquad(I.1)$$

Thus, if $d\Theta/db = (db/d\Theta)^{-1}$ can be computed, then the cross section is known.

Expressions for the deflection function can be found analytically for only a very few potentials, so that numerical methods usually must be employed. One way to solve the problem would be to integrate the equations of motion in time (i.e., Newton's law relating the acceleration to the force) to find the trajectories corresponding to various impact parameters and then to tabulate the final directions of the motion (scattering angles). This would involve integrating four coupled first-order differential equations for two coordinates and their velocities in the scattering plane, as discussed in Section 2.5 below. However, since angular momentum is conserved, the evolution of θ is related directly to the radial motion, and the problem can be reduced to a one-dimensional one, which can be solved by quadrature. This latter approach, which is simpler and more accurate, is the one we will pursue here.

To derive an appropriate expression for Θ, we begin with the conservation of angular momentum, which implies that

$$L = mvb = mr^2\frac{d\theta}{dt},\qquad(I.2)$$

is a constant of the motion. Here, $d\theta/dt$ is the angular velocity and v is the asymptotic velocity, related to the bombarding energy by $E = \frac{1}{2}mv^2$. The radial motion occurs in an effective potential that is the sum of V and the centrifugal potential, so that energy conservation implies

$$\frac{1}{2}m\left(\frac{dr}{dt}\right)^2 + \frac{L^2}{2mr^2} + V = E.\qquad(I.3)$$

If we use r as the independent variable in (I.2), rather than the time, we can write

$$\frac{d\theta}{dr} = \frac{d\theta}{dt}\left(\frac{dr}{dt}\right)^{-1} = \frac{bv}{r^2}\left(\frac{dr}{dt}\right)^{-1},\qquad(I.4)$$

and solving (I.3) for dr/dt then yields

$$\frac{d\theta}{dr} = \pm\frac{b}{r^2}\left(1 - \frac{b^2}{r^2} - \frac{V}{E}\right)^{-1/2}.\qquad(I.5)$$

Recalling that $\theta = \pi$ when $r = \infty$ on the incoming branch of the trajectory and that θ is always decreasing, this equation can be integrated immediately to give the scattering angle,

$$\Theta = \pi - 2 \int_{r_{\min}}^{\infty} \frac{b\,dr}{r^2} \left(1 - \frac{b^2}{r^2} - \frac{V}{E} \right)^{-1/2} , \qquad (I.6)$$

where $r_{\min}$ is the distance of closest approach (the turning point, determined by the outermost zero of the argument of the square root) and the factor of 2 in front of the integral accounts for the incoming and outgoing branches of the trajectory, which give equal contributions to the scattering angle.

One final transformation is useful before beginning a numerical calculation. Suppose that there exists a distance $r_{\max}$ beyond which we can safely neglect V. In this case, the integrand in (I.6) vanishes as r^{-2} for large r, so that numerical quadrature could be very inefficient. In fact, since the potential has no effect for $r > r_{\max}$, we would just be "wasting time" describing straight-line motion. To handle this situation efficiently, note that since $\Theta = 0$ when $V = 0$, Eq. (I.6) implies that

$$\pi = 2 \int_b^{\infty} \frac{b\,dr}{r^2} \left(1 - \frac{b^2}{r^2} \right)^{-1/2} , \qquad (I.7)$$

which, when substituted into (I.6), results in

$$\Theta = 2b \left[\int_b^{r_{\max}} \frac{dr}{r^2} \left(1 - \frac{b^2}{r^2} \right)^{-1/2} - \int_{r_{\min}}^{r_{\max}} \frac{dr}{r^2} \left(1 - \frac{b^2}{r^2} - \frac{V}{E} \right)^{-1/2} \right] . \quad (I.8)$$

The integrals here extend only to $r_{\max}$ since the integrands become equal when $r > r_{\max}$.

Our goal will be to study scattering by the Lennard-Jones potential (1.16), which we can safely set to zero beyond $r_{\max} = 3a$ if we are not interested in energies smaller than about

$$V(r = 3a) \approx 5 \times 10^{-3} V_0 .$$

The study is best done in the following sequence of steps:

Step 1   Before beginning *any* numerical computation, it is important to have some idea of what the results should look like. Sketch what you think the deflection function is at relatively low energies, $E \lesssim V_0$, where the peripheral collisions at large $b \lesssim r_{\max}$ will take place in a predominantly attractive potential and the more central collisions will "bounce" against the repulsive core. What happens at much higher energies, $E \gg V_0$, where the attractive pocket in V can be neglected? Note that for values of b where the deflection function has a maximum or a minimum, Eq. (I.1) shows that the cross section will be infinite, as occurs in the rainbow formed when light scatters from water drops.

Step 2   To have analytically soluble cases against which to test your program, calculate the deflection function for a square potential, where $V(r) = U_0$ for $r < r_{max}$ and vanishes for $r > r_{max}$. What happens when U_0 is negative? What happens when U_0 is positive and $E < U_0$? when $E > U_0$?

Step 3   Write a program that calculates, for a specified energy E, the deflection function by a numerical quadrature to evaluate both integrals in Eq. (I.8) at a number of equally spaced b values between 0 and r_{max}. (Note that the singularities in the integrands require some special treatment.) Check that the program is working properly and is accurate by calculating deflection functions for the square-well potential discussed in Step 2. Compare the accuracy with that of an alternative procedure in which the first integral in (I.8) is evaluated analytically, rather than numerically.

Step 4   Use your program to calculate the deflection function for scattering from the Lennard-Jones potential at selected values of E ranging from $0.1\,V_0$ to $100\,V_0$. Reconcile your answers in Step 1 with the results you obtain. Calculate the differential cross section as a function of Θ at these energies.

Step 5   If your program is working correctly, you should observe, for energies $E \lesssim V_0$, a singularity in the deflection function where Θ appears to approach $-\infty$ at some critical value of b, b_{crit}, that depends on E. This singularity, which disappears when E becomes larger than about V_0, is characteristic of "orbiting." In this phenomenon, the integrand in Eq. (I.6) has a linear, rather than a square root, singularity at the turning point, so that the scattering angle becomes logarithmically infinite. That is, the effective potential,

$$V + E\left(\frac{b}{r}\right)^2,$$

has a parabolic maximum and, when $b = b_{crit}$, the peak of this parabola is equal to the incident energy. The trajectory thus spends a very long time at the radius where this parabola peaks and the particle spirals many times around the force center. By tracing b_{crit} as a function of energy and by plotting a few of the effective potentials involved, convince yourself that this is indeed what's happening. Determine the maximum energy for which the Lennard-Jones potential exhibits orbiting, either by a solution of an appropriate set of equations involving V and its derivatives or by a systematic numerical investigation of the deflection function. If you pursue the latter approach, you might have to reconsider the treatment of the singularities in the numerical quadratures.

Chapter 2

Ordinary Differential Equations

Many of the laws of physics are most conveniently formulated in terms of differential equations. It is therefore not surprising that the numerical solution of differential equations is one of the most common tasks in modeling physical systems. The most general form of an ordinary differential equation is a set of M coupled first-order equations

$$\frac{d\mathbf{y}}{dx} = \mathbf{f}(x, \mathbf{y}) \, , \tag{2.1}$$

where x is the independent variable and $\mathbf{y}$ is a set of M dependent variables ($\mathbf{f}$ is thus an M-component vector). Differential equations of higher order can be written in this first-order form by introducing auxiliary functions. For example, the one-dimensional motion of a particle of mass m under a force field $F(z)$ is described by the second-order equation

$$m\frac{d^2 z}{dt^2} = F(z) \, . \tag{2.2}$$

If we define the momentum

$$p(t) = m\frac{dz}{dt} \, ,$$

then (2.2) becomes the two coupled first-order (Hamilton's) equations

$$\frac{dz}{dt} = \frac{p}{m} \; ; \quad \frac{dp}{dt} = F(z) \, , \tag{2.3}$$

which are in the form of (2.1). It is therefore sufficient to consider in detail only methods for first-order equations. Since the matrix structure of coupled differential equations is of the most natural form, our discussion of the case where there is only one independent variable can be generalized readily. Thus, we need be concerned only with solving

$$\frac{dy}{dx} = f(x, y) \tag{2.4}$$

for a single dependent variable $y(x)$.

In this chapter, we will discuss several methods for solving ordinary differential equations, with emphasis on the initial value problem. That is, find $y(x)$ given the value of y at some initial point, say $y(x = 0) = y_0$. This kind of problem occurs, for example, when we are given the initial position and momentum of a particle and we wish to find its subsequent motion using Eqs. (2.3). In Chapter 3, we will discuss the equally important boundary value and eigenvalue problems.

2.1 Simple methods

To repeat the basic problem, we are interested in the solution of the differential equation (2.4) with the initial condition $y(x = 0) = y_0$. More specifically, we are usually interested in the value of y at a particular value of x, say $x = 1$. The general strategy is to divide the interval $[0, 1]$ into a large number, N, of equally spaced subintervals of length $h = 1/N$ and then to develop a recursion formula relating y_n to $\{y_{n-1}, y_{n-2}, \ldots\}$, where y_n is our approximation to $y(x_n = nh)$. Such a recursion relation will then allow a step-by-step integration of the differential equation from $x = 0$ to $x = 1$.

One of the simplest algorithms is Euler's method, in which we consider Eq. (2.4) at the point x_n and replace the derivative on the left-hand side by its forward difference approximation (1.4a). Thus,

$$\frac{y_{n+1} - y_n}{h} + \mathcal{O}(h) = f(x_n, y_n), \tag{2.5}$$

so that the recursion relation expressing y_{n+1} in terms of y_n is

$$y_{n+1} = y_n + hf(x_n, y_n) + \mathcal{O}(h^2). \tag{2.6}$$

This formula has a local error (that made in taking the single step from y_n to y_{n+1}) that is $\mathcal{O}(h^2)$ since the error in (1.4a) is $\mathcal{O}(h)$. The "global" error made in finding $y(1)$ by taking N such steps in integrating from $x = 0$ to $x = 1$ is then $N\mathcal{O}(h^2) \approx \mathcal{O}(h)$. This error decreases only linearly with decreasing step size so that half as large an h (and thus twice as many steps) is required to halve the inaccuracy in the final answer. The numerical work for each of these steps is essentially a single evaluation of f.

As an example, consider the differential equation and boundary condition

$$\frac{dy}{dx} = -xy \; ; \; y(0) = 1 \; , \tag{2.7}$$

whose solution is

$$y = e^{-x^2/2} \; .$$

The following FORTRAN program integrates forward from $x = 0$ to $x = 3$ using Eq. (2.6) with the step size input, printing the result and its error as it goes along.

```
C chap2a.for
      FUNC(X,Y)=-X*Y              !dy/dx
20    PRINT *, ' Enter step size ( .le. 0 to stop)'
      READ *, H
      IF (H .LE. 0.) STOP
      NSTEP=3./H                  !number of steps to reach X=3
      Y=1.                        !y(0)=1
      DO 10 IX=0,NSTEP-1          !loop over steps
         X=IX*H                   !last X value
         Y=Y+H*FUNC(X,Y)          !new Y value from Eq 2.6
         DIFF=EXP(-0.5*(X+H)**2)-Y !compare with exact value
         PRINT *, IX,X+H,Y,DIFF
10    CONTINUE
      GOTO 20                     !start again with new value of H
      END
```

Errors in the results obtained for

$$y(1) = e^{-1/2} = 0.606531 \; , \; y(3) = e^{-9/2} = 0.011109$$

with various step sizes are shown in the first two columns of Table 2.1. As expected from (2.6), the errors decrease linearly with smaller h. However, the fractional error (error divided by y) increases with x as more steps are taken in the integration and y becomes smaller.

Table 2.1   Error in integrating $dy/dx = -xy$ with $y(0) = 1$

h	Euler's method Eq. (2.6)		Taylor series Eq. (2.10)		Implicit method Eq. (2.18)	
	$y(1)$	$y(3)$	$y(1)$	$y(3)$	$y(1)$	$y(3)$
0.500	$-.143469$	.011109	.032312	$-.006660$	$-.015691$	.001785
0.200	$-.046330$	.006519	.005126	$-.000712$	$-.002525$	.000255
0.100	$-.021625$	.003318	.001273	$-.000149$	$-.000631$	.000063
0.050	$-.010453$	.001665	.000317	$-.000034$	$-.000157$	.000016
0.020	$-.004098$	.000666	.000051	$-.000005$	$-.000025$	.000003
0.010	$-.002035$	.000333	.000013	$-.000001$	$-.000006$	.000001
0.005	$-.001014$	.000167	.000003	.000000	$-.000001$	.000000
0.002	$-.000405$	.000067	.000001	.000000	.000000	.000000
0.001	$-.000203$	.000033	.000000	.000000	.000000	.000000

■ **Exercise 2.1** A simple and often stringent test of an accurate numerical integration is to use the final value of y obtained as the initial condition to integrate backward from the final value of x to the starting point. The extent to which the resulting value of y differs from the original initial condition is then a measure of the inaccuracy. Apply this test to the example above.

Although Euler's method seems to work quite well, it is generally unsatisfactory because of its low-order accuracy. This prevents us from reducing the numerical work by using a larger value of h and so taking a smaller number of steps. This deficiency becomes apparent in the example above as we attempt to integrate to larger values of x, where some thought shows that we obtain the absurd result that $y = 0$ (exactly) for $x > h^{-1}$. One simple solution is to change the step size as we go along, making h smaller as x becomes larger, but this soon becomes quite inefficient.

Integration methods with a higher-order accuracy are usually preferable to Euler's method. They offer a much more rapid increase of accuracy with decreasing step size and hence greater accuracy for a fixed amount of numerical effort. One class of simple higher order methods can be derived from a Taylor series expansion for y_{n+1} about y_n:

$$y_{n+1} = y(x_n + h) = y_n + hy_n' + \frac{1}{2}h^2 y_n'' + \mathcal{O}(h^3) . \qquad (2.8)$$

From (2.4), we have

$$y_n' = f(x_n, y_n) , \qquad (2.9a)$$

and

$$y_n'' = \frac{df}{dx}(x_n, y_n) = \frac{\partial f}{\partial x} + \frac{\partial f}{\partial y}\frac{dy}{dx} = \frac{\partial f}{\partial x} + \frac{\partial f}{\partial y}f \,, \qquad (2.9b)$$

which, when substituted into (2.8), results in

$$y_{n+1} = y_n + hf + \frac{1}{2}h^2\left[\frac{\partial f}{\partial x} + f\frac{\partial f}{\partial y}\right] + \mathcal{O}(h^3) \,, \qquad (2.10)$$

where f and its derivatives are to be evaluated at (x_n, y_n). This recursion relation has a local error $\mathcal{O}(h^3)$ and hence a global error $\mathcal{O}(h^2)$, one order more accurate than Euler's method (2.6). It is most useful when f is known analytically and is simple enough to differentiate. If we apply Eq. (2.10) to the example (2.7), we obtain the results shown in the middle two columns of Table 2.1; the improvement over Euler's method is clear. Algorithms with an even greater accuracy can be obtained by retaining more terms in the Taylor expansion (2.8), but the algebra soon becomes prohibitive in all but the simplest cases.

2.2 Multistep and implicit methods

Another way of achieving higher accuracy is to use recursion relations that relate y_{n+1} not just to y_n, but also to points further "in the past," say y_{n-1}, y_{n-2}, ... To derive such formulas, we can integrate one step of the differential equation (2.4) *exactly* to obtain

$$y_{n+1} = y_n + \int_{x_n}^{x_{n+1}} f(x, y)\, dx \,. \qquad (2.11)$$

The problem, of course, is that we don't know f over the interval of integration. However, we can use the values of y at x_n and x_{n-1} to provide a linear extrapolation of f over the required interval:

$$f \approx \frac{(x - x_{n-1})}{h}f_n - \frac{(x - x_n)}{h}f_{n-1} + \mathcal{O}(h^2) \,, \qquad (2.12)$$

where $f_i \equiv f(x_i, y_i)$. Inserting this into (2.11) and doing the x integral then results in the Adams-Bashforth two-step method,

$$y_{n+1} = y_n + h\left(\frac{3}{2}f_n - \frac{1}{2}f_{n-1}\right) + \mathcal{O}(h^3) \,. \qquad (2.13)$$

Related higher-order methods can be derived by extrapolating with higher-degree polynomials. For example, if f is extrapolated by a cubic polynomial fitted to f_n, f_{n-1}, f_{n-2}, and f_{n-3}, the Adams-Bashforth four-step method results:

$$y_{n+1} = y_n + \frac{h}{24}(55f_n - 59f_{n-1} + 37f_{n-2} - 9f_{n-3}) + \mathcal{O}(h^5). \qquad (2.14)$$

Note that because the recursion relations (2.13) and (2.14) involve several previous steps, the value of y_0 alone is not sufficient information to get them started, and so the values of y at the first few lattice points must be obtained from some other procedure, such as the Taylor series (2.8) or the Runge-Kutta methods discussed below.

■ **Exercise 2.2** Apply the Adams-Bashforth two- and four-step algorithms to the example defined by Eq. (2.7) using Euler's method (2.6) to generate the values of y needed to start the recursion relation. Investigate the accuracy of $y(x)$ for various values of h by comparing with the analytical results and by applying the reversibility test described in Exercise 2.1.

The methods we have discussed so far are all "explicit" in that the y_{n+1} is given directly in terms of the already known value of y_n. "Implicit" methods, in which an equation must be solved to determine y_{n+1}, offer yet another means of achieving higher accuracy. Suppose we consider Eq. (2.4) at a point $x_{n+1/2} \equiv (n + \frac{1}{2})h$ mid-way between two lattice points:

$$\left.\frac{dy}{dx}\right|_{x_{n+1/2}} = f(x_{n+1/2}, y_{n+1/2}). \qquad (2.15)$$

If we then use the symmetric difference approximation for the derivative (the analog of (1.3b) with $h \rightarrow \frac{1}{2}h$) and replace $f_{n+1/2}$ by the average of its values at the two adjacent lattice points [the error in this replacement is $\mathcal{O}(h^2)$], we can write

$$\frac{y_{n+1} - y_n}{h} + \mathcal{O}(h^2) = \frac{1}{2}[f_n + f_{n+1}] + \mathcal{O}(h^2), \qquad (2.16)$$

which corresponds to the recursion relation

$$y_{n+1} = y_n + \frac{1}{2}h[f(x_n, y_n) + f(x_{n+1}, y_{n+1})] + \mathcal{O}(h^3). \qquad (2.17)$$

This is all well and good, but the appearance of y_{n+1} on both sides of this equation (an implicit equation) means that, in general, we must solve a non-trivial equation (for example, by the Newton-Raphson method discussed in Section 1.3) at each integration step; this can be very time consuming. A particular simplification occurs if f is linear in y, say $f(x, y) = g(x)y$, in which case (2.17) can be solved to give

$$y_{n+1} = \left[\frac{1 + \frac{1}{2}g(x_n)h}{1 - \frac{1}{2}g(x_{n+1})h} \right] y_n .
\tag{2.18}$$

When applied to the problem (2.7), where $g(x) = -x$, this method gives the results shown in the last two columns of Table 2.1; the quadratic behavior of the error with h is clear.

■ **Exercise 2.3** Apply the Taylor series method (2.10) and the implicit method (2.18) to the example of Eq. (2.7) and obtain the results shown in Table 2.1. Investigate the accuracy of integration to larger values of x.

The Adams-Moulton methods are both multistep and implicit. For example, the Adams-Moulton two-step method can be derived from Eq. (2.11) by using a quadratic polynomial passing through f_{n-1}, f_n, and f_{n+1},

$$f \approx \frac{(x - x_n)(x - x_{n-1})}{2h^2} f_{n+1} - \frac{(x - x_{n+1})(x - x_{n-1})}{h^2} f_n$$
$$+ \frac{(x - x_{n+1})(x - x_n)}{2h^2} f_{n-1} + \mathcal{O}(h^3) ,$$

to interpolate f over the region from x_n to x_{n+1}. The implicit recursion relation that results is

$$y_{n+1} = y_n + \frac{h}{12}(5f_{n+1} + 8f_n - f_{n-1}) + \mathcal{O}(h^4) .
\tag{2.19}$$

The corresponding three-step formula, obtained with a cubic polynomial interpolation, is

$$y_{n+1} = y_n + \frac{h}{24}(9f_{n+1} + 19f_n - 5f_{n-1} + f_{n-2}) + \mathcal{O}(h^5) .
\tag{2.20}$$

Implicit methods are rarely used by solving the implicit equation to take a step. Rather, they serve as bases for "predictor-corrector" algorithms, in which a "prediction" for y_{n+1} based only on an explicit method is then "corrected" to give a better value by using this prediction in an implicit method. Such algorithms have the advantage of allowing a continuous monitoring of the accuracy of the integration, for example by making sure that the correction is small. A commonly used predictor-corrector algorithm with local error $\mathcal{O}(h^5)$ is obtained by using the explicit Adams-Bashforth four-step method (2.14) to make the prediction, and then calculating the correction with the Adams-Moulton three-step method (2.20), using the predicted value of y_{n+1} to evaluate f_{n+1} on the right-hand side.

2.3 Runge-Kutta methods

As you might gather from the preceding section, there is quite a bit of freedom in writing down algorithms for integrating differential equations and, in fact, a large number of them exist, each having it own peculiarities and advantages. One very convenient and widely used class of methods are the Runge-Kutta algorithms, which come in varying orders of accuracy. We derive here a second-order version to give the spirit of the approach and then simply state the equations for the third- and commonly used fourth-order methods.

To derive a second-order Runge-Kutta algorithm (there are actually a whole family of them characterized by a continuous parameter), we approximate f in the integral of (2.11) by its Taylor series expansion about the mid-point of the integration interval. Thus,

$$y_{n+1} = y_n + hf(x_{n+1/2}, y_{n+1/2}) + \mathcal{O}(h^3), \qquad (2.21)$$

where the error arises from the quadratic term in the Taylor series, as the linear term integrates to zero. Although it seems as if we need to know the value of $y_{n+1/2}$ appearing in f in the right-hand side of this equation for it to be of any use, this is not quite true. Since the error term is already $\mathcal{O}(h^3)$, an approximation to y_{n+1} whose error is $\mathcal{O}(h^2)$ is good enough. This is just what is provided by the simple Euler's method, Eq. (2.6). Thus, if we define k to be an intermediate approximation to twice the difference between $y_{n+1/2}$ and y_n, the following two-step procedure gives y_{n+1} in terms of y_n:

$$k = hf(x_n, y_n); \qquad (2.22a)$$

$$y_{n+1} = y_n + hf(x_n + \frac{1}{2}h, y_n + \frac{1}{2}k) + \mathcal{O}(h^3). \qquad (2.22b)$$

This is a second-order Runge-Kutta algorithm. It embodies the general idea of substituting approximations for the values of y into the right-hand side of implicit expressions involving f. It is as accurate as the Taylor series or implicit methods (2.10) or (2.17), respectively, but places no special constraints on f, such as easy differentiability or linearity in y. It also uses the value of y at only one previous point, in contrast to the multipoint methods discussed above. However, (2.22) does require the evaluation of f twice for each step along the lattice.

Runge-Kutta schemes of higher-order can be derived in a relatively straightforward way. Any of the quadrature formulas discussed in Chapter 1 can be used to approximate the integral (2.11) by a finite sum of f values. For example, Simpson's rule yields

$$y_{n+1} = y_n + \frac{h}{6}\left[f(x_n, y_n) + 4f(x_{n+1/2}, y_{n+1/2}) + f(x_{n+1}, y_{n+1})\right]$$
$$+ \mathcal{O}(h^5) . \tag{2.23}$$

Schemes for generating successive approximations to the y's appearing in the right-hand side of a commensurate accuracy then complete the algorithms. A third-order algorithm with a local error $\mathcal{O}(h^4)$ is

$$k_1 = hf(x_n, y_n) ;$$
$$k_2 = hf(x_n + \frac{1}{2}h, y_n + \frac{1}{2}k_1) ;$$
$$k_3 = hf(x_n + h, y_n - k_1 + 2k_2) ; \tag{2.24}$$
$$y_{n+1} = y_n + \frac{1}{6}(k_1 + 4k_2 + k_3) + \mathcal{O}(h^4) .$$

It is based on (2.23) and requires three evaluations of f per step. A fourth-order algorithm, which requires f to be evaluated four times for each integration step and has a local accuracy of $\mathcal{O}(h^5)$, has been found by experience to give the best balance between accuracy and computational effort. It can be written as follows, with the k_i as intermediate variables:

$$k_1 = hf(x_n, y_n) ;$$
$$k_2 = hf(x_n + \frac{1}{2}h, y_n + \frac{1}{2}k_1) ;$$
$$k_3 = hf(x_n + \frac{1}{2}h, y_n + \frac{1}{2}k_2) ; \tag{2.25}$$
$$k_4 = hf(x_n + h, y_n + k_3) ;$$
$$y_{n+1} = y_n + \frac{1}{6}(k_1 + 2k_2 + 2k_3 + k_4) + \mathcal{O}(h^5) .$$

▪ **Exercise 2.4** Try out the second-, third-, and fourth-order Runge-Kutta methods discussed above on the problem defined by Eq. (2.7). Compare the computational effort for a given accuracy with that of other methods.

▪ **Exercise 2.5** The two coupled first-order equations

$$\frac{dy}{dt} = p \; ; \quad \frac{dp}{dt} = -4\pi^2 y \tag{2.26}$$

define simple harmonic motion with period 1. By generalizing one of the single-variable formulas given above to this two-variable case, integrate these equations with any particular initial conditions you choose and investigate the accuracy with which the system returns to its initial state at integral values of t.

2.4 Stability

A major consideration in integrating differential equations is the numerical stability of the algorithm used; i.e., the extent to which round-off or other errors in the numerical computation can be amplified, in many cases enough for this "noise" to dominate the results. To illustrate the problem, let us attempt to improve the accuracy of Euler's method and approximate the derivative in (2.4) directly by the symmetric difference approximation (1.3b). We thereby obtain the three-term recursion relation

$$y_{n+1} = y_{n-1} + 2hf(x_n, y_n) + \mathcal{O}(h^3) \, , \tag{2.27}$$

which superficially looks about as useful as either of the third-order formulas (2.10) or (2.18). However, consider what happens when this method is applied to the problem

$$\frac{dy}{dx} = -y \; ; \quad y(x = 0) = 1 \, , \tag{2.28}$$

whose solution is $y = e^{-x}$. To start the recursion relation (2.27), we need the value of y_1 as well as $y_0 = 1$. This can be obtained by using (2.10) to get

$$y_1 = 1 - h + \frac{1}{2}h^2 + \mathcal{O}(h^3) \, .$$

(This is just the Taylor series for e^{-h}.) The following FORTRAN program then uses the method (2.27) to find y for values of x up to 6 using the value of h input:

Table 2.2   Integration of $dy/dx = -y$ with $y(0) = 1$

x	Exact	Error	x	Exact	Error	x	Exact	Error
0.2	.818731	−.000269	3.3	.036883	−.000369	5.5	.004087	−.001533
0.3	.740818	−.000382	3.4	.033373	−.000005	5.6	.003698	.001618
0.4	.670320	−.000440	3.5	.030197	−.000380	5.7	.003346	−.001858
0.5	.606531	−.000517	3.6	.027324	.000061	5.8	.003028	.001989
0.6	.548812	−.000538	3.7	.024724	−.000400	5.9	.002739	−.002257
			3.8	.022371	.000133	6.0	.002479	.002439

```
C chap2b.for
10      PRINT *,' Enter value of step size ( .le. 0 to stop)'
        READ *, H
        IF (H .LE. 0) STOP
        YMINUS=1                        !Y(0)
        YZERO=1.-H+H**2/2               !Y(H)
        NSTEP=6./H
        DO 20 IX=2,NSTEP                !loop over X values
           X=IX*H                       !X at this step
           YPLUS=YMINUS-2*H*YZERO       !Y from Eq. 2.27
           YMINUS=YZERO                 !roll values
           YZERO=YPLUS
           EXACT=EXP(-X)                !analytic value at this point
           PRINT *, X,EXACT,EXACT-YZERO
20      CONTINUE
        GOTO 10                         !get another H value
        END
```

Note how the three-term recursion is implemented by keeping track of only three local variables, YPLUS, YZERO, and YMINUS.

A portion of the output of this code run for $h = 0.1$ is shown in Table 2.2. For small values of x, the numerical solution is only slightly larger than the exact value, the error being consistent with the $\mathcal{O}(h^3)$ estimate. Then, near $x = 3.5$, an oscillation begins to develop in the numerical solution, which becomes alternately higher and lower than the exact values lattice point by lattice point. This oscillation grows larger as the equation is integrated further (see values near $x = 6$), eventually overwhelming the exponentially decreasing behavior expected.

The phenomenon observed above is a symptom of an instability in the algorithm (2.27). It can be understood as follows. For the problem

(2.28), the recursion relation (2.27) reads

$$y_{n+1} = y_{n-1} - 2hy_n .\qquad(2.29)$$

We can solve this equation by assuming an exponential solution of the form $y_n = Ar^n$ where A and r are constants. Substituting into (2.29) then results in an equation for r,

$$r^2 + 2hr - 1 = 0 ,$$

the constant A being unimportant since the recursion relation is linear. The solutions of this equation are

$$r_+ = (1 + h^2)^{1/2} - h \approx 1 - h \; ; \; r_- = -(1 + h^2)^{1/2} - h \approx -(1 + h) ,$$

where we have indicated approximations valid for $h \ll 1$. The positive root is slightly less than one and corresponds to the exponentially decreasing solution we are after. However, the negative root is slightly less than -1, and so corresponds to a spurious solution

$$y_n \sim (-)^n (1 + h)^n ,$$

whose magnitude increases with n and which oscillates from lattice point to lattice point.

The general solution to the linear difference equation (2.27) is a linear combination of these two exponential solutions. Even though we might carefully arrange the initial values y_0 and y_1 so that only the decreasing solution is present for small x, numerical round-off during the recursion relation [Eq. (2.29) shows that two positive quantities are subtracted to obtain a smaller one] will introduce a small admixture of the "bad" solution that will eventually grow to dominate the results. This instability is clearly associated with the three-term nature of the recursion relation (2.29). A good rule of thumb is that instabilities and round-off problems should be watched for whenever integrating a solution that decreases strongly as the iteration proceeds; such a situation should therefore be avoided, if possible. We will see the same sort of instability phenomenon again in our discussion of second-order differential equations in Chapter 3.

■ **Exercise 2.6** Investigate the stability of several other integration methods discussed in this chapter by applying them to the problem (2.28). Can you give analytical arguments to explain the results you obtain?

2.5 Order and chaos in two-dimensional motion

A fundamental advantage of using computers in physics is the ability to treat systems that cannot be solved analytically. In the usual situation, the numerical results generated agree qualitatively with the intuition we have developed by studying soluble models and it is the quantitative values that are of real interest. However, in a few cases computer results defy our intuition (and thereby reshape it) and numerical work is then essential for a proper understanding. Surprisingly, such cases include the dynamics of simple classical systems, where the generic behavior differs *qualitatively* from that of the models covered in a traditional Mechanics course. In this example, we will study some of this surprising behavior by integrating numerically the trajectories of a particle moving in two dimensions. General discussions of these systems can be found in [He80], [Ri80], and [Ab78].

We consider a particle of unit mass moving in a potential, V, in two dimensions and assume that V is such that the particle remains confined for all times if its energy is low enough. If the momenta conjugate to the two coordinates (x, y) are (p_x, p_y), then the Hamiltonian takes the form

$$H = \frac{1}{2}(p_x^2 + p_y^2) + V(x, y) . \qquad (2.30)$$

Given any particular initial values of the coordinates and momenta, the particle's trajectory is specified by their time evolution, which is governed by four coupled first-order differential equations (Hamilton's equations):

$$\frac{dx}{dt} = \frac{\partial H}{\partial p_x} = p_x , \quad \frac{dy}{dt} = \frac{\partial H}{\partial p_y} = p_y ;$$
$$\frac{dp_x}{dt} = -\frac{\partial H}{\partial x} = -\frac{\partial V}{\partial x} , \quad \frac{dp_y}{dt} = \frac{\partial H}{\partial y} = -\frac{\partial V}{\partial y} . \qquad (2.31)$$

For any V, these equations conserve the energy, E, so that the constraint

$$H(x, y, p_x, p_y) = E$$

restricts the trajectory to lie in a three-dimensional manifold embedded in the four-dimensional phase space. Apart from this, there are very few other general statements that can be made about the evolution of the system.

One important class of two-dimensional Hamiltonians for which additional statements about the trajectories *can* be made are those that are *integrable.* For these potentials, there is a second function of the coordinates and momenta, apart from the energy, that is a constant of the motion; the trajectory is thus constrained to a two-dimensional manifold of the phase space. Two familiar kinds of integrable systems are separable and central potentials. In the separable case,

$$V(x,y) = V_x(x) + V_y(y) , \tag{2.32}$$

where the $V_{x,y}$ are two independent functions, so that the Hamiltonian separates into two parts, each involving only one coordinate and its conjugate momentum,

$$H = H_x + H_y \; ; \; H_{x,y} = \frac{1}{2}p_{x,y}^2 + V_{x,y} .$$

The motions in x and y therefore decouple from each other and each of the Hamiltonians $H_{x,y}$ is separately a constant of the motion. (Equivalently, $H_x - H_y$ is the second quantity conserved in addition to $E = H_x + H_y$.) In the case of a central potential,

$$V(x,y) = V(r) \; ; \; r = (x^2 + y^2)^{1/2} , \tag{2.33}$$

so that the angular momentum, $p_\theta = xp_y - yp_x$, is the second constant of the motion and the Hamiltonian can be written as

$$H = \frac{1}{2}p_r^2 + V(r) + \frac{p_\theta^2}{2r^2} ,$$

where p_r is the momentum conjugate to r. The additional constraint on the trajectory present in integrable systems allows the equations of motion to be "solved" by reducing the problem to one of evaluating certain integrals, much as we did for one-dimensional motion in Chapter 1. All of the familiar analytically soluble problems of classical mechanics are those that are integrable.

Although the dynamics of integrable systems are simple, it is often not at all easy to make this simplicity apparent. There is no general analytical method for deciding if there is a second constant of the motion in an arbitrary potential or for finding it if there is one. Numerical calculations are not obviously any better, as these supply only the trajectory for given initial conditions and this trajectory can be quite complicated in even familiar cases, as can be seen by recalling the Lissajous patterns

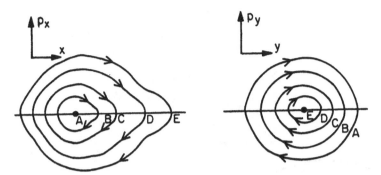

Figure 2.1 Trajectories of a particle in a two-dimensional separable potential as they appear in the (x, p_x) and (y, p_y) planes. Several trajectories corresponding to the same energy but different initial conditions are shown. Trajectories A and E are the limiting ones having vanishing E_x and E_y, respectively.

of the (x, y) trajectories that arise when the motion in both coordinates is harmonic,

$$V_x = \frac{1}{2}\omega_x^2 x^2 \ , \ \ V_y = \frac{1}{2}\omega_y^2 y^2 \ . \tag{2.34}$$

An analysis in phase space suggests one way to detect integrability from the trajectory alone. Consider, for example, the case of a separable potential. Because the motions of each of the two coordinates are independent, plots of a trajectory in the (x, p_x) and (y, p_y) planes might look as shown in Figure 2.1. Here, we have assumed that each potential $V_{x,y}$ has a single minimum value of 0 at particular values of x and y, respectively. The particle moves on a closed contour in each of these two-dimensional projections of the four-dimensional phase space, each contour looking like that for ordinary one-dimensional motion shown in Figure 1.3. The areas of these contours depend upon how much energy is associated with each coordinate (i.e., E_x and E_y) and, as we consider trajectories with the same energy but with different initial conditions, the area of one contour will shrink as the other grows. In each plane there is a limiting contour that is approached when all of the energy is in that particular coordinate. These contours are the intersections of the two-dimensional phase-space manifolds containing each of the trajectories and the (y, p_y) or (x, p_x) plot is therefore termed a "surface of section." The existence of these closed contours signals the integrability of the system.

Although we are able to construct Figure 2.1 only because we understand the integrable motion involved, a similar plot can be obtained

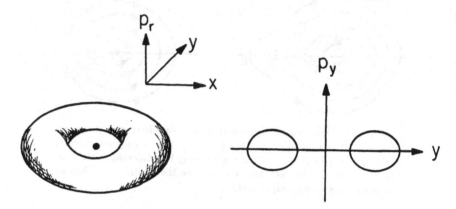

Figure 2.2 (Left) Toroidal manifold containing the trajectory of a particle in a central potential. (Right) Surface of section through this manifold at $x = 0$.

from the trajectory alone. Suppose that every time we observe one of the coordinates, say x, to pass through zero with $p_x > 0$, we plot the location of the particle on the (y, p_y) plane. In other words, crossing through the $x = 0$ plane in a positive sense triggers a "stroboscope" with which we observe the (y, p_y) variables. If the periods of the x and y motions are incommensurate (i.e., their ratio is an irrational number), then, as the trajectory proceeds, these observations will gradually trace out the full (y, p_y) contour; if the periods are commensurate (i.e., a rational ratio), then a series of discrete points around the contour will result. In this way, we can study the topology of the phase space associated with any given Hamiltonian just from the trajectories alone.

The general topology of the phase space for an integrable Hamiltonian can be illustrated by considering motion in a central potential. For fixed values of the energy and angular momentum, the radial motion is bounded between two turning points, r_{in} and r_{out}, which are the solutions of the equation

$$E - V(r) - \frac{p_\theta^2}{2r^2} = 0 \,.$$

These two radii define an annulus in the (x, y) plane to which the trajectory is confined, as shown in the left-hand side of Figure 2.2. Furthermore,

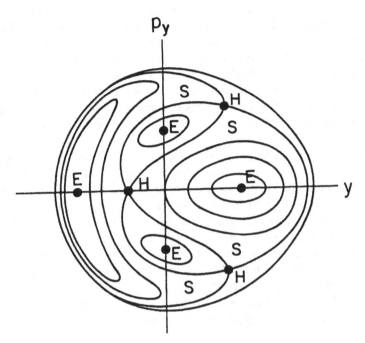

Figure 2.3 Possible features of the surface of section of a general integrable system. E and H label elliptic and hyperbolic fixed points, respectively, while the curve labeled S is a separatrix.

for a fixed value of r, energy conservation permits the radial momentum to take on only one of two values,

$$p_r = \pm\left(2E - 2V(r) - \frac{p_\theta^2}{r^2}\right)^{1/2}$$

These momenta define the two-dimensional manifold in the (x, y, p_r) space that contains the trajectory; it clearly has the topology of a torus, as shown in the left-hand side of Figure 2.2. If we were to construct a (y, p_y) surface of section by considering the $x = 0$ plane, we would obtain two closed contours, as shown in the right-hand side of Figure 2.2. (Note that $y = r$ when $x = 0$.) If the energy is fixed but the angular momentum is changed by varying the initial conditions, the dimensions of this torus change, as does the area of the contour in the surface of section.

The toroidal topology of the phase space of a central potential can be shown to be common to all integrable systems. The manifold on which the trajectory lies for given values of the constants of the motion is called an "invariant torus," and there are many of them for a given energy. The general surface of section of such tori looks like that shown in Figure 2.3. There are certain fixed points associated with trajectories that repeat themselves exactly after some period. The elliptic fixed points correspond to trajectories that are stable under small perturbations. Around each of them is a family of tori, bounded by a separatrix. The hyperbolic fixed points occur at the intersections of separatrices and are stable under perturbations along one of the axes of the hyperbola, but unstable along the other.

An interesting question is *"What happens to the tori of an integrable system under a perturbation that destroys the integrability of the Hamiltonian?"*. For small perturbations, "most" of the tori about an elliptic fixed point become slightly distorted but retain their topology (the KAM theorem due to Kolmogorov, Arnold, and Moser, [Ar68]). However, adjacent regions of phase space become "chaotic", giving surfaces of section that are a seemingly random splatter of points. Within these chaotic regions are nested yet other elliptic fixed points and other chaotic regions in a fantastic hierarchy. (See Figure 2.4.)

Large deviations from integrability must be investigated numerically. One convenient case for doing so is the potential

$$V(x,y) = \frac{1}{2}(x^2 + y^2) + x^2y - \frac{1}{3}y^3 \,, \qquad (2.35)$$

which was originally introduced by Hénon and Heiles in the study of stellar orbits through a galaxy [He64]. This potential can be thought of as a perturbed harmonic oscillator potential (a small constant multiplying the cubic terms can be absorbed through a rescaling of the coordinates and energy, so that the magnitude of the energy becomes a measure of the deviation from integrability) and has the three-fold symmetry shown in Figure 2.5. The potential is zero at the origin and becomes unbounded for large values of the coordinates. However, for energies less than 1/6, the trajectories remain confined within the equilateral triangle shown.

The FORTRAN program for Example 2, whose source code is contained in Appendix B and in the file EXMPL2.FOR, constructs surfaces of section for the Hénon-Heiles potential. The method used is to integrate the equations of motion (2.31) using the fourth-order Runge-Kutta algorithm (2.25). Initial conditions are specified by putting $x = 0$ and by giving the energy, y, and p_y; p_x is then fixed by energy conservation. As the integration proceeds, the (x,y) trajectory and the (y,p_y) surface

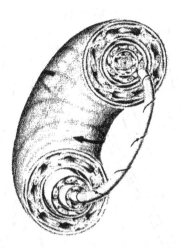

Figure 2.4 Nested tori for a slightly perturbed integrable system. Note the hierarchy of elliptic orbits interspersed with chaotic regions. A magnification of this hierarchy would show the same pattern repeated on a smaller scale and so on, *ad infinitum*. (Reproduced from [Ab78].)

of section are displayed. Points on the latter are calculated by watching for a time step during which x changes sign. When this happens, the precise location of the point on the surface of section plot is determined by switching to x as the independent variable, so that the equations of motion (2.31) become

$$\frac{dx}{dx} = 1 \;,\; \frac{dy}{dx} = \frac{1}{p_x}p_y \;;$$
$$\frac{dp_x}{dx} = -\frac{1}{p_x}\frac{\partial V}{\partial x} \;,\; \frac{dp_y}{dx} = -\frac{1}{p_x}\frac{\partial V}{\partial y} \;,$$

and then integrating one step backward in x from its value after the time step to 0 [He82]. If the value of the energy is not changed when new initial conditions are specified, all previous surface of section points can be plotted, so that plots like those in Figures 2.3 and 2.4 can be built up after some time.

The program exploits a symmetry of the Hénon-Heiles equations of motion to obtain two trajectories from one set of initial conditions. This symmetry can be seen from the equations of motion which are invariant

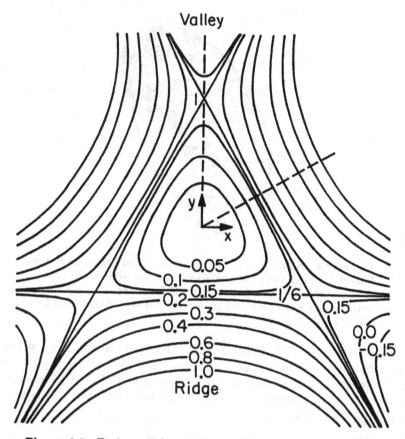

Figure 2.5   Equipotential contours of the Hénon-Heiles potential, Eq. (2.35).

if $x \to -x$ and $p_x \to -p_x$. Therefore from one trajectory we can trivially obtain a "reflected" trajectory by flipping the sign of both x and p_x at each point along the original trajectory. More specifically, if we are interested in creating the surface of section on the $x = 0$ plane we already have $x = -x$, and only the sign of p_x must be changed. Practically this means that every time we cross the $x = 0$ plane we take the corresponding y and p_y values to be on the surface of section, without consideration for the sign of p_x. If p_x is greater than zero it is a point from the initial trajectory, while if p_x is less than zero then that point belongs to the reflected trajectory.

The following exercises are aimed at improving your understanding of the physical principles and numerical methods demonstrated in this example.

■ **Exercise 2.7** One necessary (but not sufficient) check of the accuracy of the integration in Hamiltonian systems is the conservation of energy. Change the time step used and observe how this affects the accuracy. Replace the integration method by one of the other algorithms discussed in this chapter and observe the change in accuracy and efficiency. Note that we require an algorithm that is both accurate and efficient because of the very long integration times that must be considered.

■ **Exercise 2.8** Change the potential to a central one (function subroutine V) and observe the character of the (x, y) trajectories and surfaces of section for various initial conditions. Compare the results with Figure 2.2. and verify that the qualitative features don't change if you use a different central potential. Note that you will have to adjust the energy and scale of the potential so that the (x, y) trajectory does not go beyond the equilateral triangle of the Hénon-Heiles potential (Figure 2.5) or else the graphics subroutine will fail.

■ **Exercise 2.9** If the sign of the y^3 term in the Hénon-Heiles potential (2.35) is reversed, the Hamiltonian becomes integrable. Verify analytically that this is so by making a canonical transformation to the variables $x \pm y$ and showing that the potential is separable in these variables. Make the corresponding change in sign in the code and observe the character of the surfaces of section that result for various initial conditions at a given energy. (You should keep the energy below 1/12 to avoid having the trajectory become unbounded.) Verify that there are no qualitative differences if you use a different separable potential, say the harmonic one of Eq. (2.34).

■ **Exercise 2.10** Use the code to construct surfaces of section for the Hénon-Heiles potential (2.35) at energies ranging from 0.025 to 0.15 in steps of 0.025. For each energy, consider various initial conditions and integrate each trajectory long enough in time (some will require going to $t \approx 1000$) to map out the surface-of-section adequately. For each energy, see if you can find the elliptic fixed points, the tori (and tori of tori) around them, and the chaotic regions of phase space and observe how the relative proportions of each change with increasing energy. With some patience and practice, you should be able to generate a plot that resembles the schematic representation of Figure 2.4 around each elliptic trajectory.

Project II: The structure of white dwarf stars

White dwarf stars are cold objects composed largely of heavy nuclei and their associated electrons. These stars are one possible end result of the conventional nuclear processes that build the elements by binding nucleons into nuclei. They are often composed of the most stable nucleus, ^{56}Fe, with 26 protons and 30 neutrons, but, if the nucleosynthesis terminates prematurely, ^{12}C nuclei might predominate. The structure of a white dwarf star is determined by the interplay between gravity, which acts to compress the star, and the degeneracy pressure of the electrons, which tends to resist compression. In this project, we will investigate the structure of a white dwarf by integrating the equations defining this equilibrium. We will determine, among other things, the relation between the star's mass and its radius, quantities that can be determined from observation. Discussions of the physics of white dwarfs can be found in [Ch57], [Ch84], and [Sh83].

II.1 The equations of equilibrium

We assume that the star is spherically symmetric (i.e., the state of matter at any given point in the star depends only upon the distance of that point from the star's center), that it is not rotating, and that the effects of magnetic fields are not important. If the star is in mechanical (hydrostatic) equilibrium, the gravitational force on each bit of matter is balanced by the force due to spatial variation of the pressure, P. The gravitational force acting on a unit volume of matter at a radius r is

$$F_{\text{grav}} = -\frac{Gm}{r^2}\rho \,, \qquad (II.1)$$

where G is the gravitational constant, $\rho(r)$ is the mass density, and $m(r)$ is the mass of the star interior to the radius r:

$$m(r) = 4\pi \int_0^r \rho(r')r'^2\,dr' \,. \qquad (II.2)$$

The force per unit volume of matter due to the changing pressure is $-dP/dr$. When the star is in equilibrium, the net force (gravitational plus pressure) on each bit of matter vanishes, so that, using (II.1), we have

$$\frac{dP}{dr} = -\frac{Gm(r)}{r^2}\rho(r) \,. \qquad (II.3)$$

A differential relation between the mass and the density can be obtained by differentiating (II.2):

$$\frac{dm}{dr} = 4\pi r^2 \rho(r) \, . \qquad (II.4)$$

The description is completed by specifying the "equation of state," an intrinsic property of the matter giving the pressure, $P(\rho)$, required to maintain it at a given density. Upon using the identity

$$\frac{dP}{dr} = \left(\frac{d\rho}{dr}\right)\left(\frac{dP}{d\rho}\right) ,$$

Eq. (II.3) can be written as

$$\frac{d\rho}{dr} = -\left(\frac{dP}{d\rho}\right)^{-1} \frac{Gm}{r^2}\rho \, . \qquad (II.5)$$

Equations (II.4) and (II.5) are two coupled first-order differential equations that determine the structure of the star for a given equation of state. The values of the dependent variables at $r = 0$ are $\rho = \rho_c$, the central density, and $m = 0$. Integration outward in r then gives the density profile, the radius of the star, R, being determined by the point at which ρ vanishes. (On very general grounds, we expect the density to decrease with increasing distance from the center.) The total mass of the star is then $M = m(R)$. Since both R and M depend upon ρ_c, variation of this parameter allows stars of different mass to be studied.

II.2 The equation of state

We must now determine the equation of state appropriate for a white dwarf. As mentioned above, we will assume that the matter consists of large nuclei and their electrons. The nuclei, being heavy, contribute nearly all of the mass but make almost no contribution to the pressure, since they hardly move about at all. The electrons, however, contribute virtually all of the pressure but essentially none of the mass. We will be interested in densities far greater than that of ordinary matter, where the electrons are no longer bound to individual nuclei, but rather move freely through the material. A good model is then a free Fermi gas of electrons at zero temperature, treated with relativistic kinematics.

For matter at a given mass density, the number density of electrons is

$$n = Y_e \frac{\rho}{M_p} \, , \qquad (II.6)$$

where M_p is the proton mass (we neglect the small difference between the neutron and proton masses) and Y_e is the number of electrons per nucleon. If the nuclei are all ^{56}Fe, then

$$Y_e = \frac{26}{56} = 0.464 \ ,$$

while $Y_e = \frac{1}{2}$ if the nuclei are ^{12}C; electrical neutrality of the matter requires one electron for every proton.

The free Fermi gas is studied by considering a large volume V containing N electrons that occupy the lowest energy plane-wave states with momentum $p < p_f$. Remembering the two-fold spin degeneracy of each plane wave, we have

$$N = 2V \int_0^{p_f} \frac{d^3 p}{(2\pi)^3} \ , \qquad (II.7)$$

which leads to

$$p_f = (3\pi^2 n)^{1/3} \ , \qquad (II.8)$$

where $n = N/V$. (We put $\hbar = c = 1$ unless indicated explicitly.) The total energy density of these electrons is

$$\frac{E}{V} = 2 \int_0^{p_f} \frac{d^3 p}{(2\pi)^3} (p^2 + m_e^2)^{1/2} \ , \qquad (II.9)$$

which can be integrated to give

$$\frac{E}{V} = n_0 m_e x^3 \epsilon(x) \ ; \qquad (II.10a)$$

$$\epsilon(x) = \frac{3}{8x^3} \left[x(1 + 2x^2)(1 + x^2)^{1/2} - \log \left[x + (1 + x^2)^{1/2} \right] \right] \qquad (II.10b)$$

where

$$x \equiv \frac{p_f}{m_e} = \left(\frac{n}{n_0} \right)^{1/3} \ ; \quad n_0 = \frac{m_e^3}{3\pi^2} \ . \qquad (II.10c)$$

The variable x characterizes the electron density in terms of

$$n_0 = 5.89 \times 10^{29} \ \text{cm}^{-3} \ ,$$

the density at which the Fermi momentum is equal to the electron mass, m_e.

In the usual thermodynamic manner, the pressure is related to how the energy changes with volume at fixed N:

$$P = -\frac{\partial E}{\partial V} = -\frac{\partial E}{\partial x}\frac{\partial x}{\partial V} . \qquad (II.11)$$

Using (II.10c) to find

$$\frac{\partial x}{\partial V} = -\frac{x}{3V}$$

and differentiating (II.10a,b) results in

$$P = \frac{1}{3}n_0 m_e x^4 \epsilon' , \qquad (II.12)$$

where $\epsilon' = d\epsilon/dx$.

It is now straightforward to calculate $dP/d\rho$. Since P is most naturally expressed in terms of x, we must relate x to ρ. This is most easily done using (II.10c) and (II.6):

$$x = \left(\frac{n}{n_0}\right)^{1/3} = \left(\frac{\rho}{\rho_0}\right)^{1/3} ; \qquad (II.13a)$$

$$\rho_0 = \frac{M_p n_0}{Y_e} = 9.79 \times 10^5 Y_e^{-1} \text{ gm cm}^{-3} . \qquad (II.13b)$$

Thus, ρ_0 is the mass density of matter in which the electron density is n_0. Differentiating the pressure and using (II.12, II.13a) then yields, after some algebra,

$$\frac{dP}{d\rho} = \frac{dP}{dx}\frac{dx}{d\rho} = Y_e \frac{m_e}{M_p}\gamma(x) ;$$

$$\gamma(x) = \frac{1}{9x^2}\frac{d}{dx}(x^4\epsilon') = \frac{x^2}{3(1+x^2)^{1/2}} . \qquad (II.14)$$

II.3 Scaling the equations

It is often useful to reduce equations describing a physical system to dimensionless form, both for physical insight and for numerical convenience (i.e., to avoid dealing with very large or very small numbers in the computer). To do this for the equations of white dwarf structure, we introduce dimensionless radius, density, and mass variables:

$$r = R_0 \bar{r} \, , \; \rho = \rho_0 \bar{r} \, , m = M_0 \bar{m} \, , \qquad (II.15)$$

with the radius and mass scales, R_0 and M_0, to be determined for convenience. Substituting into Eqs. (II.4, II.5) and using (II.14) yields, after some rearrangement,

$$\frac{d\bar{m}}{d\bar{r}} = \left(\frac{4\pi R_0^3 \rho_0}{M_0} \right) \bar{r}^2 \bar{\rho} \; ; \qquad (II.16a)$$

$$\frac{d\bar{\rho}}{d\bar{r}} = - \left(\frac{G M_0}{R_0 Y_e (m_e / M_p)} \right) \frac{\bar{m}\bar{\rho}}{\gamma \bar{r}^2} \, . \qquad (II.16b)$$

If we now choose M_0 and R_0 so that the coefficients in parentheses in these two equations are unity, we find

$$R_0 = \left[\frac{Y_e (m_e / M_p)}{4\pi G \rho_0} \right]^{1/2} = 7.72 \times 10^8 Y_e \text{ cm} \, , \qquad (II.17a)$$

$$M_0 = 4\pi R_0^3 \rho_0 = 5.67 \times 10^{33} \, Y_e^2 \text{ gm} \, , \qquad (II.17b)$$

and the dimensionless differential equations are

$$\frac{d\bar{\rho}}{d\bar{r}} = -\frac{\bar{m}\bar{\rho}}{\gamma \bar{r}^2} \; ; \qquad (II.18a)$$

$$\frac{d\bar{m}}{d\bar{r}} = \bar{r}^2 \bar{\rho} \, . \qquad (II.18b)$$

These equations are completed by recalling that γ is given by Eq. (II.14) with $x = \bar{\rho}^{1/3}$.

The scaling quantities (II.13b) and (II.17) tell us how to relate the structures of stars with different Y_e and hence reduce the numerical work to be done. Specifically, if we have the solution for $Y_e = 1$ at some value of $\bar{\rho}_c = \rho_c / \rho_0$, then the solution at any other Y_e is obtained by scaling the density as Y_e^{-1}, the radius as Y_e, and the mass as Y_e^2.

To give some feeling for the scales (II.17), recall that the solar radius and mass are

$$R_\odot = 6.95 \times 10^{10} \text{ cm} ; \quad M_\odot = 1.98 \times 10^{33} \text{ gm} ,$$

while the density at the center of the sun is ≈ 150 gm cm^{-3}. We therefore expect white dwarf stars to have masses comparable to a solar mass, but their radii will be considerably smaller and their densities will be much higher.

II.4 Solving the equations

We are now in a position to solve the equations and determine the properties of a white dwarf. This can be done in the following sequence of steps.

■ **Step 1** Verify the steps in the derivations given above. Determine the leading behavior of ϵ and γ in the extreme non-relativistic ($x \ll 1$) and relativistic ($x \gg 1$) limits and verify that these are what you expect from simple arguments. Analyze Eqs. (II.18a,b) when $\bar{m}$ and $\bar{r}$ are finite and $\bar{\rho}$ is small to show that $\bar{\rho}$ vanishes at a finite radius; the star therefore has a well-defined surface. Determine the functional form with which $\bar{\rho}$ vanishes near the surface.

■ **Step 2** Write a program to integrate Eqs. (II.18) outward from $\bar{r} = 0$ for $Y_e = 1$. Calculate the density profiles, total masses, and radii of stars with selected values of the dimensionless central density $\bar{\rho}_c$ ranging from 10^{-1} to 10^6. By changing the radial step and the integration algorithm, verify that your solutions are accurate. Show that the total kinetic and rest energy of the electrons in the star is

$$U = \int_0^R \left(\frac{E}{V}\right) 4\pi r^2 dr ,$$

where E/V is the energy density given by (II.10) and that the total gravitational energy of the star can be written as

$$W = -\int_0^R \frac{Gm(r)}{r} \rho(r) 4\pi r^2 dr .$$

Use the scaling discussed above to cast these integrals in dimensionless form and calculate these energies for the solutions you've generated. Try to understand all of the trends you observe through simple physical reasoning.

■ **Step 3** A commonly used description of stellar structure involves a polytropic equation of state,

$$P = P_0 \left(\frac{\rho}{\rho_0} \right)^\Gamma \,,$$

where P_0, ρ_0, and Γ are constants, the latter being the adiabatic index. By suitably scaling Eqs. (II.4,5), show that a star with a polytropic equation of state obeys Eqs. (II.18) with

$$\gamma = \bar{\rho}^{\Gamma-1} = x^{3\Gamma-3} \,.$$

The degenerate electron gas therefore corresponds to $\Gamma = 4/3$ in the extreme relativistic (high-density) limit and $\Gamma = 5/3$ in the extreme non-relativistic (low-density) limit.

   Polytropic equations of state provide a simple way to study the effects of varying the equation of state (by changing Γ) and also give, for two special values of Γ, analytical solutions against which numerical solutions can be checked. For this latter purpose, it is most convenient to recast Eqs. (II.18) as a single second-order differential equation for $\bar{\rho}$ by solving (II.18a) for $\bar{m}$ and then differentiating with respect to $\bar{r}$ and using (II.18b). Show that the differential equation that results (the Lane-Emden equation, [Ch57]) is

$$\frac{1}{\bar{r}^2} \frac{d}{d\bar{r}} \left(\bar{r}^2 \bar{\rho}^{\Gamma-2} \frac{d\bar{\rho}}{d\bar{r}} \right) = -\bar{\rho} \,,$$

and that for $\Gamma = 2$ the solution is

$$\bar{\rho} = \bar{\rho}_c \frac{\sin \bar{r}}{\bar{r}} \,,$$

while for $\Gamma = 6/5$ the solution is

$$\bar{\rho} = \bar{\rho}_c \left(1 + \frac{\bar{r}^2}{3a^2} \right)^{-5/2} \; ; \; a = 5^{\frac{1}{2}} \bar{\rho}_c^{-2/5} \,.$$

Use these analytical solutions to check the code you wrote in Step 2. Then study the structure of stars with different adiabatic indexes Γ and interpret the results.

■ **Step 4** If things are working correctly, you should find that, as the central density of a white dwarf increases, its mass approaches a limiting value, (the Chandrasekhar mass, M_{Ch}) and the star becomes very small. To understand this and to get an estimate of M_{Ch}, we can follow the steps in an argument originally given by Landau in 1932. The total energy of the star is composed of the gravitational energy ($W < 0$) and the internal energy of the matter ($U > 0$). Assume that a star of a given total mass has a *constant* density profile. (This is not a very good assumption for quantitative purposes, but it is valid enough to be useful in understanding the situation.) The radius is therefore related simply to the total mass and the constant density. Calculate U and W for this density profile, assuming that the density is high enough so that the electrons can be treated in the relativistic limit. Show that both energies scale as $1/R$ and that W dominates U when the total mass exceeds a certain critical value. It then becomes energetically favorable for the star to collapse (shrink its radius to zero). Estimate the Chandrasekhar mass in this way and compare it to the result of your numerical calculation. Also verify the correctness of this argument by showing that U and W for your solutions found in Step 2 become equal and opposite as the star approaches the limiting mass.

■ **Step 5** Scale the mass-radius relation you found in Step 1 to the cases corresponding to ^{56}Fe and ^{12}C nuclei. Three white dwarf stars, Sirius B, 40 Eri B, and Stein 2051, have masses and radii (in units of the solar values) determined from observation to be (1.053 ± 0.028, 0.0074 ± 0.0006), (0.48 ± 0.02, 0.0124 ± 0.0005), and (0.50 ± 0.05 or 0.72 ± 0.08, 0.0115 ± 0.0012), respectively [Sh83]. Verify that these values are consistent with the model we have developed. What can you say about the compositions of these stars?

Chapter 3

Boundary Value and Eigenvalue Problems

Many of the important differential equations of physics can be cast in the form of a linear, second-order equation:

$$\frac{d^2y}{dx^2} + k^2(x)y = S(x) \,, \tag{3.1}$$

where S is an inhomogeneous ("driving") term and k^2 is a real function. When k^2 is positive, the solutions of the homogeneous equation (i.e., $S = 0$) are oscillatory with local wavenumber k, while when k^2 is negative, the solutions grow or decay exponentially at a local rate $(-k^2)^{\frac{1}{2}}$. For example, consider trying to find the electrostatic potential, Φ, generated by a localized charge distribution, $\rho(\mathbf{r})$. Poisson's equation is

$$\nabla^2 \Phi = -4\pi\rho \,, \tag{3.2}$$

which, for a spherically symmetric ρ and Φ, simplifies to

$$\frac{1}{r^2}\frac{d}{dr}\left(r^2\frac{d\Phi}{dr}\right) = -4\pi\rho \,. \tag{3.3}$$

The standard substitution

$$\Phi(r) = r^{-1}\phi(r)$$

then results in

$$\frac{d^2\phi}{dr^2} = -4\pi r\rho \,, \tag{3.4}$$

which is of the form (3.1) with $k^2 = 0$ and $S = -4\pi r\rho$. In a similar manner, the quantum mechanical wave function for a particle of mass m

and energy E moving in a central potential $V(r)$ can be written as

$$\Psi(\mathbf{r}) = r^{-1}R(r)Y_{LM}(\hat{\mathbf{r}}) \, ,$$

where Y_{LM} is a spherical harmonic and the radial wave function R satisfies

$$\frac{d^2R}{dr^2} + k^2(r)R = 0 \; ; \quad k^2(r) = \frac{2m}{\hbar^2}\left[E - \frac{L(L+1)\hbar^2}{2mr^2} - V(r)\right] \, . \quad (3.5)$$

This is also of the form (3.1), with $S = 0$.

The equations discussed above appear unremarkable and readily treated by the methods discussed in Chapter 2, except for two points. First, the boundary conditions imposed by the physics often appear as constraints on the dependent variable at two *separate* points of the independent variable, so that solution as an initial value problem is not obviously possible. Moreover, the Schroedinger equation (3.5) is an eigenvalue equation in which we must *find* the energies that lead to physically acceptable solutions satisfying the appropriate boundary conditions. This chapter is concerned with methods for treating such problems. We begin by deriving an integration algorithm particularly well suited to equations of the form (3.1), and then discuss boundary value and eigenvalue problems in turn.

3.1 The Numerov algorithm

There is a particularly simple and efficient method for integrating second-order differential equations having the form of (3.1). To derive this method, commonly called the Numerov or Cowling's method, we begin by approximating the second derivative in (3.1) by the three-point difference formula (1.7),

$$\frac{y_{n+1} - 2y_n + y_{n-1}}{h^2} = y_n'' + \frac{h^2}{12}y_n'''' + \mathcal{O}(h^4) \, , \quad (3.6)$$

where we have written out explicitly the $\mathcal{O}(h^2)$ "error" term, which is derived easily from the Taylor expansion (1.1, 1.2a). From the differential equation itself, we have

$$\begin{aligned}
y_n'''' &= \frac{d^2}{dx^2}(-k^2y + S)\big|_{x=x_n} \\
&= -\frac{(k^2y)_{n+1} - 2(k^2y)_n + (k^2y)_{n-1}}{h^2} \\
&\quad + \frac{S_{n+1} - 2S_n + S_{n-1}}{h^2} + \mathcal{O}(h^2) \, .
\end{aligned} \quad (3.7)$$

When this is substituted into (3.6), we can write, after some rearrangement,

$$(1 + \frac{h^2}{12}k_{n+1}^2)y_{n+1} - 2(1 - \frac{5h^2}{12}k_n^2)y_n + (1 + \frac{h^2}{12}k_{n-1}^2)y_{n-1}$$

$$= \frac{h^2}{12}(S_{n+1} + 10S_n + S_{n-1}) + \mathcal{O}(h^6) . \quad (3.8)$$

Solving this linear equation for either y_{n+1} or y_{n-1} then provides a recursion relation for integrating either forward or backward in x, with a local error $\mathcal{O}(h^6)$. Note that this is one order more accurate than the fourth-order Runge-Kutta method (2.25), which might be used to integrate the problem as two coupled first-order equations. The Numerov scheme is also more efficient, as each step requires the computation of k^2 and S at only the lattice points.

■ **Exercise 3.1**   Apply the Numerov algorithm to the problem

$$\frac{d^2y}{dx^2} = -4\pi^2 y ; \quad y(0) = 1 , \quad y'(0) = 0 .$$

Integrate from $x = 0$ to $x = 1$ with various step sizes and compare the efficiency and accuracy with some of the methods discussed in Chapter 2. Note that you will have to use some special procedure (e.g., a Taylor series) to generate the value of $y_1 \equiv y(h)$ needed to start the three-term recursion relation.

3.2 Direct integration of boundary value problems

As a concrete illustration of boundary value problems, consider trying to solve Poisson's equation (3.4) when the charge distribution is

$$\rho(r) = \frac{1}{8\pi}e^{-r} , \quad (3.9)$$

which has a total charge

$$Q = \int \rho(r)d^3r = \int_0^\infty \rho(r)4\pi r^2 \, dr = 1 .$$

The exact solution to this problem is

$$\phi(r) = 1 - \frac{1}{2}(r + 2)e^{-r}, \quad (3.10)$$

from which $\Phi = r^{-1}\phi$ follows immediately. This solution has the expected behavior at large r, $\phi \to 1$, which corresponds to $\Phi \to r^{-1}$, the Coulomb potential from a unit charge.

Suppose that we try to solve this example as an ordinary initial value problem. Since ρ has no singular behavior at the origin (e.g., there is no point charge), Φ is regular there, which implies that $\phi = r\Phi$ vanishes at $r = 0$; this is indeed the case for the explicit solution (3.10). We could then integrate (3.4) outward from the origin using the appropriate rearrangement of (3.8) (recall $k^2 = 0$ here):

$$\phi_{n+1} = 2\phi_n - \phi_{n-1} + \frac{h^2}{12}(S_{n+1} + 10S_n + S_{n-1}), \qquad (3.11)$$

with

$$S = -4\pi r\rho = -\frac{1}{2}re^{-r}.$$

However, to do so we must know the value of ϕ_1 (or, equivalently, $d\phi/dr = \Phi$ at $r = 0$) in addition to $\phi_0 = 0$. This is not a very happy situation, since ϕ_1 is part of the very function we're trying to find, and so is not known *a priori*. We will discuss below what to do in the general case, but in the present example, since we have an analytical solution, we can find $\phi_1 = \phi(r = h)$ from (3.10). The following FORTRAN program does the outward integration to $r = 20$, storing the solution in an array and printing the exact result and error as it goes along.

```
C chap3a.for
        DIMENSION PHI(0:200)              !array for the solution
        EXACT(R)=1.-(R+2)*EXP(-R)/2       !exact solution
        SOURCE(R)=-R*EXP(-R)/2            !source function
        H=.1                             !radial step
        NSTEP=20./H                      !number of points to R=20
        CONST=H**2/12                    !constant in Numerov method
        SM=0.                            !source at R=0
        SZ=SOURCE(H)                     !source at R=H
        PHI(0)=0                         !boundary condition at R=0
        PHI(1)=EXACT(H)                  !exact value at first point
        DO 10 IR=1,NSTEP-1               !loop for outward integration
          R=(IR+1)*H                     !radius at next point
          SP=SOURCE(R)                   !source at next point
                                         !Eq. 3.11
          PHI(IR+1)=2*PHI(IR)-PHI(IR-1)+CONST*(SP+10.*SZ+SM)
          SM=SZ                          !roll values
          SZ=SP
          PRINT *, R,EXACT(R),PHI(IR+1),EXACT-PHI(IR+1)
10      CONTINUE
        STOP
```

Table 3.1   Errors in solving the Poisson problem defined by Eqs. (3.4,3.9)

r	Exact $\phi(r)$	Analytical $\phi(h)$	5% Error in $\phi(h)$	Linear correction
2	0.729330	−0.000003	0.049919	−0.000016
4	0.945053	−0.000006	0.099838	−0.000011
6	0.990085	−0.000005	0.149762	−0.000007
8	0.998323	0.000001	0.199690	−0.000003
10	0.999728	0.000010	0.249622	−0.000001
12	0.999957	0.000022	0.299556	0.000002
14	0.999993	0.000036	0.349493	0.000003
16	0.999999	0.000052	0.399431	0.000000
18	1.000000	0.000065	0.449366	0.000000
20	1.000000	0.000077	0.499301	0.000000

Note how computation is minimized by keeping track of the values of S at the current and adjacent lattice points (SZ, SM, and SP) and by computing the constant $h^2/12$ appearing in the Numerov formula (3.11) outside of the integration loop.

Results generated by this program are given in the first three columns of Table 3.1, which show that the numerical solution is rather accurate for small r. All is not well, however. The error per step gets larger at large r (the error after the 20 steps from $r = 0$ to $r = 2$ is 3×10^{-6}, while during the 20 steps from $r = 18$ to $r = 20$, it grows by 1.2×10^{-5}, 4 times as much), and the solution will become quite inaccurate if continued to even larger radii. This behavior is quite surprising because the errors in the Numerov integration should be getting smaller at large r as ϕ becomes a constant.

Further symptoms of a problem can be found by considering a more general case. We then have no analytical formula to give us Φ near the origin, which we need to get the three-term recursion relation started. One way to proceed is find $\Phi(0)$ by direct numerical quadrature of the Coulomb potential,

$$\Phi(0) = \int \frac{\rho(r)}{r} d^3r = 4\pi \int_0^\infty r\rho \, dr \,,$$

perhaps using Simpson's rule. There will, however, always be some error associated with the value obtained. We can simulate such an error in the code above (suppose it is 5%) by inserting the line (just before the DO loop)

```
PHI(1)=.95*PHI(1)
```

The code then gives the errors listed in the fourth column of Table 3.1. Evidently disaster has struck, for a 5% change in the initial conditions has induced a 50% error in the solution at large r.

It is simple to understand what has happened. Solutions to the homogeneous version of (3.4),

$$\frac{d^2\phi}{dr^2} = 0 \,,$$

can be added to any particular solution of (3.4) to give yet another solution. There are two linearly independent homogeneous solutions,

$$\phi \sim r \; ; \; \Phi \sim \text{constant} \,,$$

and

$$\phi \sim \text{constant} \; ; \; \Phi \sim r^{-1} \,.$$

The general solution to (3.4) in the asymptotic region (where ρ vanishes and the equation is homogeneous) can be written as a linear combination of these two functions, but the latter, sub-dominant solution is the physical one, since we know the potential at large r is given by $\Phi \to 1/r$. Imprecision in the specification of Φ at the origin or any numerical round-off error in the integration process can introduce a small admixture of the $\phi \sim r$ solution, which eventually dominates at large r.

The cure for this difficulty is straightforward: subtract a multiple of the "bad," unphysical solution to the homogeneous equation from the numerical result to guarantee the physical behavior in the asymptotic region. It is easy to see that the "bad" results shown in the fourth column of Table 3.1 vary linearly with r for large r. The following lines of code (which can be inserted just before the STOP statement) then fit the last 10 points of the numerical solution to the form

$$\phi = mr + b$$

and subtract mr from the numerical results to guarantee the appropriate large-r behavior.

```
SLOPE=(PHI(NSTEP)-PHI(NSTEP-10))/(10*H)
DO 20 IR=1,NSTEP
  R=IR*H
  PHI(IR)=PHI(IR)-SLOPE*R
  DIFF=EXACT(R)-PHI(IR)
```

```
      PRINT *, R,EXACT(R),PHI(IR),DIFF
20    CONTINUE
```

The errors in ϕ so obtained are shown in the final column of Table 3.1; the solution is even more accurate at large r than the uncorrected one found when the exact value of PHI(1) is used to start the integration.

In this simple example, the instabilities are not too severe; satisfactory results for moderate values of r are obtained with outward integration when the exact (or reasonably accurate approximate) value of PHI(1) is used. Alternatively, it is also feasible to integrate inward, starting at large r with $\phi = Q$, independent of r. This results in a solution that often satisfies accurately the boundary condition at $r = 0$ and avoids having to perform a quadrature to determine the (approximate) starting value of PHI(1).

■ **Exercise 3.2**   Solve the problem defined by Eqs. (3.4, 3.9) by Numerov integration inward from large r using the known asymptotic behavior of ϕ for the starting values. How well does your solution satisfy the boundary condition $\phi(r = 0) = 0$?

3.3 Green's function solution of boundary value problems

When the two solutions to the homogeneous equation have very different behaviors, some extra precautions must be taken. For example, in describing the potential from a charge distribution of a multipole order $l > 0$, the monopole equation (3.4) is modified to

$$\left[\frac{d^2}{dr^2} - \frac{l(l+1)}{r^2}\right]\phi = -4\pi r \rho , \qquad (3.12)$$

which has the two homogeneous solutions

$$\phi \sim r^{l+1} ; \quad \phi \sim r^{-l} .$$

For large r, the first of these solutions is much larger than the second, so that ensuring the correct asymptotic behavior by subtracting a multiple of this dominant homogeneous solution from a particular solution we have found by outward integration is subject to large round-off errors. Inward integration is also unsatisfactory, in that the unphysical r^{-l} solution is likely to dominate at small r.

One possible way to generate an accurate solution is by combining the two methods. Inward integration can be used to obtain the potential for r greater than some intermediate radius, r_m, and outward integration can be used for the potential when $r < r_m$. As long as r_m is chosen so that neither homogeneous solution is dominant, the outer and inner potentials obtained respectively from these two integrations will match at r_m and, together, describe the entire solution. Of course, if the inner and outer potentials don't quite match, a multiple of the homogeneous solution can be added to the former to correct for any deficiencies in our knowledge of $\phi'(r = 0)$.

Sometimes the two homogeneous solutions have such different behaviors that it is impossible to find a value of r_m that permits satisfactory integration of the inner and outer potentials. Such cases can be solved by the Green's function of the homogeneous equation. To illustrate, let us consider Eq. (3.1) with the boundary condition $\phi(x = 0) = \phi(x = \infty) = 0$. Since the problem is linear, we can write the solution as

$$\phi(x) = \int_0^\infty G(x, x') S(x') \, dx' , \tag{3.13}$$

where G is a Green's function satisfying

$$\left[\frac{d^2}{dx^2} + k^2(x) \right] G(x, x') = \delta(x - x') . \tag{3.14}$$

It is clear that G satisfies the homogeneous equation for $x \neq x'$. However, the derivative of G is discontinuous at $x = x'$, as can be seen by integrating (3.14) from $x = x' - \epsilon$ to $x = x' + \epsilon$, where ϵ is an infinitesimal:

$$\left. \frac{dG}{dx} \right|_{x=x'+\epsilon} - \left. \frac{dG}{dx} \right|_{x=x'-\epsilon} = 1 . \tag{3.15}$$

The problem, of course, is to find G. This can be done by considering two solutions to the homogeneous problem, $\phi_<$ and $\phi_>$, satisfying the boundary conditions at $x = 0$ and $x = \infty$, respectively, and normalized so that their Wronskian,

$$W = \frac{d\phi_>}{dx} \phi_< - \frac{d\phi_<}{dx} \phi_> , \tag{3.16}$$

is unity. (It is easy to use the homogeneous equation to show that W is independent of x). Then the Green's function is given by

$$G(x, x') = \phi_<(x_<) \phi_>(x_>) , \tag{3.17}$$

where $x_<$ and $x_>$ are the smaller and larger of x and x', respectively. It is evident that this expression for G satisfies the homogeneous equation and the discontinuity condition (3.15). From (3.13), we then have the explicit solution

$$\phi(x) = \phi_>(x) \int_0^x \phi_<(x')S(x')dx' + \phi_<(x) \int_x^\infty \phi_>(x')S(x')\,dx' \ . \quad (3.18)$$

This expression can be evaluated by a numerical quadrature and is not subject to any of the stability problems we have seen associated with a direct integration of the inhomogeneous equation.

In the case of arbitrary k^2, the homogeneous solutions $\phi_<$ and $\phi_>$ can be found numerically by outward and inward integrations, respectively, of initial value problems and then normalized to satisfy (3.16). However, for simple forms of $k^2(x)$, they are known analytically. For example, for the problem defined by Eq. (3.12), it is easy to show that

$$\phi_<(r) = r^{l+1} \ ; \ \phi_>(r) = -\frac{1}{2l+1}r^{-l}$$

are one possible set of homogeneous solutions satisfying the appropriate boundary conditions and Eq. (3.16).

■ **Exercise 3.3** Solve the problem defined by Eqs. (3.9, 3.12) for $l = 0$ using the Green's function method. Compare your results with those obtained by direct integration and with the analytical solution.

■ **Exercise 3.4** Spherically symmetric solutions to the equation

$$(\nabla^2 - a^2)\Phi = -4\pi\rho$$

lead to the ordinary differential equation

$$\left(\frac{d^2}{dr^2} - a^2\right)\phi = -4\pi r\rho$$

with the boundary conditions $\phi(r = 0) = \phi(r \to \infty) = 0$. Here, a is a constant. Write a program to solve this problem using the Green's function when ρ is given by Eq. (3.9). Compare your numerical results for various values of a with the analytical solution,

$$\phi = \left(\frac{1}{1-a^2}\right)^2 \left(e^{-ar} - e^{-r}\left[1 + \frac{1}{2}(1 - a^2)r\right]\right) \ .$$

What happens if you try to solve this problem by integrating only inward or only outward? What happens if you try a solution by integrating inward and outward to an intermediate matching radius? How do your results change when you vary the matching radius?

3.4 Eigenvalues of the wave equation

Eigenvalue problems involving differential equations often arise in finding the normal-mode solutions of wave equations. As a simple example with which to illustrate a method of solution, we consider the normal modes of a stretched string of uniform mass density. After a suitable scaling of the physical quantities, the equation and boundary conditions defining these modes can be written as

$$\frac{d^2\phi}{dx^2} = -k^2\phi \; ; \quad \phi(x = 0) = \phi(x = 1) = 0 \; . \tag{3.19}$$

Here, $0 < x < 1$ is the scaled coordinate along the string, ϕ is the transverse displacement of the string, and k is the constant wavenumber, linearly related to the frequency of vibration. This equation is an eigenvalue equation in the sense that solutions satisfying the boundary conditions exist only for particular values of k, $\{k_n\}$, which we must find. Furthermore, it is linear and homogeneous, so that the normalization of the eigenfunction corresponding to any of the k_n, ϕ_n, is not fixed, but can be chosen for convenience.

The (un-normalized) eigenfunctions and eigenvalues of this problem are well-known analytically:

$$k_n = n\pi \; ; \quad \phi_n \sim \sin n\pi x \; , \tag{3.20}$$

where n is a positive integer. These provide a useful check of the numerical methods for solving this problem.

One suitable general strategy for numerical solution of an eigenvalue problem is an iterative one. We guess a trial eigenvalue and generate a solution by integrating the differential equation as a initial value problem. If the resulting solution does not satisfy the boundary conditions, we change the trial eigenvalue and integrate again, repeating the process until a trial eigenvalue is found for which the boundary conditions are satisfied to within a predetermined tolerance.

For the problem at hand, this strategy (known picturesquely as the "shooting" method) can be implemented as follows. For each trial value of k, we integrate forward from $x = 0$ with the initial conditions

$$\phi(x = 0) = 0 \; , \quad \phi'(x = 0) = \delta \; .$$

The number δ is arbitrary and can be chosen for convenience, since the problem we are solving is a homogeneous one and the normalization of the solutions is not specified. Upon integrating to $x = 1$, we will find, in general, a non-vanishing value of ϕ, since the trial eigenvalue will not be one of the true eigenvalues. We must then readjust k and integrate again, repeating the process until we find $\phi(x = 1) = 0$ to within a specified tolerance; we will have then found an eigenvalue and the corresponding eigenfunction.

The problem of finding a value of k for which $\phi(1)$ vanishes is a root-finding problem of the type discussed in Chapter 1. Note that the Newton-Raphson method is inappropriate since we cannot differentiate explicitly the numerically determined value of $\phi(1)$ with respect to k and the secant method could be dangerous, as there are many eigenvalues and it might be difficult to control the one to which the iterations will ultimately converge. Therefore, it is safest to use a simple search to locate an approximate eigenvalue and then, if desired, switch to the more efficient secant method.

The following FORTRAN program finds the lowest eigenvalue of stretched string problem (3.19) by the shooting method described above, printing the trial eigenvalue as it goes along. The search (beginning at line 10) is terminated when the eigenvalue is determined within a precision of 10^{-5}. The initial trial eigenvalue and the search step size are set in the second and third lines.

```
C chap3b.for
      REAL K
      K=1.                      !initial values to start search
      DK=1.
      TOLK=1.E-05               !search tolerance
      CALL INTGRT(K,PHIP)       !find PHIP at first guess for K
      PHIOLD=PHIP               !save initial PHIP value
10    CONTINUE                  !beginning of search loop
         K=K+DK                 !increment K
         CALL INTGRT(K,PHIP)    !calculate PHIP for this K
         IF (PHIP*PHIOLD .LT. 0) THEN
            K=K-DK              !if PHIP changes sign, back up
            DK=DK/2             ! and halve the step
         END IF
      IF (ABS(DK) .GT. TOLK) GOTO 10
                               !continue until DK is small enough
      EXACT=4.*ATAN(1.)         !exact value=pi
      PRINT *, ' eigenvalue, error =',K,EXACT-K
      STOP
```

```
      END
      !subroutine to calculate phi(x=1)= PHIP for value of K input
      SUBROUTINE INTGRT(K,PHIP)
      REAL K
      DATA NSTEP/100/            !setup lattice parameters
      H=1./NSTEP
      PHIM=0.                    !initial conditions
      PHIZ=.01
      CONST=(K*H)**2/12.         !constant in Numerov method
      DO 10 IX=1,NSTEP-1         !forward recursion to X=1
         PHIP=2*(1.-5.*CONST)*PHIZ -(1.+CONST)*PHIM   !Eq 3.8
         PHIP=PHIP/(1+CONST)     !the rest of Eq.3.8
         PHIM=PHIZ               !roll values
         PHIZ=PHIP
10    CONTINUE
      PRINT *, K,PHIP
      RETURN
      END
```

When run, this program generates results that converge to a value close to the exact answer, π, the error being caused by the finite integration step and the value of TOLK.

■ **Exercise 3.5**   Use the program above to find some of the higher eigenvalues. Note that the numerical inaccuracies become greater as the eigenvalue increases and the integration of the more rapidly oscillating eigenfunction becomes inaccurate. Change the search algorithm to the more efficient secant method. How close does your initial guess have to be in order to converge to a given eigenvalue? Change the code to correspond to the boundary conditions

$$\phi'(x = 0) = 0 \ , \quad \phi(x = 1) = 0 \ ,$$

and verify that the numerical eigenvalues agree with the analytical values expected.

■ **Exercise 3.6**   The wave equation in cylindrical geometry often leads to the eigenvalue problem

$$\left(\frac{d^2}{dr^2} + \frac{1}{r}\frac{d}{dr}\right)\Phi(r) = -k^2\Phi \ ; \ \ \Phi(r = 0) = 1, \ \ \Phi(r = 1) = 0 \ .$$

The analytical eigenfunctions are the regular cylindrical Bessel function of order zero, the eigenvalues being the zeros of this function:

$$k_1 = 2.404826, \quad k_2 = 5.520078, \quad k_3 = 8.653728, \quad k_4 = 11.791534, \quad \ldots \ .$$

Show that the substitution $\Phi = r^{-1/2}\phi$ changes this equation into one for which the Numerov algorithm is suitable and modify the code above to solve this problem. Compare the numerical eigenvalues with the exact values.

3.5 Stationary solutions of the one-dimensional Schroedinger equation

A rich example of the shooting method for eigenvalue problems is the task of finding the stationary quantum states of a particle of mass m moving in a one-dimensional potential $V(x)$. We'll assume that $V(x)$ has roughly the form shown in Figure 3.1: the potential becomes infinite at $x = x_{\min}$ and $x = x_{\max}$ (i.e., there are "walls" at these positions) and has a well somewhere in between. The time-independent Schroedinger equation and boundary conditions defining the stationary states are [Me68]

$$\frac{d^2\psi}{dx^2} + k^2(x)\psi(x) = 0 \ ; \quad \psi(x_{\min}) = \psi(x_{\max}) = 0 \ , \qquad (3.21)$$

which is of the form (3.1) with

$$k^2(x) = \frac{2m}{\hbar^2}[E - V(x)] \ .$$

We must find the energies E (eigenvalues) for which there is a non-zero solution to this problem. At one of these eigenvalues, we expect the eigenfunction to oscillate in the classically allowed regions where $E > V(x)$ and to behave exponentially in the classically forbidden regions where $E < V(x)$. Thus, there will be "bound" solutions with $E < 0$, which are localized within the well and decay exponentially toward the walls and "continuum" solutions with $E > 0$, which have roughly constant magnitude throughout the entire region between the walls.

The eigenvalue problem defined by Eq. (3.21) can be solved by the shooting method. Suppose that we are seeking a bound state and so take a negative trial eigenvalue. Upon integrating toward larger x from $x_{\min}$, we can generate a solution, $\psi_<$, which increases exponentially through the classically forbidden region and then oscillates beyond the left turning point in the classically allowed region (see the lower portion of Figure 3.1). If we were to continue integrating past the right turning point,

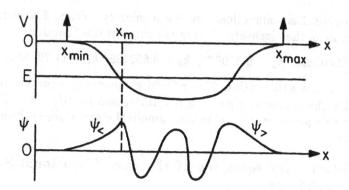

Figure 3.1 (Upper) Schematic potential used in the discussion of the one-dimensional Schroedinger equation. (Lower) Solutions $\psi_<$ and $\psi_>$ of the Schroedinger equation at an arbitrary energy $E < 0$. The left turning point is used as the matching point. When E is an eigenvalue, the derivative is continuous at the matching point.

the integration would become numerically unstable since, even at an exact eigenvalue where $\psi_<(x_{max}) = 0$, there can be an admixture of the undesirable exponentially growing solution. As a general rule, integration *into* a classically forbidden region is likely to be inaccurate. Therefore, at each energy it is wiser to generate a second solution, $\psi_>$, by integrating from x_{max} toward smaller x. To determine whether the energy is an eigenvalue, $\psi_<$ and $\psi_>$ can be compared at a matching point, x_m, chosen so that neither integration will be inaccurate. (A convenient choice for x_m is the left turning point.) Since both $\psi_<$ and $\psi_>$ satisfy a homogeneous equation, their normalizations can always be chosen so that the two functions are equal at x_m. An eigenvalue is then signaled by equality of the derivatives at x_m; i.e., the solutions match smoothly, as is invoked in analytical solutions of such problems. Thus,

$$\frac{d\psi_<}{dx}\bigg|_{x_m} - \frac{d\psi_>}{dx}\bigg|_{x_m} = 0 . \qquad (3.22)$$

If we approximate the derivatives by their simplest finite difference approximations (1.4) using the points x_m and $x_m - h$, an equivalent condition is

$$f \equiv \frac{1}{\psi}[\psi_<(x_m - h) - \psi_>(x_m - h)] = 0 , \qquad (3.23)$$

since the normalizations have been chosen to guarantee $\psi_<(x_m) = \psi_>(x_m)$. The quantity ψ in Eq. (3.23) is a convenient scale for the difference, which can be chosen to make f typically of order unity. It might be the value of $\psi_<$ at x_m or the maximum value of $\psi_<$ or $\psi_>$. Note that if there are no turning points (e.g., if $E > 0$ for a potential like that shown in Figure 3.1), then x_m can be chosen anywhere, while if there are more than two turning points, three or more homogeneous solutions, each accurate in different regions, must be patched together.

The program for Example 3, whose FORTRAN source code is given in Appendix B and in the file EXMPL3.FOR, solves for the stationary states of a one-dimensional potential by the algorithm described above on a lattice of size determined by the parameter NPTS. The potential is assumed to be of the form $V(x) = V_0 v(x)$, where the dimensionless function $v(x)$ has a minimum value of -1 and a maximum value of $+1$. (This can always be guaranteed by a suitable linear scaling of the energies.) If the coordinate is scaled by a physical length a, the Schroedinger equation (3.21) can be written as

$$\left[-\frac{1}{\gamma^2} \frac{d^2}{dx^2} + v(x) - \epsilon \right] \psi(x) = 0 \,,$$

where

$$\gamma = \left(\frac{2ma^2 V_0}{\hbar^2} \right)^{1/2}$$

is a dimensionless measure of the classical nature of the system and $\epsilon = E/V_0$ is the dimensionless energy. All eigenvalues therefore satisfy $\epsilon > -1$. The functional form of the potential defined by the program can be any one of three analytical types (square well, parabolic well, and Lennard-Jones potential); the square well may also include a square bump, and the parabolic well may be flattened symmetrically about $x = 0$. For the value of γ input, a number of states are sought using an initial trial energy and energy increment. For each state, a simple search on the energy is made to try to zero the function f defined by (3.23) and, when an approximate eigenvalue is located, the secant method is employed until $|f|$ becomes less than the value of the matching tolerance (this value may be changed at run time) For each trial eigenvalue, the Schroedinger equation is integrated forward and backward, and the two solutions are matched at the left-most turning point where the behavior of the wave function changes from oscillatory to exponential (near x_{max} if there is no such turning point). As the search proceeds, the trial eigenvalue is displayed, as is the current step in the energy, the current value of f, and the number of nodes in the trial wave function. As the solution is likely to be inaccurate when it is integrated into a classically forbidden region, the program also indicates whether or not this is the case. When an eigenvalue is found, it is indicated on a graph of the potential by a line at the appropriate level between the left-most and right-most turning points and the eigenfunction is also graphed.

A difficulty arises from the arbitrary phase of the wave function and the dependence of the sign of f on that phase. If we are arbitrary about the sign of ψ, then the sign of f is also arbitrary. However, since the search procedure uses the sign of f as a signal for crossing a root, this situation must be avoided. Therefore, to ensure that f is a continuous function of energy the sign factor is completely determined once the search is started. Often in quantum mechanics, the overall sign of the wave function is of little consequence. However, in this program the sign of $\psi(x)$ determines the sign of f which in turn signals the crossing of a root. Therefore, to ensure that f is a continuous function of energy we must choose the sign of $\psi(x)$ carefully. At the beginning of the search a sign is chosen arbitrarily (e.g., $\psi_<(h) > 0$). This sign convention continues as the energy changes until a new node appears in the trial wave function. To keep f continuous, the portion of the wave function ($\psi_<$ or $\psi_>$) that gains the node must retain the same sign between steps in the search. (By drawing a few graphs you can convince yourself that this is the case.) This new sign convention is retained until another node appears, when the new sign is chosen by the same criterion.

The following exercises can help to improve your understanding of the physical principles and numerical methods illustrated in this example.

■ **Exercise 3.7** Verify that the code gives the expected answers for the eigenvalues of the square-well and parabolic-well potentials ($\gamma = 50$ might be a convenient value to use). Observe how the discontinuity in the derivative at the matching point is smoothed out as the energy converges to each eigenvalue. Also note the increase in the number of nodes with increasing eigenvalue and, for the parabolic well, the behavior of the solution near the turning points. For states with large quantum numbers, the amplitude of the oscillations becomes larger near the turning points, consistent with the behavior of the WKB wave function, which is proportional to the inverse square-root of the classical velocity, $(E - V)^{-1/4}$. Find some solutions in these potentials also for values of γ that are small (say 10) and large (say 200), corresponding to the extreme quantum and classical limits, respectively.

■ **Exercise 3.8**   For the analytically soluble square and parabolic wells, investigate the effects of changing the integration method from the Numerov algorithm to the "naïve" one obtained by approximating the second derivative by just the three-point formula (1.7); that is, neglecting the $\mathcal{O}(h^2)$ term in (3.6).

■ **Exercise 3.9**   Change the program so that the eigenvalues are found by integrating only forward and demanding that the wave function vanish at x_{max}. Observe the problems that arise in trying to integrate an exponentially dying solution into a classically forbidden region. (It is wise to keep γ relatively small here so that the instabilities don't become *too* large.)

■ **Exercise 3.10**   When the potential is reflection symmetric about $x = 0$, the eigenfunctions will have a definite parity (symmetric or anti-symmetric about $x = 0$) and that parity will alternate as the quantum number (energy) increases. Verify that this is the case for the numerical solutions generated by the code. Can you think of a way in which the parity can be exploited to halve the numerical effort involved in finding the eigenvalues of a symmetric potential? If so, modify the code to try it out.

■ **Exercise 3.11**   If we consider a situation where $v(x) = 0$ for $x < x_{min}$ and $x > x_{max}$ (i.e., we remove the walls), then the zero boundary conditions at the ends of the lattice are inappropriate for weakly bound states ($\epsilon \lesssim 0$) since the wave function decays very slowly as $|x|$ becomes large. More appropriate boundary conditions at x_{min} and x_{max} are therefore

$$\frac{1}{\psi}\frac{d\psi}{dx} = \pm\gamma(-\epsilon)^{1/2} \ .$$

Change the code to implement these boundary conditions and observe the effect on the wave functions and energies of the states near zero energy. Note that if we were to normalize the wave function in the conventional way, the contributions from these exponential tails would have to be included. Can you derive, in the style of the Numerov algorithm, a numerical realization of these boundary conditions accurate to a high order in DX?

■ **Exercise 3.12**   Check numerically that, for a given potential, two eigenfunctions, ψ_E and ψ'_E, corresponding to different eigenvalues E and E', are orthogonal,

$$\int \psi_E(x)\psi'_E(x)\,dx = 0 \ ,$$

as is required by the general principles of quantum mechanics.

■ **Exercise 3.13**   For small, intermediate, and large values of γ, compare the exact eigenvalues of the Lennard-Jones potential with the semiclassical energies generated by the code for Example 1.

■ **Exercise 3.14**   Investigate the eigenfunctions and eigenvalues for potentials you might have encountered in learning elementary quantum mechanics: the δ-function potential, a finite square-well, double-well potentials, periodic potentials, etc. This may be done by altering the potential in subroutine POTNTL. Interpret the wave functions and eigenvalues you find. (Note that the code will sometimes have trouble finding two eigenvalues that are nearly degenerate.)

Project III: Atomic structure in the Hartree-Fock approximation

The self-consistent field approximation (Hartree-Fock) is known to be an accurate description of many of the properties of multi-electron atoms and ions. In this approximation, each electron is described by a separate single-particle wave function (as distinct from the many-electron wave function) that solves a Schroedinger-like equation. The potential appearing in this equation is that generated by the average motion of all of the other electrons, and so depends on their single-particle wave functions. The result is a set of non-linear eigenvalue equations, which can be solved by the methods introduced in this chapter. In this project, we will solve the self-consistent field equations to determine the ground-state structure of small atomic systems (e.g., the atoms and ions of the elements in the periodic table from Hydrogen through Neon). The total energies calculated can be compared directly with experimental values. The brief derivation we give here can be supplemented with the material found in [Be68], [Me68], and [We80].

III.1 Basis of the Hartree-Fock approximation

The Hamiltonian for N electrons moving about a heavy nucleus of charge Z located at the origin can be written as

$$H = \sum_{i=1}^{N} \frac{p_i^2}{2m} - \sum_{i=1}^{N} \frac{Ze^2}{r_i} + \frac{1}{2} \sum_{i \neq j=1}^{N} \frac{e^2}{r_{ij}} \,. \qquad (III.1)$$

Here, the $\{r_i\}$ are the locations of the electrons, m and $-e$ are the electron mass and charge, and $r_{ij} \equiv r_i - r_j$ is the separation between electrons i and j. The three sums in (III.1) embody the electron kinetic energy, the electron-nucleus attraction, and the inter-electron repulsion. As is appropriate to the level of accuracy of the self-consistent field approximation, we have neglected much smaller terms, such as those associated with the spin-orbit interaction, hyperfine interactions, recoil motion of the nucleus, and relativity.

A proper quantum mechanical description requires that we specify the spin state of each electron, in addition to its location. This can be done by giving its spin projection on some fixed quantization axis, $\sigma_i = \pm 1/2$. For convenience, we will use the notation $x_i \equiv (\mathbf{r}_i, \sigma_i)$ to denote all of the coordinates (space and spin) of electron i.

The self-consistent field methods are based on the Rayleigh-Ritz variational principle, which states that the ground state eigenfunction of the Hamiltonian, $\Psi(x_1, x_2, \ldots, x_N)$, is that wave function that minimizes the expectation value of H,

$$E = \langle \Psi | H | \Psi \rangle \,, \qquad (III.2)$$

subject to the constraints that Ψ obey the Pauli principle (i.e., that it be anti-symmetric under the interchange of any two of the x's) and that it be normalized to unity:

$$\int |\Psi|^2 d^N x = 1 \,. \qquad (III.3)$$

(The notation $d^N x$ means integration over all of the spatial coordinates and summation over all of the spin coordinates of the N electrons.) Furthermore, this minimum value of E *is* the ground state energy. A calculation of (III.2) for *any* normalized and anti-symmetric trial function Ψ therefore furnishes an upper bound to the ground state energy.

The Hartree-Fock approximation is based on restricting the trial wave function to be a Slater determinant:

$$\Psi(x_1, x_2, \ldots, x_N) = (N!)^{-1/2} \det \psi_\alpha(x_j) \,. \qquad (III.4)$$

Here, the $\psi_\alpha(x)$ are a set of N orthonormal single-particle wave functions; they are functions of the coordinates of only a single electron. The determinant is that of the $N \times N$ matrix formed as α and x_i each take on their N possible values, while the factor $(N!)^{-1/2}$ ensures that Ψ is normalized according to (III.3). The physical interpretation of this wave function is that each of the electrons moves independently in an orbital ψ_α under the average influence of all the other electrons. This turns out to be a good approximation to the true wave function of an atom because the smooth Coulomb interaction between the electrons averages out many of the details of their motion.

Using the properties of determinants, it is easy to see that Ψ has the required anti-symmetry under interchange of any two electrons (a determinant changes sign whenever any two of its columns are interchanged) and that Ψ is properly normalized according to Eq. (III.3) if the single-particle wave functions are orthonormal:

$$\int \psi_\alpha^*(x)\psi_{\alpha'}(x)\,dx = \delta_{\alpha\alpha'} \ . \tag{III.5}$$

Since the Hamiltonian (III.1) does not involve the electron spin variables, the spins decouple from the space degrees of freedom, so that it is useful to write each single-particle wave function as a product of space and spin functions:

$$\psi_\alpha(x) = \chi_\alpha(\mathbf{r})|\sigma_\alpha\rangle \ , \tag{III.6}$$

where $\sigma_\alpha = \pm 1/2$ is the spin projection of the orbital α. The orthonormality constraint (III.5) then takes the form

$$\delta_{\sigma_\alpha\sigma_{\alpha'}} \int \chi_\alpha^*(\mathbf{r})\chi_{\alpha'}(\mathbf{r})\,d^3r = \delta_{\alpha\alpha'} \ , \tag{III.7}$$

so that orbitals can be orthogonal by either their spin or space dependence.

The computation of the energy (III.2) using the wave function defined by (III.4–6) is straightforward but tedious. After some algebra, we have

$$E = \sum_{\alpha=1}^{N} \left\langle \alpha \left| \frac{p^2}{2m} \right| \alpha \right\rangle + \int \left[-\frac{Ze^2}{r} + \frac{1}{2}\Phi(\mathbf{r}) \right] \rho(\mathbf{r})\,d^3r$$
$$- \frac{1}{2} \sum_{\alpha,\alpha'=1}^{N} \delta_{\sigma_\alpha\sigma_{\alpha'}} \left\langle \alpha\alpha' \left| \frac{e^2}{r_{ij}} \right| \alpha'\alpha \right\rangle \ . \tag{III.8}$$

In this expression, the one-body matrix elements of the kinetic energy are

$$\left\langle \alpha \left| \frac{p^2}{2m} \right| \alpha \right\rangle = -\frac{\hbar^2}{2m} \int \chi_\alpha^*(\mathbf{r})\nabla^2\chi_\alpha(\mathbf{r})\,d^3r \ , \tag{III.9}$$

the electron density is the sum of the single-particle densities,

$$\rho(\mathbf{r}) = \sum_{\alpha=1}^{N} |\chi_\alpha(\mathbf{r})|^2 \ , \tag{III.10}$$

the electrostatic potential generated by the electrons is

$$\Phi(\mathbf{r}) = e^2 \int \frac{1}{|\mathbf{r} - \mathbf{r}'|} \rho(\mathbf{r}') \, d^3 r' \,, \qquad (III.11a)$$

so that

$$\nabla^2 \Phi = -4\pi e^2 \rho(\mathbf{r}) \,, \qquad (III.11b)$$

and the exchange matrix elements of the inter-electron repulsion are

$$\left\langle \alpha \alpha' \left| \frac{e^2}{r_{ij}} \right| \alpha' \alpha \right\rangle = e^2 \int \chi_\alpha^*(\mathbf{r}) \chi_{\alpha'}^*(\mathbf{r}') \frac{1}{|\mathbf{r} - \mathbf{r}'|} \chi_{\alpha'}(\mathbf{r}) \chi_\alpha(\mathbf{r}') \, d^3 r \, d^3 r' \,.$$
$$(III.12)$$

The interpretation of the various terms in (III.8) is straightforward. The kinetic energy is the sum of the kinetic energies of the single particle orbitals, while the electron-nucleus attraction and direct inter-electron repulsion are just what would be expected from a total charge of $-Ne$ distributed in space with density $\rho(\mathbf{r})$. The final term in (III.8) is the exchange energy, which arises from the anti-symmetry of the trial wave function (III.4). It is a sum over all pairs of orbitals with the same spin projection; pairs of orbitals with different spin projections are "distinguishable" and therefore do not contribute to this term.

The strategy of the self-consistent field approach should now be clear. The variational wave function (III.4) depends on a set of "parameters": the values of the single-particle wave functions at each point in space. Variation of these parameters so as to minimize the energy (III.8) while respecting the constraints (III.7) results in a set of Euler-Lagrange equations (the Hartree-Fock equations) that define the "best" determinental wave function and give an optimal bound on the total energy. Because these equations are somewhat complicated in detail, we consider first the two-electron problem, and then turn to situations with three or more electrons.

III.2 The two-electron problem

For two electrons that don't interact with each other, the ground state of their motion around a nucleus is the $1s^2$ configuration; i.e., both electrons are in the same real, spherically symmetric spatial state, but have opposite spin projections. It is therefore natural to take a trial wave function for the interacting system that realizes this same configuration; the corresponding two single-particle wave functions are

$$\psi(x) = \frac{1}{(4\pi)^{1/2} r} R(r) \left| \pm \frac{1}{2} \right\rangle \,, \qquad (III.13)$$

so that the many-body wave function (III.4) is

$$\Psi = \frac{1}{\sqrt{2}}\frac{1}{4\pi r_1 r_2}R(r_1)R(r_2)\left[\left|+\frac{1}{2}\right\rangle\left|-\frac{1}{2}\right\rangle - \left|-\frac{1}{2}\right\rangle\left|+\frac{1}{2}\right\rangle\right] , \quad (III.14)$$

This trial wave function is anti-symmetric under the interchange of the electron spins but is symmetric under the interchange of their space coordinates. It respects the Pauli principle, since it is antisymmetric under the interchange of all variables describing the two electrons. The normalization condition (III.5) becomes

$$\int_0^\infty R^2(r)\,dr = 1 , \quad (III.15)$$

while the energy (III.8) becomes

$$E = 2 \times \frac{\hbar^2}{2m}\int_0^\infty \left(\frac{dR}{dr}\right)^2 dr + \int_0^\infty \left[-\frac{Ze^2}{r} + \frac{1}{4}\Phi(r)\right]\rho(r)4\pi r^2\,dr ,$$
$$(III.16)$$

with (III.10) reducing to

$$\rho(r) = 2 \times \frac{1}{4\pi r^2}R^2(r) ; \quad \int_0^\infty \rho(r)4\pi r^2 dr = 2 , \quad (III.17)$$

and (III.11b) becoming

$$\frac{1}{r^2}\frac{d}{dr}\left(r^2\frac{d\Phi}{dr}\right) = -4\pi e^2 \rho . \quad (III.18)$$

Note that the exchange energy is attractive and has a magnitude of one-half of that of the direct inter-electron repulsion [resulting in a net factor of 1/4 in the final term of (III.16)] and that various factors of two have entered from the sum over the two spin projections.

A common variational treatment of the two-electron system ("poor man's Hartree-Fock") takes R to be a hydrogenic 1s orbital parametrized by an effective charge, Z^*:

$$R(r) = 2\left(\frac{Z^*}{a}\right)^{1/2}\frac{Z^* r}{a}e^{-Z^* r/a} , \quad (III.19)$$

where a is the Bohr radius. The energy (III.16) is then minimized as a function of Z^* to find an approximation to the wave function and energy. This procedure, which is detailed in many textbooks (see, for example,

[Me68]), results in

$$Z^* = Z - \frac{5}{16} \; ; \; E = -\frac{e^2}{a}\left[Z^2 - \frac{5}{8}Z + \frac{25}{256}\right] . \qquad (III.20)$$

In carrying out this minimization, it is amusing to note that the kinetic energy scales as Z^{*2}, while all of the potential energies scale as Z^*, so that, at the optimal Z^*, the kinetic energy is $-1/2$ of the potential. This is a specific case of a more general virial theorem pertaining to the Hartree-Fock approximation (see Step 1 below).

The full Hartree-Fock approximation for the two-electron problem is very much in this same variational spirit, but the most general class of normalized single-particle wave functions is considered. That is, we consider E in Eq. (III.16) to be a *functional* of R and require that it be stationary with respect to all possible norm-conserving variations of the single-particle wave function. If the normalization constraint (III.15) is enforced by the method of Lagrange multipliers, for an arbitrary variation of $\delta R(r)$ we require

$$\delta\left(E - 2\epsilon \int_0^\infty R^2 \, dr\right) = 0 , \qquad (III.21)$$

where ϵ is a Lagrange multiplier to be determined after variation so that the solution is properly normalized. The standard techniques of variational calculus then lead to

$$\int_0^\infty \delta R(r)\left[-4\frac{\hbar^2}{2m}\frac{d^2}{dr^2} - 4\frac{Ze^2}{r} + 2\Phi(r) - 4\epsilon\right]R(r)\,dr = 0 , \qquad (III.22)$$

which is satisfied if R solves the Schroedinger-like equation

$$\left[-\frac{\hbar^2}{2m}\frac{d^2}{dr^2} - \frac{Ze^2}{r} + \frac{1}{2}\Phi(r) - \epsilon\right]R(r) = 0 . \qquad (III.23)$$

Choosing ϵ (the "single-particle energy") to be an eigenvalue of the single-particle Hamiltonian appearing in (III.23) ensures that R is normalizable. Equations (III.18,23) are the two coupled non-linear differential equations in one dimension that form the Hartree-Fock approximation to the original six-dimensional Schroedinger equation. Note that only one-half of Φ appears in (III.23) since each electron interacts only with the other and not "with itself"; inclusion of the exchange term in the energy (III.16) is necessary to get this bookkeeping right.

III.3 Many-electron systems

The assumption of spherical symmetry is an enormous simplification in the two-electron problem, as it allowed us to reduce the eigenvalue problem for the single-particle wave function and the Poisson equation for the potential from three-dimensional partial differential equations to ordinary differential equations. For the two-electron problem, it is plausible (and true) that a spherically symmetric solution has the lowest energy. However, for most many-electron systems, spherical symmetry of the density and potential are by no means guaranteed. In principle, non-spherical solutions should be considered, and such "deformed" wave functions are in fact the optimal ones for describing the structure of certain nuclei.

To understand what the problem is, let us assume that the potential Φ is spherically symmetric. The solutions to the single-particle Schroedinger equation in such a potential are organized into "shells," each characterized by an orbital angular momentum, l, and a radial quantum number, n. Within each shell, all $2(2l+1)$ orbitals associated with the various values of σ_α and the projection of the orbital angular momentum, m, are degenerate. The orbitals have the form

$$\chi_\alpha(\mathbf{r}) = \frac{1}{r} R_{nl}(r) Y_{lm}(\hat{\mathbf{r}}) \; ; \quad \int_0^\infty R_{nl}^2(r)\, dr = 1 \; . \qquad (III.24)$$

However, we must decide which of these orbitals to use in constructing the Hartree-Fock determinant. Unless the number of electrons is such that all of the $2(2l+1)$ substates of a given shell are filled, the density as given by (III.10) will not be spherically symmetric. This, in turn, leads to a non-symmetric potential and a much more difficult single-particle eigenvalue equation; the general problem is therefore intrinsically three-dimensional.

A slight modification of the rigorous Hartree-Fock method (the filling or central-field approximation) is useful in generating a spherically symmetric approximation to such "open-shell" systems. The basic idea is to spread the valence electrons uniformly over the last occupied shell. For example, in discussing the neutral Carbon atom, there would be 2 electrons in the $1s$ shell, 2 electrons in the $2s$ shell, and 2 electrons spread out over the 6 orbitals of the $2p$ shell. (Note that we don't put 4 electrons in the $2p$ shell and none in the $2s$ shell since the single-particle energy of the latter is expected to be more negative.) Thus, we introduce the number of electrons in each shell, N_{nl}, which can take on integer values between 0 and $2(2l+1)$, and, using the wave functions (III.24), write the density (III.10) as

$$\rho(r) = \frac{1}{4\pi r^2} \sum_{nl} N_{nl} R_{nl}^2(r) \; ; \quad \int_0^\infty \rho(r) 4\pi r^2 \, dr = \sum_{nl} N_{nl} = N \; .$$
$$(III.25)$$

In writing this expression, we have used the identity

$$\sum_{m=-l}^{l} |Y_{lm}(\hat{\mathbf{r}})|^2 = \frac{2l+1}{4\pi} \; .$$

In the same spirit, the energy functional (III.8) can be generalized to open-shell situations as

$$E = \sum_{nl} N_{nl} \frac{\hbar^2}{2m} \int_0^\infty \left[\left(\frac{dR_{nl}}{dr} \right)^2 + \frac{l(l+1)}{r^2} R_{nl}^2 \right] dr$$
$$+ \int_0^\infty \left[-\frac{Ze^2}{r} + \frac{1}{2}\Phi(r) \right] \rho(r) 4\pi r^2 \, dr + E_{\text{ex}} \; ,$$
$$(III.26a)$$

with the exchange energy being

$$E_{\text{ex}} = -\frac{1}{4} \sum_{nln'l'} N_{nl} N_{n'l'} \sum_{\lambda=|l-l'|}^{l+l'} \begin{pmatrix} l & l' & \lambda \\ 0 & 0 & 0 \end{pmatrix}^2 I_{nl,n'l'}^{\lambda} \; . \qquad (III.26b)$$

In this expression, I is the integral

$$I_{nl,n'l'}^{\lambda} = e^2 \int_0^\infty dr \int_0^\infty dr' \, R_{nl}(r) R_{n'l'}(r') \frac{r_<^\lambda}{r_>^{\lambda+1}} R_{n'l'}(r) R_{nl}(r') \; ,$$
$$(III.27)$$

where $r_<$ and $r_>$ are the smaller and larger of r and r' and the $3-j$ symbol vanishes when $l + l' + \lambda$ is odd and otherwise has the value

$$\begin{pmatrix} l & l' & \lambda \\ 0 & 0 & 0 \end{pmatrix}^2 = \frac{(-l+l'+\lambda)!\,(l-l'+\lambda)!\,(l+l'-\lambda)!}{(l+l'+\lambda+1)!}$$
$$\times \left[\frac{p!}{(p-l)!\,(p-l')!\,(p-\lambda)!} \right]^2 ,$$

where $p = \frac{1}{2}(l + l' + \lambda)$. In deriving these expressions, we have used the multipole decomposition of the Coulomb interaction and the standard techniques of angular momentum algebra [Br68].

The Hartree-Fock equations defining the optimal radial wave functions now follow from the calculus of variations, as in the two-electron case. Lagrange multipliers ϵ_{nl} are introduced to keep each of the radial wave function normalized and, after some algebra, we have

$$\left[-\frac{\hbar^2}{2m}\frac{d^2}{dr^2} + \frac{l(l+1)\hbar^2}{2mr^2} - \frac{Ze^2}{r} + \Phi(r) - \epsilon_{nl} \right] R_{nl}(r) = -F_{nl}(r) \,, \quad (III.28a)$$

with

$$F_{nl}(r) = -\frac{e^2}{2} \sum_{n'l'} N_{n'l'} R_{n'l'}(r) \sum_{\lambda=|l-l'|}^{l+l'} \begin{pmatrix} l & l' & \lambda \\ 0 & 0 & 0 \end{pmatrix}^2 J^\lambda_{nl,n'l'} \,; \quad (III.28b)$$

$$J^\lambda_{nl,n'l'} = \frac{1}{r^{\lambda+1}} \int_0^r R_{n'l'}(r') R_{nl}(r') r'^\lambda \, dr'$$

$$+ r^\lambda \int_r^\infty \frac{R_{n'l'}(r') R_{nl}(r')}{r'^{\lambda+1}} \, dr' \,. \quad (III.28c)$$

The eigenvalue equation (III.28a) can be seen to be analogous to (III.23) for the two-electron problem, except that the exchange energy has introduced a non-locality (Fock potential) embodied in F and has coupled together the eigenvalue equations for each of the radial wave functions; it is easy to show that these two equations are equivalent when there is a single orbital with $l = 0$. It is useful to note that (III.26b, 28b) imply that the exchange energy can be also be written as

$$E_{\text{ex}} = \frac{1}{2} \sum_{nl} N_{nl} \int_0^\infty R_{nl}(r) F_{nl}(r) \, dr \,, \quad (III.29)$$

and that, by multiplying (III.28a) by R_{nl} and integrating, we can express the single particle eigenvalue as

$$\epsilon_{nl} = \frac{\hbar^2}{2m} \int_0^\infty \left[\left(\frac{dR_{nl}}{dr}\right)^2 + \frac{l(l+1)}{r^2} R_{nl}^2 \right] dr$$

$$+ \int_0^\infty \left[-\frac{Ze^2}{r} + \Phi(r) \right] R_{nl}^2(r) \, dr + \int_0^\infty R_{nl}(r) F_{nl}(r) \, dr \,. \quad (III.30)$$

III.4 Solving the equations

For the numerical solution of the Hartree-Fock equations, we must first adopt a system of units. For comparison with experimental values, it is convenient to measure all lengths in Angstroms and all energies in electron volts. If we use the constants

$$\frac{\hbar^2}{m} = 7.6359 \text{ eV-Å}^2 \; ; \; e^2 = 14.409 \text{ eV-Å} , \qquad (III.31)$$

then the Bohr radius and Rydberg have their correct values,

$$a = \frac{\hbar^2}{me^2} = 0.5299 \text{ Å} \; ; \; \text{Ry} = \frac{e^2}{2a} = 13.595 \text{ eV} . \qquad (III.32)$$

For a large atom with many electrons, the accurate solution of the Hartree-Fock equations is a considerable task. However, if we consider the ground states of systems with at most 10 electrons (requiring three shells: $1s$, $2s$, and $2p$), then the numerical work can be managed in a reasonable amount of time. A lattice of several hundred points with a radial step size of $\lesssim 0.01$ Å extending out to ≈ 3 Å should be sufficient for most cases.

The best approach to developing a program to solve the Hartree-Fock equations is to consider the two-electron problem first, for which there is only a single radial wave function solving a local eigenvalue equation, and then to consider the more complex case of several orbitals. The attack can be made through the following sequence of steps.

■ **Step 1** Verify the algebra leading to the final equations presented above for the two-electron system [Eqs. (III.16,18,23)] and for the multi-electron system [Eqs. (III.18,26,28)] and make sure that you understand the physical principles behind the derivations. Prove the virial theorem that the kinetic energy is $-1/2$ of the potential energy. This can be done by imagining that the single-particle wave functions of a solution to the Hartree-Fock equations are subject to a norm-preserving scaling transformation,

$$\chi_\alpha(\mathbf{r}) \to \tau^{3/2} \chi(\tau \mathbf{r}) ,$$

where τ is a dimensionless scaling parameter. Show that the total kinetic energy in (III.8) scales as τ^2, while all of the potential energies scale as τ. Since the energy at the Hartree-Fock solution is stationary with respect to *any* variation of the wave functions, use

$$\left. \frac{\partial E}{\partial \tau} \right|_{\tau=1} = 0$$

to prove the theorem.

■ **Step 2** Write a program to calculate the energy from (III.16) if R is known at all of the lattice points. This will require writing a subroutine that calculates Φ by solving (III.18) (you might modify the one given earlier in this chapter) and then evaluating suitable quadratures for the various terms in (III.16). Verify that your program is working by calculating the energies associated with the hydrogenic orbital (III.19) and comparing it with the analytical results [remember to normalize the wave function by the appropriate discretization of (III.15)].

■ **Step 3** Write a subroutine that uses the shooting method to solve the radial equation (III.23) for the lowest eigenvalue ϵ and corresponding normalized wave function R if the potential Φ is given at the lattice points. The zero boundary condition at the origin is easily implemented, but the boundary condition at large distances can be taken as $R(r = L) = 0$, where L is the outer end of the lattice. (Greater accuracy, particularly for weakly bound states, can be had by imposing instead an exponential boundary condition at the outer radius.) Note that the radial scale (i.e., R and the radial step size) should change with the strength of the central charge. Verify that your subroutine works by setting Φ to 0 and comparing, for $Z = 2$ and $Z = 4$, the calculated wave function, eigenvalue, and energy of the 1s orbital with the analytical hydrogenic values.

■ **Step 4** Combine the subroutines developed in Steps 2 and 3 into a code that, given a value of Z, solves the two-electron Hartree-Fock equations by iteration. An iteration scheme is as follows, the organization into subroutines being obvious:

i) "Guess" an initial wave function, say the hydrogenic one (III.19) with the appropriate value of Z^*.

ii) Solve (III.18) for the potential generated by the initial wave function and calculate the total energy of the system from Eq. (III.16).

iii) Find a new wave function and its eigenvalue by solving (III.23) and normalizing according to (III.15).

iv) Calculate the new potential and new total energy. Then go back to *ii*) and repeat *ii*) and *iii*) until the total energy has converged to within the required tolerance.

At each iteration, you should print out the eigenvalue, the total energy, and the three separate contributions to the energy appearing in (III.16); a plot of the wave function is also useful for monitoring the calculation. Note that the total energy should decrease as the iterations proceed and will converge relatively quickly to a minimum. The individual contributions to the energy will take longer to settle down, consistent with the

Table III.1: Binding energies (in eV) of small atomic systems

Z	Number of electrons, N							
	2	3	4	5	6	7	8	9
1	14.34							
2	78.88							
3	198.04	203.43						
4	371.51	389.71	399.03					
5	599.43	637.35	662.49	670.79				
6	881.83	946.30	994.17	1018.55	1029.81			
7	1218.76	1316.62	1394.07	1441.19	1471.09	1485.62		
8	1610.23	1743.31	1862.19	1939.58	1994.47	2029.58	2043.19	
9	2054.80	2239.93	2397.05	2511.27	2598.41	2661.05	2696.03	2713.45

fact that it is only the total energy that is stationary at the variational minimum, not the individual components; at convergence, the virial theorem discussed in Step 1 should be satisfied. Try beginning the iteration procedure with different single-particle wave functions and note that the converged solution is still the same. Vary the values of the lattice spacing and the boundary radius, L, and prove that your results are stable under these changes.

■ **Step 5** Use your program to solve the Hartree-Fock equations for central charges $Z = 1$-9. Compare the total energies obtained with the experimental values given in $N = 2$ column of Table III.1. (These binding energies, which are the negative of the total energies, are obtained from the measured ionization potentials of atoms and ions given in [We71].) Compare your results also with the wave functions and associated variational energies given by Eqs. (III.19,20). Note that both approximations should give upper bounds to the exact energy. Give a physical explanation for the qualitative behavior of the discrepancies as a function of Z. Can you use second-order perturbation theory to show that the discrepancy between the Hartree-Fock and exact energies should become a constant for large Z? Show that for $Z = 1$, the Hartree-Fock approximation predicts that the H^- ion is unbound in that its energy is greater than that of the H atom and so it is energetically favorable to shed the extra electron. As can be seen from Table III.1, this is not the case in the real world. In finding the $Z = 1$ solution, you might discover that convergence is quite a delicate business; it is very easy for the density to change so much from iteration to iteration that the lowest eigenvalue of the single-particle Hamiltonian becomes positive. One way to alleviate this problem is to prevent the density from changing too much from one iteration to the next, for example by averaging the new density and the old following step *ii)* above.

■ **Step 6** Modify your two-electron program to treat systems in which several orbitals are involved. It is easiest to first modify the calculation of the total energy for a given set of radial wave functions to include E_{ex}. This is most conveniently done by calculating and storing the F_{nl} of Eq. (III.28b) and using Eq. (III.29). Because of the Fock term, the eigenvalue equations (III.28a) cannot be treated by the shooting method we have discussed. However, one scheme is to treat the F_{nl} calculated from the previous set of wave functions as inhomogeneous terms in solving for the new set of wave functions. For trial values of the ϵ_{nl} calculated from (III.30) using the previous set of wave functions, (III.28a) can be solved as uncoupled inhomogeneous boundary value problems using the Green's function method of Eq. (3.18); after normalization according to (III.24), the solutions serve as a new set of wave functions. The two-electron systems can be used to check the accuracy of your modifications; for these systems you should find that the exchange energy is $-1/2$ of the direct inter-electron interaction energy and that the solutions converge to the same results as those generated by the code in Step 4. Use this Hartree-Fock code to calculate the wave functions and energies for some of the other systems listed in Table III.1 and compare your results with the experimental values; interpret what you find. A convenient set of initial wave functions are the hydrogenic orbitals, given by (III.19) for the 1*s* state and

$$R_{2s}(r) = 2\left(\frac{Z^*}{2a}\right)^{1/2}\left(1 - \frac{Z^*r}{2a}\right)\frac{Z^*r}{2a}e^{-Z^*r/2a}\,,$$

$$R_{2p}(r) = \left(\frac{2Z^*}{3a}\right)^{1/2}\left(\frac{Z^*r}{2a}\right)^2 e^{-Z^*r/2a}\,,$$

for the 2*s* and 2*p* states, respectively. The optimal common value of Z^* in these expressions should be determined for any system by minimizing the total energy.

Special Functions
and
Gaussian Quadrature

In this chapter, we discuss two loosely related topics: algorithms for computing the special functions of mathematical physics (Bessel functions, orthogonal polynomials, etc.) and efficient methods of quadrature based on orthogonal functions. In most scientific computing, large libraries supply almost all of the subroutines relevant to these tasks and so relieve the individual from the tedium of writing code. In fact, there is usually little need to know very much about how these subroutines work in detail. However, a rough idea of the methods used is useful; this is what we hope to impart in this chapter.

4.1 Special functions

The special functions of mathematical physics were largely developed long before large-scale numerical computation became feasible, when analytical methods were the rule. Nevertheless, they are still relevant today, for two reasons. One is the insight analytical solutions offer; they guide our intuition and provide a framework for the qualitative interpretation of more complicated problems. However, of particular importance to numerical work is the fact that special functions often allow part of a problem to be solved analytically and so dramatically reduce the amount of computation required for a full solution.

As an illustration, consider a one-dimensional harmonic oscillator moving under an external perturbation: its frequency, ω, is being changed with time. Suppose that the frequency has its unperturbed value, ω_0, for times before $t = 0$ and for times after $t > T$, and that we are interested in the motion for times long after the perturbation ceases. Given the oscillator's initial coordinate and velocity, one straightforward method of solution is to integrate the equations of motion,

$$\frac{dx}{dt} = v(t) \; ; \quad \frac{dv}{dt} = -\omega^2(t)x(t) \, ,$$

as an initial-value problem using one of the methods discussed in Chapter 2. However, this would be inefficient, as the motion after the perturbation stops $(t > T)$ is well understood and is readily expressed in terms of the "special" sine and cosine functions involved,

$$x(t > T) = x(T)\cos \omega_0(t - T) + \omega_0^{-1}v(T)\sin \omega_0(t - T) \, .$$

Since there are very efficient methods for computing the trigonometric functions, it is wiser to integrate numerically only the non-trivial part of the motion $(0 < t < T)$ and then to use the velocity and coordinate at $t = T$ to compute directly the sinusoidal function given above. Although this example might seem trivial, the concept of using special functions to "do part of the work" is a general one.

A useful resource in dealing with special functions is the *Handbook of Mathematical Functions* [Ab64]. This book contains the definitions and properties of most of the functions one often needs. Methods for computing them are also given, as well as tables of their values for selected arguments. These last are particularly useful for checking the accuracy of the subroutines you are using.

Recursion is a particularly simple way of computing some special functions. Many functions are labeled by an order or index and satisfy recursion relations with respect to this label. If the function can be computed explicitly for the few lowest orders, then the higher orders can be found from these formulas. As an example, consider the computation of the Legendre polynomials, $P_l(x)$, for $|x| \leq 1$ and $l = 0, 1, 2, \ldots$. These are important in the solution of wave equations in situations with a spherical symmetry. The recursion relation with respect to degree is

$$(l + 1)P_{l+1}(x) + lP_{l-1}(x) - (2l + 1)xP_l(x) = 0 \, . \tag{4.1}$$

Using the explicit values $P_0(x) = 1$ and $P_1(x) = x$, forward recursion in l yields P_l for any higher value of l required. The following FORTRAN program accomplishes this for any value for x and l input.

```
C chap4a.for
20      PRINT *, ' Enter x, 1 (1 .lt. 0 to stop)'
        READ *, X,L
        IF (L .LT. 0) THEN
            STOP
```

```
      ELSE IF (L .EQ. 0) THEN
            PL=0.                         !explicit form for L=0
      ELSE IF (L .EQ. 1) THEN
            PL=X                          !explicit form for L=1
      ELSE
            PM=1.                         !values to start recursion
            PZ=X
            DO 10 IL=1,L-1                !loop for forward recursion
            PP=((2*IL+1)*X*PZ-IL*PM)/(IL+1) !Eq. 4.1
            PM=PZ                         !roll values
            PZ=PP
10          CONTINUE
            PL=PZ
      END IF
      PRINT *,X,L,PL
      GOTO 20
      END
```

This code works with no problems, and the results agree with the values given in the tables to the arithmetic precision of the computer. We can also compute the derivatives of the Legendre polynomials with this algorithm using the relation

$$(1 - x^2)P_l' = -lxP_l + lP_{l-1} \ . \tag{4.2}$$

Other sets of orthogonal polynomials, such as Hermite and Laguerre, can be treated similarly.

As a second example, consider the cylindrical Bessel functions, $J_n(x)$ and $Y_n(x)$, which arise as the regular and irregular solutions to wave equations in cylindrical geometries. These functions satisfy the recursion relation

$$C_{n-1}(x) + C_{n+1}(x) = \frac{2n}{x}C_n(x) \ , \tag{4.3}$$

where C_n is either J_n or Y_n. To use these recursion relations in the forward direction, we need the values of C_0 and C_1. These are most easily obtained from the polynomial approximations given in [Ab64], formulas 9.4.1–3. For $|x| < 3$, we have

$$\begin{aligned} J_0(x) = 1 &- 2.2499997 \, y^2 + 1.2656208y^4 \\ &- 0.3163866y^6 + 0.0444479y^8 - 0.039444y^{10} \\ &+ 0.0002100y^{12} + \epsilon \ ; \quad |\epsilon| \leq 5 \times 10^{-8} \ , \end{aligned} \tag{4.4a}$$

Table 4.1   Forward recursion for the irregular Bessel function $Y_n(2)$

n	$Y_n(2)$
0	+0.51037
1	−0.10703
2	−0.61741
3	−1.1278
4	−2.7659
5	−9.9360
6	−46.914
7	−271.55
8	−1853.9
9	−14560.

where $y = x/3$ and

$$
Y_0(x) = \frac{2}{\pi} \log\left(\frac{1}{2}x\right) J_0(x) + 0.36746691 + 0.605593666 y^2
$$
$$
- 0.74350384 y^4 + 0.25300117 y^6 - 0.04261214 y^8
$$
$$
+ 0.00427916 y^{10} - 0.00024846 y^{12} + \epsilon ; \quad |\epsilon| \le 1.4 \times 10^{-8} ;
$$

$$(4.4b)$$

while for $x > 3$,

$$
J_0(x) = x^{-1/2} f_0 \cos\theta \qquad Y_0(x) = x^{-1/2} f_0 \sin\theta \qquad (4.4c)
$$

where

$$
f_0 = 0.79788456 - 0.00000077 y^{-1} - 0.00552740 y^{-2}
$$
$$
- 0.00009512 y^{-3} + 0.00137237 y^{-4} - 0.00072805 y^{-5} \qquad (4.4d)
$$
$$
+ 0.00014476 y^{-6} + \epsilon ; \quad |\epsilon| < 1.6 \times 10^{-8} ,
$$

and

$$
\theta = x - 0.78539816 - 0.04166397 y^{-1} - 0.00003954 y^{-2}
$$
$$
+ 0.00262573 y^{-3} - 0.00054125 y^{-4} - 0.00029333 y^{-5} \qquad (4.4e)
$$
$$
+ 0.00013558 y^{-6} + \epsilon ; \quad |\epsilon| < 7 \times 10^{-8}
$$

Similar expressions for J_1 and Y_1 are given in Sections 9.4.5–6 of [Ab64]. Note that these formulas are *not* Taylor series, but rather polynomials whose coefficients have been adjusted to best represent the Bessel functions over the intervals given.

Table 4.2   Computation of the regular Bessel function $J_n(2)$

n	Exact value	Error in forward recursion	Un-normalized backward recursion	Error in normalized backward recursion
0	$0.223891E + 00$	$0.000000E + 00$	$0.150602E - 10$	$0.000000E + 00$
1	$0.576725E + 00$	$0.000000E + 00$	$0.387940E - 10$	$0.000000E + 00$
2	$0.352834E + 00$	$0.000000E + 00$	$0.237337E - 10$	$0.000000E + 00$
3	$0.128943E + 00$	$0.000000E + 00$	$0.867350E - 11$	$0.000000E + 00$
4	$0.339957E - 01$	$-0.000002E - 01$	$0.228676E - 11$	$0.000000E - 01$
5	$0.703963E - 02$	$-0.000075E - 02$	$0.473528E - 12$	$0.000000E - 02$
6	$0.120243E - 02$	$-0.000355E - 02$	$0.808826E - 13$	$0.000000E - 02$
7	$0.174944E - 03$	$-0.020559E - 03$	$0.117678E - 13$	$0.000000E - 03$
8	$0.221795E - 04$	$-0.140363E - 03$	$0.149193E - 14$	$0.000000E - 04$
9	$0.249234E - 05$	$-0.110234E - 02$	$0.167650E - 15$	$0.000000E - 05$
10	$0.251539E - 06$	$-0.978959E - 02$	$0.169200E - 16$	$0.000000E - 06$
11	$0.230428E - 07$		$0.155000E - 17$	$0.000000E - 07$
12	$0.193270E - 08$		$0.130000E - 18$	$0.000007E - 08$
13	$0.149494E - 09$		$0.100000E - 19$	$0.000830E - 09$
14	$0.107295E - 10$		$0.000000E - 19$	$0.107295E - 10$

Let us now attempt to calculate the Y_n by forward recursion using (4.3) together with the polynomial approximations for the values of Y_0 and Y_1. Doing so leads to results that reproduce the values given in the tables of Chapter 9 of [Ab64] to within the arithmetic precision of the computer. For example, we find the results listed Table 4.1 for $x = 2$.

It is natural to try to compute the regular solutions, J_n, with the same forward recursion procedure. Using Eq. (4.4a) and its analog for J_1, we find the errors listed in the third column of Table 4.2; the exact values are given in the second column. As can be seen, forward recursion gives good results for $n < 5$, but there are gross errors for the higher values of n.

It is relatively easy to understand what is going wrong. We can think about the recursion relation (4.3) for the Bessel functions as the finite difference analog of a second-order differential equation in n. In fact, if we subtract $2C_n$ from both sides of (4.3), we obtain

$$C_{n+1} - 2C_n + C_{n-1} = 2\left(\frac{n}{x} - 1\right)C_n , \qquad (4.5)$$

which, in the limit of continuous n, we can approximate by

$$\frac{d^2C}{dn^2} = -k^2(n)C \; ; \; k^2(n) = 2\left(1 - \frac{n}{x}\right) . \tag{4.6}$$

In deriving this equation, we have used the three-point finite difference formula (1.7) for the second derivative with $h = 1$ and have identified the local wavenumber, $k^2(n)$. Equation (4.6) will have two linearly independent solutions, either both oscillatory in character (when k^2 is positive, or when $n < x$) or one exponentially growing and the other exponentially decreasing (when k^2 is negative, or $n > x$). As is clear from Table 4.1, Y_n is the solution that grows exponentially with increasing n, so that no loss of precision occurs in forward recursion. However, Table 4.2 shows that the exact values of J_n decrease rapidly with increasing n, and so precision is lost rapidly as forward recursion proceeds beyond $n = x$. This disease is the same as that encountered in Chapter 3 in integrating exponentially dying solutions of second-order differential equations; its cure is also the same: avoid using the recursion relation in the direction of decreasing values of the function.

■ **Exercise 4.1** Use Eq. (4.1) to show that recursion of the Legendre polynomials is stable in either direction.

To compute the regular cylindrical Bessel functions accurately, we can exploit the linearity of the recursion relation and use Eq. (4.3) in the direction of decreasing n. Suppose we are interested in $J_n(2)$ for $n \leq 10$. Then, choosing $J_{14} = 0$ and $J_{13} = 1 \times 10^{-20}$, an arbitrarily small number, we can recur backwards to $n = 0$. The resulting sequence of numbers will then reproduce the J_n, to within an arbitrary normalization, since, as long as we have chosen the initial value of n high enough, the required solution of the difference equation (4.3), which grows exponentially with decreasing n, will dominate for small n. The sequence can then be normalized through the identity

$$J_0(x) + 2J_2(x) + 2J_4(x) + \ldots = 1 . \tag{4.7}$$

The following FORTRAN code evaluates the regular cylindrical Bessel functions using backward recursion.

```
C chap4b.for
      REAL J(0:50)
C
```

```
100     PRINT *, ' Enter max value of n (.le. 50; .lt. 0 to stop)'
        READ *, NMAX
        IF (NMAX .LT. 0) STOP
        IF (NMAX .GT. 50) NMAX=50
        PRINT *,' Enter value of x'
        READ *,X
C
        J(NMAX)=0.                          !backward recursions
        J(NMAX-1)=1.E-20                    !initial conditions
        DO 10 N=NMAX-1,1,-1
           J(N-1)=(2*N/X)*J(N)-J(N+1)      !Eq. 4.3
10      CONTINUE
C
        SUM=J(0)                            !calculate sum in Eq. 4.7
        DO 20 N=2,NMAX,2
           SUM=SUM+2*J(N)
20      CONTINUE
C
        DO 30 N=0,NMAX                      !normalize and output
           J(N)=J(N)/SUM
           PRINT *,N,J(N)
30      CONTINUE
C
        GOTO 100
        END
```

When run for NMAX=14 and X=2, it gives the unnormalized values (i.e., after line 10) shown in the fourth column of Table 4.2 and the errors in the final values shown in the fifth column of that table. The results are surprisingly accurate, even for values of n close to 14. An alternative way of obtaining the constant with which to normalize the whole series is to calculate the value of $J_0(2)$ from the polynomial approximation (4.4a).

■ **Exercise 4.2** Run the code above for various values of NMAX at fixed X. By comparing with tabulated values, verify that the results are accurate as long as NMAX is large enough (somewhat greater than the larger of X and the maximum order desired). Change the normalization algorithm to use the approximations (4.4a,c) for J_0.

■ **Exercise 4.3** The regular and irregular spherical Bessel functions, j_l and n_l, satisfy the recursion relation

$$s_{l+1} + s_{l-1} = \frac{2l+1}{x} s_l \, ,$$

where s_l is either j_l or n_l. The explicit formulas for the few lowest orders are

$$j_0 = \frac{\sin x}{x} \; ; \; j_1 = \frac{\sin x}{x^2} - \frac{\cos x}{x} \; ; \; j_2 = \left(\frac{3}{x^3} - \frac{1}{x} \right) \sin x - \frac{3}{x^2} \cos x \, ,$$

and

$$n_0 = -\frac{\cos x}{x} \; ; \; n_1 = -\frac{\cos x}{x^2} - \frac{\sin x}{x} \; ; \; n_2 = \left(-\frac{3}{x^3} + \frac{1}{x} \right) \cos x - \frac{3}{x^2} \sin x \, .$$

At $x = 0.5$, the exact values of the functions of order 2 are

$$n_2 = -25.059923 \; ; \; j_2 = 1.6371107 \times 10^{-2} \, .$$

Show that n_2 can be calculated either by explicit evaluation or by forward recursion and convince yourself that the latter method will work for all l and x. Investigate the calculation of $j_2(0.5)$ by forward recursion, explicit evaluation, and by backward recursion and show that the first two methods can be quite inaccurate. Can you see why? Thus, even if explicit expressions for a function are available, the stability of backward recursion can make it the method of choice.

Our discussion has illustrated some pitfalls in computing the some commonly used special functions. Specific methods useful for other functions can be found in the appropriate chapters of [Ab64].

4.2 Gaussian quadrature

In Chapter 1, we discussed several methods for computing definite integrals that were most convenient when the integrand was known at a series of equally spaced lattice points. While such methods enjoy widespread use, especially when the integrand involves a numerically-generated solution to a differential equation, more efficient quadrature schemes exist if we can evaluate the integrand for arbitrary abscissae. One of the most useful of these is Gaussian quadrature.

Consider the problem of evaluating

$$I = \int_{-1}^{1} f(x) \, dx \, .$$

The formulas discussed in Chapter 1 were of the form

$$I \approx \sum_{n=1}^{N} w_n f(x_n) \,, \tag{4.8}$$

where

$$x_n = -1 + 2\frac{(n-1)}{(N-1)}$$

are the equally spaced lattice points. Here, we are referring to the "elementary" quadrature formulas [such as (1.9), (1.11), or (1.13a,b)], and not to compound formulas such as (1.12). For example, for Simpson's rule (1.11), $N = 3$ and

$$x_1 = -1 \,, \quad x_2 = 0 \,, \quad x_3 = 1 \,; \quad w_1 = w_3 = \frac{1}{3} \,, \quad w_2 = \frac{4}{3} \,.$$

From the derivation of Simpson's rule, it is clear that the formula is exact when f a polynomial of degree 3 or less, which is commensurate with the error estimate given in Eq. (1.11). More generally, if a quadrature formula based on a Taylor series uses N points, it will integrate exactly a polynomial of degree $N - 1$ (degree N if N is odd). That is, the N weights w_n can be chosen to satisfy the N linear equations

$$\int_{-1}^{1} x^p \, dx = \sum_{n=1}^{N} w_n x_n^p \,; \quad p = 0, 1, \ldots, N - 1 \,. \tag{4.9}$$

(When N is odd, the quadrature formula is also exact for the odd monomial x^N.)

A greater precision for a given amount of numerical work can be achieved if we are willing to give up the requirement of equally-spaced quadrature points. That is, we will choose the x_n in some optimal sense, subject only to the constraint that they lie within the interval $[-1, 1]$. We then have $2N$ parameters at our disposal in constructing the quadrature formula (the N x_n's and the N w_n's), and so we should be able to choose them so that Eq. (4.9) is satisfied for p ranging from 0 to $2N - 1$. That is, the quadrature formula using only N carefully chosen points can be made exact for polynomials of degree $2N - 1$ or less. This is clearly more efficient than using equally-spaced abscissae.

To see how to best choose the x_n, we consider the Legendre polynomials, which are orthogonal on the interval $[-1, 1]$:

$$\int_{-1}^{1} P_i(x) P_j(x) \, dx = \frac{2}{2i+1} \delta_{ij} \,. \tag{4.10}$$

It is easily shown that P_i is a polynomial of degree i with i roots in the interval $[-1, 1]$. Any polynomial of degree $2N - 1$ or less then can be written in the form

$$f(x) = Q(x)P_N(x) + R(x) \,,$$

where Q and R are polynomials of degree $N - 1$ or less. The exact value of the required integral (4.8) is then

$$I = \int_{-1}^{1} (Q P_N + R) \, dx = \int_{-1}^{1} R \, dx \,, \qquad (4.11)$$

where the second step follows from the orthogonality of P_N to all polynomials of degree $N - 1$ or less. If we now take the x_n to be the N zeros of P_N, then application of (4.8) gives (exactly)

$$I = \sum_{n=1}^{N} w_n \left[Q(x_n)P_N(x_n) + R(x_n) \right] = \sum_{n=1}^{N} w_n R(x_n) \,. \qquad (4.12)$$

It remains to choose the w_n so that R (a polynomial of degree $N - 1$ or less) is integrated exactly. That is, the w_n satisfy the set of linear equations (4.9) when the x_n are the zeros of P_N. It can be shown that w_n is related to the derivative of P_N at the corresponding zero. Specifically,

$$w_n = \frac{2}{(1 - x_n^2)[P_N'(x_n)]^2} \,.$$

This completes the specification of what is known as the Gauss-Legendre quadrature formula. Note that it can be applied to any definite integral between finite limits by a simple linear change of variable. That is, for an integral between limits $x = a$ and $x = b$, a change of variable to

$$t = -1 + 2 \frac{(x - a)}{(b - a)}$$

reduces the integral to the form required. Other non-linear changes of variable that make the integrand as smooth as possible will also improve the accuracy.

Other types of orthogonal polynomials provide useful Gaussian quadrature formulas when the integrand has a particular form. For example, the Laguerre polynomials, L_i, which are orthogonal on the interval $[0, \infty]$ with the weight function e^{-x}, lead to the Gauss-Laguerre quadrature formula

$$\int_0^{\infty} e^{-x} f(x) \, dx \approx \sum_{n=1}^{N} w_n f(x_n) \,, \qquad (4.13)$$

where the x_n are the roots of L_N and the w_n are related to the values of L_{N+1} at these points. Similarly, the Hermite polynomials provide Gauss-Hermite quadrature formulas for integrals of the form

$$\int_{-\infty}^{\infty} e^{-x^2} f(x)\, dx .$$

These Gaussian quadrature formulas, and many others, are given in Section 25.4 of [Ab64], which also contains tables of the abscissae and weights.

In the practical application of Gaussian quadrature formulas, one does not need to write programs to calculate the abscissae and weights. Rather, there are usually library subroutines that can be used to establish arrays containing these numbers. For example, subroutine QUAD in Example 4 (see Appendix B) establishes the Gauss-Legendre abscissae and weights for many different values of N.

As a general rule, Gaussian quadrature is the method of choice when the integrand is smooth, or can be made smooth by extracting from it a function that is the weight for a standard set of orthogonal polynomials. We must, of course, also have the ability to evaluate the integrand at the required abscissae. If the integrand varies rapidly, we can compound the basic Gaussian quadrature formula by applying it over several sub-intervals of the range of integration. Of course, when the integrand can be evaluated only at equally-spaced abscissae (such as when it is generated by integrating a differential equation), then formulas of the type discussed in Chapter 1 must be used.

As an illustration of Gaussian quadrature, consider using a 3-point Gauss-Legendre quadrature to evaluate the integral

$$I = \int_0^3 (1+t)^{\frac{1}{2}}\, dt = 4.66667 . \tag{4.14}$$

Making the change of variable to

$$x = -1 + \frac{2}{3}t$$

results in

$$I = \frac{3}{2} \int_{-1}^{1} \left(\frac{3}{2}x + \frac{5}{2} \right)^{\frac{1}{2}} dx . \tag{4.15}$$

For $N = 3$, the Gauss-Legendre abscissae and weights are

$$x_1 = -x_3 = 0.774597 , \quad x_2 = 0 ; \quad w_1 = w_3 = 0.555556 , \quad w_2 = 0.888889 .$$

Straightforward evaluation of the quadrature formula (4.8) then results in $I = 4.66683$, while a Simpson's rule evaluation of (4.14) with $h = 1.5$ gives 4.66228. Gaussian quadrature is therefore more accurate than Simpson's rule by about a factor of 27, yet requires the same number of evaluations of the integrand (three).

■ **Exercise 4.4**   Consider the integral

$$\int_{-1}^{1} (1 - x^2)^{1/2} \, dx = \frac{\pi}{2} \,.$$

Evaluate this integral using some of the quadrature formulas discussed in Chapter 1 and using Gauss-Legendre quadrature. Note that the behavior of the integrand near $x = \pm 1$ is cause for some caution. Compare the accuracy and efficiency of these various methods for different numbers of abscissae. Note that this integral can be evaluated exactly with a "one-point" Gauss-Chebyshev quadrature formula of the form

$$\int_{-1}^{1} (1 - x^2)^{1/2} f(x) \, dx = \sum_{n=1}^{N} w_n f(x_n) \,,$$

with

$$x_n = \cos \frac{n}{N+1}\pi \,; \quad w_n = \frac{\pi}{N+1} \sin^2 \frac{n}{N+1}\pi \,.$$

(See Section 25.4.40 of [Ab64].)

4.3 Born and eikonal approximations to quantum scattering

In this example, we will investigate the Born and eikonal approximations suitable for describing quantum-mechanical scattering at high energies, and in particular calculate the scattering of fast electrons (energies greater than several 10's of eV) from neutral atoms. The following project deals with the exact partial-wave solution of this problem.

Extensive discussions of the quantum theory of scattering are given in many texts (see, for example, [Me68], [Ne66], or [Wu62]); we will only review the essentials here. For particles of mass m and energy

$$E = \frac{\hbar^2}{2m}k^2 > 0 \,,$$

scattering from a central potential $V(r)$ is described by a wave function

$\Psi(\mathbf{r})$ that satisfies the Schroedinger equation,

$$-\frac{\hbar^2}{2m}\nabla^2\Psi + V\Psi = E\Psi , \qquad (4.16)$$

with the boundary condition at large distances

$$\Psi \underset{r\to\infty}{\to} e^{ikz} + f(\theta)\frac{e^{ikr}}{r} . \qquad (4.17)$$

Here, the beam is incident along the z direction and θ is the scattering angle (angle between $\mathbf{r}$ and $\hat{\mathbf{z}}$). The complex scattering amplitude f embodies the observable scattering properties and is the basic function we seek to determine. The differential cross section is given by

$$\frac{d\sigma}{d\Omega} = |f(\theta)|^2 , \qquad (4.18)$$

and the total cross section is

$$\sigma = \int d\Omega \frac{d\sigma}{d\Omega} = 2\pi \int_0^{\pi} d\theta \sin\theta |f(\theta)|^2 . \qquad (4.19)$$

In general, f is a function of both E and θ.

At this point, many elementary treatments of scattering introduce a partial-wave decomposition of Ψ, express f in terms of the phase shifts, and then proceed to discuss the radial Schroedinger equation in each partial wave, from which the phase shift and hence the exact cross section, can be calculated. We will use this method, which is most appropriate when the energies are low and only a few partial waves are important, in Project IV below. However, in this example, we will consider two approximation schemes, the Born and eikonal, which are appropriate to high-energy situations when many partial waves contribute.

Both the Born and eikonal approximations are based on an exact integral expression for the scattering amplitude derived in many advanced treatments:

$$f(\theta) = -\frac{m}{2\pi\hbar^2} \int e^{-i\mathbf{k}_f \cdot \mathbf{r}} V(r)\Psi(\mathbf{r})\, d^3r . \qquad (4.20)$$

Here, the wavenumber of the scattered particle is $\mathbf{k}_f$, so that $|\mathbf{k}_f| = k$ and $\hat{\mathbf{k}}_f \cdot \hat{\mathbf{z}} = \cos\theta$. It is also convenient to introduce the wavenumber of the incident particle, $\mathbf{k}_i = k\hat{\mathbf{z}}$.

The Born approximation (more precisely, the first Born approximation) consists of assuming that the scattering is weak, so that the full scattering wave function Ψ differs very little from the incident plane wave,

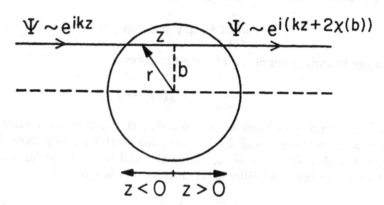

Figure 4.1   Geometry of the eikonal approximation.

$\exp(i\hat{\mathbf{k}}_i\cdot\mathbf{r})$. Making this replacement in (4.20) results in the Born scattering amplitude,

$$f_B(\theta) = -\frac{m}{2\pi\hbar^2}\int e^{-i\mathbf{q}\cdot\mathbf{r}}V(r)\,d^3r = -\frac{2m}{q\hbar^2}\int_0^\infty \sin qr\,V(r)r\,dr\,. \quad (4.21)$$

Here, we have introduced the momentum transfer, $\mathbf{q} = \mathbf{k}_f - \mathbf{k}_i$, so that

$$q = |\mathbf{k}_f - \mathbf{k}_i| = 2k\sin\frac{1}{2}\theta\,.$$

and have used the identity

$$\int e^{-i\mathbf{q}\cdot\mathbf{r}}\,d\hat{\mathbf{r}} = 4\pi j_0(qr) = 4\pi\frac{\sin qr}{qr}\,. \quad (4.22)$$

Note that the Born approximation to the scattering amplitude depends only upon q and not separately upon E and θ.

Better approximations to Ψ in Eq. (4.20) result in correspondingly better approximations to f. One possible improvement is the eikonal approximation, valid at high energies and small scattering angles. (See [Wa73] or [Ne66].) This approximation is semiclassical in nature; its essence is that each ray of the incident plane wave suffers a phase shift as it passes through the potential on a straight-line trajectory (see Figure 4.1). Since this phase shift depends upon the impact parameter of the ray, the wavefronts of the wave function are distorted after passing through the potential; it is this distortion that carries the scattering information.

To derive the eikonal approximation, we will put, without loss of generality,

$$\Psi(\mathbf{r}) = e^{i\mathbf{k}_i \cdot \mathbf{r}}\psi(\mathbf{r}) , \qquad (4.23)$$

where ψ is a slowly-varying function describing the distortion of the incident wave. Upon inserting this into the original Schroedinger equation (4.16), we obtain an equation for ψ:

$$-\frac{\hbar^2}{2m}(2i\mathbf{k}_i\cdot\nabla + \nabla^2)\psi + V\psi = 0 . \qquad (4.24)$$

If we now assume that ψ varies slowly enough so that the $\nabla^2\psi$ term can be ignored (i.e., k is very large), we have

$$ik\frac{\hbar^2}{m}\frac{\partial\psi(\mathbf{b},z)}{\partial z} = V(b,z)\psi(\mathbf{b},z) . \qquad (4.25)$$

Here, we have introduced the coordinate b in the plane transverse to the incident beam, so that

$$V(b,z) = V(r) ; \quad r = (b^2 + z^2)^{1/2} .$$

From symmetry considerations, we expect that ψ will be azimuthally symmetric and so independent of $\hat{\mathbf{b}}$. Equation (4.25) can be integrated immediately and, using the boundary condition that $\psi \to 1$ as $z \to -\infty$ since there is no distortion of the wave before the particle reaches the potential, we have

$$\psi(b,z) = e^{2i\chi(b,z)} ; \quad \chi(b,z) = -\frac{m}{2\hbar^2 k}\int_{-\infty}^{z} V(b,z')\,dz' . \qquad (4.26)$$

Having obtained the eikonal approximation to the scattering wave function, we can now obtain the eikonal scattering amplitude, f_e. Inserting Eq. (4.23) into (4.20), we have

$$f_e = -\frac{m}{2\pi\hbar^2}\int d^2b \int_{-\infty}^{\infty} dz\, e^{-i\mathbf{q}\cdot\mathbf{r}}V(b,z)\psi(b,z) . \qquad (4.27)$$

Using (4.25), we can relate $V\psi$ directly to $\partial\psi/\partial z$. Furthermore, if we restrict our consideration to relatively small scattering angles, so that $q_z \approx 0$, then the z integral in (4.27) can be done immediately and, using (4.26) for ψ, we obtain

$$f_e = -\frac{ik}{2\pi}\int d^2b\, e^{-i\mathbf{q}\cdot\mathbf{b}}(e^{2i\chi(b)} - 1) , \qquad (4.28)$$

with the "profile function"

$$\chi(b) = \chi(b, z = \infty) = -\frac{m}{2\hbar^2 k} \int_{-\infty}^{\infty} V(b, z)\, dz \, . \qquad (4.29)$$

Since χ is azimuthally symmetric, we can perform the azimuthal integration in (4.28) and obtain our final expression for the eikonal scattering amplitude,

$$f_e = -ik \int_0^{\infty} b\, db\, J_0(qb)(e^{2i\chi(b)} - 1) \, . \qquad (4.30)$$

In deriving this expression, we have used the identity [compare with Eq. (4.22)]

$$J_0(qb) = \frac{1}{2\pi} \int_0^{2\pi} e^{-iqb \cos \phi}\, d\phi \, .$$

Note that in contrast to f_B, f_e depends upon both E (through k) and q.

An important property of the exact scattering amplitude is the optical theorem, which relates the total cross section to the imaginary part of the forward scattering amplitude. After a bit of algebra, one can show that f_e satisfies this relation in the limit that the incident momentum becomes large compared to the length scale over which the potential varies:

$$\sigma = \frac{4\pi}{k} \mathrm{Im} f(q = 0) = 8\pi \int_0^{\infty} b\, db\, \sin^2 \chi(b) \, . \qquad (4.31)$$

The Born approximation cannot lead to a scattering amplitude that respects this relation, as Eq. (4.21) shows that f_B is purely real. It is also easy to show that, in the extreme high-energy limit, where $k \to \infty$ and χ becomes small, the Born and eikonal amplitudes become equal (see Exercise 4.5). The eikonal formula (4.30) also can be related to the usual partial wave expression for f (see Exercise 4.6).

One practical application of the approximations discussed above is in the calculation of the scattering of high-energy electrons from neutral atoms. In general, this is a complicated multi-channel scattering problem since there can be reactions leading to final states in which the atom is excited. However, as the reaction probabilities are small in comparison to elastic scattering, for many purposes the problem can be modeled by the scattering of an electron from a central potential. This potential represents the combined influence of the attraction of the central nuclear charge (Z) and the screening of this attraction by the Z atomic electrons. For a neutral target atom, the potential vanishes at large distances faster than r^{-1}. A very accurate approximation to this potential can be had by solving for the self-consistent Hartree-Fock potential of the neutral atom, as was done in Project III. However, a much simpler estimate can be obtained using an approximation to the Thomas-Fermi model of the atom given by Lenz and Jensen [Go49]:

$$V = -\frac{Ze^2}{r}e^{-x}(1 + x + b_2x^2 + b_3x^3 + b_4x^4) ; \qquad (4.32a)$$

with

$$e^2 = 14.409 ; \quad b_2 = 0.3344 ; \quad b_3 = 0.0485 ; \quad b_4 = 2.647 \times 10^{-3} ; \qquad (4.32b)$$

and

$$x = 4.5397Z^{1/6}r^{1/2} . \qquad (4.32c)$$

Here, the potential is measured in eV and the radius is measured in Å. Note that there is a possible problem with this potential, since it is singular as r^{-1} at the origin, and so leads to a divergent expression for χ at $b = 0$. However, if the potential is regularized by taking it to be a constant within some small radius r_{min}, (say the radius of the atoms 1s shell), then the calculated cross section will be unaffected except at momentum transfers large enough so that $qr_{min} \gg 1$.

Our goal is to compute the Born and eikonal approximations to the differential and total cross sections for a given central potential at a specified incident energy, and in particular for the potential (4.32). To do this, we must compute the integrals (4.21), (4.29), and (4.30) defining the scattering amplitudes, as well as the integral (4.19) for the total cross section. The FORTRAN program for Example 4, whose source code is given in Appendix B, as well as in the file EXMPL4.FOR, does these calculations and graphs the results on a semi-log plot; the total cross section given by the optical theorem, Eq. (4.31), is also calculated.

The incident particle is assumed to have the mass of the electron, and, as is appropriate for atomic systems, all lengths are measured in Å and all energies in eV. The potential can be chosen to be a square well of radius 2 Å, a Gaussian well of the form

$$V(r) = -V_0 e^{-2r^2} ,$$

or the Lenz-Jensen potential (4.32). All potentials are assumed to vanish beyond 2 Å. Furthermore, the r^{-1} singularity in the Lenz-Jensen potential is cutoff inside the radius of the 1s shell of the target atom.

Because the differential cross sections become very peaked in the forward direction at the high energies where the Born and eikonal approximations are valid, the integration over $\cos \theta$ in Eq. (4.19) is divided into two regions for a more accurate integration of the forward peak. One of these extends from $\theta = 0$ to $\theta = \theta_{cut}$, where

$$q_{cut} R_{cut} = 2\pi = 2k R_{cut} \sin \frac{1}{2}\theta_{cut}$$

and R_{cut} is 1 Å for the Lenz-Jensen potential and 2 Å for either the square-or Gaussian-well potentials, and the other extends from θ_{cut} to π. All integrals are done by Gauss-Legendre quadrature using the same number of points, and the Bessel function of order zero required by Eq. (4.30) is evaluated using the approximations (4.4).

The following exercises are aimed at improving your understanding of this program and the physics it describes.

■ **Exercise 4.5**   Verify the algebra in the derivations above. Show that in the limit of very high energies, where χ is small, so that $\sin \chi \approx \chi$, the Born and eikonal results are identical. Also prove that the eikonal amplitude satisfies the optical theorem (4.31) in the limit where the incident momentum becomes large in comparison with the length scale of the potential.

■ **Exercise 4.6**   Show that if the conventional expression for f in terms of a sum over partial waves [Eq. (IV.4) below] is approximated by an integral over l (or, equivalently, over $b = l/k$) and the small-θ/large-l approximation

$$P_l(\cos \theta) \approx J_0(l\theta)$$

is used, Eq. (4.30) results, with the identification $\chi(b) = \delta_l$. Investigate, either numerically or analytically, the validity of this relation between the Bessel function of order zero and the Legendre polynomials.

■ **Exercise 4.7**   Test the Born approximation cross sections generated by the code by comparing the numerical values with the analytical Born results for a square or Gaussian well of depth 20 eV and for varying incident energies from 1 eV to 10 keV. Verify for these cases that $\chi(b)$ as computed by the code has the expected values. Investigate the variation of the numerical results with changes in the number of quadrature points. (Note that only particular values of N are allowed by the subroutine generating the Gauss-Legendre abscissae and weights.)

■ **Exercise 4.8**   Fix the depth of a square well at 20 eV and calculate for various incident energies to get a feeling for how the differential and total cross sections vary. Compare the square well cross sections with those of a Gaussian well of comparable depth and explain the differences. Show that the Born and eikonal results approach each other at high energies for either potential and that the optical relation for the eikonal amplitude becomes better satisfied at higher energies.

■ **Exercise 4.9**   Using the Lenz-Jensen potential and a fixed charge and incident energy, say $Z = 50$ and $E = 1000$ eV, investigate the sensitivity of the calculation to the number of quadrature points used and to the small-r regularization of the potential. Calculate the cross sections for various Z ranging from 20 to 100 and for E ranging from 10 eV to 10 keV; interpret the trends you observe.

■ **Exercise 4.10**   Use the program constructed in Project I to calculate the classical differential cross section for electrons scattering from the Lenz-Jensen potential for various Z and incident energies. Compare the results with the Born and eikonal cross sections. Can you establish an analytical connection between the classical description and the quantum approximations? (See [Ne66] for a detailed discussion.)

Project IV: Partial wave solution of quantum scattering

In this project, we will use the method of partial waves to solve the quantum scattering problem for a particle incident on a central potential, and in particular consider the low-energy scattering of electrons from neutral atoms. The strategy of the method is to employ a partial wave expansion of the scattering wave function to decompose the three-dimensional Schroedinger equation (4.16) into a set of uncoupled one-dimensional ordinary differential equations for the radial wave functions; each of these is then solved as a boundary-value problem to determine the phase shift, and hence the scattering amplitude.

IV.1 Partial wave decomposition of the wave function

The standard partial wave decomposition of the scattering wave function Ψ is

$$\Psi(\mathbf{r}) = \sum_{l=0}^{\infty} (2l + 1)i^l e^{i\delta_l} \frac{R_l(r)}{kr} P_l(\cos\theta), \qquad (IV.1)$$

When this expansion is substituted into the Schroedinger equation (4.16), the radial wave functions R_l are found to satisfy the radial differential equations

$$\left[-\frac{\hbar^2}{2m}\frac{d^2}{dr^2} + V(r) + \frac{l(l+1)\hbar^2}{2mr^2} - E\right]R_l(r) = 0 . \qquad (IV.2)$$

Although this is the same equation as that satisfied by a bound state wave function, the boundary conditions are different. In particular, R vanishes at the origin, but it has the large-r asymptotic behavior

$$R_l \to kr[\cos \delta_l j_l(kr) - \sin \delta_l n_l(kr)] , \qquad (IV.3)$$

where j_l and n_l are the regular and irregular spherical Bessel functions of order l. (See Exercise 4.3.)

The scattering amplitude f is related to the phase shifts δ_l by

$$f(\theta) = \frac{1}{k}\sum_{l=0}^{\infty}(2l+1)e^{i\delta_l}\sin \delta_l P_l(\cos \theta) , \qquad (IV.4)$$

and the total cross section is easily found from Eq. (4.19) and the orthogonality of the Legendre polynomials, (4.10):

$$\sigma = \frac{4\pi}{k^2}\sum_{l=0}^{\infty}(2l+1)\sin^2 \delta_l . \qquad (IV.5)$$

Although the sums in (IV.4,5) extend over all l, they are in practice limited to only a finite number of partial waves. This is because, for large l, the repulsive centrifugal potential in (IV.2) is effective in keeping the particle outside the range of the potential and so the phase shift is very small. If the potential is negligible beyond a radius r_{max} an estimate of the highest partial wave that is important, l_{max}, can be had by setting the turning point at this radius. Thus,

$$\frac{l_{max}(l_{max}+1)\hbar^2}{2mr_{max}^2} = E ,$$

which leads to $l_{max} \sim kr_{max}$. This estimate is usually slightly low, since penetration of the centrifugal barrier leads to non-vanishing phase shifts in partial waves somewhat higher than this.

IV.2 Finding the phase shifts

To find the phase shift in a given partial wave, we must solve the radial equation (IV.2), using, for example, the Numerov method discussed in Chapter 3. Although the boundary conditions specified by (IV.3) and the vanishing of R at the origin are non-local, we can still integrate the equation as an initial value problem. This is because the equation is linear, so that the boundary condition at large r can be satisfied simply by appropriately normalizing the solution.

If we put $R_l(r = 0) = 0$ and take the value at the next lattice point, $R_l(r = h)$, to be any convenient small number (R will generally rise rapidly through the centrifugal barrier and so we must avoid overflows), it would seem that we were ready to use the Numerov method. However, $k(r = 0)$ is infinite. The solution is to use Eq. (1.7) for $R_l''(h)$, along with the known values $R_l(0)$, $R_l(h)$, and $k(h)$ to find $R_l(2h)$. Now we can integrate outward in r to a radius $r^{(1)} > r_{max}$. (See Figure IV.1.) Here, V vanishes and R must be a linear combination of the free solutions, $krj_l(kr)$ and $krn_l(kr)$:

$$R_l^{(1)} = Akr^{(1)}[\cos \delta_l j_l(kr^{(1)}) - \sin \delta_l n_l(kr^{(1)})] . \qquad (IV.6)$$

Although the constant A depends upon the value chosen for $R(r = h)$, it is largely irrelevant for our purposes; however, it must be kept small enough so that overflows are avoided.

Knowing only $R_l^{(1)}$ does not allow us to solve for the two unknowns, A and δ_l. However, if we continue integrating to a larger radius $r^{(2)} > r^{(1)}$, then we also have

$$R_l^{(2)} = Akr^{(2)}[\cos \delta_l j_l(kr^{(2)}) - \sin \delta_l n_l(kr^{(2)})] . \qquad (IV.7)$$

Equations (IV.6,7) can then be solved for δ_l. After a bit of algebra, we have

$$\tan \delta_l = \frac{Gj_l^{(1)} - j_l^{(2)}}{Gn_l^{(1)} - n_l^{(2)}} ; \quad G = \frac{r^{(1)} R_l^{(2)}}{r^{(2)} R_l^{(1)}} , \qquad (IV.8)$$

where $j_l^{(1)} = j_l(kr^{(1)})$, etc. Note that this equation determines δ_l only within a multiple of π, although this does not affect the physical observables [see Eqs. (IV.4,5)]. The correct number of π's at a given energy can be determined by comparing the number of nodes in R and in the free solution, krj_l, which occur for $r < r_{max}$. With the conventional definitions we have used, the phase shift in each partial wave vanishes at high energies and approaches $N_l\pi$ at zero energy, where N_l is the number of bound states in the potential in the l'th partial wave.

IV.3 Solving the equations

Our goal in a numerical treatment of this problem is to investigate the scattering of electrons with energies from 0.5 eV to 10 keV from the potential (4.32); a radius $r_{max} = 2$ Å is reasonable. For the charge Z

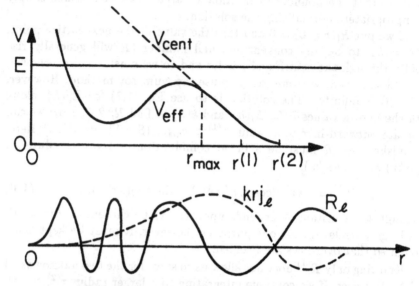

Figure IV.1   (Upper) Schematic centrifugal potential (dashed line) and effective total potential (solid line) in a partial wave $l > 0$. The incident energy is also shown (horizontal line), as are the range of the potential (r_{max}) and the two matching radii, $r^{(1)}$ and $r^{(2)}$. (Lower) Schematic scattering wave function, R_l (solid line), and free wave function, $kr j_l$ (dashed line), for the same situation.

and energy E specified, the program should calculate the phase shift and plot the radial wave function for each important partial wave, and then sum the contributions from each l to find the total cross section and the differential cross section as a function of θ from $0°$ to $180°$, say in $5°$ steps. This program can be constructed and exploited in the following sequence of steps.

■ **Step 1**   Write a subroutine that computes the values of the Legendre polynomials for the degrees and angles required and stores them in an array. [See Eq. (4.1).] Also write a subroutine that computes j_l and n_l for a given value of x. (See Exercise 4.3). Check that these routines are working correctly by comparing their results with tabulated values.

■ **Step 2**   Write a subroutine that calculates the phase shift in a specified partial wave. Use the Numerov method to integrate to $r^{(1)}$ and then

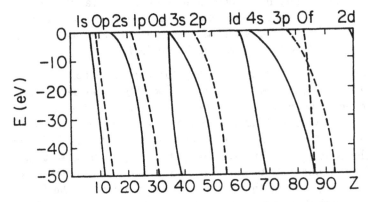

Figure IV.2   Weakly-bound levels of the Lenz-Jensen potential. Levels with a negative parity (p and f) are shown by dashed curves.

to $r^{(2)}$ and then determine the phase shift from Eq. (IV.8). Note that if $r^{(2)}$ is too close to $r^{(1)}$, problems with numerical precision may arise (both the numerator and denominator of (IV.8) vanish), while too great a distance between the matching radii leads to wasted computation. Check that your subroutine is working properly by verifying that the calculated phase shift vanishes when $V(r)$ is set to 0 and that it is independent of the choice of the matching radii, as long as they are greater than r_{max}. Also check that the phase shifts you find for a square-well potential agree with those that you can calculate analytically. Note that since quite a bit of numerical work is involved in the calculation of the Lenz-Jensen potential (4.32), it is most efficient if the values of V at the lattice points are stored in an array to be used in the integration of all important partial waves.

■ **Step 3**   Write a main program that, for a given charge and energy, calls the subroutines you have constructed in performing the sum over partial waves to calculate the total and differential cross sections. Verify that your estimate of l_{max} is reasonable in that all non-vanishing phase shifts are computed but that no computational effort is wasted on partial waves for which the phase shift is negligible.

■ **Step 4**   Study the cross section as a function of energy for several Z, say 20, 50, and 100. Show that, at low energies, resonances occurring in particular partial waves make the angular distributions quite complex and the total cross sections very energy-dependent, but that all quantities become smooth and monotonic at high energies. Tabulations of experimental data with which to compare your results are given in [Ki71]. Also compare your high energy results with the Born and eikonal approximations generated by the program in Example 4.

- **Step 5** To understand the low-energy resonances, consider Figure IV.2, which gives the energies of weakly bound levels in the Lenz-Jensen potential as a function of Z. By studying the cross section at a fixed energy of 5 eV as a function of Z, show that the regions near $Z = 46$ and $Z = 59$ reflect the extensions of the $2p$ and $1d$ bound states into the continuum. By examining the $Z = 59$ cross section, angular distribution, phase shifts, and radial wave functions as functions of energy near $E = 5$ eV, show that the resonance is indeed in the $l = 2$ partial wave. Note that since the Lenz-Jensen potential is based on the Thomas-Fermi approximation, it is not expected to be accurate in the outer region of the atom. The resonance energies and widths you find therefore differ quantitatively from those that are found in experiment.

- **Step 6** Use the Hartree-Fock code from Project III to generate the direct potential, Φ, for the neutral Ne atom and modify your program to calculate the scattering of electrons from this potential. Note that when the energy of the incident electron is large, it is easily "distinguishable" from the electrons in the atom, and so any effects of the Fock (exchange) potential will be small. Compare your results with those produced with the Lenz-Jensen potential. A further discussion of the electron-atom scattering problem can be found in [Bo74].

Chapter 5

Matrix
Operations

Linearization is a common assumption or approximation in describing physical processes and so linear systems of equations are ubiquitous in computational physics. Indeed, the matrix manipulations associated with finding eigenvalues or with solving simultaneous linear equations are often the bulk of the work involved in solving many physical problems. In this chapter, we will discuss briefly two of the more non-trivial matrix operations: inversion and diagonalization. Our treatment here will be confined largely to "direct" methods appropriate for "dense" matrices (where most of the elements are non-zero) of dimension less than several hundred; iterative methods for treating the very large sparse matrices that arise in the discretization of ordinary and partial differential equations will be discussed in the following two chapters. As is the case with the special functions of the previous chapter, a variety of library subroutines employing several different methods for solving matrix problems are usually available on any large computer. Our discussions here are therefore limited to selected basic methods, to give a flavor of what has to be done. More detailed treatments can be found in many texts, for example [Ac70] and [Bu81].

5.1 Matrix inversion
Let us consider the problem of inverting a square $(N \times N)$ matrix $\mathbf{A}$. This might arise, for example, in solving the linear system

$$\mathbf{A}\mathbf{x} = \mathbf{b} , \qquad (5.1)$$

for the unknown vector $\mathbf{x}$, where $\mathbf{b}$ is a known vector. The most natural way of doing this is to follow our training in elementary algebra and use Cramer's rule; i.e., to evaluate $\mathbf{A}^{-1}$ as being proportional to the transpose of the matrix of cofactors of $\mathbf{A}$. To do this, we would have to calculate

N^2 determinants of matrices of dimension $(N - 1) \times (N - 1)$. If these were done in the most naive way, in which we would evaluate

$$\det \mathbf{A} = \sum_P (-)^P A_{1P1} A_{2P2} \dots A_{NPN} , \qquad (5.2)$$

for the determinant of the $N \times N$ matrix $\mathbf{A}$, where P is one of the $N!$ permutations of the N columns, Pi is the i'th element of P, and $(-)^P$ is the signature of P, then the numerical work would be horrendous. For example, of order $N!$ multiplications are required to evaluate (5.2). For $N = 20$, this is some 2×10^{18} multiplications. On the fastest computer currently available, 10^8 multiplications per second might be possible, so that some 10^3 years are required just to invert this one 20×20 matrix! Clearly, this is not the way to evaluate determinants (or to invert matrices, as it turns out).

One of the simplest practical methods for evaluating $\mathbf{A}^{-1}$ is the Gauss-Jordan method. Here, the idea is to consider a class of elementary row operations on the matrix $\mathbf{A}$. These involve multiplying a particular row of $\mathbf{A}$ by a constant, interchanging two rows, or adding a multiple of one row to another. Each of these three operations can be represented by left-multiplying $\mathbf{A}$ by a simple matrix, $\mathbf{T}$. For example, when $N = 3$, the matrices

$$\begin{bmatrix} 1 & 0 & 0 \\ 0 & 1 & 0 \\ 0 & 0 & 2 \end{bmatrix}, \quad \begin{bmatrix} 0 & 1 & 0 \\ 1 & 0 & 0 \\ 0 & 0 & 1 \end{bmatrix}, \text{ and } \begin{bmatrix} 1 & 0 & 0 \\ 0 & 1 & 0 \\ -1/2 & 0 & 1 \end{bmatrix} \qquad (5.3)$$

will multiply the third row by 2, interchange the first and second rows, and subtract one-half of the first row from the third row, respectively. The Gauss-Jordan strategy is to find a sequence of such operations,

$$\mathbf{T} = \dots \mathbf{T}_3 \mathbf{T}_2 \mathbf{T}_1$$

which, when applied to $\mathbf{A}$, reduces it to the unit matrix. That is,

$$\mathbf{TA} = (\dots \mathbf{T}_3 \mathbf{T}_2 \mathbf{T}_1)\mathbf{A} = \mathbf{I} , \qquad (5.4)$$

so that $\mathbf{T} = \mathbf{A}^{-1}$ is the required inverse. Equivalently, the same sequence of operations applied to the unit matrix yields the inverse of $\mathbf{A}$.

A practical method to find the appropriate sequence of row operations is best illustrated by example. Consider the 3×3 matrix $\mathbf{A}$ and the unit matrix:

$$\mathbf{A} = \begin{bmatrix} 1 & 2 & 1 \\ 4 & 2 & 2 \\ 2 & 4 & 1 \end{bmatrix} ; \mathbf{I} = \begin{bmatrix} 1 & 0 & 0 \\ 0 & 1 & 0 \\ 0 & 0 & 1 \end{bmatrix} . \qquad (5.5a)$$

We first apply transformations to zero all but the first element in the first column of **A**. Subtracting 4 times the first row from the second yields

$$\mathbf{TA} = \begin{bmatrix} 1 & 2 & 1 \\ 0 & -6 & -2 \\ 2 & 4 & 1 \end{bmatrix} \; ; \; \mathbf{TI} = \begin{bmatrix} 1 & 0 & 0 \\ -4 & 1 & 0 \\ 0 & 0 & 1 \end{bmatrix} , \qquad (5.5b)$$

and then subtracting twice the first row from the third yields

$$\mathbf{TA} = \begin{bmatrix} 1 & 2 & 1 \\ 0 & -6 & -2 \\ 0 & 0 & -1 \end{bmatrix} \; ; \; \mathbf{TI} = \begin{bmatrix} 1 & 0 & 0 \\ -4 & 1 & 0 \\ -2 & 0 & 1 \end{bmatrix} . \qquad (5.5c)$$

We now go to work on the second column, where, by adding 1/3 of the second row to the first, we can zero all but the second element:

$$\mathbf{TA} = \begin{bmatrix} 1 & 0 & 1/3 \\ 0 & -6 & -2 \\ 0 & 0 & -1 \end{bmatrix} \; ; \; \mathbf{TI} = \begin{bmatrix} -1/3 & 1/3 & 0 \\ -4 & 1 & 0 \\ -2 & 0 & 1 \end{bmatrix} , \qquad (5.5d)$$

and, multiplying the second row by $-1/6$ yields

$$\mathbf{TA} = \begin{bmatrix} 1 & 0 & 1/3 \\ 0 & 1 & 1/3 \\ 0 & 0 & -1 \end{bmatrix} \; ; \; \mathbf{TI} = \begin{bmatrix} -1/3 & 1/3 & 0 \\ 2/3 & -1/6 & 0 \\ -2 & 0 & 1 \end{bmatrix} . \qquad (5.5e)$$

Finally, we can reduce the third column of **TA** to the required form by adding 1/3 of the third row to the first and second rows, and then multiplying the third row by -1:

$$\mathbf{TA} = \begin{bmatrix} 1 & 0 & 0 \\ 0 & 1 & 0 \\ 0 & 0 & 1 \end{bmatrix} \; ; \; \mathbf{TI} = \begin{bmatrix} -1 & 1/3 & 1/3 \\ 0 & -1/6 & 1/3 \\ 2 & 0 & -1 \end{bmatrix} . \qquad (5.5f)$$

This finds the required inverse.

A moment's thought shows how this algorithm can be generalized to an $N \times N$ matrix and that, when this is done, it requires, for large N, of order N^3 multiplications and additions. Thus, it is computationally tractable, as long as N is not too large.

In practice, several subtleties are important. For example, it might happen that, at some point in the procedure, the diagonal element of **TA** vanishes in the column we are working on. In this case, an interchange of two rows will bring a non-vanishing value to this "pivot" element (if no such row interchange can do this, then the matrix **A** is singular). In fact, numerical accuracy requires that rows be interchanged to bring into the pivot position that element in the column being worked on that has the largest absolute value. Problems associated with numerical round-off can also arise during the inversion if elements of the matrix differ greatly in magnitude. For this reason, it is often useful to scale the rows or columns so that all entries have roughly the same magnitude ("equilibration"). Various special cases (such as when **A** is symmetric or when we are only interested in solving (5.1) for **x** and not for A^{-1} itself) can result in reductions of the numerical work. For example, in the latter case, we can apply **T** only to the vector **b**, rather than to the whole unit matrix. Finally, if we are interested in computing only the determinant of **A**, successive row transformations, which have a simple and easily calculable effect on the determinant, can be used to bring **TA** to lower or upper diagonal form (all elements vanishing above or below the diagonal, respectively), and then the determinant can be evaluated simply as the product of the diagonal elements.

■ **Exercise 5.1** Use Eq. (5.2) to show that interchanging the rows of a matrix changes the sign of its determinant, that adding a multiple of one row to another leaves its determinant unchanged, and that multiplying one row by a constant multiplies the determinant by that same constant.

■ **Exercise 5.2** Write (or just flowchart) a subroutine that will use the Gauss-Jordan method to find the inverse of a given matrix. Incorporate pivoting as described above to improve the numerical accuracy.

5.2 Eigenvalues of a tri-diagonal matrix

We turn now to the problem of finding the eigenvalues and eigenvectors of an $N \times N$ matrix **A**; that is, the task of finding the N scalars λ_n and their associated N-component vectors ϕ_n that satisfy

$$A\phi_n = \lambda_n \phi_n \, . \tag{5.6}$$

Equivalently, the eigenvalues are zeros of the N'th degree characteristic polynomial of **A**:

$$P_A(\lambda) \equiv \det(A - \lambda I) = \prod_{n=1}^{N} (\lambda_n - \lambda) \, . \tag{5.7}$$

For simplicity, we will restrict our discussion to matrices **A** that are real and symmetric, so that the eigenvalues are always real and the eigenvectors can be chosen to be orthonormal; this is the most common type of matrix arising in modeling physical systems. We will also consider only cases where all or many of the eigenvalues are required, and possibly their associated eigenvectors. If only a few of the largest or smallest eigenvalues of a large matrix are needed, then the iterative methods described in Section 7.4, which involve successive applications of A or its inverse, can be more efficient.

The general strategy for diagonalizing A is to reduce the task to the tractable one of finding the eigenvalues and eigenvectors of a symmetric tri-diagonal matrix; that is, one in which all elements are zero except those on or neighboring the diagonal. This can always be accomplished by applying a suitable sequence of orthogonal transformations to the original matrix, as described in the next section. For now, we assume that A has been brought to a tri-diagonal form and discuss a strategy for finding its eigenvalues.

To find the eigenvalues of a symmetric tri-diagonal matrix, we must find the roots of the characteristic polynomial, (5.7). This polynomial is given in terms of the elements of A by the determinant

$$
P_A(\lambda) = \begin{vmatrix}
A_{11} - \lambda & A_{12} & & & & \\
A_{21} & A_{22} - \lambda & A_{23} & & & \\
& A_{32} & A_{33} - \lambda & & & \\
& & A_{43} & & & \\
& & & \cdots & & \\
& & & & A_{N-1N-1} - \lambda & A_{N-1N} \\
& & & & A_{NN-1} & A_{NN} - \lambda
\end{vmatrix} ,
$$

$$(5.8)$$

where all elements not shown explicitly are zero and where the symmetry of A implies that $A_{nm} = A_{mn}$. To find the zeros of P_A, any of the root finding strategies discussed in Section 1.3 can be employed, providing that we can find the numerical value of P_A for a given value of λ. This latter can be done conveniently in a recursive manner. Let $P_n(\lambda)$ be the value of the $n \times n$ sub-determinant of (5.8) formed from the first n rows and columns. Clearly, P_n is a polynomial of degree n, $P_N = P_A$ (the polynomial we are after), and

$$
P_1(\lambda) = A_{11} - \lambda \ ; \quad P_2(\lambda) = (A_{22} - \lambda)P_1(\lambda) - A_{12}^2 \ . \tag{5.9}
$$

Moreover, by expanding the determinant for P_n in terms of the minors of the n'th column, it is easy to derive the recursion relation

$$
P_n(\lambda) = (A_{nn} - \lambda)P_{n-1}(\lambda) - A_{nn-1}^2 P_{n-2}(\lambda) \ . \tag{5.10}
$$

This, together with the starting values (5.9), allows an efficient evaluation of P_A.

■ **Exercise 5.3**   Prove Eq. (5.10) above.

Several features of the problem above help considerably in finding the roots of P_A. If an algorithm like Newton's method is used, it is quite easy to differentiate (5.10) once or twice with respect to λ and so derive recursion relations allowing the simple evaluation of the first or second derivatives of $P_A(\lambda)$. More importantly, it is possible to show that the number of times the sign changes in the sequence

$$1, \ P_1(\lambda), \ P_2(\lambda), \dots, P_N(\lambda)$$

is equal to the number of eigenvalues less than λ. This fact is useful in several ways: it is a means of making sure that no roots are skipped in the search for eigenvalues, it provides a simple way of localizing a root initially, and it can be used with the simple search procedure to locate the root accurately, although perhaps not in the most efficient way possible.

To make a systematic search for all of the eigenvalues (roots of P_A), it is essential to know where to begin looking. If we are working through finding the entire sequence of eigenvalues, a natural guess for the next-highest one is some distance above that eigenvalue just found. Some guidance in how far above and in how to estimate the lowest eigenvalue can be found in Gerschgorin's bounds on the eigenvalues. It is quite simple to show that

$$\lambda_n \geq \min_i \left\{ A_{ii} - \sum_{j \neq i} |A_{ij}| \right\}, \tag{5.11a}$$

and that

$$\lambda_n \leq \max_i \left\{ A_{ii} + \sum_{j \neq i} |A_{ij}| \right\} \tag{5.11b}$$

for all n. For the tri-diagonal forms under consideration here, the sums over j in these expressions involve only two terms. The lower bound, Eq. (5.11a), is a convenient place to begin a search for the lowest eigenvalue, and the difference between the upper and lower bounds gives some measure of the average spacing between eigenvalues.

■ **Exercise 5.4** Write a subroutine that finds all of the eigenvalues of a tri-diagonal matrix using the procedure described above, together with any root-finding algorithm you find convenient. Test your subroutine on an $N \times N$ tri-diagonal matrix of the form

$$A_{nn} = -2 \; ; \; A_{nn-1} = A_{n-1n} = +1 \; ,$$

whose eigenvalues are known analytically to be

$$\lambda_n = -4 \sin^2 \left(\frac{n\pi}{2(N+1)} \right) \; .$$

5.3 Reduction to tri-diagonal form

To apply the method for finding eigenvalues discussed in the previous section, we must reduce a general real symmetric matrix A to tri-diagonal form. That is, we must find an orthogonal $N \times N$ matrix, O, satisfying

$$O^t O = O O^t = I \; , \tag{5.12}$$

where O^t is the transpose of O, such that $O^t A O$ is tri-diagonal. Elementary considerations of linear algebra imply that the eigenvalues of the transformed matrix are identical to those of the original matrix. The problem, of course, is to find the precise form of O for any particular matrix A. We discuss in this section two methods for effecting such a transformation.

The Householder method is a common and convenient strategy for reducing a matrix to tri-diagonal form. It takes the matrix O to be the product of $N - 2$ orthogonal matrices,

$$O = O_1 O_2 \ldots O_{N-2} \; , \tag{5.13}$$

each of which successively transforms one row and column of A into the required form. (Only $N - 2$ transformations are required as the last two rows and columns are already in tri-diagonal form.) A simple method can be applied to find each of the O_n.

To be more explicit about how the Householder algorithm works, let us consider finding the first orthogonal transformation, O_1, which we will choose to annihilate most of the first row and column of A; that is,

$$O_1^t A O_1 = \begin{bmatrix} A_{11} & k^{(1)} & 0 & \cdots & 0 \\ k^{(1)} & A_{22}^{(2)} & A_{23}^{(2)} & \cdots & A_{2N}^{(2)} \\ 0 & A_{32}^{(2)} & A_{33}^{(2)} & \cdots & A_{3N}^{(2)} \\ \vdots & \vdots & \vdots & \vdots & \vdots \\ 0 & A_{N2}^{(2)} & A_{N3}^{(2)} & \cdots & A_{NN}^{(2)} \end{bmatrix} \; . \tag{5.14}$$

Here, $k^{(1)}$ is a possibly non-vanishing element and the matrix $\mathbf{A}^{(2)}$ is the result of applying the transformation $\mathbf{O}_1$ to the last $N-1$ rows and columns of the original matrix $\mathbf{A}$. Once $\mathbf{O}_1$ is found and applied (by methods discussed below), we can choose $\mathbf{O}_2$ so that it effects the same kind of transformation on the matrix $\mathbf{A}^{(2)}$; that is, it annihilates most of the first row and column of that matrix and transforms its last $N-2$ rows and columns:

$$\mathbf{O}_2^t \mathbf{A}^{(2)} \mathbf{O}_2 = \begin{bmatrix} A_{22}^{(2)} & k^{(2)} & 0 & \cdots & 0 \\ k^{(2)} & A_{33}^{(3)} & A_{34}^{(3)} & \cdots & A_{N3}^{(3)} \\ 0 & A_{43}^{(3)} & A_{44}^{(3)} & \cdots & A_{4N}^{(3)} \\ \vdots & \vdots & \vdots & \vdots & \vdots \\ 0 & A_{N3}^{(3)} & A_{N4}^{(3)} & \cdots & A_{NN}^{(3)} \end{bmatrix} . \tag{5.15}$$

Continued orthogonal transformations of decreasing dimension defined in this way will transform the original matrix into a tri-diagonal one after a total of $N-2$ transformations, the diagonal elements of the transformed matrix being

$$A_{11}, \; A_{22}^{(2)}, \; A_{33}^{(3)}, \ldots, A_{N-1\,N-1}^{(N-1)}, \; A_{NN}^{(N-1)} ,$$

and the off-diagonal elements being

$$k^{(1)}, \; k^{(2)}, \ldots, k^{(N-1)} .$$

It remains, of course, to find the precise form of each of the $\mathbf{O}_n$. We illustrate the procedure with $\mathbf{O}_1$, which we take to be of the form

$$\mathbf{O} = \begin{bmatrix} 1 & \mathbf{0}^t \\ \mathbf{0} & \mathbf{P} \end{bmatrix} , \tag{5.16}$$

where $\mathbf{0}$ is an $N-1$-dimensional column vector of zeros, $\mathbf{0}^t$ is a similar row vector, and $\mathbf{P}$ is an $(N-1) \times (N-1)$ symmetric matrix satisfying

$$\mathbf{P}^2 = \mathbf{I} \tag{5.17}$$

if $\mathbf{O}$ is to be orthogonal. (In these expressions and the following, for simplicity we have dropped all superscripts and subscripts which indicate that it is the first orthogonal transformation of the sequence (5.13) which is being discussed.) A choice for $\mathbf{P}$ that satisfies (5.17) is

$$\mathbf{P} = \mathbf{I} - 2\mathbf{u}\mathbf{u}^t . \tag{5.18}$$

In this expression, I is the $N - 1$-dimensional unit matrix, u is an $N - 1$-dimensional unit vector satisfying

$$u^t u = 1 , \qquad (5.19)$$

and the last term involves the *outer* product of u with itself. Each element of (5.18) therefore reads

$$P_{nm} = \delta_{nm} - 2u_n u_m ,$$

where n and m range from 1 to $N - 1$. (It should cause no confusion that the indices on P and u range from 1 to $N - 1$, while those on the full matrices O and A range from 1 to N.)

Under a transformation of the form (5.16), A becomes

$$O^t A O = \begin{bmatrix} A_{11} & (P\alpha)^t \\ P\alpha & A^{(2)} \end{bmatrix} , \qquad (5.20)$$

where we have defined an $N - 1$-dimensional vector α as all elements but the first in the first column of A; that is, $\alpha_i = A_{i+1,1}$ for i ranging from 1 to $N - 1$. Upon comparing (5.20) with the desired form (5.14), it is apparent that the action of P on this vector must yield

$$P\alpha \equiv \alpha - 2u(u^t \alpha) = k , \qquad (5.21)$$

where k is the vector

$$k = [k, 0, 0, \ldots, 0]^t .$$

Equation (5.21) is what we must solve to determine u, and hence the required transformation, P. To do so, we must first find the scalar k, which is easily done by taking the scalar product of (5.21) with its transpose and using the idempotence of P [Eq. (5.17)]:

$$k^t k = k^2 = (P\alpha)^t (P\alpha) = \alpha^t \alpha \equiv \alpha^2 = \sum_{i=2}^{N} A_{i1}^2 , \qquad (5.22)$$

so that $k = \pm\alpha$. Having found k (we will discuss the choice of the sign in a moment), we can then rearrange (5.21) as

$$\alpha - k = 2u(u^t \alpha) . \qquad (5.23)$$

Upon taking the scalar product of this equation with itself, using $\alpha^t k = \pm A_{21}\alpha$, and recalling (5.22), we have

$$2(u^t \alpha)^2 = (\alpha^2 \mp A_{21}\alpha) , \qquad (5.24)$$

so that we can then solve (5.21) for u:

$$u = \frac{\alpha - k}{2(u^t \alpha)} \, . \tag{5.25}$$

This completes the steps necessary to find P. To recapitulate, we first solve (5.22) for k, then calculate the square of the scalar product (5.24), form the vector u according to (5.25), and then finally P according to (5.18). In evaluating (5.24), considerations of numerical stability recommend taking that sign which makes the right-hand side largest. Note that we need evaluate only the square of the scalar product as the vector u enters P bilinearly. Note also that a full matrix product need not be performed in evaluating $A^{(2)}$. Indeed, from (5.18,20) we have

$$\begin{aligned} A^{(2)} &= (1 - 2uu^t)A(1 - 2uu^t) \\ &= A - 2u(Au)^t - 2(Au)u^t + 4uu^t(u^tAu) \, , \end{aligned} \tag{5.26}$$

where the symbol A in this expression stands for the square symmetric matrix formed from the last $N - 1$ rows and columns of the original matrix. Thus, to evaluate $A^{(2)}$, once we have the vector u we need only calculate the vector Au and the scalar u^tAu. Finally, we note that numerical stability during the successive transformations is improved if the diagonal element having largest absolute magnitude is in the upper left-hand corner of the sub-matrix being transformed. This can always be arranged by a suitable interchange of rows and columns (which leaves the eigenvalues invariant) after each of the orthogonal transformations is applied.

■ **Exercise 5.5** By writing either a detailed flowchart or an actual subroutine, convince yourself that you understand the Householder algorithm described above.

With an algorithm for transforming A to tri-diagonal form, we have specified completely how to find the eigenvalues of a real symmetric matrix. Once these are in hand, the eigenvectors are a relatively simple matter. The method of choice, inverse vector iteration, works as follows. Let $\phi_n^{(1)}$ be any guess for the eigenvector associated with λ_n. This guess can be refined by evaluating

$$\phi_n^{(2)} = [A - (\lambda_n + \epsilon)I]^{-1} \phi_n^{(1)} \, . \tag{5.27}$$

Here, ϵ is a small, non-zero scalar that allows the matrix to be inverted. It is easy to see that this operation enhances that component of $\phi_n^{(1)}$ along the true eigenvector at the expense of the spurious components. Normalization of $\phi_n^{(2)}$, followed by repeated refinements according to (5.27) converges quickly to the required eigenvector, often in only two iterations.

An alternative to the Householder method is the Lanczos algorithm [Wh77], which is most suitable when we are interested in many of the lowest eigenvalues of very large matrices. The strategy here is to construct iteratively a set of orthonormal basis vectors, $\{\psi_n\}$, in which A is explicitly tri-diagonal. To begin the construction, we choose an arbitrary first vector in the basis, ψ_1, normalized so that $\psi_1^t \psi_1 = 1$. We then form the second vector in the basis as

$$\psi_2 = C_2(A\psi_1 - A_{11}\psi_1)\,, \qquad (5.28)$$

where $A_{11} = \psi_1^t A \psi_1$ (it is *not* the element of A in the first row and column), and C_2 is a normalization chosen to ensure that $\psi_2^t \psi_2 = 1$:

$$C_2 = [(A\psi_1)^t(A\psi_1) - (A_{11})^2]^{-1/2}\,. \qquad (5.29)$$

It is easy to show that $\psi_2^t \psi_1 = 0$. Subsequent vectors in the basis are then constructed recursively as

$$\psi_{n+1} = C_{n+1}(A\psi_n - A_{nn}\psi_n - A_{nn-1}\psi_{n-1})\,, \qquad (5.30)$$

with

$$C_{n+1} = [(A\psi_n)^t(A\psi_n) - (A_{nn})^2 - (A_{nn-1})^2]^{-1/2}\,. \qquad (5.31)$$

Thus, each successive ψ_{n+1} is that unit vector which is coupled to ψ_n by A and which is orthogonal to both ψ_n and ψ_{n-1}. The matrix A is explicitly tri-diagonal in this basis since Eq. (5.30) shows that when A acts on ψ_n, it yields only terms proportional to ψ_n, ψ_{n-1} and ψ_{n+1}. Continuing the recursion (5.30) until ψ_N is generated completes the basis and the representation of A in it.

The Lanczos method is well-suited to large matrices, as only the ability to apply A to a vector is required, and only the vectors ψ_n, ψ_{n-1}, and $A\psi_n$ need be stored at any given time. We must, of course, also be careful to choose ψ_1 so that it is not an eigenvector of A. However, the Lanczos method is not appropriate for finding all of the eigenvalues of a large matrix, as round-off errors in the orthogonalizations of (5.30) will accumulate as the recursive generation of the basis proceeds; i.e., the scalar products of basis vectors with large n with those having small n will not vanish identically.

The real utility of the Lanczos algorithm is in cases where many, but not all, of the eigenvalues are required. Suppose, for example, that we are interested in the 10 lowest eigenvalues of a matrix of dimension 1000. What is done is to generate recursively some number of states in the basis greater than the number of eigenvalues being sought (say 25), and then to find the 10 lowest eigenvalues and eigenvectors of that limited sub-matrix. If an arbitrary linear combination (say the sum) of these eigenvectors is then used as the initial vector in constructing a new basis of dimension 25, it can be shown that iterations of this process (generating a limited basis, diagonalizing the corresponding tri-diagonal matrix, and using the normalized sum of lowest eigenvectors found as the first vector in the next basis) will converge to the required eigenvalues and eigenvectors.

■ **Exercise 5.6** Write a subroutine that uses the Lanczos method to generate a complete basis in which a symmetric input matrix **A** is tri-diagonal. The program should output the diagonal and off-diagonal elements of **A** in this basis, as well as the basis vectors themselves. Show by explicit example that if the dimension of the matrix is too large, round-off errors cause the basis generated to be inaccurate. Try curing this problem by doing the computation in double-precision arithmetic and observe the results.

■ **Exercise 5.7** A simple limit of the Lanczos procedure for generating and truncating the basis is when we are interested in only the lowest eigenvalue and retain only ψ_1 and ψ_2 to form a 2×2 matrix to be diagonalized. Show that the lower eigenvalue of this matrix is always less than or equal to A_{11}, so that iterations of the procedure lead to a monotonically decreasing estimate for the lowest eigenvalue of **A**. This procedure is closely related to the time-evolution algorithm discussed in Section 7.4.

5.4 Determining nuclear charge densities

The distribution of electrical charge within the atomic nucleus is one of the most basic aspects of nuclear structure. The interactions of electrons and muons with the nucleus can be used to determine this distribution with great precision, as these particles interact almost exclusively through the well understood electromagnetic interaction (for a general discussion, see [Fo66]). In this example, we will explore how the experimentally determined cross sections for the elastic scattering of high-energy (several 100 MeV) electrons from nuclei can be analyzed to determine the nuclear charge distribution; the method relies on the solution of a set of linear equations by matrix inversion.

To illustrate the basic idea, we begin by considering the scattering of non-relativistic electrons of momentum k and energy E from a localized charge distribution, $\rho(\mathbf{r})$, that contains Z protons, so that

$$\int d\mathbf{r}\,\rho(\mathbf{r}) = Z , \qquad (5.32)$$

and that is fixed in space. The electrons interact with this charge distribution through the Coulomb potential it generates, $V(\mathbf{r})$, which satisfies Poisson's equation,

$$\nabla^2 V = 4\pi\alpha\rho , \qquad (5.33)$$

where $\alpha = 1/137.036$ is the fine structure constant. (We henceforth work in units where $\hbar = c = 1$, unless explicitly indicated, and note that $\hbar c = 197.329$ MeV-fm.)

The Born approximation to quantum scattering discussed in the previous chapter illustrates how electron scattering cross sections carry information about the charge distribution, although precision work requires a better approximation to the scattering, as discussed below. Equations (4.18, 4.21) show that the Born cross section for scattering through an angle θ is proportional to the square of the Fourier transform of the potential at the momentum transfer $\mathbf{q}$, where $q = 2k \sin \frac{1}{2}\theta$. The Fourier transform of (5.33) results in

$$V(\mathbf{q}) = -\frac{4\pi\alpha}{q^2}\rho(\mathbf{q}) , \qquad (5.34)$$

where

$$\rho(\mathbf{q}) = \int d\mathbf{r}\,e^{-i\mathbf{q}\cdot\mathbf{r}}\rho(\mathbf{r}) \qquad (5.35)$$

is the Fourier transform of the charge density. Thus, the differential cross section can be written as (we use σ as a shorthand notation for $d\sigma/d\Omega$):

$$\sigma = \sigma_{\text{ruth}}|F(\mathbf{q})|^2 . \qquad (5.36)$$

Here, the Rutherford cross section for scattering from a point charge of strength Z is

$$\sigma_{\text{ruth}} = \frac{4Z^2\alpha^2 m^2}{q^4} , \qquad (5.37)$$

m is the electron mass, and the nuclear "form factor" is $F(\mathbf{q}) = Z^{-1}\rho(\mathbf{q})$. For the spin-0 nuclei we will be considering, the charge density is spherically symmetric, so that $\rho(\mathbf{r}) = \rho(r)$, and

$$F(\mathbf{q}) = F(q) = \frac{4\pi}{Zq}\int_0^\infty dr\,r \sin qr\rho(r) . \qquad (5.38)$$

Equation (5.36) illustrates how electron scattering is used to study nuclear structure. Deviations of the measured cross section from the Rutherford value are a direct measure of the Fourier transform of the nuclear charge density. It also shows that scatterings at high momentum transfers are needed to probe the nucleus on a fine spatial scale. As nuclear sizes are typically several fermis (10^{-13} cm), a spatial resolution better than 1 fm is desirable, implying momentum transfers of several fm^{-1} and so beam energies of several hundred MeV. (Recall that $2k$ is the maximum momentum transfer that can be achieved with a beam of momentum k.)

At energies of several hundred MeV, the electron is highly relativistic (the electron mass is only 0.511 MeV) and so the discussion above must be redone beginning with the Dirac equation rather than the Schroedinger equation. One trivial change in the final expressions that result is that ultra-relativistic kinematics are used for the electron (i.e., its momentum is proportional to its energy), so that the momentum transfer can be written as

$$q = 2E \sin \frac{1}{2}\theta . \tag{5.39}$$

The only other modification to Eq. (5.36) is that the Rutherford cross section is replaced by the Mott cross section,

$$\sigma = \sigma_{\text{mott}} |F(q)|^2 \; ; \; \sigma_{\text{mott}} = \frac{4Z^2\alpha^2 E^2 \cos^2 \frac{1}{2}\theta}{q^4} , \tag{5.40}$$

where the additional factor of $\cos \frac{1}{2}\theta$ in the scattering amplitude arises from the spin-$\frac{1}{2}$ nature of the electron.

Let us now consider how measured cross sections are to be analyzed to determine the charge density. Suppose that we have a set of I experimental values of the cross section for elastic scattering of electrons from a particular nucleus at a variety of momentum transfers (e.g., angular distributions at one or several beam energies). Let these values be σ_i^e and their statistical uncertainties by Δ_i. Suppose also that we parametrize the nuclear charge density by a set of N parameters, C_n, so that $\rho(r) = \rho(r; \{C_n\})$. Of course, the C's must be chosen so that the normalization constraint (5.32) is satisfied. That is,

$$Z(\{C_n\}) \equiv 4\pi \int_0^\infty dr\, r^2 \rho(r; \{C_n\}) = Z . \tag{5.41}$$

A specific choice for this parametrization will be discussed shortly.

The usual methods of data analysis [Be69] state that the "best" values of the parameters C_n implied by the data are those that minimize

$$\chi^2 \equiv \sum_{i=1}^{I} \frac{(\sigma_i^e - \sigma_i^t)^2}{\Delta_i^2} , \qquad (5.42)$$

subject to the normalization constraint (5.41) above. Here, σ_i^t is the "theoretical" cross section calculated from the appropriate Mott cross section and nuclear form factor; it depends parametrically upon the C's. This minimum value of χ^2 measures the quality of the fit, a satisfactory value being about the number of degrees of freedom (the number of data points less the number of parameters).

There are several computational strategies for finding the parameters that minimize χ^2 or, equivalently, that satisfy the N non-linear equations

$$\frac{\partial \chi^2}{\partial C_n} = 0 . \qquad (5.43)$$

This is a specific example of the commonly encountered problem of minimizing a non-linear function of several parameters; it is often fraught with difficulties [Ac70], not the least of which is finding a local minimum, rather than the global one usually sought. Most strategies are based on an iterative refinement of a "guess" for the optimal C_n, and, if this guess is close to the required solution, there is usually no problem. One commonly used approach is to compute the direction of the N-dimensional gradient [left-hand side of (5.43)] either analytically or numerically at the current guess and then to generate the next guess by stepping some distance in C-space directly away from this direction. Alternatively, a multi-dimensional generalization of the Newton-Raphson method, Eq. (1.14), can be used. The simple approach we will adopt here is based on a local linearization of σ_i^t about the current guess, C_n^0. For small variations of the C's about this point,

$$C_n = C_n^0 + \delta C_n ,$$

the theoretical cross sections can be expanded as

$$\sigma_i^t = \sigma_i^0 + \sum_{n=1}^{N} W_{in} \delta C_n ; \ W_{in} \equiv \frac{\partial \sigma_i^t}{\partial C_n} ,$$

so that χ^2 is given by

$$\chi^2 = \sum_{i=1}^{I} (\sigma_i^e - \sigma_i^0 - \sum_{n=1}^{N} W_{in} \delta C_n)^2 / \Delta_i^2 , \qquad (5.44)$$

where σ_i^0 is the theoretical cross section corresponding to the current guess, C^0. The charge normalization constraint (5.41) can be linearized similarly as

$$Z(\{C_n\}) = Z + \sum_{n=1}^{N} \frac{\partial Z}{\partial C_n} \delta C_n \, , \tag{5.45}$$

where we have assumed that the current guess is normalized properly, so that $Z(\{C_n^0\}) = Z$.

Equations for the δC's which make χ^2 stationary can be had by requiring

$$\frac{\partial}{\partial \delta C_n}(\chi^2 - 2\lambda Z) = 0 \, . \tag{5.46}$$

Here, we have introduced a Lagrange multiplier, 2λ, to ensure the proper charge normalization. Some simple algebra using Eqs. (5.44,45) then results in the linear equations

$$\sum_{m=1}^{N} a_{nm}\delta C_m - \frac{\partial Z}{\partial C_n}\lambda = b_n \, , \tag{5.47a}$$

$$\sum_{m=1}^{N} \frac{\partial Z}{\partial C_m}\delta C_m = 0 \, , \tag{5.47b}$$

with

$$a_{nm} = \sum_{i=1}^{I} \frac{W_{in}W_{im}}{\Delta_i^2} \, ; \quad b_n = \sum_{i=1}^{I} \frac{(\sigma_i^e - \sigma_i^0)W_{in}}{\Delta_i^2} \, . \tag{5.48}$$

For a given guess for the C's, Eqs. (5.47) can be solved to determine what change in the C's will reduce χ^2, subject to the validity of the linearization. The new improved values can then be used as the next guess, the process being continued until χ^2 converges to a minimum. To solve Eqs. (5.47), it is convenient to define the $(N+1) \times (N+1)$ matrix

$$A_{nm} = A_{mn} = a_{nm} \, ; \quad A_{n,N+1} = A_{N+1,n} = \frac{\partial Z}{\partial C_n} \, ; \quad A_{N+1,N+1} = 0 \, , \tag{5.49a}$$

and the $N+1$-component vector

$$B_n = b_n \, ; \quad B_{N+1} = 0 \, , \tag{5.49b}$$

and identify $\lambda \equiv \delta C_{N+1}$. (The indices m and n in these expressions range from 1 to N.) Equations (5.47) can therefore be written as

$$\sum_{m=1}^{N+1} A_{nm}\delta C_m = B_n \,, \tag{5.50}$$

which is simply solved by inversion of the symmetric matrix $\mathbf{A}$:

$$\delta C_n = \sum_{m=1}^{N+1} A_{nm}^{-1}B_m = \sum_{m=1}^{N} A_{nm}^{-1}b_m \,. \tag{5.51}$$

After the process described above converges to the optimal values to the C's, we can then enquire into the precision with which the parameters are determined. This can be related to the matrix $\mathbf{A}$ defined above as follows. Let us consider an ensemble of data sets in which each experimental cross section fluctuates independently about its mean value. That is,

$$\langle \delta\sigma_i^e \delta\sigma_j^e \rangle = \Delta_i^2 \delta_{ij} \,, \tag{5.52}$$

where $\langle \ldots \rangle$ denotes ensemble average, and $\delta\sigma_i^e$ is the deviation of σ_i^e from its mean value. Such fluctuations lead to fluctuations of the C's about their optimal value. Since the b_n are linearly related to the experimental cross sections by (5.48), these fluctuations are given by (5.51) as

$$\delta C_n = \sum_{m=1}^{N} A_{nm}^{-1}\delta b_m = \sum_{m=1}^{N} A_{nm}^{-1}\sum_{i=1}^{I} \frac{\delta\sigma_i^e W_{im}}{\Delta_i^2} \,. \tag{5.53}$$

Upon using this equation twice (for δC_n and δC_m) and taking the ensemble average, we have, after some algebra,

$$\langle \delta C_n \delta C_m \rangle = A_{nm}^{-1} \,. \tag{5.54}$$

The uncertainties in the density parameters can be translated into a more physical statement about the correlated uncertainties in the density itself by expanding the parametrized density about the optimal parameters. In particular, the correlated uncertainties in the densities at radii r and r' can be written as

$$\langle \delta\rho(r)\delta\rho(r') \rangle = \sum_{n=1}^{N}\sum_{m=1}^{N} \frac{\partial\rho(r)}{\partial C_n}\frac{\partial\rho(r')}{\partial C_m}\langle \delta C_n \delta C_m \rangle \,. \tag{5.55}$$

Thus, $\langle \delta\rho(r)\delta\rho(r) \rangle^{1/2}$ can be taken as a measure of the precision with which $\rho(r)$ is determined, although it should be remembered that the uncertainties in the density determined at different spatial points are correlated through (5.55).

We now turn to the parametrization of $\rho(r)$. On general grounds (for example, the validity of treating the nucleus as a sharp-surface, incompressible liquid drop), we expect the charge density to be fairly uniform within the nuclear interior and to fall to zero within a relatively thin surface region. This picture was largely confirmed by early electron scattering experiments [Ho57], although the measured cross sections extended only to momentum transfers less than about $1.5 \, \mathrm{fm}^{-1}$, which is barely enough resolution to see the nuclear surface. Such data could therefore yield only limited information, and a suitable form for the density with only three parameters was found to be of the form of a Fermi function:

$$\rho(r) = \frac{\rho_0}{1 + e^{4.4(r-R_0)/t}} \, , \qquad (5.56)$$

as illustrated in Figure 5.1. Since it turns out that $4.4R_0/t \gg 1$, the parameter ρ_0 is essentially the central density [i.e., $\rho(r = 0)$] the parameter R_0 is the radius where the charge density drops to one-half of its central value, and t, the surface thickness, is a measure of the distance over which this drop occurs. These three parameters are, of course, not all independent, as ρ_0 can be related to R_0 and t by the normalization constraint (5.32). Systematic studies of electron scattering from a number of nuclei showed that, for a nucleus with A nucleons, suitable values were $t = 2.4 \, \mathrm{fm}$ and $R_0 = 1.07A^{1/3} \, \mathrm{fm}$.

More modern studies of electron scattering use higher energy beams, and so cover momentum transfers to approximately $4 \, \mathrm{fm}^{-1}$. A typical example of the range and quality of the data is shown in Figure 5.2. Such data have led to a number of "model-independent" analyses [Fr75], in which a very flexible parametrization of the charge is used. One of these [Si74] takes the charge density to be the sum of many (of order 10) Gaussian "lumps", and adjusts the location of each lump and the amount of charge it carries to best fit the data. An alternative [Fr73], which we pursue here, assumes that the density vanishes outside a radius R, considerably beyond the nuclear surface, and expands the density as a finite Fourier series of N terms. That is, for $r < R$, we take

$$r\rho(r) = \sum_{n=1}^{N} C_n \sin\left(\frac{n\pi r}{R}\right) , \qquad (5.57)$$

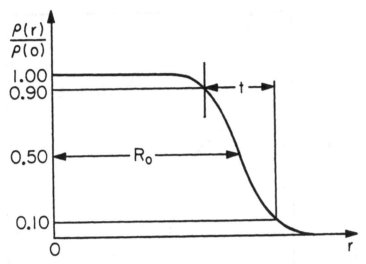

Figure 5.1   The Fermi function parametrization of the nuclear charge density. The density falls to one-half of its central value at the radius R_0; the distance over which the density drops from 90% to 10% of its central value is t. (After [Ho57].)

and for $r > R$, we take $\rho(r) = 0$. Equations (5.32) and (5.38), together with some straightforward algebra, show that

$$Z(\{C_n\}) = 4R^2 \sum_{n=1}^{N} \frac{(-)^{n+1}}{n} C_n \, , \qquad (5.58)$$

and that

$$F(q) = \frac{4\pi R^2}{Z} \sum_{n=1}^{N} C_n (-)^n \frac{n\pi}{(qR)^2 - n^2\pi^2} \frac{\sin qR}{qR} \, . \qquad (5.59)$$

These expressions, together with Eqs. (5.38,40), can be used to derive explicit forms for W_{in}, and hence for the χ^2 minimization procedure described above. It is useful to note that the n'th term in the Fourier expansion of the density gives a contribution to the form factor that is peaked near $q_n \equiv n\pi/R$. Hence, truncation of the series at N implies little control over Fourier components with q larger than $q_N = N\pi/R$. This maximum wavenumber contained in the parametrization must be commensurate with the maximum momentum transfer covered by the experimental data, q_{max}. If $q_N < q_{max}$, then data at larger momentum

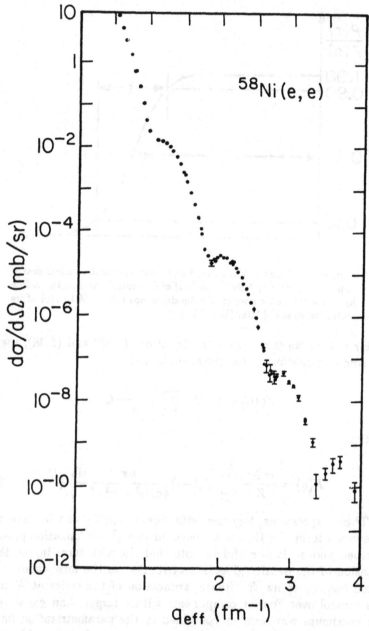

Figure 5.2 Experimental cross sections for the elastic scattering of electrons from ^{58}Ni at an energy of 449.8 MeV. (From [Ca80].)

transfers will not be well described, while if $q_N > q_{max}$, then the C_n's with $n > q_{max}R/\pi$ will not be well-determined by the fit.

We now return to the adequacy of the Born approximation for describing the experimental data. For charge distributions of the shape expected (roughly like those in Figure 5.1), the relatively sharp nuclear surface induces zeros in $F(q)$. That is, there are particular momentum transfers (or equivalently, scattering angles) where the Born scattering amplitude vanishes. At these values of q, corrections associated with the distortion of the plane wave of the incident electron by the nuclear Coulomb potential become relatively more important, so that the experimental data has these zeros largely filled in, often to the extent that they become merely shoulders in a rapidly falling cross section (see Figure 5.2). A precision determination of the nuclear charge density from the measured cross sections therefore requires a more sophisticated description of the scattering of the electrons from the Coulomb potential.

The rigorous solution to this problem involves solving the Dirac equation for the electron in the Coulomb potential generated by the assumed charge distribution [Ye54, Fr73]. This results in a calculation very similar to that done in Project IV: a partial wave decomposition, solution of the radial equation in each partial wave to determine the phase-shift, and then summation over partial waves to determine the scattering amplitude at each scattering angle. While such a calculation is certainly possible, it is too long for interactive programs. However, the eikonal approximation to Dirac scattering can be sufficiently accurate and results in a reasonable amount of numerical work.

The eikonal approximation for ultra-relativistic electrons scattering from a charge distribution results in final expressions very similar to those derived in Section 4.3 [Ba64]. The cross section is related to a complex scattering amplitude, f, by

$$\sigma = \cos^2 \frac{1}{2}\theta |f|^2 , \qquad (5.60)$$

and the scattering amplitude is given by the Fourier-Bessel transform [compare Eq. (4.30)]

$$f = -ik \int_0^\infty J_0(qb)[e^{2i\chi(b)} - 1]b \, db , \qquad (5.61)$$

where the profile function is the integral of the potential along the straight-line trajectory at impact parameter b:

$$\chi = -\frac{1}{2} \int_{-\infty}^\infty V(r) \, dz .$$

For the particular parametrization (5.57), ρ vanishes beyond R, and $V(r)$ is simply $-Z\alpha/r$ in this region. It is therefore convenient to split the impact parameter integral (5.61) into regions $b < R$ and $b > R$, so that the latter integral can be done analytically. This results in

$$f = f_{outer} + f_{inner} = f_{outer} - ik \int_0^R J_0(qb)[e^{2i\chi(b)} - 1]b\,db\,, \qquad (5.62)$$

where

$$f_{outer} = \frac{ik}{q^2} \Big[-2i\alpha ZX J_0(X) X^{2i\alpha Z} S_{-2i\alpha Z, -1}(X)$$
$$+ XJ_1(X) X^{2i\alpha Z} S_{1-2i\alpha Z, 0}(X) - XJ_1(X) \Big]\,, \qquad (5.63)$$

with $X = qR$. In this expression, $S_{\mu\nu}$ is a Lommel function (essentially the incomplete integral of the product of a Bessel function and a power), which has the useful asymptotic expansion for large X

$$S_{\mu\nu}(X) \sim X^{\mu-1} \Big[1 - \frac{(\mu-1)^2 - \nu^2}{X^2}$$
$$+ \frac{\{(\mu-1)^2 - \nu^2\}\{(\mu-3)^2 - \nu^2\}}{X^4} - \cdots \Big]\,. \qquad (5.64)$$

Note that since we will typically have $R \gtrsim 8$ fm and $q \lesssim 1$ fm^{-1}, X will be 8 or more, and so the number of terms shown should be quite sufficient.

The interior contribution to the scattering amplitude, f_{inner}, involves the potential for $r < R$ and so depends upon the detailed form of the nuclear charge density. Given the simple relation between the potential and density through Poisson's equation (5.33), it should not be too surprising that it is possible to express $\chi(b < R)$ directly in terms of the density:

$$\chi(b) = -Z\alpha \log\Big(\frac{b}{R}\Big) - 4\pi\alpha \int_b^R r^2 \rho(r)\phi(b/r)\,dr\,, \qquad (5.65)$$

where

$$\phi(y) \equiv \log\Big(\frac{1 + (1-y^2)^{1/2}}{y}\Big) - (1-y^2)^{1/2}\,.$$

The parametrization (5.57) then allows the profile function to be written in the form

$$\chi(b) = -Z\alpha \log\Big(\frac{b}{R}\Big) + \sum_{n=1}^N C_n \chi_n(b)\,; \qquad (5.66a)$$

$$\chi_n(b) = -4\pi\alpha R^2 \int_{b/R}^1 z\,dz \sin(n\pi z)\phi(b/zR)\,, \qquad (5.66b)$$

with the change of integration variable $z = r/R$.

Equations (5.61–64,66) complete the specification of the eikonal cross section in terms of the nuclear charge density. It is also easy to show that the quantities W_{in} are given by

$$W_{in} = 2\cos^2 \frac{1}{2}\theta_i \mathrm{Re}\left(f_i^* \frac{\partial f_i}{\partial C_n} \right) ; \tag{5.67a}$$

$$\frac{\partial f_i}{\partial C_n} = 2k \int_0^R J_0(q_i b)e^{2i\chi(b)}\chi_n(b)b\,db . \tag{5.67b}$$

Several fine points remain to be discussed before turning to the program. One is that the nucleus is not infinitely heavy, and so recoils when struck by the electron. To correct for this, the cross sections calculated above, which have all been for an electron impinging on a static charge distribution, must be divided by the factor

$$\eta \equiv 1 + 2\frac{E}{M} \sin^2 \frac{1}{2}\theta ,$$

where M is the target mass (roughly $A \times 940$ MeV). Note that this correction vanishes as M becomes large compared to E. Second, because of the Coulomb attraction, electrons at the nucleus have a slightly higher momentum than the beam value. In the eikonal treatment, this can be approximately corrected for by replacing q in all of the formulas by

$$q_{\mathrm{eff}} = q\left(1 - \frac{\bar{V}}{E}\right) , \tag{5.68}$$

and by multiplying the eikonal scattering amplitude by $(q_{\mathrm{eff}}/q)^2$. Here, $\bar{V}$ is a suitable average of the Coulomb potential over the nuclear volume, conveniently given by

$$\bar{V} = -\frac{4}{3}\frac{Z\alpha\hbar c}{R_0} ,$$

where R_0 is the nuclear radius. Similarly, in the Born approximation (5.40), the nuclear form factor is to be evaluated at q_{eff}, but the Mott cross section is still to be evaluated at q.

The FORTRAN program for Example 5, whose listing can be found in Appendix B and in the file EXMPL5.FOR, uses the method described above to analyze electron scattering cross sections for the nuclei ^{40}Ca $(Z = 20)$, ^{58}Ni $(Z = 28)$, and ^{208}Pb $(Z = 82)$ to determine their charge densities. Many different data sets for each target, taken at a variety of beam energies and angles (see, for example, [Fr77]), have been converted for equivalent cross sections at one beam energy [Ca80]. The integrals (5.62,66b,67b) are evaluated by 20-point Gauss-Legendre quadrature and the matrix inversion required by (5.51) is performed by an efficient implementation of Gauss-Jordan elimination [Ru63].

After requesting the target nucleus, the boundary radius R, and the number of sine functions N, the program iterates the procedure described above, displaying at each iteration plots of the fit, fractional error in the fit, and the charge density of the nucleus, together with the values of the C's and the density. The program also allows for initial iterations with fewer sines than are ultimately required, the number then being increased on request. This technique improves the convergence of the more rapidly oscillating components of the density as it reduces the effective dimension of the χ^2 search by first converging the smoother density components. Note that, at each iteration, only points at an effective momentum transfer smaller than that described accurately by the current number of sines are used in calculating χ^2.

The following exercises will be useful in understanding the physical and numerical principles important in this calculation.

■ **Exercise 5.8** To check that the program is working properly, use the eikonal formulas and Fermi-function charge density described above to generate a set of "pseudo-data" cross sections ranging from q between 0.5 and 4 fm^{-1}. Make sure that these cross sections are given at intervals typical of the experimental data and also assign errors to them typical of the data. Then verify the extent to which the correct (i.e., Fermi-function) charge distribution results when these pseudo-data are used as input to the program. One potential source of error you should investigate is the "completeness error". That is, the ability of a finite Fourier series to reproduce densities of the form expected. This can be studied by changing the number of sines used in the fit.

■ **Exercise 5.9** Run the code with the actual experimental data to determine the charge densities for the three nuclei treated. Study the quality of the fit and the errors in the density extracted as functions of the number of expansion terms used. Also verify that the fitting procedure converges to the same final solution if the initial guess for the density is varied within reasonable limits. Note that the converged solutions do not have a uniform interior density, but rather show "wiggles" due to the specific shell-model structure of each nucleus.

■ **Exercise 5.10** Extend the program to calculate the moments of the density,

$$M_k \equiv \left[\frac{4\pi}{Z} \int_0^\infty r^{2+k} \rho(r) \, dr \right]^{1/k},$$

and their uncertainties for integer k ranging from 2 to 5. This can be done conveniently by using Eq. (5.57) to express these moments directly in terms of the C_n. Verify that the size of these moments is consistent with an $A^{1/3}$ scaling of the nuclear radius.

■ **Exercise 5.11** A simple model for the doubly-magic nucleus ^{40}Ca is a Slater determinant constructed by putting four nucleons (neutron and proton, spin up and down) in the 10 lowest orbitals of a spherically-symmetric harmonic oscillator potential. Show that the charge density predicted by this model is

$$\rho(r) = \frac{1}{\pi^{3/2} r_0^3} e^{-(r/r_0)^2} \left[5 + 4 \left(\frac{r}{r_0} \right)^4 \right],$$

where $r_0 = (\hbar/m\omega)^{1/2}$, m being the nucleon mass and ω being the harmonic oscillator frequency. Determine the value of r_0 (and hence ω) required to reproduce the experimental value of the root-mean-square radius (M_2), and then compare the detailed form of ρ predicted by this model with that obtained from the data.

■ **Exercise 5.12** Modify the program so that it calculates the Born, rather than eikonal, cross section and show that the former is grossly inadequate to describe the data by comparing the Born cross sections for a Fermi distribution with the experimental data.

■ **Exercise 5.13** Modify the program to calculate the "off-diagonal" uncertainties in the density determined using Eq. (5.55) and show that the correlation in the uncertainties largely decreases as the separation of the points increases.

Project V: A schematic shell model

The assumption of particles moving independently in a potential is a common approximation in treating the quantum mechanics of many-particle systems. For example, this picture underlies the Hartree-Fock method for treating the many electrons in an atom (see Project III), the shell model for treating the many nucleons in a nucleus, and many models for treating the quarks in a hadron. However, it is only a rough approximation

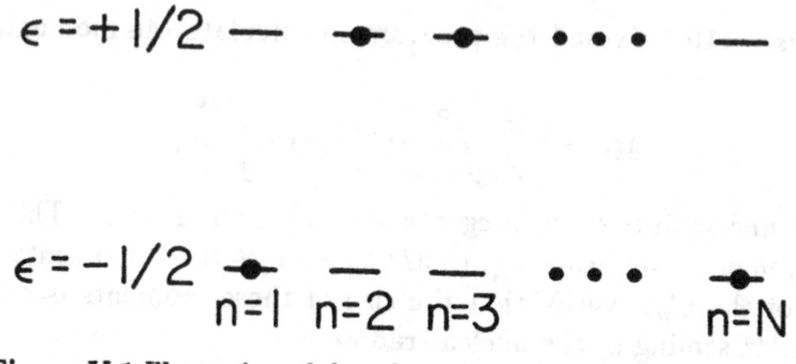

$$\epsilon = +1/2 \quad \text{—} \quad \bullet \quad \bullet \quad \bullet \bullet \bullet \quad \text{—}$$

$$\epsilon = -1/2 \underset{\substack{\text{n=1 \quad n=2 \quad n=3}}}{\bullet \quad \text{—} \quad \text{—}} \quad \bullet \bullet \bullet \quad \underset{\text{n=N}}{\bullet}$$

Figure V.1 Illustration of the orbitals in the schematic shell model.

in many cases and a detailed description of the exact quantum eigenstates often requires consideration of the "residual" interactions between the particles making up the system. In atoms and hadrons, the residual interactions are not strong enough to induce qualitative changes in the spectrum expected from simply placing the particles in different orbitals. In nuclei, however, the coherence and strength of the residual interactions can lead to a "collective" behavior in which many nucleons participate in the excitations and the character of the eigenstates is quite different from that expected naively.

Realistic calculations of the effects of residual interactions are quite complex and involve the diagonalization of large matrices expressing the way in which the interaction moves particles among the orbitals [Mc80]. However, the phenomenon of collectivity can be illustrated by a schematic shell model introduced by by Lipkin *et al.* [Li65], which we consider in this project. The model is non-trivial, yet simple enough to be soluble with only a modest numerical diagonalization. It has therefore served as a testing ground for approximations to be used in treating many-body systems and is also the prototype for more sophisticated group-theoretic models of nuclear spectra [Ar81].

V.1 Definition of the model
The schematic model consists of N distinguishable particles labeled by $n = 1, 2, \ldots, N$. Each of these can occupy one of two orbitals, a lower and an upper, having energies $+1/2$ and $-1/2$ and distinguished by $s = -1$ and $s = +1$, respectively (see Figure V.1). There are then 2^N states in the model, each defined by which particles are "up" and which are "down", and each having an unperturbed energy (in the absence of the residual interaction) equal to one-half of the difference between the number up and the number down.

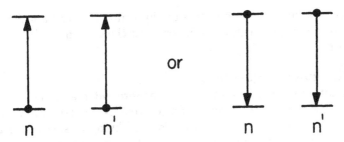

Figure V.2 Illustration of the residual interaction in the schematic model.

The residual interaction in the schematic model is one that changes simultaneously the states of any pair of particles; it therefore couples states of the unperturbed system. In particular, it is convenient to take a residual interaction of strength $-V$ that promotes any two particles that are "down" to "up" or that drops any two that are "up" to "down", as illustrated in Figure V.2. An interaction that simultaneously raises one particle and lowers another can also be included, but introduces no qualitatively new features.

To define the model precisely, it is easiest to use the language of second quantization and introduce the operators $a_{ns}^\dagger$ and their adjoints, a_{ns}, which respectively create and annihilate a particle in the orbital ns. As the model is constructed so that no more than one particle can occupy each orbital, it is convenient to endow these operators with the usual fermion anti-commutation relations,

$$\{a_{ns}^\dagger, a_{n's'}^\dagger\} = 0 , \tag{V.1a}$$

$$\{a_{ns}, a_{n's'}\} = 0 , \tag{V.1b}$$

$$\{a_{ns}^\dagger, a_{n's'}\} = \delta_{nn'}\delta_{ss'} . \tag{V.1c}$$

The operator $a_{ns}^\dagger a_{ns}$ then counts the number of particles in the orbital ns (i.e., it has eigenvalues 0 and 1), while the operators $a_{n1}^\dagger a_{n-1}$ and $a_{n-1}^\dagger a_{n1}$ respectively raise or lower the n'th particle. In terms of these operators, the Hamiltonian for the model can be written as

$$H = \frac{1}{2} \sum_{n=1}^{N} (a_{n1}^\dagger a_{n1} - a_{n-1}^\dagger a_{n-1})$$
$$- \frac{1}{2} V \sum_{n=1}^{N} \sum_{n'=1}^{N} (a_{n1}^\dagger a_{n-1} a_{n'1}^\dagger a_{n'-1} + a_{n-1}^\dagger a_{n1} a_{n'-1}^\dagger a_{n'1}) , \tag{V.2}$$

where we need not worry about the unphysical $n' = n$ term in the residual interaction since an attempt to raise or lower the same particle twice yields a vanishing state.

V.2 The exact eigenstates

The Hamiltonian described above can be represented as a matrix coupling the 2^N states among themselves. Apart from symmetry and a vanishing of many of its elements, this matrix is not obviously special, so that its numerical diagonalization to find the exact eigenstates and eigenvalues is a substantial problem for all but the smallest values of N. Fortunately, the relatively simple structure of the problem allows it to be expressed in the familiar language of the group SU(2) (the quantum angular momentum operators), a transformation that carries us a good deal of the way to an exact diagonalization and also affords some insight into the problem.

Let us define the operators

$$J_z = \frac{1}{2} \sum_{n=1}^{N} (a_{n1}^\dagger a_{n1} - a_{n-1}^\dagger a_{n-1}) , \qquad (V.3a)$$

$$J_+ = \sum_{n=1}^{N} a_{n1}^\dagger a_{n-1} , \quad J_- = (J_+)^\dagger = \sum_{n=1}^{N} a_{n-1}^\dagger a_{n1} . \qquad (V.3b)$$

Thus, J_z measures (one-half of) the difference between the number of "up" particles and the number of "down" particles, while J_+ and J_- coherently raise or lower all of the particles. Using these operators, the Hamiltonian (V.2) can be written as

$$H = J_z - \frac{1}{2}V(J_+^2 + J_-^2) = J_z - V(J_x^2 - J_y^2) , \qquad (V.4)$$

where we have introduced the operators

$$J_x = \frac{1}{2}(J_+ + J_-) , \quad J_y = -\frac{1}{2}i(J_+ - J_-) . \qquad (V.5)$$

Using the fundamental anti-commutation rules (V.1), it is easy to show that the three operators J_z, $J_\pm$ satisfy the commutation rules of a quantum angular momentum:

$$[J_z, J_\pm] = \pm J_\pm , \quad [J_+, J_-] = 2J_z . \qquad (V.6)$$

Thus, although these "quasi-spin" operators have nothing to do with a physical angular momentum, all of the techniques and experience in dealing with quantum spin operators can be applied immediately to this problem.

We can begin by realizing that since the total quasi-spin operator

$$J^2 = J_x^2 + J_y^2 + J_z^2 \qquad (V.7)$$

commutes with $J_{x,y,z}$, it commutes with the Hamiltonian and thus each eigenstate of H can be labeled by its total quasi-spin, j, where $j(j+1)$ is the eigenvalue of J^2. The eigenstates therefore can be classified into non-degenerate multiplets of $2j+1$ states each, all states in a multiplet having the same j; the Hamiltonian has a non-vanishing matrix element between two states only if they belong to the same multiplet. However, since H does not commute with J_z, its eigenstates will not be simultaneously eigenstates of this latter operator.

To see what values of j are allowed, we can classify the 2^N states of the model (not the eigenstates of H) according to their eigenvalues of J_z. The one state with the largest eigenvalue of J_z has all particles up, corresponding to an eigenvalue $m = \frac{1}{2}N$. We therefore expect one multiplet of $2 \cdot \frac{1}{2}N + 1 = N + 1$ states corresponding to $j = \frac{1}{2}N$. Turning next to states with $m = \frac{1}{2}N - 1$, we see that there are N states, corresponding to one of the N particles down and all of the rest up. One linear combination of these N states (the totally symmetric one) belongs to the $j = \frac{1}{2}N$ multiplet, implying that there are $N - 1$ multiplets with $j = \frac{1}{2}N - 1$. Continuing in a similar way, there are $\frac{1}{2}N(N - 1)$ states with $m = \frac{1}{2}N - 2$ (two particles down, the rest up), of which one linear combination belongs to the $j = \frac{1}{2}N$ multiplet and $N - 1$ linear combinations belong to the $j = \frac{1}{2}N - 1$ multiplets. There are thus $\frac{1}{2}N(N - 3)$ multiplets with $j = \frac{1}{2}N - 2$. By continuing in this way, we can classify the 2^N states of the model into multiplets with j running from $\frac{1}{2}N$ down to 0 or $\frac{1}{2}$, if N is even or odd, respectively.

Because H involves only the quasi-spin operators, its action within a multiplet depends only upon the value of j involved. For a given N, the spectrum of one multiplet therefore serves for all with the same j and, in fact, also serves for multiplets with this j in systems of larger N. We can therefore restrict our attention to the multiplet with the largest value of j, $\frac{1}{2}N$.

By these considerations, we have reduced the problem of diagonalizing the full Hamiltonian to one of diagonalizing (V.4) in the $N + 1$ eigenstates of J_z belonging to the multiplet with $j = \frac{1}{2}N$. We can label these states as $|m\rangle$, where m runs from $-j$ to j in integer steps. Using the usual formulas for the action of the angular momentum raising and lowering operators,

$$J_\pm |m\rangle = C_m^\pm |m \pm 1\rangle \; ; \; C_m^\pm = [j(j+1) - m(m \pm 1)]^{1/2} , \qquad (V.8)$$

we can write the elements of H in this basis as

$$\langle m'|H|m \rangle = m\delta_{m'm} - \frac{1}{2}V[C_m^+ C_{m+1}^+ \delta_{m'm+2} + C_m^- C_{m-1}^- \delta_{m'm-2}] . \quad (V.9)$$

Thus, H is tri-diagonal in this basis and in fact separates into two un-coupled problems involving the states $m = -j, -j+2, \ldots, +j$ and $m = -j+1, -j+3, \ldots, +j-1$ when N is even (j is integral) and the states $m = -j, -j+2, \ldots, +j-1$ and $m = -j+1, -j+3, \ldots, +j$ when N is odd (j is half-integral). For small values of j, the resulting matrices can be diagonalized analytically, while for larger systems, the numerical methods discussed in this chapter can be employed to find the eigenvalues and eigenvectors.

The quasi-spin method allows us to make one other statement about the exact solution of the model. Let

$$R = e^{i\pi(J_x + J_y)/2^{1/2}}$$

be the unitary operator effecting a rotation in quasi-spin space by an angle of π about the axis $(\hat{x} + \hat{y})/2^{1/2}$. It is easy to see that this rotation transforms the quasi-spin operators as

$$RJ_x R^\dagger = J_y \; ; \; RJ_y R^\dagger = J_x \; ; \; RJ_z R^\dagger = -J_z , \quad (V.10)$$

so that the Hamiltonian (V.4) transforms as

$$RHR^\dagger = -H . \quad (V.11)$$

Hence, if ψ is an eigenstate of H with energy E, then $R\psi$ is an eigenstate of H with energy $-E$. Thus, the eigenvalues of H come in pairs of equal magnitude and opposite sign (or are 0). This also allows us to see that if N is even, at least one of the eigenvalues will be zero.

V.3 Approximate eigenstates

We turn now to approximate methods for solving the schematic model and consider first ordinary Rayleigh-Schroedinger perturbation theory. If we treat the J_z term in (V.4) as the unperturbed Hamiltonian and the $J_x^2 - J_y^2$ term as the perturbation, then the unperturbed eigenstates are $|m\rangle$ with unperturbed energies $E_m^{(0)} = m$ and the perturbation series is an expansion in powers of V. Since $\langle m|J_\pm|m \rangle = 0$, the energies are unperturbed in first order and are perturbed in second order as

$$\Delta E_m^{(2)} = \frac{\langle m+2|\frac{1}{2}VJ_+^2|m \rangle^2}{E_m^{(0)} - E_{m+2}^{(0)}} + \frac{\langle m-2|\frac{1}{2}VJ_-^2|m \rangle^2}{E_m^{(0)} - E_{m-2}^{(0)}} . \quad (V.12)$$

Using Eq. (V.9) for the matrix elements involved, explicit expressions for the second-order energies can be derived; the fourth-order terms can also be done with a bit of patient algebra.

While the weak-coupling (small-V) limit can be treated by straightforward perturbation theory, an approximation for the strong-coupling (large-V) situation is not so obvious. One appealing approach is the semiclassical method, valid in the limit of large N [Ka79, Sh80]. The discussion begins by considering the equations of motion for the time-dependence of the expectation values of the quasi-spin operators:

$$i\frac{d}{dt}\langle J_i \rangle = \langle [J_i, H] \rangle , \qquad (V.13)$$

where J_i is any one of the three components of the quasi-spin. Using the Hamiltonian (V.4) and the commutation rules (V.6), it is easy to write out Eq. (V.13) for each of the three components:

$$\frac{d}{dt}\langle J_x \rangle = -\langle J_y \rangle + V\langle J_z J_y + J_y J_z \rangle , \qquad (V.14a)$$

$$\frac{d}{dt}\langle J_y \rangle = \langle J_x \rangle + V\langle J_z J_x + J_x J_z \rangle , \qquad (V.14b)$$

$$\frac{d}{dt}\langle J_z \rangle = -2V\langle J_y J_x + J_x J_y \rangle . \qquad (V.14c)$$

Unfortunately, these equations do not form a closed set, as the time derivatives of expectation values of single quasi-spin operators depend upon expectation values of bilinear products of these operators. However, if we ignore the fact that the J's are operators, and put, for example,

$$\langle J_x J_y \rangle = \langle J_x \rangle \langle J_y \rangle ,$$

then a closed set of equations results:

$$\frac{d}{dt}\langle J_x \rangle = -\langle J_y \rangle + 2V\langle J_z \rangle \langle J_y \rangle , \qquad (V.15a)$$

$$\frac{d}{dt}\langle J_y \rangle = \langle J_x \rangle + 2V\langle J_z \rangle \langle J_x \rangle , \qquad (V.15b)$$

$$\frac{d}{dt}\langle J_z \rangle = -4V\langle J_y \rangle \langle J_x \rangle . \qquad (V.15c)$$

It is easy to show that Eqs. (V.15) conserve

$$J^2 \equiv \langle J_x \rangle^2 + \langle J_y \rangle^2 + \langle J_z \rangle^2 ,$$

so that a convenient parametrization is in terms of the usual polar angles,

$$\langle J_z \rangle = J\cos\theta , \quad \langle J_x \rangle = J\sin\theta\cos\phi , \quad \langle J_y \rangle = J\sin\theta\sin\phi , \qquad (V.16)$$

where

$$J = [j(j+1)]^{1/2} \approx \frac{1}{2}N \ .$$

If we further define the variables

$$p \equiv \frac{1}{2}N \cos \theta \ , \quad q \equiv \phi + \frac{1}{4}\pi \ , \qquad\qquad (V.17)$$

so that $|p| \leq \frac{1}{2}N$ and $0 \leq q \leq 2\pi$, then, upon introducing the convenient coupling constant

$$\chi \equiv NV \approx 2JV \ ,$$

Eqs. (V.15) can be written as

$$\frac{dp}{dt} = 2\frac{\chi}{N}\left(\frac{N^2}{4} - p^2\right)\cos 2q = -\frac{\partial E(p,q)}{\partial q} \ , \qquad (V.18a)$$

$$\frac{dq}{dt} = 1 + 2\frac{\chi}{N}p \sin 2q = \frac{\partial E(p,q)}{\partial p} \ . \qquad\qquad (V.18b)$$

Here, E is the expectation value of the Hamiltonian

$$E(p,q) \equiv \langle H \rangle = \langle J_z \rangle - V(\langle J_x \rangle^2 - \langle J_y \rangle^2) \qquad (V.19a)$$

$$= p - \frac{\chi}{N}\left(\frac{N^2}{4} - p^2\right)\sin 2q \ . \qquad (V.19b)$$

By these manipulations, we have transformed the quasi-spin problem into one of a time-dependent classical system with "momentum" p and "coordinate" q satisfying equations of motion (V.18). As these equations are of the canonical form, p and q are canonically conjugate and (V.19b) can be identified as the Hamiltonian for this classical system. Our derivation has been heuristic, but not rigorous. A more careful calculation defines the classical Hamiltonian as the expectation value of the quantum Hamiltonian between coherent states [Ya82]. However, the only modification this procedure gives to our results is the redefinition:

$$\chi = (N-1)V \ .$$

Note that the Hamiltonian so derived is a variational one, therefore the quantum ground state energy will be *lower* than the minima in the classical energy.

We can now infer properties of the quantum eigenstates by analyzing this classical system. Using Eq. (V.19b), we can calculate the trajectories in phase-space; i.e., the value of p for a given q and $E = \langle H \rangle$. Equivalently, we can consider contour lines of E in the (p,q) plane. In an analysis of

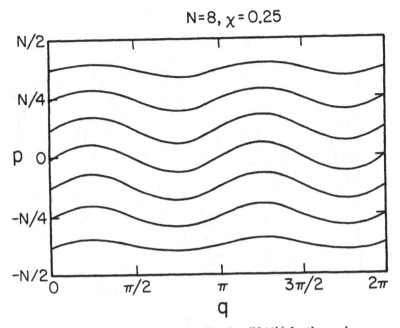

Figure V.3　Contours of the Hamiltonian (V.19b) for the weak coupling case $N = 8$ and $\chi = 0.25$. The contour passing through ($p = 0$, $q = 0$) has zero energy. The upper and lower boundary lines $p = \pm\frac{1}{2}N$ are contours at energy $\pm\frac{1}{2}N$, respectively. Successive contours shown differ in energy by 1. (From [Ka79].)

these contours, it is useful to distinguish between weak ($\chi < 1$) and strong ($\chi > 1$) coupling cases.

Figure V.3 shows a typical weak-coupling case, $N = 8$ and $\chi = 0.25$. The trajectory with the lowest allowed energy, which we can identify with the ground state of the quantum system, has $p = -\frac{1}{2}N$, $E = -\frac{1}{2}N$, and q satisfying the differential equation (V.18b). Note that this trajectory ranges over all values of q.

In the strong coupling case, $\chi > 1$, the trajectories look as shown in Figure V.4, which corresponds to $N = 8$ and $\chi = 2.5$. There are now minima in the energy surface at $p = -\frac{1}{2}N/\chi$, $q = \pi/4$, $5\pi/4$, which, using the equations of motion (V.18), can be seen to correspond to stationary points (i.e., time-independent solutions of the equations of motion). The energy at these minima is $E = -N(\chi + \chi^{-1})/4$, which we can identify as an estimate of the ground state energy. There is therefore a "phase transition" at $\chi = 1$ where the ground state changes from one where

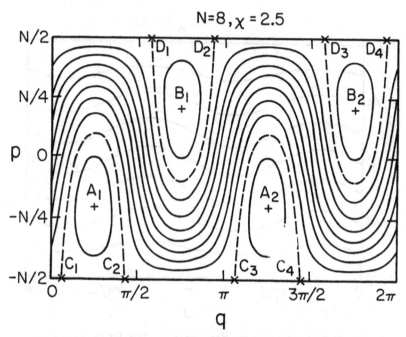

Figure V.4 Similar to Figure V.3 for the strong coupling case $N = 8$, $\chi = 2.5$. The points $A_{1,2}$, $B_{1,2}$ are minima and maxima, respectively, and C_{1-4}, D_{1-4} are saddle points. Contours through the saddle points, which have energy $\pm\frac{1}{2}N$, are drawn as dashed lines; these separate trajectories that are localized and extended in q. Successive contours differ in energy by 1. (From [Ka79].)

$p = -\frac{1}{2}N$ to one where $p > -\frac{1}{2}N$, the energy being continuous across this transition. This qualitative change in the nature of the ground state is a direct consequence of the coherence and strength of the interactions among the particles. It should also be noted that, in direct analogy with the usual quantum double-well problem, the presence of two degenerate minima in Figure V.4 suggests that the ground and first-excited states will be nearly degenerate, the splitting between them caused only by "tunneling" between the wells.

The semiclassical analysis can be extended to predict the energies of excited states as well, essentially by applying the analog of the Bohr-Sommerfeld quantization rule given in Eqs. (1.20,21) [Sh80]. One finds that N controls the quantum nature of the problem, with the system becoming more "classical" as the number of particles increases. In the weak coupling case, all trajectories extend over the full range of q. However, for strong coupling, there are two types of trajectories: those that vibrate around the energy minima or maxima (and are hence confined to limited regions of q) and those that extend throughout all q. It is easy to show from (V.19b) that the transitions between these two behaviors occur at $|E| = J$ (e.g., the dashed lines in Figure V.4). In analogy with the usual quantum double-well, we expect states associated with the confined trajectories to occur in nearly degenerate pairs, the degeneracy being broken more strongly as the energy increases above the minima or decreases below the maxima.

V.4 Solving the model

The simple semiclassical picture we have discussed above can be tested by an analysis of the exact eigenstates of the model. This can be carried out through the following sequence of steps.

■ **Step 1** Verify the algebra in the discussion above. In particular, show that Eq. (V.4) is a valid representation of the Hamiltonian. Also evaluate the second-order perturbation to the energies, Eq. (V.12).

■ **Step 2** Verify the correctness of Eqs. (V.14,18,19) and the discussion of the energy contours in the strong and weak coupling cases. In the strong coupling case, linearize the equations of motion about the minima or maxima and show that the frequency of harmonic motion about these points is $[2(\chi^2 - 1)]^{1/2}$. This is the spacing expected between the pairs of nearly degenerate states in the exact spectrum.

■ **Step 3** Write a program that finds the eigenvalues of the tri-diagonal Hamiltonian matrix (V.9), perhaps by modifying that written for Exercise 5.4. Note that the numerical work can be reduced by treating basis states with even and odd m separately and by using the symmetry of the spectrum about $E = 0$. In addition, you can save yourself some tedious bookkeeping concerning the size of the bases by taking N even only.

■ **Step 4** Use the program written in Step 3 to calculate the spectrum of the model for selected values of N between 8 and 40 and for values of χ between 0.1 and 5. At weak coupling, compare your results with the perturbation estimates (V.12). For strong couplings, compare your ground state energy with the semiclassical estimate and verify the expected pairwise degeneracy of the low-lying states. Also compare the excitation energies of these states with your estimate in Step 2. How does the accuracy of the semi-classical estimate change with N?

■ **Step 5**   Write a program that takes the eigenvalues found in Step 2 and solves for the eigenvectors by inverse vector iteration [Eq. (5.27)]. For selected values of N and χ, compute the expectation value of J_z for each eigenvector. In a semiclassical interpretation, this expectation value can be identified with the time-average of p over the associated trajectory in phase space. Verify that $\langle J_z \rangle$ changes through the spectrum in accord with what you expect from Figures V.3, V.4.

Chapter 6

Elliptic
Partial
Differential
Equations

Partial differential equations are involved in the description of virtually every physical situation where quantities vary in space or in space and time. These include phenomena as diverse as diffusion, electromagnetic waves, hydrodynamics, and quantum mechanics (Schroedinger waves). In all but the simplest cases, these equations cannot be solved analytically and so numerical methods must be employed for quantitative results. In a typical numerical treatment, the dependent variables (such as temperature or electrical potential) are described by their values at discrete points (a lattice) of the independent variables (e.g., space and time) and, by appropriate discretization, the partial differential equation is reduced to a large set of difference equations. Although these difference equations then can be solved, in principle, by the direct matrix methods discussed in Chapter 5, the large size of the matrices involved (dimension comparable to the number of lattice points, often more than several thousand) makes such an approach impractical. Fortunately, the locality of the original equations (i.e., they involve only low-order derivatives of the dependent variables) makes the resulting difference equations "sparse" in the sense that most of the elements of the matrices involved vanish. For such matrices, iterative methods of inversion and diagonalization can be very efficient. These methods are the subject of this and the following chapter.

Most of the physically important partial differential equations are of second order and can be classified into three types: parabolic, elliptic, or hyperbolic. Roughly speaking, parabolic equations involve only a first-order derivative in one variable, but have second order derivatives in the remaining variables. Examples are the diffusion equation and the time-dependent Schroedinger equation, which are first order in time, but second order in space. Elliptic equations involve second order derivatives in each of the independent variables, each derivative having the same sign when all terms in the equation are grouped on one side. This class includes Poisson's equation for the electrostatic potential and the time-independent Schroedinger equation, both in two or more spatial variables. Finally, there are hyperbolic equations, which involve second derivatives of opposite sign, such as the wave equation describing the vibrations of a stretched string. In this chapter, we will discuss some numerical methods appropriate for elliptic equations, reserving the discussion of parabolic equations for Chapter 7. Hyperbolic equations often can be treated by methods similar to those we discuss, although unique difficulties do arise [Ri67].

For concreteness in our discussion, we will consider particular forms of elliptic boundary value and eigenvalue problems for a field ϕ in two spatial dimensions (x, y). The boundary value problem involves the equation

$$-\left[\frac{\partial^2}{\partial x^2} + \frac{\partial^2}{\partial y^2}\right]\phi = S(x, y) . \tag{6.1}$$

Although this is not the most general elliptic form, it nevertheless covers a wide variety of situations. For example, in an electrostatics problem, ϕ is the potential and S is related to the charge density [compare Eq. (3.2)], while in a steady-state heat diffusion problem, ϕ is the temperature, and S is the local rate of heat generation or loss. Our discussion can be generalized straightforwardly to other elliptic cases, such as those involving three spatial dimensions or a spatially varying diffusion or dielectric constant.

Of course, Eq. (6.1) by itself is an ill-posed problem, as some sort of boundary conditions are required. These we will take to be of the Dirichlet type; that is, ϕ is specified on some large closed curve in the (x, y) plane (conveniently, the unit square) and perhaps on some additional curves within it (see Figure 6.1). The boundary value problem is then to use (6.1) to find ϕ everywhere within the unit square. Other classes of boundary value problems, such as Neumann (where the normal derivative of ϕ is specified on the surfaces) and mixed (where a linear combination of ϕ and its normal derivative is specified), can be handled by very similar methods.

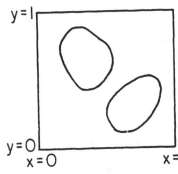
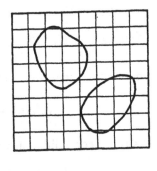

Figure 6.1 (Left) A two-dimensional boundary value problem with Dirichlet boundary conditions. Values of ϕ are specified on the edges of the unit square and on the surfaces within. (Right) Discretization of the problem on a uniform cartesian lattice.

The eigenvalue problems we will be interested in might involve an equation of the form

$$-\left[\frac{\partial^2}{\partial x^2} + \frac{\partial^2}{\partial y^2}\right]\phi + V(x,y)\phi = \epsilon\phi , \qquad (6.2)$$

together with a set of Dirichlet boundary conditions. This might arise, for example, as the time-independent Schroedinger equation, where ϕ is the wave function, V proportional to the potential, and ϵ is related to the eigenvalue. Such an equation might also be used to describe the fields in an acoustic or electromagnetic waveguide, where ϵ is then related to the square of the cut-off frequency. The eigenvalue problem then consists of finding the values ϵ_λ and the associated eigenfunctions ϕ_λ for which Eq. (6.2) and the boundary conditions are satisfied. As methods for solving such problems are closely related to those for solving a related parabolic equation, we will defer their discussion to Chapter 7.

6.1 Discretization and the variational principle

Our first step is to cast Eq. (6.1) in a form suitable for numerical treatment. To do so, we define a lattice of points covering the region of interest in the (x,y) plane. For convenience, we take the lattice spacing, h, to be uniform and equal in both directions, so that the unit square in covered by $N \times N$ lattice points (see Figure 6.1). These points can be labeled by indices (i,j), each of which runs from 0 to N, so that the coordinates of the point (i,j) are $(x_i = ih, y_j = jh)$. If we then define $\phi_{ij} = \phi(x_i, y_j)$, and similarly for S_{ij}, it is straightforward to apply the three-point difference approximation (1.7) for the second derivative in each direction and so approximate (6.1) as

$$-\left[\frac{\phi_{i+1j} + \phi_{i-1j} - 2\phi_{ij}}{h^2} + \frac{\phi_{ij+1} + \phi_{ij-1} - 2\phi_{ij}}{h^2}\right] = S_{ij} , \qquad (6.3a)$$

or, in a more convenient notation,

$$-\left[(\delta_i^2\phi)_{ij} + (\delta_j^2\phi)_{ij}\right] = h^2 S_{ij} . \qquad (6.3b)$$

Here, δ_i^2 is the second-difference operator in the i index,

$$(\delta_i^2\phi)_{ij} \equiv \phi_{i+1j} + \phi_{i-1j} - 2\phi_{ij} ,$$

and δ_j^2 is defined similarly.

Although Eq. (6.3) is the equation we will be solving, it is useful to derive it in a different way, based on a variational principle. Such an approach is handy in cases where the coordinates are not cartesian, as discussed in Section 6.3 below, or when more accurate difference formulas are required. It also is guaranteed to lead to symmetric (or hermitian) difference equations, an often useful property. The variational method also affords some insight into how the solution algorithm works. A good review of this approach can be found in [Ad84].

Consider the quantity E, defined to be a functional of the field ϕ of the form

$$E = \int_0^1 dx \int_0^1 dy \left[\frac{1}{2}(\nabla\phi)^2 - S\phi\right] . \qquad (6.4)$$

In some situations, E has a physical interpretation. For example, in electrostatics, $-\nabla\phi$ is the electric field and S is the charge density, so that E is indeed the total energy of the system. However, in other situations, such as the steady-state diffusion equation, E should be viewed simply as a useful quantity.

It is easy to show that, at a solution to (6.1), E is stationary under all variations $\delta\phi$ that respect the Dirichlet boundary conditions imposed. Indeed, the variation is

$$\delta E = \int_0^1 dx \int_0^1 dy \left[\nabla\phi \cdot \nabla\delta\phi - S\delta\phi\right] , \qquad (6.5)$$

which, upon integrating the second derivative term by parts, becomes

$$\delta E = \int_C dl\, \delta\phi \mathbf{n} \cdot \nabla\phi + \int_0^1 dx \int_0^1 dy\, \delta\phi[-\nabla^2\phi - S] , \qquad (6.6)$$

where the line integral is over the boundary of the region of interest (C) and n is the unit vector normal to the boundary. Since we consider only variations that respect the boundary conditions, $\delta\phi$ vanishes on C, so that the line integral does as well. Demanding that δE be zero for all such variations then implies that ϕ satisfies (6.1). This then furnishes a variational principle for our boundary value problem.

To derive a discrete approximation to the partial differential equation based on this variational principle, we first approximate E in terms of the values of the field at the lattice points and then vary with respect to them. The simplest approximation to E is to employ the two-point difference formula to approximate each first derivative in $(\nabla\phi)^2$ at the points halfway between the lattice points and to use the trapezoidal rule for the integrals. This leads to

$$E = \frac{1}{2}\sum_{i=1}^{N}\sum_{j=1}^{N}[(\phi_{ij}-\phi_{i-1j})^2+(\phi_{ij}-\phi_{ij-1})^2]-h^2\sum_{i=1}^{N-1}\sum_{j=1}^{N-1}S_{ij}\phi_{ij}\ . \quad (6.7)$$

Putting

$$\frac{\partial E}{\partial\phi_{ij}} = 0$$

for all ij then leads to the difference equation derived previously, (6.3). Of course, a more accurate discretization can be obtained by using better approximations for the first derivatives and for the integrals, taking care that the accuracies of both are commensurate. It is also easy to show from (6.7) that not only is E stationary at the solution, but that it is a minimum as well.

■ **Exercise 6.1** Show that the vanishing of the derivatives of Eq. (6.7) with respect to the values of ϕ at the lattice points lead to the difference Eq. (6.3). Prove, or argue heuristically, that E is a minimum when ϕ is the correct solution.

■ **Exercise 6.2** Use the differentiation formulas given in Table 1.2 to derive discretizations that are more accurate than (6.3). Can you see how the boundary conditions are to be incorporated in these higher-order formulas? What are the corresponding discretizations of E analogous to (6.7)? (These are not trivial; see [Fl78].)

We must now discuss where the boundary conditions enter the set of linear equations (6.3). Unless the coordinate system is well adapted to the geometry of the surfaces on which the boundary conditions are imposed (e.g., the surfaces are straight lines in cartesian coordinates or arcs in cylindrical or spherical coordinates), the lattice points will only roughly describe the geometry (see Fig. 6.1). One can always improve the accuracy by using a non-uniform lattice spacing and placing more points in the regions near the surfaces or by transforming to a coordinate system in which the boundary conditions are expressed more naturally. In any event, the boundary conditions will then provide the values of the ϕ_{ij} at some subset of lattice points. At a point far away from one of the boundaries, the boundary conditions do not enter (6.3) directly. However, consider (6.3) at a point just next to a boundary, say $(i, N - 1)$. Since ϕ_{iN} is specified as part of the boundary conditions (as it is on the whole border of the unit square), we can rewrite (6.3b) as

$$4\phi_{iN-1} - \phi_{i+1N-1} - \phi_{i-1N-1} - \phi_{iN-2} = h^2 S_{iN-1} + \phi_{iN} ; \qquad (6.8a)$$

that is, ϕ_{iN} enters not as an unknown, but rather as an inhomogeneous, known term. Similarly, if a Neumann boundary condition were imposed at a surface, say $\partial\phi/\partial y = g(x)$ at $y = 1$ or, equivalently, $j = N$, then this could be approximated by the discrete boundary condition

$$\phi_{iN} - \phi_{iN-1} = hg_i ,$$

which means that at $j = N - 1$, Eq. (6.3) would become

$$3\phi_{iN-1} - \phi_{i+1N-1} - \phi_{i-1N-1} - \phi_{iN-2} = h^2 S_{iN-1} + hg_i . \qquad (6.8b)$$

These considerations, and a bit more thought, show that the discrete approximation to the differential equation (6.1) is equivalent to a system of linear equations for the unknown values of ϕ at the interior points. In a matrix notation, this can be written as

$$M\phi = s , \qquad (6.9)$$

where M is the matrix appearing in the linear system (6.3) and the inhomogeneous term s is proportional to S at the interior points and is linearly related to the specified values of ϕ or its derivatives on the boundaries. In any sort of practical situation there are a very large number of these equations (some 2500 if $N = 50$, say), so that solution by direct inversion of M is impractical. Fortunately, since the discrete approximation to the Laplacian involves only neighboring points, most of the elements of M vanish (it is sparse) and there are then efficient iterative techniques for solving (6.9). We begin their discussion by considering an analogous, but simpler, one-dimensional boundary value problem, and then return to the two-dimensional case.

6.2 An iterative method for boundary value problems

The one-dimensional boundary value problem analogous to the two-dimensional problem we have been discussing can be written as

$$-\frac{d^2\phi}{dx^2} = S(x) , \qquad (6.10)$$

with $\phi(0)$ and $\phi(1)$ specified. The related variational principle involves

$$E = \int_0^1 dx \left[\frac{1}{2} \left(\frac{d\phi}{dx} \right)^2 - S\phi \right] , \qquad (6.11)$$

which can be discretized on a uniform lattice of spacing $h = 1/N$ as

$$E = \frac{1}{2h} \sum_{i=1}^{N} (\phi_i - \phi_{i-1})^2 - h \sum_{i=1}^{N-1} S_i \phi_i . \qquad (6.12)$$

When varied with respect to ϕ_i, this yields the difference equation

$$2\phi_i - \phi_{i+1} - \phi_{i-1} = h^2 S_i , \qquad (6.13)$$

which is, of course, just the naive discretization of Eq. (6.10).

Methods of solving the boundary value problem by integrating forward and backward in x were discussed in Chapter 3, but we can also consider (6.13), together with the known values of ϕ_0 and ϕ_N, as a set of linear equations. For a modest number of points (say $N \lesssim 100$), the linear system above can be solved by the direct methods discussed in Chapter 5 and, in fact, a very efficient special direct method exists for such "tri-diagonal" systems, as discussed in Chapter 7. However, to illustrate the iterative methods appropriate for the large sparse matrices of elliptic partial differential equations in two or more dimensions, we begin by rewriting (6.13) in a "solved" form for ϕ_i:

$$\phi_i = \frac{1}{2}[\phi_{i+1} + \phi_{i-1} + h^2 S_i] . \qquad (6.14)$$

Although this equation is not manifestly useful, since we don't know the ϕ's on the right-hand side, it can be interpreted as giving us an "improved" value for ϕ_i based on the values of ϕ at the neighboring points. Hence the strategy (Gauss-Seidel iteration) is to guess some initial solution and then to sweep systematically through the lattice (say from left to right), successively replacing ϕ at each point by an improved value. Note that the most "current" values of the $\phi_{i\pm1}$ are to be used in the right-hand side of Eq. (6.14). By repeating this sweep many times, an initial guess for ϕ can be "relaxed" to the correct solution.

To investigate the convergence of this procedure, we generalize Eq. (6.14) so that at each step of the relaxation ϕ_i is replaced by a linear mixture of its old value and the "improved" one given by (6.14):

$$\phi_i \to \phi_i' = (1 - \omega)\phi_i + \omega\frac{1}{2}[\phi_{i+1} + \phi_{i-1} + h^2 S_i] . \qquad (6.15)$$

Here, ω is a parameter that can be adjusted to control the rate of relaxation: "over-relaxation" corresponds to $\omega < 1$, while "under-relaxation" means $\omega > 1$. The optimal value of ω that maximizes the rate of relaxation will be discussed below. To see that (6.15) results in an "improvement" in the solution, we calculate the change in the energy functional (6.12), remembering that all ϕ's except ϕ_i are to be held fixed. After some algebra, one finds

$$E' - E = -\frac{\omega(2 - \omega)}{2h}\left[\frac{1}{2}(\phi_{i+1} + \phi_{i-1} + h^2 S_i) - \phi_i\right]^2 \leq 0 , \qquad (6.16)$$

so that, as long as $0 < \omega < 2$, the energy never increases, and should thus converge to the required minimum value as the sweeps proceed. (The existence of other, spurious minima of the energy would imply that the linear system (6.13) is not well posed.)

■ **Exercise 6.3**   Use Eqs. (6.12) and (6.15) to prove Eq. (6.16).

As an example of this relaxation method, let us consider the one-dimensional boundary-value problem of the form (6.10) with

$$S(x) = 12x^2 ; \quad \phi(0) = \phi(1) = 0 .$$

The exact solution is

$$\phi(x) = x(1 - x^3)$$

and the energy is

$$E = -\frac{9}{14} = -0.64286 .$$

The following FORTRAN code implements the relaxation algorithm and prints out the energy after each sweep of the 21-point lattice. An initial guess of $\phi_i = 0$ is used, which is clearly quite far from the truth.

```
C chap6a.for
      PARAMETER (NSTEP=20)                   !lattice size
      DIMENSION PHI(0:NSTEP),S(0:NSTEP) !field and h**2*source
50    PRINT *,' Enter omega (.le. 0 to stop)'
      READ *, OMEGA                          !relaxation parameter
      IF (OMEGA .LE. 0) STOP
```

```
C
      H=1./NSTEP
      DO 10 IX=0,NSTEP
        X=IX*H
        S(IX)=H*H*12*X*X          !fill h**2*source array
        PHI(IX)=0.                !initial guess for field
10    CONTINUE
C
      DO 20 ITER=1,500            !relaxation loop
        DO 15 IX=1,NSTEP-1        !sweep through lattice
          PHIP=(PHI(IX-1)+PHI(IX+1)+S(IX))/2    !Eq. 6.14
          PHI(IX)=(1.-OMEGA)*PHI(IX)+OMEGA*PHIP !Eq. 6.15
15      CONTINUE
        IF (MOD(ITER-1,20) .EQ. 0) THEN !sometimes calc energy
          E=0.                   !Eq. (6.11) for E
          DO 30 IX=1,NSTEP
            E=E+(((PHI(IX)-PHI(IX-1))/H)**2)/2
            E=E-S(IX)*PHI(IX)/H**2
30        CONTINUE
          E=E*H
          PRINT *,' iteration = ',iter,'  energy = ',E
        END IF
20    CONTINUE
      GOTO 50
      END
```

Results for the energy as a function of iteration number are shown in Table 6.1 for three different values of ω. Despite the rather poor initial guess for ϕ, the iterations converge and the converged energy is independent of the relaxation parameter, but differs somewhat from the exact answer due to discretization errors (i.e., h not vanishingly small); the discrepancy can be reduced, of course, by increasing N. A detailed examination of the solution indicates good agreement with the analytical result. Note that the rate of convergence clearly depends upon ω. A general analysis [Wa66] shows that the best choice for the relaxation parameter depends upon the lattice size and on the geometry of the problem; it is usually greater than 1. The optimal value can be determined empirically by examining the convergence of the solution for only a few iterations before choosing a value to be used for many iterations.

Table 6.1 Convergence of the energy functional during relaxation of a 1-D boundary value problem

Iteration	$\omega = 0.5$	$\omega = 1.0$	$\omega = 1.5$
1	−0.01943	−0.04959	−0.09459
21	−0.24267	−0.44024	−0.60688
41	−0.36297	−0.56343	−0.63700
61	−0.44207	−0.61036	−0.63831
81	−0.49732	−0.62795	−0.63836
101	−0.53678	−0.63450	−0.63837
121	−0.56517	−0.63693	−0.63837
141	−0.58563	−0.63783	−0.63837
161	−0.60037	−0.63817	−0.63837
181	−0.61100	−0.63829	−0.63837
201	−0.61866	−0.63834	−0.63837
221	−0.62418	−0.63836	−0.63837
241	−0.62815	−0.63836	−0.63837
⋮	⋮	⋮	⋮
381	−0.63734	−0.63837	−0.63837
401	−0.63763	−0.63837	−0.63837

■ **Exercise 6.4** Use the code above to verify that the energy approaches the analytical value as the lattice is made finer. Investigate the accuracy of the solution at each of the lattice points and note that the energy can be considerably more accurate than the solution itself; this is a natural consequence of the minimization of E at the correct solution. Use one of the higher-order discretizations you derived in Exercise 6.2 to solve the problem.

The application of the relaxation scheme described above to two- (or even three-) dimensional problems is now straightforward. Upon solving (6.3a) for ϕ_{ij}, we can generate the analogue of (6.15):

$$\phi_{ij} \rightarrow \phi'_{ij} = (1-\omega)\phi_{ij} + \frac{\omega}{4}[\phi_{i+1j} + \phi_{i-1j} + \phi_{ij+1} + \phi_{ij-1} + h^2 S_{ij}]. \quad (6.17)$$

If this algorithm is applied successively to each point in the lattice, say sweeping the rows in order from top to bottom and each row from left to right, one can show that the energy functional (6.7) always decreases (if ω is within the proper range) and that there will be convergence to the required solution.

Several considerations can serve to enhance this convergence in practice. First, starting from a good guess at the solution (perhaps one with similar, but simpler, boundary conditions) will reduce the number of iterations required. Second, an optimal value of the relaxation parameter should be used, either estimated analytically or determined empirically, as described above. Third, it may sometimes be more efficient to concentrate the relaxation process, for several iterations, in some sub-area of the lattice where the trial solution is known to be particularly poor, thus not wasting effort on already-relaxed parts of the solution. Finally, one can always do a calculation on a relatively coarse lattice that relaxes with a small amount of numerical work, and then interpolate the solution found onto a finer lattice to be used as the starting guess for further iterations.

6.3 More on discretization

It is often the case that the energy functional defining a physical problem has a form more complicated than the simple "integral of the square of the derivative" that we have been considering so far. For example, in an electrostatics problem with spatially-varying dielectric properties or in a diffusion problem with a spatially-varying diffusion coefficient, the boundary-value problem (6.1) is modified to

$$-\nabla \cdot D\nabla\phi = S(x,y) \,, \tag{6.18}$$

where $D(x,y)$ is the dielectric constant or diffusion coefficient, and the corresponding energy functional is [compare Eq. (6.4)]

$$E = \int_0^1 dx \int_0^1 dy \left[\frac{1}{2}D(\nabla\phi)^2 - S\phi\right] \,. \tag{6.19}$$

Although it is possible to discretize Eq. (6.18) directly, it should be evident from the previous discussion that a far better procedure is to discretize (6.19) first and then to differentiate with respect to the field variables to obtain the difference equations to be solved.

To see how this works out in detail, consider the analog of the one-dimensional problem defined by (6.11),

$$E = \int_0^1 dx \left[\frac{1}{2}D(x)\left(\frac{d\phi}{dx}\right)^2 - S\phi\right] \,, \tag{6.20}$$

The discretization analogous to (6.12) is

$$E = \frac{1}{2h}\sum_{i=1}^N D_{i-\frac{1}{2}}(\phi_i - \phi_{i-1})^2 - h\sum_{i=1}^{N-1} S_i\phi_i \,, \tag{6.21}$$

where $D_{i-1/2}$ is the diffusion constant at the half-lattice points. This might be known directly if we have an explicit formula for $D(x)$, or it might be approximated with appropriate accuracy by $\frac{1}{2}(D_i + D_{i-1})$. Note that, in either case, we have taken care to center the differencing properly. Variation of this equation then leads directly to the corresponding difference equations [compare Eq. (6.13)],

$$(D_{i+1/2} + D_{i-1/2})\phi_i - D_{i+1/2}\phi_{i+1} - D_{i-1/2}\phi_{i-1} = h^2 S_i \ . \qquad (6.22)$$

These can then be solved straightforwardly by the relaxation technique described above.

A problem treated in cylindrical or spherical coordinates presents very much the same kind of situation. For example, when the diffusion or dielectric properties are independent of space, the energy functional in cylindrical coordinates will involve

$$E = \int_0^\infty dr\, r \left[\frac{1}{2} \left(\frac{d\phi}{dr} \right)^2 - S\phi \right], \qquad (6.23)$$

where r is the cylindrical radius. (We suppress here the integrations over the other coordinates.) This is of the form (6.20), with $D(r) = r$ and an additional factor of r appearing in the source integral. Discretization on a lattice $r_i = hi$ in analogy to (6.21) then leads to the analog of (6.22),

$$2r_i\phi_i - r_{i+1/2}\phi_{i+1} - r_{i-1/2}\phi_{i-1} = h^2 r_i S_i \ . \qquad (6.24)$$

At $i = 0$, this equation just tells us that $\phi_1 = \phi_{-1}$, or equivalently, that $\partial\phi/\partial r = 0$ at $r = 0$. This is to be expected, as, in the electrostatics language, Gauss's law allows no radial electric field at $r = 0$. At $i = 1$, Eq. (6.24) gives an equation involving three unknowns, ϕ_0, ϕ_1, and ϕ_2, but putting $\phi_0 = \phi_1$ as a rough approximation to the zero-derivative boundary condition gives an equation involving only two unknowns, which is what we expect at a boundary on the basis of our experience with the cartesian problems discussed above.

A more elegant discretization of problems with cylindrical symmetry naturally incorporates the zero-derivative boundary condition at $r = 0$ by working on a lattice defined by $r_i = (i - \frac{1}{2})h$. In this case, Eq. (6.24) is still valid, but for $i = 1$ the coefficient of the term involving ϕ_{i-1} vanishes, giving directly an equation with only two unknowns, ϕ_1 and ϕ_2.

■ **Exercise 6.5** Verify that variation of Eq. (6.21) leads to Eq. (6.22). Write down explicitly the discretizations of Eq. (6.23) on the $r_i = ih$ and $r_i = (i - \frac{1}{2})h$ lattices and verify their variation leads to (6.24) in either case. Why can the contribution to the energy functional from the region between $r = 0$ and $r = \frac{1}{2}h$ be safely neglected on the $r_i = (i - \frac{1}{2})h$ lattice to the same order as the accuracy of approximation used for the derivative?

6.4 Elliptic equations in two dimensions

The application of relaxation methods to elliptic boundary value problems is illustrated by the program for Example 6, which solves Laplace's equation, $\nabla^2 \phi = 0$, on a uniform rectangular lattice of unit spacing with Dirichlet boundary conditions specified on the lattice borders and on selected points within the lattice. This situation might describe the steady-state temperature distribution within a plate whose edges and certain interior regions are held at specified temperatures, or it might describe an electrostatics problem specified by a number of equipotential surfaces. During the iterations, the potential is displayed, as is the energy functional (6.7) to monitor convergence. The code's source listing can be found in Appendix B and in the file EXMPL6.FOR.

The following exercises, phrased in the language of electrostatics, might help you to understand better the physical and numerical principles illustrated by this example.

■ **Exercise 6.6** Verify that the solutions corresponding to particular boundary conditions in the interior of the lattice agree with your intuition. You might try fixing a single interior point to a potential different from that of the boundary, fixing two symmetrically located points to different potentials, or fixing a whole line to a given potential. Other possibilities include constructing a "Faraday cage" (a closed region bounded by a surface at fixed potential), studying a quadrupole pattern of boundary conditions, or calculating the capacitance of various configurations of conductors. Study of what happens when you increase or decrease the size of the lattice is also interesting.

■ **Exercise 6.7** For a given set of boundary conditions, investigate the effects of under-relaxation and over-relaxation on the convergence of the relaxation process. Change the discretization of the Laplacian to a higher-order formula as per Exercise 6.2 and observe any the changes in the solution or the efficiency with which it is approached.

■ **Exercise 6.8** Modify the code to solve Poisson's equation; i.e., allow for a charge density to be specified throughout the lattice. Use this to study the solutions for certain simple charge distributions. For example, you might try computing the potential between two (line) charges as a function of their separation and comparing it with your analytical expectations.

■ **Exercise 6.9** Modify the code to use Neumann boundary conditions on selected borders and interior regions of the lattice. Study the solution for simple geometries and compare it with what you expect.

■ **Exercise 6.10** Modify the code to allow for a spatially-varying dielectric constant. (Note that you must change both the relaxation formula and the energy functional.) Study the solutions for selected simple geometries of the dielectric constant (e.g., a half-space filled with dielectric) and simple boundary conditions.

■ **Exercise 6.11** An alternative to the Dirichlet boundary conditions are periodic boundary conditions, in which the potentials on the left and right and top and bottom borders of the rectangle are constrained to be equal, but otherwise arbitrary. That is,

$$\phi_{i1} = \phi_{iN} \; ; \;\; \phi_{1j} = \phi_{Nj}$$

for all i and j. This might correspond to the situation in a crystal, where the charge density is periodic in space. Modify the code to incorporate these boundary conditions into Poisson's equation and verify that the solutions look as expected for simple charge distributions.

■ **Exercise 6.12** Change the relaxation formula and energy functional to treat situations with an azimuthal symmetry by re-interpreting one of the coordinates as the cylindrical radius while retaining the other as a cartesian coordinate. Use the resulting program to model a capacitor made of two circular disks, and in particular calculate the capacitance and potential field for varying separations between the disks. Compare your results with your expectations for very large or very small separations.

Project VI: Steady-state hydrodynamics in two dimensions

The description of the flow of fluids is one of the richest and most challenging problems that can be treated on a computer. The non-linearity of the equations and the complexity of phenomena they describe (e.g., turbulence) sometimes make computational fluid dynamics more of an art than a science, and several book-length treatments are required to cover the field adequately (see, for example, [Ro76]). In this project, we will consider one relatively simple situation that can be treated by the relaxation methods for elliptic equations described in this chapter and that will serve to give some idea of the problems involved. This situation is the time-independent flow of a viscous, incompressible fluid past an object. For simplicity, we will take the object to be translationally invariant in one direction transverse to the flow, so that the fluid has a non-trivial motion only in two-coordinates, (x, y); this might describe a rod or beam placed in a steady flow of water with incident speed V_0. We will also consider only the case where the cross section of this rod is a rectangle with dimensions $2W$ transverse to the flow and T along the flow (see Figure VI.1). This will greatly simplify the coding needed to treat the boundary conditions, while still allowing the physics to be apparent. We begin with an exposition of the basic equations and their discretization, follow with a brief discussion of the boundary conditions, and then give some guidance in writing the program and in extracting some understanding from it.

VI.1 The equations and their discretization

In describing the flow of a fluid through space, at least two fields are important: ρ, the mass density, and $\mathbf{V}$, the velocity, of the fluid element at each point in space. These are related through two fundamental equations of hydrodynamics [La59]:

$$\frac{\partial \rho}{\partial t} + \nabla \cdot \rho \mathbf{V} = 0 \; ; \qquad (VI.1a)$$

$$\frac{\partial \mathbf{V}}{\partial t} = -(\mathbf{V} \cdot \nabla)\mathbf{V} - \frac{1}{\rho}\nabla P + \nu\nabla^2 \mathbf{V} \; . \qquad (VI.1b)$$

The first of these (the continuity equation) expresses the conservation of mass, and states that the density can change at a point in space only due to a net in- or out-flow of matter. The second equation (Navier-Stokes) expresses the conservation of momentum, and states that the velocity changes in response to convection, $(\mathbf{V} \cdot \nabla)\mathbf{V}$, spatial variations in the pressure, ∇P, and viscous forces $\nu\nabla^2\mathbf{V}$, where ν is the kinematic viscosity, assumed constant in our discussion. In general, the pressure is given in terms of the density and temperature through an "equation of state", and when the temperature varies as well, an additional equation embodying the conservation of energy is also required. We will assume that the temperature is constant throughout the fluid.

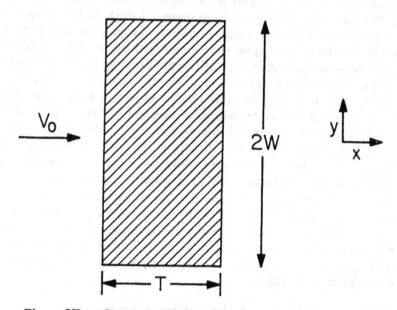

Figure VI.1   Geometry of the two-dimensional flow past a plate to be treated in this project.

We will be interested in studying time-independent flows, so that all time derivatives can be set to zero in these equations. Furthermore, we will assume that the fluid is incompressible, so that the density is constant (this is a good approximation for water under many conditions). Equations (VI.1) then become

$$\nabla \cdot \mathbf{V} = 0 ; \tag{$VI.2a$}$$

$$(\mathbf{V} \cdot \nabla)\mathbf{V} = -\frac{1}{\rho}\nabla P + \nu\nabla^2\mathbf{V} . \tag{$VI.2b$}$$

For two-dimensional flow, these equations can be written explicitly in terms of the x and y components of the velocity field, denoted by u and v, respectively:

$$\frac{\partial u}{\partial x} + \frac{\partial v}{\partial y} = 0 ; \tag{$VI.3a$}$$

$$u\frac{\partial u}{\partial x} + v\frac{\partial u}{\partial y} = -\frac{1}{\rho}\frac{\partial P}{\partial x} + \nu\nabla^2 u ; \tag{$VI.3b$}$$

$$u\frac{\partial v}{\partial x} + v\frac{\partial v}{\partial y} = -\frac{1}{\rho}\frac{\partial P}{\partial y} + \nu\nabla^2 v . \tag{$VI.3c$}$$

Here, as usual,

$$\nabla^2 = \frac{\partial^2}{\partial x^2} + \frac{\partial^2}{\partial y^2} \ .$$

Equations (VI.3) are three scalar equations for the fields u, v, and P. While these equations could be solved directly, it is more convenient for two-dimensional problems to replace the velocity fields by two equivalent scalar fields: the stream function, $\psi(x,y)$, and the vorticity, $\zeta(x,y)$. The first of these is introduced as a convenient way of satisfying the continuity equation (VI.3a). The stream function is defined so that

$$u = \frac{\partial \psi}{\partial y} \ ; \quad v = -\frac{\partial \psi}{\partial x} \ . \qquad (VI.4)$$

It is easily verified that this definition satisfies the continuity equations (VI.3a) for any function ψ and that such a ψ exists for all flows that satisfy the continuity condition (VI.2a). Furthermore, one can see that $(\mathbf{V} \cdot \nabla)\psi = 0$, so that $\mathbf{V}$ is tangent to contour lines of constant ψ, the "stream lines".

The vorticity is defined as

$$\zeta = \frac{\partial u}{\partial y} - \frac{\partial v}{\partial x} \ , \qquad (VI.5)$$

which is seen to be (the negative of) the curl of the velocity field. From the definitions (VI.4), it follows that ζ is related to the stream function ψ by

$$\nabla^2 \psi = \zeta \ . \qquad (VI.6)$$

An equation for ζ can be derived by differentiating (VI.3b) with respect to y and (VI.3c) with respect to x. Upon subtracting one from the other and invoking the continuity equation (VI.3a) and the definitions (VI.4), one finds, after some algebra,

$$\nu \nabla^2 \zeta = \left[\frac{\partial \psi}{\partial y} \frac{\partial \zeta}{\partial x} - \frac{\partial \psi}{\partial x} \frac{\partial \zeta}{\partial y} \right] \ . \qquad (VI.7)$$

Finally, an equation for the pressure can be derived by differentiating (VI.3b) with respect to x and adding it to the derivative of (VI.3c) with respect to y. Upon expressing all velocity fields in terms of the stream function, one finds, after a bit of rearranging,

$$\nabla^2 P = 2\rho \left[\left(\frac{\partial^2 \psi}{\partial x^2} \right) \left(\frac{\partial^2 \psi}{\partial y^2} \right) - \left(\frac{\partial^2 \psi}{\partial x \partial y} \right)^2 \right] \ . \qquad (VI.8)$$

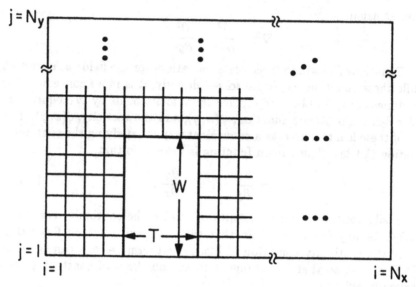

Figure VI.2 The lattice to be used in calculating the fluid flow past the plate illustrated in Figure VI.1.

Equations (VI.6–8) are a set of non-linear elliptic equations equivalent to the original hydrodynamic equations (VI.3). They are particularly convenient in that if it is just the velocity field we want, only the two equations (VI.6,7) need be solved simultaneously, since neither involves the pressure. If we do happen to want the pressure, it can be obtained by solving (VI.8) for P after we have found ψ and ζ.

To solve Eqs. (VI.6,7) numerically, we introduce a two-dimensional lattice of uniform spacing h having N_x and N_y points in the x and y directions, respectively, and use the indices i and j to measure these coordinates. (See Figure VI.2.) Note that since the centerline of the rectangle is a line of symmetry, we need only treat the upper half-plane, $y > 0$. Moreover, it is convenient to place the lattice so that the edges of the plate are on lattice points. The location of the plate relative to the up- and downstream edges of the lattice is arbitrary, although it is wise to place the plate far enough forward to allow the "wake" to develop behind it, yet not too close to the upstream edge, whose boundary conditions can influence the flow pattern spuriously. It is also convenient to scale the equations by measuring all lengths in units of h and all velocities in units of the incident fluid velocity, V_0. The stream function is then measured in units of $V_0 h$, while the vorticity is in units of V_0/h, and the pressure is conveniently scaled by ρV_0^2. Upon differencing (VI.6) in the usual way, we have

$$(\delta_i^2 \psi)_{ij} + (\delta_j^2 \psi)_{ij} = \zeta_{ij} , \qquad (VI.9)$$

where δ^2 is the symmetric second-difference, as in Eq. (6.3b). Similarly, (VI.7) can be differenced as

$$(\delta_i^2 \zeta)_{ij} + (\delta_j^2 \zeta)_{ij} = \frac{R}{4} \left[(\delta_j \psi)_{ij} (\delta_i \zeta)_{ij} - (\delta_i \psi)_{ij} (\delta_j \zeta)_{ij} \right] , \qquad (VI.10)$$

where the symmetric first-difference operator is

$$(\delta_i \psi)_{ij} \equiv \psi_{i+1j} - \psi_{i-1j}$$

and similarly for δ_j. The lattice Reynolds number, $R = V_0 h / \nu$, is a dimensionless measure of the strength of the viscous forces or, equivalently, of the speed of the incident stream. It is related to the physical Reynolds number Re by replacing the mesh spacing by the width of the rectangle: $Re = 2WV_0/\nu$. Finally, the pressure equation (VI.8) can be differenced as

$$\left[(\delta_i^2 P)_{ij} + (\delta_j^2 P)_{ij} \right] = 2[(\delta_i^2 \psi)_{ij} (\delta_j^2 \psi)_{ij} - \frac{1}{16} (\delta_i \delta_j \psi)_{ij}^2] . \qquad (VI.11)$$

VI.2 Boundary conditions

In order for the elliptic problems (VI.6–8) to be well posed, we must specify either the values or the normal derivatives of the stream function, the vorticity, and the pressure at all boundaries of the lattice shown in Figure VI.2. These boundaries can be classified into three groups, in the notation of Figure VI.3:

    i) the centerline boundaries (A and E);

    ii) the boundaries contiguous with the rest of the fluid (F, G, and H);

    iii) the boundaries of the plate itself (B, C, and D).

We treat ψ and ζ on each of these in turn, followed by the pressure boundary conditions. Throughout this discussion, we use unscaled (i.e., physical) quantities. It also helps to recall that in the freely flowing fluid (i.e., no obstruction), the solution is $u = V_0$, $v = 0$, so that $\psi = V_0 y$ and $\zeta = 0$.

The boundary conditions on the centerline surfaces A and E are determined by symmetry. The y component of the velocity, v, must vanish here, so that the x derivative of ψ vanishes. It follows that A and E are

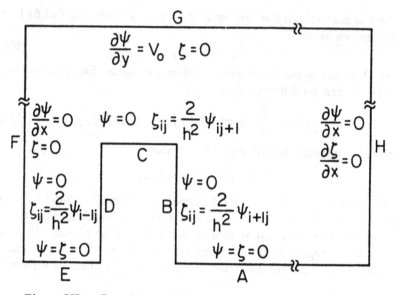

Figure VI.3 Boundary conditions to be imposed on the stream function and vorticity.

each stream lines. Moreover, since the normal velocity (and hence the tangential derivative of ψ) also vanishes on B, C, and D, the entire surface ABCDE is a single stream line, and we may arbitrarily put $\psi = 0$ on it. Note that the velocities depend only upon the derivatives of ψ, so that the physical description is invariant to adding a spatially independent constant to the stream function; the choice of ψ on this streamline fixes the constant. From symmetry, we can also conclude that the vorticity vanishes on A and E.

The boundary conditions on the upstream surface F are also fairly straightforward. This surface is contiguous with the smoothly flowing incident fluid, so that specifying

$$v = -\frac{\partial \psi}{\partial x} = 0 \; ; \; \zeta = 0 \qquad \text{on F} ,$$

as is the case far upstream, seems reasonable. Similarly, if the lattice is large enough, we may expect the upper boundary G to be in free flow, so that fixing

$$u = \frac{\partial \psi}{\partial y} = V_0 \; ; \; \zeta = 0 \qquad \text{on G} ,$$

is one appropriate choice. The downstream boundary H is much more ambiguous and, as long as it is sufficiently far from the plate, there should be many plausible choices. However, the boundary conditions on a boundary that approaches the plate can influence the shape of the solution found. One convenient choice is to say that nothing changes beyond the lattice boundary, so that

$$\frac{\partial \psi}{\partial x} = \frac{\partial \zeta}{\partial x} = 0 \quad \text{on H}.$$

At the walls of the plate (B, C, and D), one of the correct boundary conditions is that the normal velocity of the fluid is zero. This we have used already by requiring that this surface be a stream line. However, the other boundary condition appropriate for viscous flow is that the tangential velocity be zero. Implementing this by setting the normal derivative of ψ to zero would be an overspecification of the elliptic problem for ψ. Instead, the "no-slip" boundary condition is imposed on the vorticity. Consider a point ij on the upper surface of the plate, C. We can write the stream function at the point on lattice spacing above this one, $ij + 1$, as a Taylor series in y:

$$\psi_{ij+1} = \psi_{ij} + h\frac{\partial \psi}{\partial y}\bigg|_{ij} + \frac{h^2}{2}\frac{\partial^2 \psi}{\partial y^2}\bigg|_{ij} + \cdots . \qquad (VI.12)$$

Since, at the wall,

$$\frac{\partial \psi}{\partial y} = u = 0$$

and

$$\frac{\partial^2 \psi}{\partial y^2} = \frac{\partial u}{\partial y} = \zeta ,$$

because $\partial v/\partial x = 0$, (VI.12) can be reduced to

$$\zeta_{ij} = 2\frac{\psi_{ij+1} - \psi_{ij}}{h^2} \quad \text{on C}, \qquad (VI.13a)$$

which provides a Dirichlet boundary condition for ζ. (This is the general form of the boundary condition; recall that we had previously specified $\psi_{ij} = 0$ on the plate boundaries.) The same kind of arguments can be applied to the surfaces B and D to yield

$$\zeta_{ij} = 2\frac{\psi_{i+1j} - \psi_{ij}}{h^2} \quad \text{on B}; \qquad (VI.13b)$$

$$\zeta_{ij} = 2\frac{\psi_{i-1j} - \psi_{ij}}{h^2} \quad \text{on D}. \qquad (VI.13c)$$

Note that there is an ambiguity at the "corners", where surfaces B and C and D and C intersect, as here the vorticity can be computed in two ways (horizontal or vertical difference of ψ). In practice, there are several ways of resolving this: use one form and verify that the other gives similar values (a check on the accuracy of the calculation), use the average of the two methods, or use the horizontal value when relaxing the point just to the right or left of the corner and the vertical value when relaxing the point just above the corner.

The boundary conditions for the pressure on all surfaces are of the Neumann type, and follow from Eqs. (VI.3). We leave their explicit finite-difference form in terms of ψ and ζ as an exercise. Note that from symmetry, $\partial P/\partial y = 0$ on the centerlines A and E.

VI.3 Solving the equations

Our general strategy will be to solve the coupled non-linear elliptic equations (VI.9,10) for the stream function and vorticity using the relaxation methods discussed in this chapter. One possible iteration scheme goes as follows. We begin by choosing trial values corresponding to the free-flowing solution: $\psi = y$ and $\zeta = 0$. We then perform one relaxation sweep of (VI.9) to get an improved value of ψ. The Dirichlet boundary conditions for ζ on the walls of the plate are then computed from (VI.13) and then one relaxation sweep of (VI.10) is performed. With the new value of ζ so obtained, we go back to a sweep of (VI.9) and repeat the cycle many times until convergence. If required, we can then solve the pressure equation (VI.11) with Neumann boundary conditions determined from (VI.3). This iteration scheme can be implemented in the following sequence of steps:

■ **Step 1** Write a section of code to execute one relaxation sweep of (VI.9) for ψ given ζ and subject to the boundary conditions discussed above. Be sure to allow for an arbitrary relaxation parameter. This code can be adapted from that for Example 6; in particular, the technique of using letters on the screen as a crude contour plot can be useful in displaying ψ.

■ **Step 2** Write a section of code that calculates the boundary conditions for ζ on the plate walls if ψ is known.

■ **Step 3** Write a section of code that does one relaxation sweep of (VI.10) for ζ if ψ is given. Build in the boundary conditions discussed above and be sure to allow for an arbitrary relaxation parameter. If you have a contour plotting routine and a sufficiently high-resolution terminal, it is useful to display ψ and ζ on the screen together.

■ **Step 4** Combine the sections of code you wrote in the previous three steps into a program that executes a number of iterations of the coupled ψ-ζ problem. Test the convergence of this scheme on several coarse lattices for several different Reynolds numbers and choices of relaxation parameters.

■ **Step 5** Calculate the flow past several plates by running your program to convergence on some large lattices (say 70 × 24) for several increasing values of the lattice Reynolds number. A typical size of the plate might be $W = 8h$ and $T = 8h$, while lattice Reynolds numbers might run from 0.1 to 8. For the larger Reynolds numbers, you might find instabilities in the relaxation procedure due to the nonlinearity of the equations. These can be suppressed by using relaxation parameters as small as 0.1 and by using as trial solutions the flow patterns obtained at smaller Reynolds numbers. Verify that the flow around the plate is smooth at small Reynolds numbers but that at larger velocities the flow separates from the back edge of the plate and a small vortex (eddy) develops behind it. Check also that your solutions are accurate by running two cases with different lattice Reynolds numbers but the same physical Reynolds number.

■ **Step 6** Two physically interesting quantities that can be computed from the flow pattern are the net pressure and viscous forces per unit area of the plate, F_P and F_v. By symmetry, these act in the x direction and are measured conveniently in terms of the flow of momentum incident on the face of the plate per unit area, $2W\rho V_0^2$. The pressure force is given by:

$$F_P = \int_D P\,dy - \int_B P\,dy \,, \qquad (VI.14)$$

where the integrals are over the entire front and back surfaces of the plate. This shows clearly that it is only the relative values of the pressure over the plate surfaces that matter. These can be obtained from the flow solution using Eqs. (VI.3b,c). Consider, for example, the front face of the plate. Using (VI.3c), together with (VI.4,6) and the fact that the velocities vanish at the plate surface, we can write the derivative of the pressure along the front face as:

$$\frac{\partial P}{\partial y} = -\nu\rho\frac{\partial \zeta}{\partial x} \,. \qquad (VI.15)$$

Hence, ζ from the flow solution can be integrated to find the (relative) pressure along the front face. Similar expressions can be derived for the top and back faces, so that the pressure (apart from an irrelevant additive constant) can be computed on all faces by integrating up the front face, across the top, and down the back face. The net viscous force per unit area of the plate is due only to the flow past the top (and bottom) surfaces and is given by

$$F_v = 2\rho\nu \int_C \frac{\partial u}{\partial y} \, dx \, . \qquad (VI.16a)$$

However, since the boundary conditions on the top surface of the plate require $\partial v/\partial x = 0$, this can be written as

$$F_v = 2\rho\nu \int_C \zeta \, dx \, , \qquad (VI.16b)$$

which is conveniently evaluated from the flow solution. Investigate the variation of the viscous and pressure forces with Reynolds number and compare your results with what you expect.

■ **Step 7** Three other geometries can be investigated with only minor modifications of the code you've constructed. One of these is a "jump", in which the thickness of the plate is increased until its downstream edge meets the downstream border of the lattice (H). Another is a pair of rectangular plates placed symmetrically about a centerline, so that the fluid can flow between them, as well around them (a crude nozzle). Finally, two plates, one behind the other, can also be calculated. Modify your code to treat each of these cases and explore the flow patterns at various Reynolds numbers.

Chapter 7

Parabolic
Partial Differential
Equations

Typical parabolic partial differential equations one encounters in physical situations are the diffusion equation

$$\frac{\partial \phi}{\partial t} = \nabla \cdot (D\nabla\phi) + S , \qquad (7.1)$$

where D is the (possibly space dependent) diffusion constant and S is a source function, and the time-dependent Schroedinger equation

$$i\hbar\frac{\partial \phi}{\partial t} = -\frac{\hbar^2}{2m}\nabla^2\phi + V\phi , \qquad (7.2)$$

where V is the potential. In contrast to the boundary value problems encountered in the previous chapter, these problems are generally of the initial value type. That is, we are given the field ϕ at an initial time and seek to find it at a later time, the evolution being subject to certain spatial boundary conditions (e.g., the Schroedinger wave function vanishes at very large distances or the temperature or heat flux is specified on some surfaces). The methods by which such problems are solved on the computer are basically straightforward, although a few subtleties are involved. We will also see that they provide a natural way of solving elliptic eigenvalue problems, particularly if it is only the few lowest or highest eigenvalues that are required.

7.1 Naive discretization and instabilities
We begin our discussion by treating diffusion in one dimension with a uniform diffusion constant. We take x to vary between 0 and 1 and assume Dirichlet boundary conditions that specify the value of the field

at the end points of this interval. After appropriate scaling, the analogue of (7.1) is

$$\frac{\partial \phi}{\partial t} = \frac{\partial^2 \phi}{\partial x^2} + S(x,t) . \tag{7.3}$$

As usual, we approximate the spatial derivatives by finite differences on a uniform lattice of $N+1$ points having spacing $h = 1/N$, while the time derivative is approximated by the simplest first-order difference formula assuming a time step Δt. Using the superscript n to label the time step (that is, $\phi^n \equiv \phi(t_n)$, $t_n = n\Delta t$), we can approximate (7.3) as

$$\frac{\phi_i^{n+1} - \phi_i^n}{\Delta t} = \frac{1}{h^2}(\delta^2 \phi^n)_i + S_i^n , \tag{7.4}$$

At $i = 1$ and $i = N - 1$, this equation involves the Dirichlet boundary conditions specifying ϕ_0 and ϕ_N.

Equation (7.4) is an *explicit* differencing scheme since, given ϕ at one time, it is straightforward to solve for ϕ at the next time. Indeed, in an obvious matrix notation, we have:

$$\phi^{n+1} = (1 - H\Delta t)\phi^n + S^n\Delta t , \tag{7.5}$$

where the action of the operator H is defined by

$$(H\phi)_i \equiv -\frac{1}{h^2}(\delta^2 \phi)_i .$$

As an example of how this explicit scheme might be applied, consider the case where $S = 0$ with the boundary conditions $\phi(0) = \phi(1) = 0$. Suppose that we are given the initial condition of a Gaussian centered about $x = 1/2$,

$$\phi(x, t = 0) = e^{-20(x-1/2)^2} - e^{-20(x-3/2)^2} - e^{-20(x+1/2)^2} ,$$

where the latter two "image" Gaussians approximately ensure the boundary conditions at $x = 1$ and $x = 0$, respectively, and we seek to find ϕ at later times. The following FORTRAN code applies (7.5) to do this on a lattice with $N = 25$.

```
C chap7a.for
      PARAMETER (NSTEP=25)
      DIMENSION PHI(0:NSTEP)      !array for solution
                                  !form of the analytical  solution
      GAUSS(X,T)=EXP(-20.*(X-.5)**2/(1.+80*T))/SQRT(1+80*T)
      EXACT(X,T)=GAUSS(X,T)-GAUSS(X-1.,T)-GAUSS(X+1.,T)
      H=1./NSTEP
```

```
                              !get input from terminal
50      PRINT *, ' Enter time step and total time (0 to stop)'
        READ *,DT,TIME
        IF (DT .EQ. 0.) STOP
        NITER=TIME/DT
        DTH=DT/H**2
                              !setup initial conditions
        T=0.
        PHI(0)=0.
        PHI(NSTEP)=0.
        DO 10 IX=1,NSTEP-1
           PHI(IX)=EXACT(IX*H,T)
10      CONTINUE
                              !loop over time steps
        DO 20 ITER=1,NITER
           POLD=0.            !old PHI at last point
           DO 30 IX=1,NSTEP-1 !new PHI at this point
              PNEW=PHI(IX)+DTH*(POLD+PHI(IX+1)-2*PHI(IX))
              POLD=PHI(IX)         !roll POLD value
              PHI(IX)=PNEW         !store new value
30         CONTINUE
           IF (MOD(ITER,10) .EQ. 0) THEN  !output every 10 steps
              PRINT *, ' iteration = ', ITER, ' time = ',ITER*DT
              T=ITER*DT
              DO 40 IX=1,NSTEP-1
                 DIFF=PHI(IX)-EXACT(IX*H,T)
                 PRINT *, ' phi = ', PHI(IX),' error = ', DIFF
40            CONTINUE
           END IF
20      CONTINUE
        GOTO 50
        END
```

Typical results from this code are shown in Figure 7.1. Things seem to be working fine for a time step of 0.00075 and the results agree reasonably well with the analytical solution of a spreading Gaussian,

$$\phi(x,t) = \tau^{-1/2}\left[e^{-20(x-1/2)^2/\tau} - e^{-20(x-3/2)^2/\tau} - e^{-20(x+1/2)^2/\tau}\right];$$
$$\tau = 1 + 80t\,.$$

This time step is, however, quite small compared to the natural time scale of the solution, $t \approx 0.01$, so that many steps are required before anything interesting happens. If we try to increase the time step, even to only 0.001, things go very wrong: an unphysical instability develops in the numerical solution, which quickly acquires violent oscillations from one lattice point to another soon after $t = 0.02$.

It is easy to understand what is happening here. Let the set of states ψ_λ be the eigenfunctions of the discrete operator H with eigenvalues ϵ_λ. Since H is an hermitian operator, the eigenvalues are real and the eigenvectors can be chosen to be orthonormal. We can expand the solution at any time in this basis as

$$\phi^n = \sum_\lambda \phi_\lambda^n \psi_\lambda \ .$$

The exact time evolution is given by

$$\phi^n = e^{-nH\Delta t}\phi^0 \ ,$$

so that each component of the solution should evolve as

$$\phi_\lambda^n = e^{-n\epsilon_\lambda \Delta t}\phi_\lambda^0 \ .$$

This corresponds to the correct behavior of the diffusion equation, where short-wavelength components (with larger eigenvalues) disappear more rapidly as the solution "smooths out". However, (7.5) shows that the explicit scheme will evolve the expansion coefficients as

$$\phi_\lambda^n = (1 - \epsilon_\lambda \Delta t)^n \phi_\lambda^0 \ . \tag{7.6}$$

As long as Δt is chosen to be small enough, the factor in (7.6) approximates $e^{-n\epsilon_\lambda \Delta t}$ and the short-wavelength components damp with time. However, if the time step is too large, one or more of the quantities $1 - \epsilon_\lambda \Delta t$ has an absolute value larger than unity. The corresponding components of the initial solution, even if present only due to very small numerical round-off errors, are then amplified with each time step, and soon grow to dominate.

To quantify this limit on Δt, we have some guidance in that the eigenvalues of H are known analytically in this simple model problem. It is easily verified that the functions

$$(\psi_\lambda)_i = \sin \frac{\lambda \pi i}{N}$$

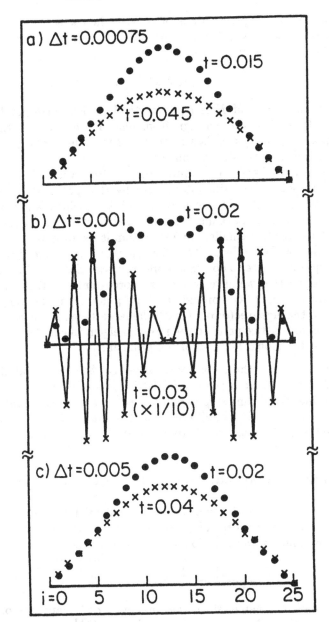

Figure 7.1　Results for the one-dimensional diffusion of a Gaussian; a fixed lattice spacing $h = 0.04$ is used in all calculations. a) and b) result from the explicit algorithm (7.5) while c) is the implicit algorithm (7.8).

are (un-normalized) eigenfunctions of H with the correct boundary conditions on a lattice of $N+1$ points for $\lambda = 1, 2, \ldots, N-1$ and that the associated eigenvalues are

$$\epsilon_\lambda = \frac{4}{h^2} \sin^2 \frac{\lambda\pi}{2N} .$$

The largest eigenvalue of H is $\epsilon_{N-1} \approx 4h^{-2}$, which corresponds to an eigenvector that alternates in sign from one lattice point to the next. Requiring $|1 - \epsilon_{N-1}\Delta t| < 1$ then restricts Δt to be less than $\frac{1}{2}h^2$, which is 0.0008 in the example we are considering.

The question of stability is quite distinct from that of accuracy, as the limit imposed on the time step is set by the spatial step used and not by the characteristic time scale of the solution, which is much larger. The explicit scheme we have discussed is unsatisfactory, as the instability forces us to use a much smaller time step than is required to describe the evolution adequately. Indeed, the situation gets even worse if we try to use a finer spatial lattice to obtain a more accurate solution. Although the restriction on Δt that we have derived is rigorous only for the simple case we have considered, it does provide a useful guide for more complicated situations, as the eigenvector of H with the largest eigenvalue will always oscillate from one lattice point to the next; its eigenvalue is therefore quite insensitive to the global features of the problem.

■ **Exercise 7.1** Use the code above to verify that the instability sets in for smaller time steps if the spatial lattice is made finer and that the largest possible stable time step is roughly $\frac{1}{2}h^2$. Show that the instability is present for other initial conditions, and that round-off causes even the exact lowest eigenfunction of H to be unstable under numerical time evolution.

7.2 Implicit schemes and the inversion of tri-diagonal matrices

One way around the instability of the explicit algorithm described above is to retain the general form of (7.4), but to replace the second space derivative by that of the solution at the new time. That is, (7.4) is modified to

$$\frac{\phi_i^{n+1} - \phi_i^n}{\Delta t} = \frac{1}{h^2}(\delta^2 \phi^{n+1})_i + S_i^n . \tag{7.7}$$

This is an *implicit* scheme, since the unknown, ϕ^{n+1}, appears on both side of the equation. We can, of course, solve for ϕ^{n+1} as

$$\phi^{n+1} = \frac{1}{1 + H\Delta t}[\phi^n + S^n \Delta t] . \tag{7.8}$$

This scheme is equivalent to (7.5) to lowest order in Δt. However, it is much better in that larger time steps can be used, as the operator $(1 + H\Delta t)^{-1}$ has eigenvalues $(1 + \epsilon_\lambda \Delta t)^{-1}$, all of whose moduli are less than 1 for any Δt. All components of the solution therefore decrease with each time step, as they should. Although this decrease is inaccurate (i.e., not exponential) for the most rapidly oscillating components, such components should not be large in the initial conditions if the spatial discretization is accurate. In any event, there is no amplification, which implies stability. For the slowly varying components of the solution corresponding to small eigenvalues, the evolution closely approximates the exponential at each time step.

Note that if we had tried to be more accurate than (7.7) and had used the average of the second space derivatives at the two time steps involved,

$$\frac{\phi_i^{n+1} - \phi_i^n}{\Delta t} = \frac{1}{h^2}(\delta^2 \frac{1}{2}[\phi^{n+1} + \phi^n])_i + S_i^n , \qquad (7.9)$$

so that the time evolution is effected by

$$\phi^{n+1} = \frac{1}{1 + \frac{1}{2}H\Delta t}[(1 - \frac{1}{2}H\Delta t)\phi^n + S^n \Delta t] , \qquad (7.10)$$

this would have been almost as good as the implicit scheme, because the components of the solution are diminished by factors whose absolute values are less than one.

A potential drawback of the implicit scheme (7.8) is that it requires the solution of a set of linear equations (albeit tri-diagonal) at each time step to find ϕ^{n+1}; this is equivalent to the application of the inverse of the matrix $1 + H\Delta t$ to the vector appearing in brackets. Since the inverse itself is time-independent, it might be found only once at the beginning of the calculation and then used for all times, but application still requires of order N^2 operations if done directly. Fortunately, the following algorithm (Gaussian elimination and back-substitution, [Va62]) provides a very efficient solution (of order N operations) of a tri-diagonal system of equations such as that posed by Eq. (7.8).

Let us consider trying to solve the tri-diagonal linear system of equations $\mathbf{A}\phi = \mathbf{b}$ for the unknowns ϕ_i:

$$A_i^- \phi_{i-1} + A_i^0 \phi_i + A_i^+ \phi_{i+1} = b_i . \qquad (7.11)$$

Here, the $A_i^{\pm,0}$ are the only non-vanishing elements of $\mathbf{A}$ and the b_i are known quantities. This is the form of the problem posed by the evaluation of (7.8) for ϕ^{n+1}, where ϕ_0 and ϕ_N are given by the Dirichlet boundary conditions. In particular,

$$b_i = \phi_i^n + S_i^n \Delta t \;,\;\; A_i^0 = 1 + \frac{2\Delta t}{h^2} \;,\;\; A_i^\pm = -\frac{\Delta t}{h^2} \;.$$

To solve this system of equations, we assume that the solution satisfies a one-term forward recursion relation of the form

$$\phi_{i+1} = \alpha_i \phi_i + \beta_i \;, \tag{7.12}$$

where the α_i and β_i are coefficients to be determined. Substituting this into (7.11), we have

$$A_i^- \phi_{i-1} + A_i^0 \phi_i + A_i^+(\alpha_i \phi_i + \beta_i) = b_i \;, \tag{7.13}$$

which can solved for ϕ_i to yield

$$\phi_i = \gamma_i A_i^- \phi_{i-1} + \gamma_i(A_i^+ \beta_i - b_i) \;, \tag{7.14}$$

with

$$\gamma_i = -\frac{1}{A_i^0 + A_i^+ \alpha_i} \;. \tag{7.15}$$

Upon comparing Eq. (7.14) with (7.12), we can identify the following backward recursion relations for the α's and β's:

$$\alpha_{i-1} = \gamma_i A_i^- \;; \tag{7.16a}$$

$$\beta_{i-1} = \gamma_i(A_i^+ \beta_i - b_i) \;. \tag{7.16b}$$

The strategy to solve the system should now be clear. We will use the recursion relations (7.15,16) in a backwards sweep of the lattice to determine the α_i and β_i for i running from $N-2$ down to 0. The starting values to be used are

$$\alpha_{N-1} = 0 \;,\;\; \beta_{N-1} = \phi_N \;,$$

which will guarantee the correct value of ϕ at the last lattice point. Having determined these coefficients, we can then use the recursion relation (7.12) in a forward sweep from $i = 0$ to $N - 1$ to determine the solution, with the starting value ϕ_0 known from the boundary conditions. We have then determined the solution in only two sweeps of the lattice, involving of order N arithmetic operations.

The following code implements the algorithm (7.8) for the model diffusion problem we have been considering. Results are shown in Figure 7.1c, where it is clear that accurate solutions can be obtained with a much larger time step than can be used with the explicit scheme; the increase in numerical effort per time step is only about a factor of two. Note that the α_i and γ_i are independent of b, so that, as the inversion must be done at every time step, it is more efficient to compute these coefficients only once and store them at the beginning of the calculation (DO loop 15); only the β_i then need be computed for each inversion (DO loop 30).

```
C chap7b.for
      PARAMETER (NSTEP=25)    !lattice parameter
      DIMENSION PHI(0:NSTEP)  !array for solution
      DIMENSION ALPHA(0:NSTEP),BETA(0:NSTEP),GAMMA(0:NSTEP)
                             !analytical solution
      GAUSS(X,T)=EXP(-20.*(X-.5)**2/(1.+80*T))/SQRT(1+80*T)
      EXACT(X,T)=GAUSS(X,T)-GAUSS(X-1.,T)-GAUSS(X+1.,T)
      H=1./NSTEP
                             !obtain input from terminal
50    PRINT *, ' Enter time step and total time (0 to stop)'
      READ *,DT,TIME
      IF (DT .EQ. 0.) STOP
      NITER=TIME/DT
      DTH=DT/H**2
                             !initial conditions
      T=0.
      PHI(0)=0.
      PHI(NSTEP)=0.
      DO 10 IX=1,NSTEP-1
         PHI(IX)=EXACT(IX*H,T)
10    CONTINUE
                             !find ALPHA and GAMMA once
      AP=-DTH                !A coefficients
      AZ=1.+2*DTH
      ALPHA(NSTEP-1)=0.      !starting values
      GAMMA(NSTEP-1)=-1./AZ
      DO 15 IX=NSTEP-1,1,-1  !backward sweep
         ALPHA(IX-1)=GAMMA(IX)*AP
         GAMMA(IX-1)=-1./(AZ+AP*ALPHA(IX-1))
15    CONTINUE
                             !time loop
      DO 20 ITER=1,NITER
         BETA(NSTEP-1)=PHI(NSTEP)  !find BETA at this time
```

```
          DO 25 IX=NSTEP-1,1,-1         !backward sweep
            BETA(IX-1)=GAMMA(IX)*(AP*BETA(IX)-PHI(IX))
   25     CONTINUE
                                        !find new PHI
          PHI(0)=0.                     !initial value
          DO 30 IX=0,NSTEP-1            !forward sweep
            PHI(IX+1)=ALPHA(IX)*PHI(IX)+BETA(IX)
   30     CONTINUE
                                        !output every 10th time step
          IF (MOD(ITER,10) .EQ. 0) THEN
            PRINT *, ' iteration = ', ITER, ' time = ',ITER*DT
            T=ITER*DT
            DO 40 IX=1,NSTEP-1
              DIFF=PHI(IX)-EXACT(IX*H,T)
              PRINT *, ' phi = ', PHI(IX),' error = ', DIFF
   40       CONTINUE
          END IF
   20     CONTINUE
          GOTO 50
          END
```

■ **Exercise 7.2** Use the code above to investigate the accuracy and stability of the implicit algorithm for various values of Δt and for various lattice spacings. Study the evolution of various initial conditions and verify that they correspond to your intuition. Incorporate sources or sinks along the lattice and study the solutions that arise when ϕ vanishes everywhere at $t = 0$.

■ **Exercise 7.3** Modify the code above to apply the algorithm (7.10) and study its stability properties. Explore the effects of taking different linear combinations of ϕ^n and ϕ^{n+1} on the right-hand side of Eq. (7.9) (e.g., $\frac{3}{4}\phi^n + \frac{1}{4}\phi^{n+1}$).

■ **Exercise 7.4** A simple way to impose the Neumann boundary condition

$$\frac{\partial \phi}{\partial x}\bigg|_{x=0} = g$$

is to require that $\phi_1 = \phi_0 + hg$. Show that this implies that the forward recursion of (7.12) is to be started with

$$\phi_0 = \left(\frac{hg - \beta_0}{\alpha_0 - 1}\right).$$

What is the analogous expression implied by the more accurate constraint $\phi_1 = \phi_{-1} + 2hg$? What are the initial conditions for the backward recursion of the α and β coefficients if Neumann boundary conditions are imposed at the right-hand edge of the lattice? Modify the code above to incorporate the boundary condition that ϕ have vanishing derivative at $x = 0$ (as is appropriate if an insulator is present in a heat conduction problem) and observe its effect on the solutions. Show that the inversion scheme for tri-diagonal matrices discussed above cannot be applied if periodic boundary conditions are required (i.e., $\phi_N = \phi_0$).

■ **Exercise 7.5** Solve the boundary value problem posed in Exercise 3.4 by discretization and inversion of the resulting tri-diagonal matrix. Compare your solution with those obtained by the Green's function method and with the analytical result.

7.3 Diffusion and boundary value problems in two dimensions

The discussion above shows that diffusion in one dimension is best handled by an implicit method and that the required inversion of a tri-diagonal matrix is a relatively simple task. It is therefore natural to attempt to extend this approach to two or more spatial dimensions. For the two-dimensional diffusion equation,

$$\frac{\partial \phi}{\partial t} = \nabla^2 \phi \,,$$

the discretization is straightforward and, following our development for the one-dimensional problem, the time evolution should be effected by

$$\phi^{n+1} = \frac{1}{1 + H\Delta t}\phi^n \,, \tag{7.17}$$

where

$$(H\phi)_{ij} \equiv -\frac{1}{h^2}\left[(\delta_i^2 \phi)_{ij} + (\delta_j^2 \phi)_{ij}\right] .$$

Unfortunately, while H is very sparse, it is not tri-diagonal, so that the algorithm that worked so well in one dimension does not apply; some thought shows that H cannot be put into a tri-diagonal form by any permutation of its rows or columns. However, the fact that H can be written as a sum of operators that separately involve differences only in the i or j indices:

$$H = H_i + H_j \;;\; H_{i,j} = -\frac{1}{h^2}\delta_{i,j}^2 \,, \tag{7.18}$$

means that an expression equivalent to (7.17) through order Δt is

$$\phi^{n+1} = \frac{1}{1 + H_i \Delta t} \frac{1}{1 + H_j \Delta t} \phi^n \, . \tag{7.19}$$

This can now be evaluated exactly, as each of the required inversions involves a tri-diagonal matrix. In particular, if we define the auxiliary function $\phi^{n+1/2}$, we can write

$$\phi^{n+1/2} = \frac{1}{1 + H_j \Delta t} \phi^n \; ; \; \phi^{n+1} = \frac{1}{1 + H_i \Delta t} \phi^{n+1/2} \, .$$

Thus, $(1 + H_j \Delta t)^{-1}$ is applied by forward and backward sweeps of the lattice in the j direction, independently for each value of i. The application of $(1 + H_i \Delta t)^{-1}$ is then carried out by forward and backward sweeps in the i direction, independently for each value of j. This "alternating-direction" scheme is easily seen to be stable for all values of the time step and is generalized straightforwardly to three dimensions.

The ability to invert a tri-diagonal matrix exactly and the idea of treating separately each of the second derivatives of the Laplacian also provides a class of iterative alternating-direction methods [Wa66] for solving the elliptic boundary value problems discussed in Chapter 6. The matrix involved is written as the sum of several parts, each containing second differences in only one of the lattice indices. In two dimensions, for example, we seek to solve [see Eq. (6.3b)]

$$(H_i + H_j)\phi = s \, . \tag{7.20}$$

If we add a term $\omega\phi$ to both sides of this equation, where ω is a constant discussed below, the resulting expression can be solved for ϕ in two different ways:

$$\phi = \frac{1}{\omega + H_i}[s - (H_j - \omega)\phi] \; ; \tag{7.21a}$$

$$\phi = \frac{1}{\omega + H_j}[s - (H_i - \omega)\phi] \, . \tag{7.21b}$$

This pair of equations forms the basis for an iterative method of solution: they are solved in turn, ϕ on the right-hand sides being taken as the previous iterate; the solution involves only the inversion of tri-diagonal matrices. The optimal choice of the "acceleration parameter". ω, depends on the matrices involved and can even be taken to vary from one iteration to the next to improve convergence. A rather complicated analysis is required to find the optimal values [Wa66]. However, a good rule of thumb is to take

$$\omega = (\alpha\beta)^{1/2} \ , \qquad (7.22)$$

where α and β are, respectively, lower and upper bounds to the eigenvalues of H_i and H_j. In general, these alternating direction methods are much more efficient than the simple relaxation scheme discussed in Chapter 6, although they are slightly more complicated to program. Note that there is a slight complication when boundary conditions are also specified in the interior of the lattice.

7.4 Iterative methods for eigenvalue problems

Our analysis of the diffusion equation (7.1) shows that the net result of time evolution is to enhance those components of the solution with smaller eigenvalues of H relative to those with larger eigenvalues. Indeed, for very long times it is only that component with the lowest eigenvalue that is significant, although it has very small amplitude. This situation suggests a scheme for finding the lowest eigenvalue of an elliptic operator, as defined by (6.2): guess a trial eigenvector and subject it to a fictitious time evolution that will "filter" it to the eigenvector having lowest eigenvalue. Since we are dealing with a linear problem, the relentlessly decreasing or increasing magnitude of the solution can be avoided by renormalizing continuously as time proceeds.

To make the discussion concrete, consider the time-independent Schroedinger equation in one dimension with $\hbar = 2m = 1$. The eigenvalue problem is

$$\left[-\frac{d^2}{dx^2} + V(x)\right]\phi = \epsilon\phi \ , \qquad (7.23)$$

with the normalization condition

$$\int dx \ \phi^2 = 1$$

(ϕ can always be chosen to be real if V is). The corresponding energy functional is

$$E = \int dx \left[\left(\frac{d\phi}{dx}\right)^2 + V(x)\phi^2(x)\right] \ . \qquad (7.24)$$

On general grounds, we know that E is stationary at an eigenfunction with respect to variations of ϕ that respect the normalization condition and that the value of E at this eigenfunction is the associated eigenvalue.

To derive a discrete approximation to this problem, we discretize E as

$$E = \sum_i h \left[\frac{(\phi_i - \phi_{i-1})^2}{h^2} + V_i \phi_i^2 \right] , \qquad (7.25)$$

and the normalization constraint takes the form

$$\sum_i h \phi_i^2 = 1 . \qquad (7.26)$$

Variation with respect to ϕ_i gives the eigenvalue problem

$$(H\phi)_i \equiv -\frac{1}{h^2}(\delta^2 \phi)_i + V_i \phi_i = \epsilon \phi_i , \qquad (7.27)$$

with ϵ entering as a Lagrange multiplier ensuring the normalization. (Compare with the derivation of the Hartree-Fock equations given in Project III.)

We can interpret (7.27) as defining the problem of finding the real eigenvalues and eigenvectors of a (large) symmetric tri-diagonal matrix representing the operator H. Although direct methods for solving this problem were discussed in Chapter 5, they cannot be applied to the very large banded matrices that arise in two- and three-dimensional problems. However, in such cases, one is usually interested in the few highest or lowest eigenvalues of the problem, and for these the diffusion analogy is appropriate. Thus, we consider the problem

$$\frac{\partial \phi}{\partial \tau} = -H\phi ,$$

where τ is a "fake" time. For convenience, we suppose that things have been arranged so that the lowest eigenvalue of H is positive definite. (This can be guaranteed by shifting H by a spatially-independent constant chosen so that the resultant V_i is positive for all i.)

To solve this "fake" diffusion problem, any of the algorithms discussed above can be employed. The simplest is the explicit scheme analogous to (7.5):

$$\phi^{n+1} \sim (1 - H\Delta\tau)\phi^n , \qquad (7.28)$$

where $\Delta\tau$ is a small, positive parameter. Here, the symbol $\sim$ is used to indicate that ϕ^{n+1} is to be normalized to unity according to (7.26). For an initial guess, we can choose ϕ^0 to be anything not orthogonal to the exact eigenfunction, although guessing something close to the solution will speed the convergence. At each step in this refining process, computation of the energy from (7.25) furnishes an estimate of the true eigenvalue.

As an example, consider the problem where

$$V = 0 \ , \ \phi(0) = \phi(1) = 0 \ ;$$

this corresponds to a free particle in hard-walled box of unit length. The analytical solutions for the normalized eigenfunctions are

$$\psi_\lambda = 2^{1/2} \sin \lambda \pi x \ ,$$

and the associated eigenvalues are

$$\epsilon_\lambda = \lambda^2 \pi^2 \ .$$

Here, λ is a non-zero integer. The following FORTRAN program implements the scheme (7.28) on a lattice of 21 points and calculates the energy (7.25) at every fourth iteration. The initial trial function $\phi \sim x(1-x)$ is used, which roughly approximates the shape of the exact ground state.

```
C chap7c.for
        PARAMETER (NSTEP=20)        !lattice parameter
        DIMENSION PHI(0:NSTEP)      !solution array
        H=1./NSTEP
                                    !obtain input from screen
100     PRINT *, ' Enter step size ( .eq. 0 to stop)'
        READ *, DT
        IF (DT .EQ. 0.) STOP
        DTH=DT/H**2
                                    !initial guess for PHI
        DO 10 IX=0,NSTEP
          X=IX*H
          PHI(IX)=X*(1.-X)
10      CONTINUE
        CALL NORM(PHI,NSTEP)        !normalize guess
                                    !iteration loop
        DO 20 ITER=1,100
          POLD=0.
          DO 30 IX=1,NSTEP-1    !apply (1-H*DT)
            PNEW=PHI(IX)+DTH*(POLD+PHI(IX+1)-2*PHI(IX))
            POLD=PHI(IX)
            PHI(IX)=PNEW
30        CONTINUE
          CALL NORM(PHI,NSTEP) !normalize
                               !calculate Energy, Eq. (7.25)
```

```
          IF (MOD(ITER,4) .EQ. 0) THEN
            E=0.
            DO 50 IX=1,NSTEP
              E=E+(PHI(IX)-PHI(IX-1))**2
50          CONTINUE
            E=E/(H)
            PRINT *,' iteration = ', ITER, '  energy = ', E
          END IF
20      CONTINUE
        GOTO 100
        END
C
C
        SUBROUTINE NORM(PHI,NSTEP)
        DIMENSION PHI(0:NSTEP)
        XNORM=0.
        DO 10 IX=0,NSTEP
          XNORM=XNORM+PHI(IX)**2
10      CONTINUE
        XNORM=SQRT(NSTEP/XNORM)
        DO 20 IX=0,NSTEP
          PHI(IX)=PHI(IX)*XNORM
20      CONTINUE
        RETURN
        END
```

The results generated by the code above for several different values of $\Delta\tau$ are shown in Table 7.1. Note that for values of $\Delta\tau$ smaller than the stability limit, $\frac{1}{2}h^2 \approx 0.00125$, the energy converges to the expected answer and does so more rapidly for larger $\Delta\tau$. However, if $\Delta\tau$ becomes too large, the large-eigenvalue components of trial eigenfunction are amplified rather than diminished, and the energy obtained finally corresponds to the *largest* eigenvalue of H, the exact finite difference value of which is [see the discussion following Eq. (7.6)]

$$1600 \sin^2 \left(\frac{19\pi}{40} \right) = 1590.15 \, .$$

This phenomenon then suggests how to find the eigenvalue of an operator with the largest absolute value: simply apply H to a trial function many times.

Table 7.1   Convergence of the lowest eigenvalue of the square-well problem. The analytic result is 9.86960; the exact finite-difference value is 9.84933.

Iteration	$\Delta\tau = .0005$	$\Delta\tau = .0010$	$\Delta\tau = .0015$
4	9.93254	9.90648	9.88909
8	9.90772	9.87841	9.86569
12	9.89109	9.86441	9.87311
16	9.87946	9.85719	10.05357
20	9.87116	9.85342	12.34984
24	9.86519	9.85146	42.96950
28	9.86086	9.85044	379.69290
32	9.85773	9.84991	1301.501
36	9.85544	9.84963	1565.746
40	9.85378	9.84949	1587.891
44	9.85257	9.84941	1589.690
48	9.85169	9.84937	1589.966
52	9.85105	9.84935	1590.061

Although the procedure outlined above works, it is unsatisfactory in that the limitation on the size of $\Delta\tau$ caused by the lattice spacing often results in having to iterate many times to refine a poor trial function. This can be alleviated to some extent by choosing a good trial function. Even better is to use an implicit scheme [such as (7.8)] that does not amplify the large-eigenvalue components for any value of $\Delta\tau$. Another possibility is to use $\exp(-H\Delta\tau)$ to refine the trial function. While exact calculation of this matrix can be difficult if large dimensions are involved, it can be well-represented by its series expansion through a finite number of terms. This series is easy to evaluate since it involves only applying H many times to a trial state and generally a larger $\Delta\tau$ can be used than if only the first order approximation to the exponential, (7.28), is employed.

We have shown so far how the method we have discussed can be used to find the lowest or highest eigenvalue of an operator or matrix. To see how to find other eigenvalues and their associated eigenfunctions, consider the problem of finding the second-lowest eigenvalue. We first find the lowest eigenvalue and eigenfunction by the method described above. A trial function for the second eigenfunction is then guessed and refined in the same way, taking care, however, that at each stage of the refinement the solution remain orthogonal to the lowest eigenfunction already found.

This can be done by continuously projecting out that component of the solution not orthogonal to the lowest eigenfunction. (This projection is not required when there is some symmetry, such as reflection symmetry, that distinguishes the two lowest solutions and that is preserved by the refinement algorithm.) Having found, in this way, the second-lowest eigenfunction, the third lowest can be found similarly, taking care that during *its* refinement, it remains orthogonal to both of the eigenfunctions with lower eigenvalues. This process cannot be applied to find more than the few lowest (or highest) eigenvectors, however, as numerical round-off errors in the orthogonalizations to the many previously-found eigenvectors soon grow to dominate.

Although the methods described above have been illustrated by a one-dimensional example, it should be clear that they can be applied directly to find the eigenvalues and eigenfunctions of elliptic operators in two or more dimensions, for example via Eq. (7.19).

■ **Exercise 7.6**   Extend the code given above to find the eigenfunctions and eigenvalues of the first two excited states of each parity in the one-dimensional square-well. Compare your results with the exact solutions and with the analytical finite-difference values.

■ **Exercise 7.7**   Write a program (or modify that given above) to find the few lowest and few highest eigenvalues and eigenfunctions of the Laplacian operator in a two-dimensional region consisting of a square with a square hole cut out of its center. Investigate how your results vary as functions of the size of the hole.

7.5 The time-dependent Schroedinger equation

The numerical treatment of the time-dependent Schroedinger equation for a particle moving in one dimension provides a good illustration of the power of the techniques discussed above and some striking examples of the operation of quantum mechanics [Go67]. We consider the problem of finding the evolution of the (complex) wave function ϕ under Eq. (7.2), given its value at some initial time. After spatial discretization in the usual way, we have the parabolic problem

$$\frac{\partial \phi_i}{\partial t} = -i(H\phi)_i , \tag{7.29}$$

where H is the operator defined in (7.27) and we have put $\hbar = 1$. To discretize the time evolution, we have the possibility of employing the analogue of any of the three methods discussed above for the diffusion equation. The explicit method related to (7.5), in which the evolution is effected by $(1 - iH\Delta t)$, is unstable for *any* value of Δt because the eigenvalues of this operator,

$$(1 - i\epsilon_\lambda \Delta t) ,$$

have moduli

$$(1 + \epsilon_\lambda^2 \Delta t^2)^{1/2}$$

greater than unity. The implicit scheme analogous to (7.8), in contrast, is stable for all Δt, as the moduli of the eigenvalues of the evolution operator,

$$(1 + \epsilon_\lambda^2 \Delta t^2)^{-1/2} ,$$

are always less than one. This is still unsatisfactory, though, as the numerical evolution then does not have the important unitarity property of the exact evolution; the norm of the wave function continually decreases with time. Fortunately, the analogue of (7.10) turns out to be very suitable. It is

$$\phi^{n+1} = \left(\frac{1 - i\frac{1}{2}H\Delta t}{1 + i\frac{1}{2}H\Delta t} \right) \phi^n . \tag{7.30}$$

This evolution operator is manifestly unitary (recall that H is hermitian) with eigenvalues of unit modulus, so that the norm of ϕ computed according to (7.26) is the same from one time to the next. (Of course, the square of the real wave function in (7.26) is to be replaced by the modulus squared of the complex wave function.) The algorithm (7.30) also has the desirable feature that it approximates the exact exponential evolution operator, $\exp(-iH\Delta t)$, through second order in Δt, which is one more power than would have been supposed naively.

For actual numerical computation, it is efficient to rewrite (7.30) as

$$\phi^{n+1} = \left(\frac{2}{1 + i\frac{1}{2}H\Delta t} - 1 \right) \phi^n \equiv \chi - \phi^n . \tag{7.31}$$

This form eliminates the sweep of the lattice required to apply the numerator of (7.30). To find, at each time step, the intermediate function χ defined by

$$(1 + i\frac{1}{2}H\Delta t)\chi = 2\phi^n ,$$

we write this equation explicitly as

$$-\frac{i\Delta t}{2h^2}\chi_{j+1} + \left(1 + \frac{i\Delta t}{h^2} + \frac{i\Delta t}{2}V_j \right)\chi_j - \frac{i\Delta t}{2h^2}\chi_{j-1} = 2\phi_j^n , \tag{7.32}$$

which, upon dividing by $-i\Delta t/2h^2$ becomes

$$\chi_{j+1} + \left(-2 + \frac{2ih^2}{\Delta t} - h^2 V_j\right)\chi_j + \chi_{j-1} = \frac{4ih^2}{\Delta t}\phi_j^n . \qquad (7.33)$$

This has the form of (7.11), and can therefore be solved by the two-sweep method discussed in connection with that equation.

The FORTRAN program for Example 7, whose source code is contained in Appendix B and in the file EXMPL7.FOR, uses the method described above to solve the time-dependent Schroedinger equation. Several analytical forms of the potential (square well or barrier, Gaussian well or barrier, potential step, or parabolic well) can be defined on a point lattice whose size is set by the parameter NPTS. An initial Gaussian or Lorentzian wavepacket of specified average position, average momentum, and spatial width can then be set up on the lattice and evolved under the boundary condition that ϕ vanish at the lattice boundaries. The probability density, $|\phi|^2$, and potential are displayed at the requested frequency, while the total probabilities and average positions of those portions of the wavepacket to the left and right of a specified point are displayed at every time step. The time evolution is continued until the user requests termination of the program.

The following exercises will be useful in improving your understanding of this example.

■ **Exercise 7.8** Test the accuracy of the integration by integrating forward for some time interval, changing the sign of the time step, and then continuing the integration for an equal time interval to see if you return to the initial wavepacket. Note that the energy calculated in subroutine ENERGY is conserved independent of the time step. Verify that this should be so using Eq. 7.30.

■ **Exercise 7.9** Verify that wavepackets in the absence of any potential and wavepackets in a parabolic potential well behave as you expect them to.

■ **Exercise 7.10** Send wavepackets of varying widths and incident energies at barriers and wells of various sizes and shapes. Interpret all features that you observe during the time evolution. For square-well and step potentials, compare the fractions of the initial probability transmitted and reflected with the analytical values of the usual transmission and reflection coefficients. Set up a resonance situation by considering scattering from potentials with a "pocket" in them and observe the separation of the initial wavepacket into prompt and delayed components. Set up a "double-well" tunneling situation and observe the evolution.

■ **Exercise 7.11** Replace the evolution algorithm (7.30) by the unstable explicit method or the non-unitary implicit method and observe the effects.

■ **Exercise 7.12** Replace the vanishing Dirichlet boundary condition at one lattice end by a zero-derivative Neumann boundary condition and observe what happens when a wavepacket approaches.

■ **Exercise 7.13** In two- and three-dimensional calculations, the large number of lattice points forces the use of as large a lattice spacing as is possible, so that higher-order approximations to the spatial derivative become necessary [Fl78]. Replace the "three-point" formula for the spatial second derivative by the more accurate "five-point" one listed in Table 1.2. Develop an algorithm to invert the penta-diagonal matrix involved in the time evolution. Implement this in the code and observe any changes in the results or the computational efficiency.

Project VII: Self-organization in chemical reactions

Recent work in several branches of physics has shown that the solutions of non-linear equations can display a rich variety of phenomena. Among these is "pattern selection", in which stable, non-trivial patterns in space and/or time emerge spontaneously from structureless initial conditions. In this project, we will investigate analytically and numerically a model of chemical reactions, the "Brusselator", whose solutions exhibit behavior that is very similar to the striking phenomena observed in actual chemical systems [Wi74]. Our discussion follows that of [Ni77] and [Bo76].

VII.1 Description of the model

We consider a network of chemical reactions in which reagent species A and B are converted into product species D and E through intermediates X and Y:

$$A \rightarrow X \; ; \qquad (VII.1a)$$

$$B + X \rightarrow Y + D \; ; \qquad (VII.1b)$$

$$2X + Y \rightarrow 3X \; ; \qquad (VII.1c)$$

$$X \rightarrow E \; . \qquad (VII.1d)$$

We assume that the concentrations of A and B are fixed and are spatially uniform and that the species D and E are "dead" in the sense of being chemically inert or being removed from the reaction volume. We also assume that the processes (VII.1a–d) are sufficiently exoergic to make the reverse reactions negligible. Under these conditions, we can write the following equations for the evolution of the concentrations of X and Y:

$$\frac{\partial X}{\partial t} = k_a A - k_b BX + k_c X^2 Y - k_d X + D_X \nabla^2 X \; ; \quad (VII.2a)$$

$$\frac{\partial Y}{\partial t} = k_b BX - k_c X^2 Y + D_Y \nabla^2 Y \; . \quad\quad\quad (VII.2b)$$

Here, k_{a-d} are the rate constants for the reactions (VII.1a–d), respectively, and $D_{X,Y}$ are the diffusion constants of the species X, Y.

It is convenient to scale the non-linear diffusion equations (VII.2) to a dimensionless form. If we measure time in units of k_d^{-1}, space in units of l, a characteristic size of the reaction volume, X and Y in units of $(k_d/k_c)^{\frac{1}{2}}$, A in units of $(k_d^3/k_a^2 k_c)^{\frac{1}{2}}$, B in units of k_d/k_b, and $D_{X,Y}$ in units of $k_d l^2$, then Eqs. (VII.2) become

$$\frac{\partial X}{\partial t} = A - (B+1)X + X^2 Y + D_X \nabla^2 X \; ; \quad (VII.3a)$$

$$\frac{\partial Y}{\partial t} = BX - X^2 Y + D_Y \nabla^2 Y \; . \quad\quad\quad (VII.3b)$$

From this scaling, it is clear that the constants A and B can be of order unity, although $D_{X,Y}$ might be considerably smaller, depending upon the value of l.

One trivial solution to these equations can be found by assuming that X and Y are independent of space and time. Setting all derivatives to zero yields a set of algebraic equations that can be solved for the equilibrium point. After a bit of algebra, we find that equilibrium is at

$$X = X_0 = A \, , \; Y = Y_0 = \frac{B}{A} \, .$$

To completely specify the model, we must give the spatial boundary conditions. Of the various possible choices, two are of particular physical interest. These are the "no-flux" boundary conditions, in which the normal derivatives of X and Y are required to vanish on the surface of the reaction volume (as might the case if the chemistry takes place in a closed vessel) and the "fixed" boundary conditions, where X and Y are required to have their equilibrium values on the surface of the reaction volume.

VII.2 Linear stability analysis

To get some insight into the behavior of this system before beginning to compute, it is useful to analyze the behavior of small perturbations about the equilibrium state. To do so, we put

$$X = X_0 + \delta X(r,t) \; ; \qquad\qquad (VII.4a)$$

$$Y = Y_0 + \delta Y(r,t) \; , \qquad\qquad (VII.4b)$$

where r represents the spatial variables and δX, δY are small quantities dependent upon space and time. Inserting this into (VII.3) and linearizing in the small quantities, we have

$$\frac{\partial \delta X}{\partial t} = (2X_0Y_0 - B - 1 + D_X\nabla^2)\delta X + X_0^2\delta Y \; ; \quad (VII.5a)$$

$$\frac{\partial \delta Y}{\partial t} = (B - 2X_0Y_0)\delta X + (D_Y\nabla^2 - X_0^2)\delta Y \; . \qquad (VII.5b)$$

We now specialize to the no-flux boundary condition in one dimension. We can then expand δX as

$$\delta X(r,t) = \sum_{m=0}^{\infty} \delta X_m e^{\omega_m t} \cos m\pi r \; , \qquad (VII.6)$$

and similarly for δY. Here, $0 \leq r \leq 1$ is the spatial coordinate and the ω_m indicate the stability of each normal mode. In particular, $Re\,\omega_m < 0$ indicates a stable mode that damps in time, while $Re\,\omega_m > 0$ indicates an unstable perturbation that grows; a complex ω_m indicates a mode that also oscillates as it grows or damps. For the fixed boundary conditions, the expansion analogous to (VII.6) involves sin $m\pi r$ rather than cos $m\pi r$.

Upon introducing the Fourier expansion into the linearized equations (VII.5a,b) and equating the coefficients of each spatial mode, we obtain the following homogeneous eigenvalue equations:

$$\begin{aligned}\omega_m\delta X_m &= (2X_0Y_0 - B - 1 - D_X m^2\pi^2)\delta X_m + X_0^2\delta Y_m \; ; \\ \omega_m\delta Y_m &= (B - 2X_0Y_0)\delta X_m - (X_0^2 + D_Y m^2\pi^2)\delta Y_m \; .\end{aligned} \qquad (VII.7)$$

It is easy to see that these hold for both types of boundary conditions and that the eigenvalues then satisfy the characteristic equation

$$\omega_m^2 + (\beta_m - \alpha_m)\omega_m + A^2 B - \alpha_m\beta_m = 0 \; , \qquad (VII.8)$$

with

$$\alpha_m = B - 1 - m^2\pi^2 D_X \; ; \quad \beta_m = A^2 + m^2\pi^2 D_Y \; . \qquad (VII.9)$$

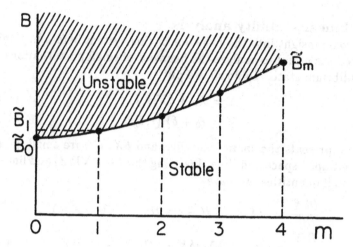

Figure VII.1 Stability of the uniform equilibrium state with respect to oscillatory behavior. The $m = 0$ mode is absent for fixed boundary conditions. (From [Ni77].)

(We have here written X_0 and Y_0 explicitly in terms of A and B). The roots of this equation are

$$\omega_m^{\pm} = \frac{1}{2}\{\alpha_m - \beta_m \pm [(\alpha_m + \beta_m)^2 - 4A^2B]^{1/2}\} \,. \qquad (VII.10)$$

A detailed analysis of Eqs. (VII.9,10) reveals several interesting aspects about the stability of the uniform equilibrium state. It is easy to see that if m becomes large for fixed values of A, B, and $D_{X,Y}$, then the $\omega_m^{\pm}$ become very negative; modes with large wavenumbers are therefore stable. Let us now imagine that A and $D_{X,Y}$ are fixed and B is allowed to vary. Then, since there are complex roots only if the discriminant in (VII.10) is negative, we can conclude that there will be oscillating modes only if

$$(A - \Delta_m^{1/2})^2 < B < (A + \Delta_m^{1/2})^2 \,, \qquad (VII.11)$$

where

$$\Delta_m = 1 + m^2\pi^2(D_X - D_Y) \,.$$

Since we must have $\Delta_m > 0$, this implies that there are oscillations only when

$$D_Y - D_X < \frac{1}{m^2\pi^2} \,.$$

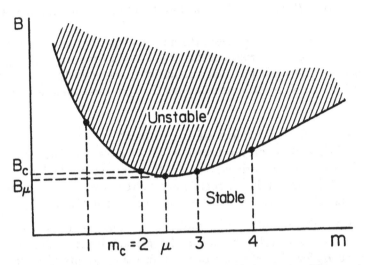

Figure VII.2 Stability of the uniform equilibrium state with respect to exponentially growing perturbations (From [Ni77].)

Furthermore, if there is a complex eigenvalue, its real part will be positive (an unstable mode) only if

$$B > \tilde{B}_m \equiv 1 + A^2 + m^2\pi^2(D_X + D_Y).$$

The situation is summarized in Figure VII.1.

Modes with real, positive frequencies are present only if $\alpha_m\beta_m - A^2B > 0$, which implies that

$$B > B_m \equiv 1 + A^2\left[\frac{D_X}{D_Y} + \frac{1}{D_Y m^2\pi^2}\right] + D_X m^2\pi^2. \qquad (VII.12)$$

If we imagine m to be a continuous variable, then it is easy to see that B_m has a minimum value of

$$B_\mu = \left[1 + A\left(\frac{D_X}{D_Y}\right)^{1/2}\right]^2$$

at

$$m = \mu = \left[\frac{A}{\pi^2(D_X D_Y)^{1/2}}\right]^{1/2}.$$

Of course, since the physical values of m are restricted to integers, as B is increased from 0, the non-oscillatory mode m_c that first becomes unstable at $B = B_c$ is that which is closest to the minimum, as shown in Figure VII.2.

To summarize this discussion, the uniform equilibrium state is unstable with respect to perturbations that oscillate in time when $B > \tilde{B}_0$ or $\tilde{B}_1$ for no-flux and fixed boundary conditions, respectively. Similarly, it is unstable with respect to perturbations that grow exponentially in time when $B > B_c$.

VII.3 Numerical solution of the model

Although the stability analysis presented above gives some hint at the behavior of the system for various ranges of the parameters, the full richness of the model in the non-linear regime can be revealed only by numerical experiments. These can be carried out through the following steps.

■ **Step 1**   Verify the algebra in the stability analysis discussed above. Assume a fixed value of $A = 2$ in a one-dimensional situation and calculate the $\tilde{B}_m$ and B_c for a set of values of the diffusion constants of order 10^{-3}. How do your results change if the reaction takes place in the unit square in two space dimensions?

■ **Step 2**   Write a program that solves the non-linear diffusion equations (VII.3a,b) in one dimension on the interval [0,1] for the case of no-flux boundary conditions; a reasonable number of spatial points might be between 25 and 100. The diffusion terms should be treated implicitly to prevent unphysical numerical instabilities, while the reaction terms can be treated explicitly. Have your program plot out X and Y at each time step.

■ **Step 3**   Use your one-dimensional program to investigate the behavior of the solutions for different values of B, $D_{X,Y}$. A reasonable place to start might be $D_X = 1 \times 10^{-3}$, $D_Y = 4 \times 10^{-3}$; the linear stability analysis should then give some guidance as to what values of B are interesting. Investigate initial conditions corresponding to smooth sinusoidal and point-wise random perturbations of the uniform equilibrium configuration (the latter can be generated with the help of function RANNOS, found at the end of Appendix D.3, or your compiler's intrinsic random number generator). Verify that you can find cases in which the system relaxes back to the uniform state, in which it asymptotically approaches a time-independent solution with a non-trivial spatial variation (dissipative structure), and in which it approaches a space- and time-dependent oscillating solution. Throughout, make sure that your time step is small enough to allow an accurate integration of the equations.

■ **Step 4** Extend your one-dimensional code to solve the Brusselator with no-flux boundary conditions in two space dimensions using, for example, an alternating-direction algorithm like Eq. (7.19). Plot your results for X at each time. (The technique of displaying an array of characters, as in Example 6, can be useful.) Investigate parameter ranges and initial conditions as in Step 3.

Chapter 8

Monte Carlo
Methods

Systems with a large number of degrees of freedom are often of interest in physics. Among these are the many atoms in a chunk of condensed matter, the many electrons in an atom, or the infinitely many values of a quantum field at all points in a region of space-time. The description of such systems often involves (or can be reduced to) the evaluation of integrals of very high dimension. For example, the classical partition function for a gas of A atoms at a temperature $1/\beta$ interacting through a pair-wise potential v is proportional to the $3A$-dimensional integral

$$Z = \int d^3 r_1 \ldots d^3 r_A \, \exp\left[-\beta \sum_{i<j} v(r_{ij})\right] . \qquad (8.1)$$

The straightforward evaluation of an integral like this by one of the quadrature formulas discussed in Chapters 1 or 4 is completely out of the question except for the very smallest values of A. To see why, suppose that the quadrature allows each coordinate to take on 10 different values (not a very fine discretization), so that the integrand must be evaluated at 10^{3A} points. For a modest value of $A = 20$ and a very fast computer capable of some 10^7 evaluations per second, this would take some 10^{53} seconds, more than 10^{34} times the age of the universe! Of course, tricks like exploiting the permutation symmetry of the integrand can reduce this estimate considerably, but it still should be clear that direct quadrature is hopeless.

The Monte Carlo methods discussed in this chapter are ways of efficiently evaluating integrals of high dimension. The name "Monte Carlo" arises from the random or "chance" character of the method and the famous casino in Monaco. The essential idea is not to evaluate the integrand at every one of a large number of quadrature points, but rather at only a representative random sampling of abscissae. This is analogous to predicting the results of an election on the basis of a poll of a small number of voters. Although it is by no means obvious that anything sensible can come out of random numbers in a computer, the Monte Carlo strategy turns out to be very appropriate for a broad class of problems in statistical and quantum mechanics. More detailed presentations of the method than that given here can be found in [Ha64] and [Ka85].

8.1 The basic Monte Carlo strategy

Even though the real power of Monte Carlo methods is in evaluating multi-dimensional integrals, it is easiest to illustrate the basic ideas in a one-dimensional situation. Suppose that we have to evaluate the integral

$$I = \int_0^1 f(x)\,dx$$

for some particular function f. Chapters 1 and 4 discussed several different quadrature formulas that employed values of f at very particular values of x (e.g., equally spaced). However, an alternative way of evaluating I is to think about it as the average of f over the interval $[0,1]$. In this light, a plausible quadrature formula is

$$I \approx \frac{1}{N} \sum_{i=1}^{N} f(x_i) \,. \tag{8.2}$$

Here, the average of f is evaluated by considering its values at N abscissae, $\{x_i\}$, chosen at random with equal probability anywhere within the interval $[0,1]$. We discuss below methods for generating such "random" numbers, but for now it is sufficient to suppose that there is a computer function that provides as many of them as are required, one after the other. Note that there is no such FORTRAN-77 function, but many compilers provide their own (e.g., RAN in VAX FORTRAN) and we provide function RANNOS (found in the file UTIL.FOR and listed at the end of Appendix D.3).

To estimate the uncertainty associated with this quadrature formula, we can consider $f_i \equiv f(x_i)$ as a random variable and invoke the central limit theorem for large N. From the usual laws of statistics, we have

$$\sigma_I^2 \approx \frac{1}{N}\sigma_f^2 = \frac{1}{N}\left[\frac{1}{N}\sum_{i=1}^{N} f_i^2 - \left(\frac{1}{N}\sum_{i=1}^{N} f_i\right)^2\right], \tag{8.3}$$

where σ_f^2 is the variance in f; i.e., a measure of the extent to which f deviates from its average value over the region of integration.

Equation (8.3) reveals two very important aspects of Monte Carlo quadrature. First, the uncertainty in the estimate of the integral, σ_I, decreases as $N^{-1/2}$. Hence, if more points are used, we will get a more precise answer, although the error decreases very slowly with the number of points (a factor of four more numerical work is required to halve the uncertainty in the answer). This is to be contrasted with a method like the trapezoidal rule, where Eqs. (1.8,1.9) show that the error scales like N^{-2}, which affords a much greater accuracy for a given amount of numerical work. (This advantage vanishes in multi-dimensional cases, as we discuss below.) The second important point to realize from (8.3) is that the precision is greater if σ_f is smaller; that is, if f is as smooth as possible. One limit of this is when f is a constant, in which case we need its value at only one point to define its average. To see the other limit, consider a situation in which f is zero except for a very narrow peak about some value of x. If the x_i have an equal probability to lie anywhere between 0 and 1, it is probable that all but a few of them will lie outside the peak of f, and that only these few of the f_i will be non-zero; this will lead to a poorly defined estimate of I.

As an example of a Monte Carlo evaluation of an integral, consider

$$\int_0^1 \frac{dx}{1+x^2} = \frac{\pi}{4} = 0.78540 . \tag{8.4}$$

The following FORTRAN program calculates this integral for the value of N input, together with an estimate of the precision of the quadrature.

```
C chap8a.for
      INTEGER SEED              !seed for random number generator
      DATA SEED/987654321/
      DATA EXACT/.78540/        !analytic answer
      FUNC(X)=1./(1.+X**2)      !function to integrate
C
10    PRINT *, ' Enter number of points (0 to stop)'
      READ *, N
      IF (N .EQ. 0) STOP
C
      SUMF=0.                   !zero sums
      SUMF2=0.
      DO 20 IX=1,N              !loop over samples
         FX=FUNC(RAN(SEED))
         SUMF=SUMF+FX           !add contributions to sums
```

Table 8.1 Monte Carlo evaluation of the integral (8.4) using two different weight functions, $w(x)$. The exact value is 0.78540.

N	$w(x) = 1$ I	σ_I	$w(x) = \frac{1}{3}(4 - 2x)$ I	σ_I
10	0.81491	0.04638	0.79982	0.00418
20	0.73535	0.03392	0.79071	0.00392
50	0.79606	0.02259	0.78472	0.00258
100	0.79513	0.01632	0.78838	0.00194
200	0.78677	0.01108	0.78529	0.00140
500	0.78242	0.00719	0.78428	0.00091
1000	0.78809	0.00508	0.78524	0.00064
2000	0.78790	0.00363	0.78648	0.00045
5000	0.78963	0.00227	0.78530	0.00028

```
          SUMF2=SUMF2+FX**2
20     CONTINUE
       FAVE=SUMF/N                    !averages
       F2AVE=SUMF2/N
       SIGMA=SQRT((F2AVE-FAVE**2)/N)    !error
       PRINT *,
   +   ' integral =',FAVE,' +- ',SIGMA,'  error = ',EXACT-FAVE
       GOTO 10
       END
```

Results of running this program for various values of N are given in the first three columns of Table 8.1. The calculated result is equal to the exact value within a few (usually less than one) standard deviations and the quadrature becomes more precise as N increases.

Since Eq. (8.3) shows that the uncertainty in a Monte Carlo quadrature is proportional to the variance of the integrand, it is easy to devise a general scheme for reducing the variance and improving the efficiency of the method. Let us imagine multiplying and dividing the integrand by a positive weight function $w(x)$, normalized so that

$$\int_0^1 dx\, w(x) = 1 .$$

The integral can then be written as

$$I = \int_0^1 dx\, w(x)\frac{f(x)}{w(x)} \ . \tag{8.5}$$

If we now make a change of variable from x to

$$y(x) = \int_0^x dx'\, w(x') \ , \tag{8.6}$$

so that

$$\frac{dy}{dx} = w(x) \ ; \ y(x = 0) = 0 \ ; \ y(x = 1) = 1 \ ,$$

then the integral becomes

$$I = \int_0^1 dy\, \frac{f\big(x(y)\big)}{w\big(x(y)\big)} \ . \tag{8.7}$$

The Monte Carlo evaluation of this integral proceeds as above, namely averaging the values of f/w at a random sampling of points uniformly distributed in y over the interval $[0,1]$:

$$I \approx \frac{1}{N}\sum_{i=1}^{N} \frac{f\big(x(y_i)\big)}{w\big(x(y_i)\big)} \ . \tag{8.8}$$

The potential benefit of the change of variable should now be clear. If we choose a w that behaves approximately as f does (i.e., it is large where f is large and small where f is small), then the integrand in (8.7), f/w, can be made very smooth, with a consequent reduction in the variance of the Monte Carlo estimate (8.8). This benefit is, of course, contingent upon being able to find an appropriate w and upon being able to invert the relation (8.6) to obtain $x(y)$.

A more general way of understanding why a change of variable is potentially useful is to realize that the uniform distribution of points in y implies that the distribution of points in x is $dy/dx = w(x)$. This means that points are concentrated about the most "important" values of x where w (and hopefully f) is large, and that little computing power is spent on calculating the integrand for "unimportant" values of x where w and f are small.

As an example of how a change of variable can improve the efficiency of Monte Carlo quadrature, we consider again the integral (8.4). A good choice for a weight function is

$$w(x) = \frac{1}{3}(4 - 2x) \ ,$$

which is positive definite, decreases monotonically over the range of integration (as does f), and is normalized correctly. Moreover, since $f/w = 3/4$ at both $x = 0$ and $x = 1$, w well approximates the behavior of f. According to (8.6), the new integration variable is

$$y = \frac{1}{3}x(4 - x) \, ,$$

which can be inverted to give

$$x = 2 - (4 - 3y)^{1/2} \, .$$

The following FORTRAN code then evaluates I according to (8.8).

```
C chap8b.for
      INTEGER SEED                 !seed for random number generator
      DATA SEED/987654321/
      DATA EXACT/.78540/           !analytic value
      FUNC(X)=1./(1.+X**2)         !function to integrate
      WEIGHT(X)=(4.-2*X)/3.        !weight function
      XX(Y)=2.-SQRT(4.-3.*Y)       !X as a function of Y
C
10    PRINT *, ' Enter number of points (0 to stop)'
      READ *, N
      IF (N .EQ. 0) STOP
C
      SUMF=0.                      !zero sums
      SUMF2=0.
      DO 20 IX=1,N                 !loop over samples
         Y=RAN(SEED)
         X=XX(Y)
         FX=FUNC(X)/WEIGHT(X)      !integrand
         SUMF=SUMF+FX              !add contributions to sum
         SUMF2=SUMF2+FX**2
20    CONTINUE
      FAVE=SUMF/N                  !averages
      F2AVE=SUMF2/N
      SIGMA=SQRT((F2AVE-FAVE**2)/N)  !error
      PRINT *,
     + ' integral =',FAVE,' +- ',SIGMA,'  error = ',EXACT-FAVE
      GOTO 10
      END
```

Results of running this code for various values of N are shown in the last two columns of Table 8.1, where the improvement over the $w = 1$ case treated previously is evident.

The one-dimensional discussion above can be readily generalized to d-dimensional integrals of the form $I = \int d^d x \, f(\mathbf{x})$. The analog of (8.2) is

$$I \approx \frac{1}{N} \sum_{i=1}^{N} f(\mathbf{x}_i) \,, \tag{8.9}$$

with the several components of the random points $\mathbf{x}_i$ to be chosen independently. Thus, the following FORTRAN program calculates

$$\pi = 4 \int_0^1 dx_1 \int_0^1 dx_2 \, \theta(1 - x_1^2 - x_2^2) \,;$$

that is, it compares the area of a quadrant of the unit circle to that of the unit square (θ is the unit step function):

```
C chap8c.for
      INTEGER SEED              !seed for random number generator
      DATA SEED/3274927/
      EXACT=4.*ATAN(1.)         !pi
C
20    PRINT *, ' Enter number of points (0 to stop)'
      READ *, N
      IF (N .EQ. 0) STOP
C
      ICOUNT=0                  !zero count
      DO 10 IX=1,N             !loop over samples
         X=RAN(SEED)
         Y=RAN(SEED)
         IF ((X**2+Y**2) .LE. 1.) ICOUNT=ICOUNT+1
10    CONTINUE
      PI4=REAL(ICOUNT)/N        !pi/4
      SIGMA=SQRT(PI4*(1.-PI4)/N)   !error
      PRINT *,
     + ' pi = ', 4*PI4,' +- ',4*SIGMA,'   error = ',EXACT-4*PI4
      GOTO 20
      END
```

When run, this program generated the satisfactory result of 3.1424 ± 0.0232.

■ **Exercise 8.1** Verify that the error estimate used (two lines after line 10) of this program is that given by Eq. (8.3).

The change of variable discussed above can also be generalized to many dimensions. For a weight function $w(\mathbf{x})$ normalized so that its integral over the region of integration is unity, the appropriate new variable of integration is $\mathbf{y}$, where the Jacobian is $|\partial \mathbf{y}/\partial \mathbf{x}| = w(\mathbf{x})$. It is generally very difficult (if not impossible) to construct $\mathbf{x}(\mathbf{y})$ explicitly, so that it is more convenient to think about the change of variable in the multi-dimensional case in the sense discussed above; i.e., it distributes the points $\mathbf{x}_i(\mathbf{y}_i)$ with distribution w. Various practical methods for doing this for arbitrary w will be discussed below.

Although the results were satisfactory in the examples given above, Monte Carlo quadrature does not appear to be particularly efficient. Even with the "good" choice of w, the results in Table 8.1 show a precision of only about 10^{-4} for $N = 5000$, whereas the conventional trapezoidal formula with 5000 points is accurate to better than 10^{-5}. However, consider evaluating a multi-dimensional integral such as (8.1). Suppose that we are willing to invest a given amount of numerical work (say to evaluate the integrand N times), and wish to compare the efficiencies of conventional and Monte Carlo quadratures. In a conventional quadrature, say a multi-dimensional analog of the trapezoidal rule, if there are a total of N points, then each dimension of a d-dimensional integral is broken up into $\sim N^{1/d}$ intervals of spacing $h \sim N^{-1/d}$. The analog of (1.9) shows that the error in the integral over each cell of volume h^d in the integration region is $O(h^{d+2})$, so that the total error in the conventional quadrature is

$$NO(h^{d+2}) = O(N^{-2/d}) \; ;$$

for large d, this decreases very slowly with increasing N. On the other hand, Eq. (8.3) above shows that the uncertainty of a Monte Carlo quadrature decreases as $N^{-1/2}$, independent of d. Assuming that the prefactors in these estimates are all of order unity, we see that Monte Carlo quadrature is more efficient when $d \gtrsim 4$. Of course, this estimate depends in detail upon the conventional quadrature scheme we use or how good a weight function is used in the Monte Carlo scheme, but the basic point is the very different way in which the two errors scale with increasing N for large d.

8.2 Generating random variables with a specified distribution

The discussion above shows that Monte Carlo quadrature involves two basic operations: generating abscissae randomly distributed over the integration volume with a specified distribution $w(x)$ (which may perhaps be unity) and then evaluating the function f/w at these abscissae. The second operation is straightforward, but it is not obvious how to generate "random" numbers on a deterministic computer. In this section, we will cover a few of the standard methods used.

The generation of uniformly distributed random numbers is the computer operation that underlies any treatment of a more complicated distribution. There are numerous methods for performing this basic task and for checking that the results do indeed correspond to the uniform random numbers required. One of the most common algorithms, and in fact that used in the VAX FORTRAN RAN function, is a "linear congruential" method, which "grows" a whole sequence of random numbers from a "seed" number. The current member of the sequence is multiplied by a first "magic" number, incremented by a second magic number, and then the sum is taken modulo a third magic number. Thus,

$$x_{i+1} = (ax_i + c)\mathrm{mod}\,m \,, \tag{8.10}$$

where i is the sequence label and a, c, and m are the magic numbers. The latter are often very large, with their precise values depending upon the word length of the computer being used. Note that the x_i cannot be truly "random" as they arise from the seed in a well-defined, deterministic algorithm; indeed, two sequences grown from the the same seed are identical. For this reason, they are often termed "pseudo-random". Nevertheless, for many practical purposes pseudo-random numbers generated in this way can be used as if they were truly random. A good discussion of uniform random number generators and of the tests that can be applied to determine if they are working properly can be found in [Kn69].

We have already seen how to choose one-dimensional random variables distributed as a specified weight function $w(x)$. According to the discussion in the previous section, the procedure is to choose y, the incomplete integral of w, uniformly, and then to find x by inverting Eq. (8.6). Thus, if we are faced with evaluating the integral

$$I = \int_0^\infty dx\, e^{-x} g(x) \,,$$

with g a relatively smooth function, it is sensible to generate x between 0 and ∞ with distribution e^{-x}, and then to average g over these values. According to (8.5), we have

$$y = 1 - e^{-x} \; , \quad x = -\log(1 - y) \, .$$

The following FORTRAN subroutine generates values of X with the required distribution.

```
c chap8d.for
      SUBROUTINE LOGDST(X)
      INTEGER SEED
      DATA SEED /98347927/
c
      X=-LOG(1.-RAN(SEED))
      RETURN
      END
```

■ **Exercise 8.2** Verify that the code above generates X with the required distribution, for example by generating and histogramming a large number of values of X. Use these values to evaluate I and its uncertainty for $g(x) = x$, x^2, and x^3 and compare your results with the analytical values. Can you understand the trend in the uncertainties calculated?

While the method of generating the incomplete integral is infallible, it requires that we be able to find $x(y)$, which can be done analytically only for a relatively small class of functions. For example, if we had wanted to use

$$w(x) = \frac{6}{5}(1 - \frac{1}{2}x^2)$$

in our efforts to evaluate Eq. (8.4), we would have been faced with solving a cubic equation to find $y(x)$. While this is certainly possible, choices of w that might follow more closely the behavior of f generally will lead to more complicated integrals that cannot be inverted analytically. However, it is possible to do the integral and inversion numerically. Let us imagine tabulating the values $x^{(j)}$ for which the incomplete integral of w takes on a series of uniformly spaced values $y^{(j)} \equiv j/M$, $j = 0, 1, \ldots, M$ that span the interval $[0, 1]$. Thus,

$$y^{(j)} \equiv \frac{j}{M} = \int_0^{x^{(j)}} dx' w(x') \, . \tag{8.11}$$

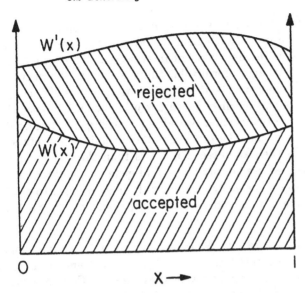

Figure 8.1 Illustration of the von Neumann rejection method for generating random numbers distributed according to a given distribution $w(x)$.

Values of $x^{(j)}$ with j an integer chosen from the set $0, 1, \ldots, M$ with equal probability will then approximate the required distribution. (Some special treatment is required to handle the end-points $j = 0$ and $j = M$ properly.) To generate the x^j, is, of course, the problem. This can be done by integrating the differential equation $dy/dx = w(x)$ through the simple discretization

$$\frac{y^{(j+1)} - y^{(j)}}{x^{(j+1)} - x^{(j)}} = w(x^{(j)}) .$$

Since $y^{(j+1)} - y^{(j)} = 1/M$, we have the convenient recursion relation

$$x^{(j+1)} = x^{(j)} + \frac{1}{M w(x^{(j)})} , \qquad (8.12)$$

which can be used with the starting value $x^{(0)} = 0$.

Another convenient method for generating one- (or multi-) dimensional random variables is von Neumann rejection, whose geometrical basis is illustrated in Figure 8.1. Suppose we are interested in generating x between 0 and 1 with distribution $w(x)$ and let $w'(x)$ be a positive function such that $w'(x) > w(x)$ over the region of integration. Note that this means that the definite integral of w' is greater than 1. A convenient, but not always useful, choice for w' is any constant greater than the maximum value of w in the region of integration. If we generate points in two dimensions that uniformly fill the area under the curve $w'(x)$ and then "accept" for use only those points that are under $w(x)$, then the accepted points will be distributed according to w. Practically, what is done is to choose two random variables, x_i and η, the former distributed proportional to w' and the latter distributed uniformly between 0 and 1. The value x_i is then accepted if η is less than the ratio $w(x_i)/w'(x_i)$; if a point is rejected, we simply go on and generate another pair of x_i and η. This technique is clearly efficient only if w' is close to w throughout the entire range of integration; otherwise, much time is wasted rejecting useless points.

■ **Exercise 8.3** Use von Neumann rejection to sample points in the interval $[0, 1]$ distributed as $w(x) = \frac{6}{5}(1 - \frac{1}{2}x^2)$ and evaluate the integral (8.4) and its uncertainty with these points. Compare the efficiency of your calculation for various choices of w'.

The Gaussian (or normal) distribution with zero mean and unit variance,

$$w(x) = (2\pi)^{-1/2} e^{-x^2/2} ,$$

plays a central role in probability theory and so is often needed in Monte Carlo calculations. It is possible, but not too efficient, to generate this distribution by inverting its incomplete integral using polynomial approximations to the error function. A more "clever" method is based on the central limit theorem, which states that the sum of a large number of uniformly distributed random numbers will approach a Gaussian distribution. Since the mean and variance of the uniform distribution on the interval $[0,1]$ are 1/2 and 1/12, respectively, the sum of 12 uniformly distributed numbers will have a mean value of 6 and a variance of 1 and will very closely approximate a Gaussian distribution. Hence, the following FORTRAN subroutine returns a Gaussian random variable, GAUSS, with zero mean and unit variance:

```
C chap8e.for
C subroutine to generate the normally distributed number GAUSS
      SUBROUTINE DIST(GAUSS)
      INTEGER SEED
      DATA SEED/39249187/
C
```

```
      GAUSS=0.
      DO 10 I=1,12
      GAUSS=GAUSS+RAN(SEED)
10    CONTINUE
      GAUSS=GAUSS-6.
      RETURN
      END
```

■ **Exercise 8.4** By histogramming a large number of the values of GAUSS produced by the subroutine above, convince yourself that their distribution is very close to normal. Can you derive a quantitative estimate of the extent to which this distribution deviates from a Gaussian? Compare with the results of summing 6 or 24 uniformly distributed random numbers. (Note that these latter sums have means of 3 and 12 and variances of $\frac{1}{2}$ and 2, respectively.)

Another efficient method for generating normally distributed variables is to consider a Gaussian distribution in two dimensions (x_1, x_2), for which the number of points in a differential area is proportional to

$$e^{-\frac{1}{2}(x_1^2 + x_2^2)} \, dx_1 \, dx_2 \ .$$

In terms of the usual polar coordinates

$$r = (x_1^2 + x_2^2)^{1/2} \ , \quad \theta = \tan^{-1} \frac{x_2}{x_1} \ ,$$

the distribution is

$$e^{-\frac{1}{2}r^2} r \, dr \, d\theta \ ,$$

or, if $u = \frac{1}{2}r^2$, the distribution is

$$e^{-u} \, du \, d\theta \ .$$

Hence, if we generate u between 0 and ∞ with an exponential distribution and θ uniformly between 0 and 2π, then the corresponding values of

$$x_1 = (2u)^{1/2} \cos \theta \ , \quad x_2 = (2u)^{1/2} \sin \theta$$

will be distributed normally. Thus, the following FORTRAN subroutine returns two normally distributed random variables, GAUSS1 and GAUSS2:

```
C chap8f.for
C returns 2 normally distributed numbers, GAUSS1 and GAUSS2
      SUBROUTINE DIST(GAUSS1,GAUSS2)
      INTEGER SEED
```

```
          DATA SEED/39249187/
          DATA PI/3.1415926/
    C

          TWOU=-2.*LOG(1.-RAN(SEED))
          RADIUS=SQRT(TWOU)
          THETA=2*PI*RAN(SEED)
          GAUSS1=RADIUS*COS(THETA)
          GAUSS2=RADIUS*SIN(THETA)
          RETURN
          END
```

To generate a Gaussian distribution with mean $\bar{x}$ and variance σ^2,

$$w(x) = (2\pi\sigma^2)^{-1/2} e^{-\frac{1}{2}(x-\bar{x})^2/\sigma^2} ,$$

we need only take the values generated by this subroutine, multiply them by σ, and then increment them by $\bar{x}$.

8.3 The algorithm of Metropolis *et al.*

Although the methods we have discussed above for generating random numbers according to a specified distribution can be very efficient, it is difficult or impossible to generalize them to sample a complicated weight function in many dimensions, and so an alternative approach is required. One very general way to produce random variables with a given probability distribution of arbitrary form is known as the Metropolis, Rosenbluth, Rosenbluth, Teller, and Teller algorithm [Me53]. As it requires only the ability to calculate the weight function for a given value of the integration variables, the algorithm has been applied widely in statistical mechanics problems, where the weight function of the canonical ensemble can be a very complicated function of the coordinates of the system [see Eq. (8.1)] and so cannot be sampled conveniently by other methods. However, it is not without its drawbacks.

Although the algorithm of Metropolis *et al.* can be implemented in a variety of ways, we begin by describing one simple realization. Suppose that we want to generate a set of points in a (possibly multi-dimensional) space of variables $\mathbf{X}$ distributed with probability density $w(\mathbf{X})$. The Metropolis algorithm generates a sequence of points, $\mathbf{X}_0$, $\mathbf{X}_1, \ldots$, as those visited successively by a random walker moving through $\mathbf{X}$ space; as the walk becomes longer and longer, the points it connects approximate more closely the desired distribution.

The rules by which the random walk proceeds through configuration space are as follows. Suppose that the walker is at a point $\mathbf{X}_n$ in the sequence. To generate X_{n+1}, it makes a trial step to a new point X_t. This new point can be chosen in any convenient manner, for example uniformly at random within a multi-dimensional cube of small side δ about X_n. This trial step is then "accepted" or "rejected" according to the ratio

$$r = \frac{w(X_t)}{w(X_n)} .$$

If r is larger than one, then the step is accepted (i.e., we put $X_{n+1} = X_t$), while if r is less than one, the step is accepted with probability r. This latter is conveniently accomplished by comparing r with a random number η uniformly distributed in the interval [0,1] and accepting the step if $\eta < r$. If the trial step is not accepted, then it is rejected, and we put $X_{n+1} = X_n$. This generates X_{n+1}, and we may then proceed to generate X_{n+2} by the same process, making a trial step from X_{n+1}. Any arbitrary point, X_0, can be used as the starting point for the random walk.

The following subroutine illustrates the application of the Metropolis algorithm to sample a two-dimensional distribution in the variables X1 and X2. Each call to the subroutine executes another step of the random walk and returns the next values of X1 and X2; the main program must initialize these variables, as well as set the value of DELTA and define the distribution WEIGHT(X1,X2).

```
C chap8g.for
C subroutine to take a step in the Metropolis algorithm
      SUBROUTINE METROP(X1,X2,WEIGHT,DELTA)
      EXTERNAL WEIGHT
      INTEGER SEED
      DATA SEED/39249187/
            !take a trial step in square about (X1,X2)
      X1T=X1+DELTA*(2*RAN(SEED)-1)
      X2T=X2+DELTA*(2*RAN(SEED)-1)
      RATIO=WEIGHT(X1T,X2T)/WEIGHT(X1,X2)
      IF (RATIO .GT. RAN(SEED)) THEN   !step accepted
         X1=X1T
         X2=X2T
      END IF
      RETURN
      END
```

This code could be made more efficient by saving the weight function at the current point of the random walk, so that it need not be computed again when deciding whether or not to accept the trial step; the evaluation of w is often the most time-consuming part of a Monte Carlo calculation using the Metropolis algorithm.

To prove that the algorithm described above does indeed generate a sequence of points distributed according to w, let us consider a large number of walkers starting from different initial points and moving independently through $\mathbf{X}$ space. If $N_n(\mathbf{X})$ is the density of these walkers at $\mathbf{X}$ after n steps, then the net number of walkers moving from point $\mathbf{X}$ to point $\mathbf{Y}$ in the next step is

$$\Delta N(\mathbf{X}) = N_n(\mathbf{X})P(\mathbf{X} \to \mathbf{Y}) - N_n(\mathbf{Y})P(\mathbf{Y} \to \mathbf{X})$$

$$= N_n(\mathbf{Y})P(\mathbf{X} \to \mathbf{Y})\left[\frac{N_n(\mathbf{X})}{N_n(\mathbf{Y})} - \frac{P(\mathbf{Y} \to \mathbf{X})}{P(\mathbf{X} \to \mathbf{Y})}\right]. \qquad (8.13)$$

Here, $P(\mathbf{X} \to \mathbf{Y})$ is the probability that a walker will make a transition to $\mathbf{Y}$ if it is at $\mathbf{X}$. This equation shows that there is equilibrium (no net change in population) when

$$\frac{N_n(\mathbf{X})}{N_n(\mathbf{Y})} = \frac{N_e(\mathbf{X})}{N_e(\mathbf{Y})} \equiv \frac{P(\mathbf{Y} \to \mathbf{X})}{P(\mathbf{X} \to \mathbf{Y})}, \qquad (8.14)$$

and that changes in $N(\mathbf{X})$ when the system is not in equilibrium tend to drive it toward equilibrium (i.e., $\Delta N(\mathbf{X})$ is positive if there are "too many" walkers at $\mathbf{X}$, or if $N_n(\mathbf{X})/N_n(\mathbf{Y})$ is greater than its equilibrium value). Hence it is plausible (and can be proved) that, after a large number of steps, the population of the walkers will settle down to its equilibrium distribution, N_e.

It remains to show that the transition probabilities of the Metropolis algorithm lead to an equilibrium distribution of walkers $N_e(\mathbf{X}) \sim w(\mathbf{X})$. The probability of making a step from $\mathbf{X}$ to $\mathbf{Y}$ is

$$P(\mathbf{X} \to \mathbf{Y}) = T(\mathbf{X} \to \mathbf{Y})A(\mathbf{X} \to \mathbf{Y}),$$

where T is the probability of making a trial step from $\mathbf{X}$ to $\mathbf{Y}$ and A is the probability of accepting that step. If $\mathbf{Y}$ can be reached from $\mathbf{X}$ in a single step (i.e., if it is within a cube of side δ centered about $\mathbf{X}$), then

$$T(\mathbf{X} \to \mathbf{Y}) = T(\mathbf{Y} \to \mathbf{X}),$$

so that the equilibrium distribution of the Metropolis random walkers satisfies

$$\frac{N_e(\mathbf{X})}{N_e(\mathbf{Y})} = \frac{A(\mathbf{Y} \to \mathbf{X})}{A(\mathbf{X} \to \mathbf{Y})}. \qquad (8.15)$$

If $w(\mathbf{X}) > w(\mathbf{Y})$, then $A(\mathbf{Y} \to \mathbf{X}) = 1$ and

$$A(\mathbf{X} \to \mathbf{Y}) = \frac{w(\mathbf{Y})}{w(\mathbf{X})} \, ,$$

while if $w(\mathbf{X}) < w(\mathbf{Y})$ then

$$A(\mathbf{Y} \to \mathbf{X}) = \frac{w(\mathbf{X})}{w(\mathbf{Y})}$$

and $A(\mathbf{X} \to \mathbf{Y}) = 1$. Hence, in either case, the equilibrium population of Metropolis walkers satisfies

$$\frac{N_e(\mathbf{X})}{N_e(\mathbf{Y})} = \frac{w(\mathbf{X})}{w(\mathbf{Y})} \, ,$$

so that the walkers are indeed distributed with the correct distribution.

Note that although we made the discussion concrete by choosing $\mathbf{X}_t$ in the neighborhood of $\mathbf{X}_n$, we can use any transition and acceptance rules that satisfy

$$\frac{w(\mathbf{X})}{w(\mathbf{Y})} = \frac{T(\mathbf{Y} \to \mathbf{X})A(\mathbf{Y} \to \mathbf{X})}{T(\mathbf{X} \to \mathbf{Y})A(\mathbf{X} \to \mathbf{Y})} \, . \qquad (8.16)$$

Indeed, one limiting choice is $T(\mathbf{X} \to \mathbf{Y}) = w(\mathbf{Y})$, independent of $\mathbf{X}$, and $A = 1$. This is the most efficient choice, as no trial steps are "wasted" through rejection. However, this choice is somewhat impractical, because if we knew how to sample w to take the trial step, we wouldn't need to use the algorithm to begin with.

An obvious question is "If trial steps are to be taken within a neighborhood of $\mathbf{X}_n$, how do we choose the step size, δ?" To answer this, suppose that $\mathbf{X}_n$ is at a maximum of w, the most likely place for it to be. If δ is large, then $w(\mathbf{X}_t)$ will likely be very much smaller than $w(\mathbf{X}_n)$ and most trial steps will be rejected, leading to an inefficient sampling of w. If δ is very small, most trial steps will be accepted, but the random walker will never move very far, and so also lead to a poor sampling of the distribution. A good rule of thumb is that the size of the trial step should be chosen so that about half of the trial steps are accepted.

One bane of applying the Metropolis algorithm to sample a distribution is that the points that make up the random walk, $\mathbf{X}_0, \mathbf{X}_1, \ldots$ are not independent of one another, simply from the way in which they were generated; that is, $\mathbf{X}_{n+1}$ is likely to be in the neighborhood of $\mathbf{X}_n$. Thus, while the points might be distributed properly as the walk becomes very long, they are not statistically independent of one another, and some care must be taken in using them to calculate integrals. For example, if we calculate

$$I = \frac{\int d\mathbf{X}\, w(\mathbf{X}) f(\mathbf{X})}{\int d\mathbf{X}\, w(\mathbf{X})}$$

by averaging the values of f over the points of the random walk, the usual estimate of the variance, Eq. (8.3), is invalid because the $f(\mathbf{X}_i)$ are not statistically independent. This can be quantified by calculating the auto-correlation function

$$C(k) = \frac{\langle f_i f_{i+k} \rangle - \langle f_i \rangle^2}{\langle f_i^2 \rangle - \langle f_i \rangle^2} \, . \tag{8.17}$$

Here, $\langle \ldots \rangle$ indicates average over the random walk; e.g.,

$$\langle f_i f_{i+k} \rangle = \frac{1}{N-k} \sum_{i=1}^{N-k} f(\mathbf{X}_i) f(\mathbf{X}_{i+k}) \, .$$

Of course, $C(0) = 1$, but the non-vanishing of C for $k \neq 0$ means that the f's are not independent. What can be done in practice is to compute the integral and its variance using points along the random walk separated by a fixed interval, the interval being chosen so that there is effectively no correlation between the points used. An appropriate sampling interval can be estimated from the value of k for which C becomes small (say $\lesssim 0.1$).

Another issue in applying the Metropolis algorithm is where to start the random walk; i.e., what to take for $\mathbf{X}_0$. In principle, any location is suitable and the results will be independent of this choice, as the walker will "thermalize" after some number of steps. In practice, an appropriate starting point is a probable one, where w is large. Some number of thermalization steps then can be taken before actual sampling begins to remove any dependence on the starting point.

■ **Exercise 8.5** Use the algorithm of Metropolis *et al.* to sample the normal distribution in one dimension. For various trial step sizes, study the acceptance ratio (fraction of trial steps accepted), the correlation function (and hence the appropriate sampling frequency), and the overall computational efficiency. Use the random variables you generate to calculate

$$\int_{-\infty}^{\infty} dx\, x^2 e^{-\frac{1}{2}x^2}$$

and estimate the uncertainty in your answer. Study how your results depend upon where the random walker is started and on how many thermalization steps you take before beginning the sampling. Compare the efficiency of the Metropolis algorithm with that of a calculation that uses one of the methods discussed in Section 8.2 to generate the normal distribution directly.

8.4 The Ising model in two dimensions

Models in which the degrees of freedom reside on a lattice and interact locally arise in several areas of condensed matter physics and field theory. The simplest of these is the Ising model [Hu63], which can be taken as a crude description of a magnetic material or a binary alloy. In this example, we will use Monte Carlo methods to calculate the thermodynamic properties of this model.

If we speak in the magnetic language, the Ising model consists of a set of spin degrees of freedom interacting with each other and with an external magnetic field. These might represent the magnetic moments of the atoms in a solid. We will consider in particular a model in two spatial dimensions, where the spin variables are located on the sites of an $N_x \times N_y$ square lattice. The spins can therefore be labeled as S_{ij}, where i, j are the indices for the two spatial directions, or as S_α, where α is a generic site label. Each of these spin variables can be either "up" ($S_\alpha = +1$) or "down" ($S_\alpha = -1$). This mimics the spin-1/2 situation, although note that we take the spins to be classical degrees of freedom and do not impose the angular momentum commutation rules characteristic of a quantum description. (Doing so would correspond to the Heisenberg model.)

The Hamiltonian for the system is conventionally written as

$$H = -J \sum_{\langle \alpha\beta \rangle} S_\alpha S_\beta - B \sum_\alpha S_\alpha. \qquad (8.18)$$

Here, the notation $\langle \alpha\beta \rangle$ means that the sum is over nearest-neighbor *pairs* of spins; these interact with a strength J (see Figure 8.2). Thus, the spin at site ij interacts with the spins at $i \pm 1j$ and $ij \pm 1$. (We assume periodic boundary conditions on the lattice, so that, for example, the lower neighbors of the spins with $i = N_x$ are those with $i = 1$ and the left-hand neighbors of those with $j = 1$ are those with $j = N_y$; the lattice therefore has the topology of a torus.) When J is positive, the energy is lower if a spin is in the same direction as its neighbors (ferromagnetism), while when J is negative, a spin will tend to be anti-aligned with its

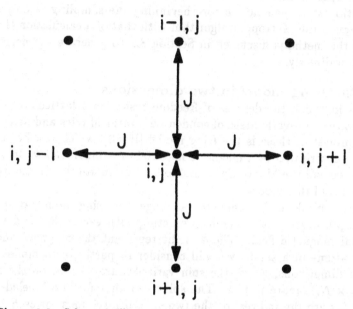

Figure 8.2   Schematic illustration of the two-dimensional Ising model.

neighbors (anti-ferromagnetism). The term involving B represents the interaction of the spins with an external magnetic field, which tends to align all spins in the same direction.

We will be interested in the thermodynamics of this system. In this case, it is convenient to measure the coupling energies J and B in units of the temperature, so that heating the system corresponds to decreasing these couplings. Configurations of the system are specified by giving the values of all $N_x \times N_y \equiv N_s$ spin variables and the weighting of any one of the 2^{N_s} spin configurations, S, in the canonical ensemble is

$$w(\mathbf{S}) = \frac{e^{-H(\mathbf{S})}}{Z} , \tag{8.19}$$

where the partition function is

$$Z(J, B) = \sum_{\mathbf{S}} e^{-H(\mathbf{S})} . \tag{8.20}$$

The thermodynamic quantities we will be interested are the magnetization

$$M = \frac{\partial \log Z}{\partial B} = \sum_{\mathbf{S}} w(\mathbf{S}) \left(\sum_{\alpha} S_{\alpha} \right) , \qquad (8.21a)$$

the susceptibility

$$\chi = \frac{\partial M}{\partial B} = \sum_{\mathbf{S}} w(\mathbf{S}) \left(\sum_{\alpha} S_{\alpha} \right)^2 - M^2 , \qquad (8.21b)$$

the energy

$$E = \sum_{\mathbf{S}} w(\mathbf{S}) H(\mathbf{S}) , \qquad (8.21c)$$

and the specific heat at constant field,

$$C_B = \sum_{\mathbf{S}} w(\mathbf{S}) H^2(\mathbf{S}) - E^2 . \qquad (8.21d)$$

In the limit of an infinitely large lattice ($N_{x,y} \to \infty$), it is possible to solve the Ising model exactly; discussions of the solution, originally due to Onsager, can be found in [Hu63] and [Mc73]. The expressions are simplest at $B = 0$. In this limit, the energy is given by

$$E = -N_s J (\coth 2J) \left[1 + \frac{2}{\pi} \kappa' K_1(\kappa) \right] , \qquad (8.22a)$$

and the specific heat is

$$C_B = N_s \frac{2}{\pi} (J \coth 2J)^2 \left(2K_1(\kappa) - 2E_1(\kappa) - (1 - \kappa') \left[\frac{\pi}{2} + \kappa' K_1(\kappa) \right] \right) , \qquad (8.22b)$$

while the magnetization is given by

$$M = \pm N_s \frac{(1 + z^2)^{1/4} (1 - 6z^2 + z^4)^{1/8}}{(1 - z^2)^{1/2}} \qquad (8.22c)$$

for $J > J_c$ and vanishes for $J < J_c$. In these expressions,

$$\kappa = 2 \frac{\sinh 2J}{\cosh^2 2J} \leq 1 , \quad \kappa' = 2 \tanh^2 2J - 1 ,$$

the complete elliptic integrals of the first and second kinds are

$$K_1(\kappa) \equiv \int_0^{\pi/2} \frac{d\phi}{(1 - \kappa^2 \sin^2 \phi)^{1/2}} , \quad E_1(\kappa) \equiv \int_0^{\pi/2} d\phi \, (1 - \kappa^2 \sin^2 \phi)^{1/2} ,$$

$z = e^{-2J}$, and $J_c = 0.4406868$ is the critical value of J for which $\kappa = 1$, where K_1 has a logarithmic singularity. Thus, all thermodynamic functions are singular at this coupling, strongly suggesting a phase transition. This is confirmed by the behavior of the magnetization, which vanishes below the critical coupling (or above the critical temperature), and can take on one of two equal and opposite values above this coupling.

A numerical solution of the Ising model is useful both as an illustration of the techniques we have been discussing and because it can be generalized readily to more complicated Hamiltonians [Fo63]. Because of the large number of terms involved, a direct evaluation of the sums in Eqs. (8.21) is out of the question. (For even a modest 16 × 16 lattice, there are $2^{256} \approx 10^{77}$ different configurations.) Hence, it is most efficient to generate spin configurations S with probability $w(\mathbf{S})$ using the Metropolis algorithm and then to average the required observables over these configurations. To implement the Metropolis algorithm, we could make our trial step from S to $\mathbf{S}_t$ by changing all of the spins randomly. This would, however, bring us to a configuration very different from S, and so there would be a high probability of rejection. It is therefore better to take smaller steps, and so we consider trial configurations that differ from the previous one only by the flipping of one spin. This is done by sweeping systematically through the lattice and considering whether or not to flip each spin, one at a time. Hence, we consider two configurations, S and $\mathbf{S}_t$, differing only by the flipping of one spin, $S_\alpha \equiv S_{ij}$. Acceptance of this trial step depends upon the ratio of the weight functions,

$$r = \frac{w(\mathbf{S}_t)}{w(\mathbf{S})} = e^{-H(\mathbf{S}_t)+H(\mathbf{S})} \ .$$

Specifically, if $r > 1$ or if $r < 1$ but larger than a uniformly distributed random number between 0 and 1, then the spin S_α is flipped; otherwise, it is not. From (8.18), it is clear that only terms involving S_{ij} will contribute to r, so that after some algebra, we have

$$r = e^{-2S_\alpha(Jf+B)} \ ; \quad f = S_{i+1j} + S_{i-1j} + S_{ij+1} + S_{ij-1} \ .$$

Here, f is the sum of the four spins neighboring the one being flipped. Because f can take on only 5 different values, 0, ±2, ±4, only 10 different values of r can ever arise (there are two possible values of S_α); these can be conveniently calculated and stored in a table before the calculation begins so that exponentials need not be calculated repeatedly. Note that if we had used trial configurations that involved flipping several spins, the calculation of r would have been much more complicated.

The program for Example 8, whose source listing is given in Appendix B and in the file EXMPL8.FOR, performs a Monte Carlo simulation of the Ising model using the algorithm just described. An initially random configuration of spins is used to start the Metropolis random walk and the lattice is shown after every NFREQ sweep, if requested. Thermalization sweeps (no calculation of the observables) are allowed for; these permit the random walk to "settle down" before observables are accumulated. Values of the energy, magnetization, susceptibility, and specific heat per spin are displayed as the calculation proceeds, as is the fraction of the trial steps accepted.

One feature of this program requires further explanation. This is a simple technique used to monitor the sweep-to-sweep correlations in the observables inherent in the Metropolis algorithm. The basic observables (energy and magnetization) are computed every NFREQ sweeps. These values are then binned into "groups" with NSIZE members. For each group, the means and standard deviations of the energy and magnetization are calculated. As more groups are generated, their means are combined into a grand average. One way to compute the uncertainty in this grand average is to treat the group means as independent measurements and to use Eq. (8.3). Alternatively, the uncertainty can be obtained by averaging the standard deviations of the groups in quadrature. If the sampling frequency is sufficiently large, these two estimates will agree. However, if the sampling frequency is too small and there are significant correlations in the successive measurements, the values within each group will be too narrowly distributed, and the second estimate of the uncertainty will be considerably smaller than the first. These two estimates of the uncertainty for the grand average of the energy and magnetization per spin are therefore displayed. Note that this technique is not so easily implemented for the specific heat and susceptibility, as they are themselves fluctuations in the energy and magnetization [Eqs. (8.21b,d)], and so only the uncertainties in their grand averages computed by the first method are displayed.

The following exercises will be useful in better understanding this example:

■ **Exercise 8.6** When $J = 0$, the Hamiltonian (8.18) reduces to that for independent spins in an external magnetic field, a problem soluble by elementary means. For this case, obtain analytical expressions for the thermodynamic observables and verify that these are reproduced by the program.

■ **Exercise 8.7** Use Eqs. (8.22) to calculate and graph the exact energy, specific heat, and magnetization per spin for an infinite lattice at $B = 0$ and for J running from 0 to 0.8.

■ **Exercise 8.8** Modify the code to compute the sweep-to-sweep correlation functions for the energy and the magnetization using (8.17). Using runs for a 16×16 lattice for $B = 0$ and for several values of J between 0.1 and 0.6, estimate the proper sampling frequency at each coupling strength. Show that the two estimates of the uncertainties in the energy and magnetization agree when a proper sampling frequency is used and that they disagree when the samples are taken too often (a reasonable group size is 10). Also show that the sweep-to-sweep correlations become stronger when the system is close to the phase transition (critical slowing down).

■ **Exercise 8.9** Run the code to obtain results for 8×8, 16×16, and 32×32 lattices at $B = 0$ for a sequence of ferromagnetic couplings from 0.1 to 0.6; pay particular attention to the region near the expected phase transition. Compare your results with the exact behavior of the infinite lattice and show that the finite size smooths out the singularities in the thermodynamic observables. Notice that the size of the magnetic domains becomes very large near the critical coupling.

■ **Exercise 8.10** Use the code to explore the thermodynamics of the model for finite B and for anti-ferromagnetic couplings ($J < 0$). Also consider simulations of a model in which a given spin S_{ij} interacts with its neighbors ferromagnetically and with its diagonal neighbors S_{i-1j-1}, S_{i+1j-1}, S_{i-1j+1}, S_{i+1j+1} anti-ferromagnetically.

■ **Exercise 8.11** The "heat bath" algorithm is an alternative to the Metropolis algorithm for sampling the canonical ensemble. In this method, the particular spin being considered is set to -1 with probability $1/(1+g)$ and to $+1$ with probability $g/(1+g)$, where $g = \exp[2(Jf + B)]$. This can be interpreted as placing the spin in equilibrium with a heat bath at the specified temperature. Verify that this algorithm corresponds to putting $A = 1$ and taking

$$T(\mathbf{S} \to \mathbf{S'}) = \frac{w(\mathbf{S'})}{w(\mathbf{S'}) + w(\mathbf{S})}$$

in Eq. (8.16) and so leads to a correct sampling of spin configurations. Modify the code to use the heat-bath algorithm and compare its efficiency with that of the conventional Metropolis algorithm.

Project VIII: Quantum Monte Carlo for the H_2 molecule

In this project, we will consider Monte Carlo methods that can be used to calculate the *exact* properties of the ground states of quantum many-body systems. These methods are based on the formal similarity between the Schroedinger equation in imaginary time and a multi-dimensional diffusion equation (recall Section 7.4). Since the latter can be handled by Monte Carlo methods (ordinary diffusion arises as the result of many random microscopic collisions of the diffusing particles), these same techniques can also be applied to the Schroedinger equation. However, the Monte Carlo method is no panacea: it can make exact statements about only the ground state energy, it requires an already well-chosen variational wave function for the ground state, and it is generally intractable for fermion systems. However, there have been successful applications of these techniques to liquid ^{4}He [Ka81], the electron gas [Ce80], small molecules [Re82], and lattice gauge theories [Ch84]. More detailed discussions can be found in these references, as well as in [Ce79], which contains a good general overview.

VIII.1 Statement of the problem

The specific problem we will treat is the structure of the H_2 molecule: two protons bound by two electrons. This will be done within the context of the accurate Born-Oppenheimer approximation, which is based on the notion that the heavy protons move slowly compared to the much lighter electrons. The potential governing the protons' motion at a separation S, $U(S)$, is then the sum of the inter-proton electrostatic repulsion and the eigenvalue, $E_0(S)$, of the two-electron Schroedinger equation:

$$U(S) = \frac{e^2}{S} + E_0(S) . \qquad (VIII.1)$$

The electronic eigenvalue is determined by the Schroedinger equation

$$H(S)\Psi_0(\mathbf{r}_1,\mathbf{r}_2;S) \equiv [K + V(S)]\Psi_0 = E_0(S)\Psi_0(\mathbf{r}_1,\mathbf{r}_2;S) . \qquad (VIII.2)$$

Here, the electronic wave function, Ψ_0, is a function of the space coordinates of the two electrons, $\mathbf{r}_{1,2}$ and depends parametrically upon the inter-proton separation. If we are interested in the electronic ground state of the molecule and are willing to neglect small interactions involving the electrons' spin, then we can assume that the electrons are in an anti-symmetric spin-singlet state; the Pauli principle then requires that Ψ_0 be symmetric under the interchange of $\mathbf{r}_1$ and $\mathbf{r}_2$. Thus, even though the electrons are two fermions, the spatial equation satisfied by their wave

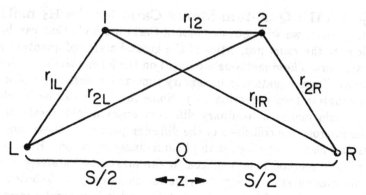

Figure VIII.1   Coordinates used in describing the H_2 molecule.

function is analogous to that for two bosons; the ground state wave function Ψ_0 will therefore have no nodes, and can be chosen to be positive everywhere.

The electron kinetic energy appearing in (VIII.2) is

$$K = -\frac{\hbar^2}{2m}(\nabla_1^2 + \nabla_2^2) \,, \qquad (VIII.3a)$$

where m is the electron mass, while the potential V involves the attraction of the electrons by each nucleus and the inter-electron repulsion:

$$V = -e^2\left[\frac{1}{r_{1L}} + \frac{1}{r_{2L}} + \frac{1}{r_{1R}} + \frac{1}{r_{2R}}\right] + \frac{e^2}{r_{12}} \,. \qquad (VIII.3b)$$

If we place the protons at locations $\pm\frac{1}{2}S$ on the z-axis, then the distance between electron 1 and the left or right proton is

$$r_{1L,R} = |\mathbf{r}_1 \pm \frac{1}{2}S\hat{\mathbf{z}}| \,,$$

and the distances between electron 2 and the protons, $r_{2L,R}$ are given similarly. The interelectron distance is $r_{12} = |\mathbf{r}_1 - \mathbf{r}_2|$. (See Figure VIII.1.)

Our goal in this project is therefore to solve the six-dimensional partial differential eigenvalue equation (VIII.2) for the lowest eigenvalue E_0 at each S, and so trace out the potential $U(S)$ via Eq. (VIII.1). This potential should look much like the 6–12 or Morse potentials studied in Chapter 1; the depth and location of the minimum and the curvature of U about this minimum are related to observable properties of the H_2 spectrum. We will also be able to calculate the exact ground state energies of two-electron atoms, $E_0(S = 0)$. (These same systems were treated in the Hartree-Fock approximation in Project III.)

VIII.2 Variational Monte Carlo and the trial wave function

We begin by discussing a variational Monte Carlo solution to the eigenvalue problem. If $\Phi(\mathbf{r})$ is any trial wave function (which we can choose to be real) not orthogonal to the exact ground state Ψ_0, then an upper bound to the electronic eigenvalue is the variational energy

$$E_v = \frac{\langle \Phi | H | \Phi \rangle}{\langle \Phi | \Phi \rangle} \qquad (VIII.4a)$$

$$= \frac{\int d\mathbf{r}\, \Phi^2(\mathbf{r}) \left(\frac{1}{\Phi} H \Phi(\mathbf{r}) \right)}{\int d\mathbf{r}\, \Phi^2(\mathbf{r})} \,, \qquad (VIII.4b)$$

where we have used $\mathbf{r}$ as a shorthand notation for the six coordinates of the problem, $\mathbf{r}_{1,2}$, and have written (VIII.4b) in a somewhat unconventional way. Note that this last equation can be interpreted as giving the variational energy as the average of a "local energy",

$$\epsilon(\mathbf{r}) \equiv \Phi^{-1} H \Phi(\mathbf{r}) = \Phi^{-1} K \Phi(\mathbf{r}) + V(\mathbf{r})$$

$$= -\frac{\hbar^2}{2m} \Phi^{-1} \sum_{i=1,2} \nabla_i^2 \Phi + V(\mathbf{r}) \,, \qquad (VIII.5)$$

over all of the six-dimensional $\mathbf{r}$-space with weighting $w(\mathbf{r}) \sim \Phi^2(\mathbf{r})$. Hence, to evaluate the variational energy by a Monte Carlo quadrature, all we need do is generate configurations (values of $\mathbf{r}_{1,2}$) distributed according to Φ^2 (by the Metropolis algorithm, for example), and then average ϵ over these configurations. It should be clear that this method can be generalized readily to systems involving a larger number of coordinates.

The choice of Φ is constrained by the requirements that it be simple enough to allow a convenient evaluation of Φ^2 and ϵ, yet also be a good approximation to the true ground state, Ψ_0. Indeed, if we are fortunate enough to choose a Φ that is the exact solution, then ϵ is independent of $\mathbf{r}$, so that the Monte Carlo quadrature gives the exact energy with zero variance. Thus, not only will the variational bound become better as Φ better approximates Ψ, but the variance in the Monte Carlo quadrature will become smaller. The trial wave function should also be symmetric under the interchange of the two electrons' spatial coordinates, but it need not be normalized properly, as the normalization cancels in evaluating (VIII.4).

One plausible choice for the trial wave function is a correlated product of molecular orbitals:

$$\Phi(\mathbf{r}_1, \mathbf{r}_2) = \phi(\mathbf{r}_1)\phi(\mathbf{r}_2)f(r_{12}) \,. \qquad (VIII.6a)$$

Here, the first two factors are an independent-particle wave function placing each electron in a molecular orbital in which it is shared equally between the two protons. A simple choice for the molecular orbital is the symmetric linear combination of atomic orbitals centered about each proton,

$$\phi(\mathbf{r}_i) = e^{-r_{iL}/a} + e^{-r_{iR}/a} \, , \qquad (VIII.6b)$$

with the variational parameter a to be determined below. The final factor in the trial wave function, f, expresses the correlation between the two electrons due to their Coulomb repulsion. That is, we expect f to be small when r_{12} is small and to approach a large constant value as the electrons become well separated. A convenient and reasonable choice is

$$f(r) = \exp\left[\frac{r}{\alpha(1 + \beta r)}\right] \, , \qquad (VIII.6c)$$

where α and β are additional positive variational parameters. Note that β controls the distance over which the trial wave function "heals" to its uncorrelated value as the two electrons separate.

The singularity of the Coulomb potential at short distances places additional constraints on the trial wave function. If one of the electrons (say 1) approaches one of the nuclei (say the left one) while the other electron remains fixed, then the potential term in ϵ becomes large and negative, since r_{1L} becomes small. This must be cancelled by a corresponding positive divergence in the kinetic energy term if we are to keep ϵ smooth and have a small variance in the Monte Carlo quadrature. Thus, the trial wave function should have a "cusp" at $r_{1L} = 0$, which means that the molecular orbital should satisfy

$$\lim_{r_{1L} \to 0}\left[-\frac{\hbar^2}{2m}\frac{1}{\phi(\mathbf{r}_{1L})}\nabla_1^2\phi(\mathbf{r}_{1L}) - \frac{e^2}{r_{1L}}\right] = \text{finite terms.} \qquad (VIII.7)$$

Similar conditions must also be satisfied whenever any one of the distances r_{1R}, $r_{2R,L}$, or r_{12} vanishes. Using (VIII.6) and a bit of algebra, it is easy to see that these constraints imply that a satisfies the transcendental equation

$$a = \frac{a_0}{(1 + e^{-S/a})} \qquad (VIII.8)$$

and that $\alpha = 2a_0$, where $a_0 = \hbar^2/me^2$ is the Bohr radius. Thus, β is the only variational parameter at our disposal. Note that these Coulomb cusp conditions would not have been as important if we were to use some other quadrature method to evaluate the variational energy, as then each of the divergences in ϵ would be cancelled by a geometrical r^2 weighting in the associated distance.

With the trial wave function specified by Eqs. (VIII.6), explicit expressions can be worked out for $\epsilon(\mathbf{r})$ in terms of the values and derivatives of ϕ and f. Note that this form of the trial function is not appropriate for very large proton separations, as it contains a finite amplitude to find the two electrons close to the same proton.

VIII.3 Monte Carlo evaluation of the exact energy

We now turn to a Monte Carlo method for evaluating the *exact* electronic eigenvalue, E_0. It is based upon evolution of the imaginary-time Schroedinger equation to refine the trial wave function to the exact ground state. In particular, the latter can be obtained by applying the operator $\exp(-Ht/\hbar)$ to the trial state Φ and considering the long-time limit $t \to \infty$. (See the discussion concerning eigenvalues of elliptic operators given in Section 7.4 for further details.) Thus, we define

$$\Psi(\mathbf{r},t) = \exp\left[\int_0^t E_n(t')\,dt'/\hbar\right] e^{-Ht/\hbar}\Phi(\mathbf{r})\,, \qquad (VIII.9)$$

where $E_n(t')$ is an as-yet-undetermined c-number function. Note that as long as $\langle \Psi_0|\Phi\rangle \neq 0$, $\Psi(t)$ will approach the (un-normalized) exact ground state Ψ_0 as t becomes large. Our method (Path Integral Monte Carlo) is equivalent to evaluating the path integral representation of (VIII.9) numerically. An alternative method (Green's Function Monte Carlo, [Ce79]) would be to filter with the inverse of the shifted Hamiltonian, instead of the exponential in (VIII.9).

To compute the exact ground state energy E_0, we consider a slight generalization of the variational energy (VIII.4) using the hermiticity of H:

$$E(t) = \frac{\langle\Phi|H|\Psi(t)\rangle}{\langle\Phi|\Psi(t)\rangle} = \frac{\int d\mathbf{r}\,\Phi(\mathbf{r})\Psi(\mathbf{r},t)\epsilon(\mathbf{r})}{\int d\mathbf{r}\,\Phi(\mathbf{r})\Psi(\mathbf{r},t)}\,. \qquad (VIII.10)$$

It should be clear that $E(t=0) = E_v$, the variational energy associated with Φ, that $E(t \to \infty) = E_0$, and that $E(t)$ is independent of the function $E_n(t')$.

Equation (VIII.10) expresses the exact energy in a form suitable for Monte Carlo evaluation. To see this, we define $G(\mathbf{r},t) = \Phi(\mathbf{r})\Psi(\mathbf{r},t)$, so that (VIII.10) becomes

$$E(t) = \frac{\int d\mathbf{r}\,G(\mathbf{r},t)\epsilon(\mathbf{r})}{\int d\mathbf{r}\,G(\mathbf{r},t)}\,. \qquad (VIII.11)$$

The exact energy $E(t)$ is thus the average of the local energy ϵ over the distribution $G(\mathbf{r}, t)$. (Note that for the problem we are considering, G is positive definite since neither Φ nor Ψ has any nodes.) A Monte Carlo evaluation of E therefore requires an "ensemble" of N configurations $\{\mathbf{r}_1, \ldots, \mathbf{r}_N\}$ distributed according to $G(\mathbf{r}, t)$, through which $E(t)$ can be estimated by

$$E(t) \approx \frac{1}{N} \sum_{i=1}^{N} \epsilon(\mathbf{r}_i) , \qquad (VIII.12)$$

with an expression for the variance in analogy with (8.3). Note that if $\Phi(\mathbf{r})$ is in fact the *exact* ground state Ψ_0, then $\epsilon(\mathbf{r}) = E_0$, independent of $\mathbf{r}$, and so $E(t) = E_0$ with zero variance.

Of course, the expressions above are of no practical use without a way to generate the ensemble of configurations. At $t = 0, G(\mathbf{r}, t) = |\Phi(\mathbf{r})|^2$, so that a method like that of Metropolis *et al.* can be used to generate the initial ensemble, typically having $N \approx 30$ members. To evolve the ensemble in time, note that since

$$\hbar \frac{\partial \Psi}{\partial t} = (E_n - H)\Psi ,$$

G satisfies the evolution equation

$$\frac{\partial G}{\partial t} = \hbar^{-1} \left(E_n(t) - \Phi(\mathbf{r}) H \frac{1}{\Phi(\mathbf{r})} \right) G(\mathbf{r}, t) \qquad (VIII.13a)$$

$$= \frac{\hbar}{2m} \frac{\partial^2 G(\mathbf{r}, t)}{\partial \mathbf{r}^2} - \frac{\partial}{\partial \mathbf{r}} \cdot [\mathbf{D}(\mathbf{r}) G(\mathbf{r}, t)]$$
$$- \hbar^{-1} [\epsilon(\mathbf{r}) - E_n(t)] G(\mathbf{r}, t) . \qquad (VIII.13b)$$

We have used here an obvious notation for the spatial derivatives.

Equation (VIII.13b) can be interpreted as a diffusion equation for G, with a drift function

$$\mathbf{D}(\mathbf{r}) = \frac{\hbar}{m} \frac{1}{\Phi(\mathbf{r})} \frac{\partial \Phi(\mathbf{r})}{\partial \mathbf{r}} = \frac{\hbar}{m} \frac{\partial \log \Phi}{\partial \mathbf{r}} . \qquad (VIII.14)$$

It shows that the kinetic energy acts to diffuse G, that D tends to keep G confined to regions where Φ is large, and that the "source" increases G where $\epsilon(\mathbf{r})$ is smallest.

The evolution of G over a short time from t to $t + \Delta t$ can be represented through order Δt by the integral kernel

$$G(\mathbf{r}, t + \Delta t) = \int d\mathbf{r}' \, P(\mathbf{r}, \mathbf{r}'; \Delta t) G(\mathbf{r}', t) , \qquad (VIII.15)$$

where

$$P(\mathbf{r}, \mathbf{r}'; \Delta t) = \exp\left\{-[\epsilon(\mathbf{r}) - E_n(t)]\Delta t/\hbar\right\}$$

$$\times \left(\frac{m}{2\pi\hbar\Delta t}\right)^3 \exp\left\{\frac{-[\mathbf{r} - \mathbf{r}' - \mathbf{D}(\mathbf{r}')\Delta t]^2}{2\hbar\Delta t/m}\right\} \qquad (VIII.16)$$

This kernel (which is positive definite) can be interpreted as the conditional probability for a configuration to evolve from $\mathbf{r}'$ at time t to $\mathbf{r}$ at time $t + \Delta t$. This probability contains a factor associated with the kinetic energy, which acts to diffuse the system about $\mathbf{r}'$ through a normalized Gaussian probability distribution with variance $\hbar\Delta t/m$ and mean $\mathbf{D}\Delta t$. The other factor in P is associated with the local energy, which acts to keep the system in regions of space where ϵ is most negative by enhancing the probability for jumps to occur to such locations. The quantum mechanical structure of the exact ground state is thus determined by the balance between these two competing tendencies.

The algorithm for evolving the ensemble should now be evident. A configuration at time t at the point $\mathbf{r}'$ generates a contribution to $G(\mathbf{r}, t + \Delta t)$ equal to $P(\mathbf{r}, \mathbf{r}'; \Delta t)$. This is realized by placing in the new ensemble a configuration $\mathbf{r}$ chosen according to the distribution

$$\exp\left\{\frac{-[\mathbf{r} - \mathbf{r}' - \mathbf{D}(\mathbf{r})\Delta t]^2}{2\hbar\Delta t/m}\right\},$$

and then weighting the importance of this configuration by

$$\exp\left\{-[\epsilon(\mathbf{r}) - E_n(t)]\Delta t/\hbar\right\}.$$

One way of effecting this weighting in practice is to replicate or delete the configuration in the new ensemble with probabilities given by this latter function. In this case, N fluctuates from time step to time step but can be held roughly constant by continuous adjustment of E_n. Indeed, to keep $\int d\mathbf{r}\, G(\mathbf{r}, t)$ (and hence N) constant, $E_n(t)$ should be equal to $E(t)$, so that E_n also furnishes an estimate of $E(t)$. An alternative way of effecting the weighting is to assign an *importance* (i.e., a weight W_i) to each configuration in the ensemble when averaging ϵ to compute the energy. That is, the summand in (VIII.12) is modified to $W_i\epsilon(\mathbf{r}_i)$. At each time step, W_i for each configuration is multiplied by the first factor in (VIII.16), with E_n readjusted to keep average weight of each configuration in the ensemble equal to 1. This latter method requires less bookkeeping than does that of replicating and deleting, but can be inefficient, as a configuration that happens to acquire a very small weight is still followed during the evolution.

In summary, the method is as follows. The system is described by an ensemble of configurations, each having a relative weight in describing the properties of the system and initially distributed in r according to the trial function $|\Phi(\mathbf{r})|^2$. The overall efficiency of the method is closely related to the accuracy of this trial function. To evolve the ensemble in time, each member is moved in r with a shifted Gaussian probability function [the second factor in Eq. (VIII.16)] and its weight is multiplied by the first factor in Eq. (VIII.16). The quantity $E_n(t)$, which is adjusted after each time step to keep the average weight of the ensemble equal to 1, provides an estimate of the energy, as does the weighted average of $\epsilon(\mathbf{r})$ over the ensemble at any time. Furthermore, once the total evolution time is sufficiently large to filter the trial wave function, continued evolution generates independent ensembles distributed according to the exact ground state wave function, which allows the statistics to be improved to any required accuracy.

Note that since the ensemble moves through configuration space at a rate determined by Δt, which must be sufficiently small so that Eqs. (VIII.15,16) are accurate, the estimates of the energy at successive time steps will not be statistically independent. Therefore, in forming averages and computing variances, care must be taken to use ensembles only at intervals of t sufficiently large that the values are uncorrelated. Such intervals are conveniently determined by examining the autocorrelation functions of the estimate. Alternatively, one can use the method of binning the values into groups, as discussed in connection with the Ising calculation in Section 8.4. It should also be noted that the finite time step requires that calculations be done for several different values of Δt and the results extrapolated to the $\Delta t = 0$ limit. A useful way to set a scale for Δt is to note that the average step the Gaussian distribution in (VIII.16) is roughly $(\hbar \Delta t/m)^{1/2}$; this must be small compared to the length scales in the wave function.

Expectation values of observables other than the energy in the exact ground state are not simply given by this method. To see this, consider the ground state expectation value of some observable A that does not commute with the Hamiltonian. Then

$$\langle \Psi_0 | A | \Psi_0 \rangle \equiv \lim_{t \to \infty} \frac{\langle \Psi(t) | A | \Psi(t) \rangle}{\langle \Psi(t) | \Psi(t) \rangle}$$

$$= \frac{\int d\mathbf{r} \, \Psi^2(\mathbf{r}, t) \Psi^{-1} A \Psi(\mathbf{r}, t)}{\int d\mathbf{r} \, \Psi^2(\mathbf{r}, t)} . \qquad (VIII.17)$$

Evaluation of this integral requires an ensemble of configurations distributed as $|\Psi|^2$. However, the diffusion process described above generates an ensemble distributed as $G = \Phi\Psi$, which is not what is required. Although this ensemble can be used, through a rather complicated algorithm, to calculate exact ground state expectation values, a good estimate can be had by using it directly to calculate the first term in

$$\langle \Psi_0|A|\Psi_0 \rangle \approx \lim_{t\to\infty} 2\frac{\langle \Phi|A|\Psi(t) \rangle}{\langle \Phi|\Psi(t) \rangle} - \frac{\langle \Phi|A|\Phi \rangle}{\langle \Phi|\Phi \rangle} , \qquad (VIII.18)$$

the second term being evaluated easily with an ensemble distributed according to Φ^2. Thus, this expression provides a way of perturbatively correcting the expectation value in the trial state; some algebra shows that it is accurate to second order in the error in the trial function, $\Psi_0 - \Phi$.

VIII.4 Solving the problem
The Monte Carlo methods described above can be applied to solve for the properties of the H_2 molecule through the following sequence of steps.

■ **Step 1** Verify that Eq. (VIII.8) and the choice $\alpha = 2a_0$ imply that the wave function (VIII.6) satisfies the Coulomb cusp condition. Derive explicit analytical expressions for $\epsilon(r)$ [Eq. (VIII.5)] and $D(r)$ [Eq. (VIII.14)] for the wave function (VIII.6).

■ **Step 2** Write a code that uses Monte Carlo quadrature to calculate the variational energy estimate associated with the trial wave function (VIII.6). You will need subroutines to evaluate Φ^2 and ϵ for a given configuration. Sample Φ^2 using the Metropolis algorithm. Study the auto-correlation function of ϵ to determine the minimum acceptable sampling frequency along the random walk.

■ **Step 3** For various inter-proton separations, S, find the parameter β that minimizes the electron eigenvalue and so determine the variational potential for the H_2 molecule. Verify that the uncertainties in your results behave with β as expected. Your value at $S = 0$ can be compared with the variational results obtained in Project III using scaled hydrogenic and full Hartree-Fock wave functions.

■ **Step 4** Verify the validity of Eqs. (VIII.15,16) by making a Taylor expansion of $G(r',t)$ about r and then doing the resulting Gaussian r' integrals. (You will also have to make the approximation $D(r') \approx D(r)$, which is accurate to $O(\Delta t)$.)

■ **Step 5** Test the Path Integral Monte Carlo scheme for finding the exact ground state energy of a Hamiltonian on the simple situation of a particle in a one-dimensional harmonic-oscillator potential. Write a program that assumes a Gaussian trial function with an incorrect width and verify that the time evolution refines the energy in the expected way. An ensemble of 20–30 members should be sufficient. Investigate how the quality of your results varies with the error in the trial wave function and with the time step you use.

■ **Step 6** Combine the programs you wrote in Steps 2 and 5 into one that determines the exact eigenvalue of the ground state of the two-electron problem and so determine the exact H_2 molecular potential at various separations. In particular, determine the location of the minimum in the potential and compare with the empirical values given in Section 1.4. Use the trial functions determined in Step 2 and verify that your exact energies are always lower than the variational values. Also verify that your exact results are independent of the precise choice of β and that they extrapolate smoothly as a function of Δt. Determine the binding energy of the He atom by considering $S = 0$ and compare with the exact value given in the discussion of Project III.

Appendix A

How to Use
the Programs

This appendix contains the information needed to execute the Computational Physics codes. To follow the instructions below, you must be familiar with your computer's operating system, FORTRAN compiler, linker, editor, and graphics package. If you are unfamiliar with any of these, consult your computer system documentation before proceeding. You may also find it useful to have a FORTRAN language reference (e.g., [Ko87]) handy.

A.1 Installation
Before anything else can be done, you must install the codes by downloading them from diskette to your computer or transferring them over the Internet network. Network transfer is much quicker and free (see Appendix E for details), but requires software and network connections that are not universally available. The other modes of distribution are IBM-PC or Macintosh formatted diskettes; an order form for these is in the back of this book.

For those users who purchase the diskettes and are planning to run the codes on an IBM-PC or Mac the installation procedure is simply to copy the codes into a directory on the hard drive. However, if you are not planning to run the codes on an IBM compatible or Mac, then downloading may not be simple and will depend on the details of your hardware. The general scheme is to find a PC or Mac connected via modem or network to your machine, and then to use a file transfer program (e.g., KERMIT) to transfer the *Computational Physics* files. You might even find it necessary to go through one or more intermediate machines to get to yours. If you are inexperienced with file transfer, consult a local expert. Once you download the files on your machine, we suggest you back them up *immediately* on media that can be read directly by that machine.

231

A.2 Files

There are five categories of files included on the *Computational Physics* diskette: text codes, common utility programs, physics programs, data files, and include files. These were written in accordance with FORTRAN-77 standards (with two exceptions noted below) and therefore should run on any machine with a FORTRAN-77 compiler. The listings for the examples are in Appendix B, the projects in Appendix C, and the common utility codes in Appendix D.

1) The text codes are the short codes that appear in the text. They are named CHAPnx.FOR, where 'n' indicates the chapter number, and 'x' is a letter to differentiate codes within the same chapter. Due to lack of space on the diskette, only the longer of these codes are included; the others can be entered from your keyboard. These require no editing to run, with the exception of those in Chapter 8 which require a random number generator that is non-standard FORTRAN-77. If your compiler does not provide one, use the function RANNOS, which is in file UTIL.FOR.

2) The first five common utility FORTRAN codes are called by, and therefore must be linked with, each of the physics programs. UTIL.FOR contains the routines for menuing, I/O, file opening/closing, etc. SETUP.FOR contains all the variables and routines that are hardware and compiler dependent, e.g., screen length and terminal unit number. You may need to edit this file to get the most efficient and attractive output. (See section A.3.) The three remaining utility codes are graphics files: GRAPHIT.HI, GRAPHIT.LO, and GRAPHIT.BLK, which provide the interface between the physics code and your graphics package. These three files contain the same subroutines, but for different types of graphics packages. You will need to link in only one of these. GRAPHIT.HI calls a high-level package, e.g., a package that can draw axes with one call; GRAPHIT.LO calls a low-level package, e.g., a package that is only capable of drawing primitives such as lines and circles and so requires many calls to draw the axes; and GRAPHIT.BLK contains empty subroutines for a non-existent package. This last file is included so that there are no unsatisfied calls at link time. If you want to produce plots, see section A.5.

The last utility code, STRIP.FOR, strips all '!'-delimited comments out of FORTRAN codes. The '!' is not standard but we have used it for clear documentation.

3) The sixteen physics codes for the examples and projects are named EXMPLn.FOR and PROJn.FOR, where 'n' indicates the chapter. With two exceptions, these do not require editing to run (see A.3).

4) The three data files all have the .DAT extension and contain data to be read into EXMPL5.FOR at run time. They require no editing.

5) The include files contain common blocks and variable type declarations that must be included in the FORTRAN codes to compile them. (We have *not* conformed to the default typing supported by FORTRAN, e.g., not all variables starting with the letters 'I' through 'N' are integers).

The file name extension indicates in which codes the file must be included. Those that have the ALL extension must be included in all the programs except STRIP.FOR and the text codes. They contain variables for graphics (GRFDAT.ALL and UISGRF.ALL), menuing (MENU.ALL), and I/O (IO.ALL). Those that have the extension En or Pn (where 'n' indicates the chapter) belong to a particular example or project. The file name indicates the function of the variables included in the common block. The MAP files contain variables to map the menu items to the program parameters. There are only seven of these. The PARAM files contain all the parameters that are constant during one calculation. There are sixteen of these. The one exception to these naming conventions is the file UISGRF.ALL. This file is *only* used with GRAPHIT.LO. Each include file is listed with the most closely related code: those with En or Pn extensions are listed with the corresponding example or project in Appendix B or C, GRFDAT.ALL and UISGRF.ALL are in Appendix D.4, MENU.ALL is in Appendix D.3, and IO.ALL is in Appendix D.2.

All of these files are included in the programs by the compiler with the INCLUDE statement. This is the other non-standard statement in the codes. We have used it both because many compilers support it, and because it is a significant aid in program development and maintenance.

A.3 Compilation

To prepare the codes for compilation is a matter of simple (but possibly tedious) editing.

1) First, if your FORTRAN compiler does not support the INCLUDE statement, you must edit each of the physics and utility programs to include the common block files. (See the previous section for a description of these files.) Before doing so, check your compiler manual to see if it provides this statement or one with a different name but serving the same purpose (e.g., USE). In some compilers, this command is treated as a FORTRAN statement, while in others, it is a compiler directive. If such a statement is provided, simply replace INCLUDE with your compiler's equivalent statement. Note that the include file names must be *exactly* the same in the

program and in the directory. If your operating system is case sensitive (e.g., UNIX), the names must be in the same case as well.

2) The other FORTRAN-77 non-standard grammar in the codes is the '!' comment delimiter. If your compiler does not accept this character, then you must compile, link, and run the program STRIP to change all of the files (this includes utility codes, physics programs, and common blocks, but not data files). This program is itself completely standard, but you must set TERM equal to the unit number connected to your keyboard. STRIP will prompt you for input and output file names. When giving names for the stripped versions of the include files, remember that the name of these files must match exactly that given in the program INCLUDE statement. Therefore, this process is easier (especially if your operating system does not allow version numbers) if you keep the stripped and unstripped versions of the code in different directories.

A sample session on the VAX is given below. Note that the '$' is the VAX/VMS prompt, the unstripped files are in the directory old: the stripped versions are placed in directory new:, and the words in brackets [ ] are comments, not to be typed in.

```
$ fortran strip.for                          [compile STRIP.FOR]
$ link strip.obj                             [link STRIP.OBJ]
$ run strip.exe                              [execute STRIP.EXE]
ENTER NAME OF FORTRAN PROGRAM                 [STRIP prompts for input]
old:exmpl1.for                               [user input]
ENTER NAME OF NEW FILE                        [STRIP prompts for output]
new:exmpl1.for                               [user input]
DO YOU WISH TO STANDARDIZE ANOTHER FILE? [Y]  [STRIP prompt]
                                             [pressing Return accepts Y]
ENTER NAME OF FORTRAN PROGRAM                 [and so on...]
old:param.e1
ENTER NAME OF NEW FILE
new:param.e1
DO YOU WISH TO STANDARDIZE ANOTHER FILE? [Y]
```

3) The next step is to edit the subroutine SETUP in the file SETUP.FOR. Instructions are given in that file. You must edit in constant values for variables that control the I/O. For example, you will need to know how many lines there are on your terminal (usually 24), unit numbers for I/O to the screen (these are 5 and 6 on a VAX), unit numbers for output files, and your own preference for default output (e.g., will you usually want graphics sent to the terminal?)

4) This last step is optional, and involves the three other routines in SETUP.FOR that are also hardware dependent. These routines come with all the lines commented out, except SUBROUTINE, RETURN, and END. Subroutine CLEAR clears the screen by sending escape characters to the terminal. The appropriate sequence for your terminal is located somewhere in the terminal manual—probably in one of the appendices. This routine simply keeps the screen from looking too cluttered; it is not essential. Code for VT200, Tektronix 4010, and PST screens is provided and can be used as a template for other screens. The last two routines (GMODE and TMODE) switch the terminal between graphics and text mode and vice versa. These are only applicable if you have a terminal with both graphics and text screens (e.g., GO-235 and HDS3200 terminals) and plan to use graphics. Again, the appropriate escape sequences are in the terminal manual.

With these changes completed, you can now compile the physics program, UTIL.FOR, SETUP.FOR, and GRAPHIT.BLK; link them together; and execute. For example, on a VAX you would type:

```
$ fortran exmpl1.for,setup.for,util.for,graphit.blk
$ link exmpl1.obj,setup.obj,util.obj,graphit.obj
$ run exmpl1.exe
```

During compilation, you must be sure that the INCLUDE files are present in the default directory.

A.4 Execution

Background: Before running any code, you should read the relevant chapter and the preface to the code in Appendix B (for examples) or C (for projects). The text provides general information about the algorithms and physics, the Appendix provides details that will be of interest only if you run these codes.

Execution: Here we discuss the conventions that are followed, more or less closely, by all of the programs; exceptions are noted in Appendices B and C. All of the programs are designed to be run interactively. Each begins with a title page describing the physical problem to be investigated and the output that will be produced. Next, the main menu is displayed, giving you the choice of entering parameter values, examining parameter values, running the program, or terminating the program. If you choose to run the program, execution proceeds and output is sent to the requested devices. For many of the programs, the calculation has a definite end (e.g., finding the cross section at a given energy), while for other programs (e.g., those involving time evolution) there is no clear end. In the second case, you are asked (at a frequency that you specify) whether you wish to continue the run or return to the main menu. When the calculation is finished, all values are zeroed (except the default parameters), and the main menu is re-displayed, giving you the opportunity to redo the calculation with a new set of parameters or to end execution.

Input: There are two times when the program waits for input. First, the program pauses for you to examine output to the screen (e.g., the title page, a page of results, a plot); the execution is resumed when you press the Return key. Second, the program will pause for you to enter a parameter value. All parameter values have default values displayed in brackets []; to accept the default value, press "return" at the prompt. If you choose to input a new value, but use the incorrect format, the program reprompts for input. If you input a value outside the range set in the menu, the program prompts for input, stating the range of values allowed.

Real numbers must be input in E9.2 format. This is a compromise in order to allow default values. The user input functions (GETFLT and GETINT) accept default values by reading the input into a string variable and checking for zero length. If the string is not empty, the parameter value must be read from the string which acts as an internal unit. Therefore, the read must be formatted since list-directed reads from internal files are not standard. Note that some compilers do not insist on the exponential (i.e., 1.23 and 1.23E+0 are both understood), or do allow list-directed reads to internal units, so the formatting restrictions can be relaxed. Fortunately, integers may be entered in the "natural" way, as left justified numerals with no decimal. This is because the program tacks a decimal point after the last non-blank character in the input string, and then reads the number from the string using F7.0 format.

In many programs you will be prompted for input outside of the main menu. This happens for several reasons. First, if the input parameters failed to pass a check, the program re-prompts for input. Second, guidance may be needed in choosing parameter values (e.g., the range of partial waves in PROJ4 depends on the input energy), and the program prompts for these values after reasonable default values have been calculated. Third, calculations can often allow the change of parameters during the run (e.g., the time step can be altered during time evolution), and the program prompts for these values during a run.

Please note that our "defensive programming" has not been airtight. It is possible to crash the program or get meaningless results by entering inappropriate parameters. The default parameters for each code are listed in Appendix B or C in brackets []. See A.8 for details on changing default values.

Output: To accommodate a variety of situations, there are several choices for run-time output. Text output can be displayed on the screen and/or sent to a file. Graphics output can be displayed on the screen and/or sent to a hardcopy device. The data used to generate graphics can also be sent to a file. You may choose the type of output by changing the default values in SETUP.FOR or by altering I/O parameters from the menu at run-time.

A.5 Graphics

One drawback to programming in FORTRAN is that there are no standard FORTRAN graphics. This puts you in one of three situations: 1) you have no graphics package at all or 2) your graphics package is executed independently of the program that creates the data (i.e., it reads in the data from a file) or 3) your graphics package can be called from a FORTRAN program.

If you have no graphics package, you should edit SETUP.FOR and set GRFTRM=0, GRFHRD=0, and GRFFIL=0 so that no graphics are produced. Because the graphics output greatly enhances many of these programs, you might wish to consider purchasing such a package.

For those users whose graphics package must read the data from a file, the programs will (on request) output graphics data to a file. If you edit SETUP.FOR so that GRFFIL=1, GRFHRD=0, and GRFTRM=0 graphics data will always be sent to a file, but never to the screen or hardcopy device.

Finally, for those who have a FORTRAN callable package, you will have to modify our codes to work with your package. To facilitate this task, we provide a set of routines (contained in GRAPHIT.HI and GRAPHIT.LO) to act as an interface between the physics programs and a general graphics package. The physics programs call the routines in the GRAPHIT files, which in turn call graphics package routines. This way, you only need to modify one GRAPHIT code instead of all of the physics codes. The code is documented for ease of modification. GRAPHIT.HI calls the high level graphics package CA-DISSPLA; GRAPHIT.LO calls the low-level package UIS graphics (for the VAXstation). If you have either one of these packages, you can produce graphics with no further work, although you should read Appendix D.4 for information specific to your package. Also, routines for GKS on Sun workstations (written by Michel Vallières at Drexel University) are available over the network (see Appendix E). Otherwise, use either GRAPHIT.HI or GRAPHIT.LO, whichever is more appropriate, as a template to create routines to call your graphics package.

A.6 Program Structure

To facilitate code modification, we have attempted to conform to general "good" programming practices, such as those discussed in [Ke78]. The code for each program has been formatted to make it easier to read. To the extent possible, variable names have been chosen mnemonically and loops have been indented, as have sections of the codes controlled by IF statements. Running comments on the right-hand side of many statement lines indicate what computation is being done. To keep calls to subroutines as uncluttered, but as informative, as possible, variables that are constant for one run are passed in common blocks, while variables that change are passed in calls. All variables are explicitly typed. This gives a check on typographical errors (using a VAX FORTRAN compiler switch) and provides a dictionary for variable names.

The programs are organized in subroutines, each performing limited and well defined tasks. The subroutines that perform the calculations are at the beginning of the code, and within those routines the most important lines are listed in boldface. The bookkeeping and I/O routines are listed at the end, and we have tried to keep these uniform from chapter to chapter. A header describes the purpose of the subroutine and the variables it uses. These are five types: INPUT and OUTPUT variables (which are passed to the subroutine in the call statement), LOCAL variables (which are used only within that subroutine), FUNCTIONS (which are defined by FORTRAN FUNCTION routines), and GLOBAL variables (which are passed in common blocks). Below we list the purpose of the subroutines that are common to all the codes, while specific details for each code precede the code listing in Appendices B and C.

INIT   displays header screen, initializes constants (e.g., PI = 3.1415926), and sets up the arrays needed for the menu. It is in this routine that default values and limits for input parameters are set.

PARAM   calls the menuing routine to obtain input parameters and then calculates any parameters that are functions of other input parameters. Files are opened and closed here, and the program stops in this routine, if so requested. Checks are performed to ensure that parameter values are reasonable.

PRMOUT   outputs parameters to the terminal or file.

TXTOUT   outputs text data to a terminal or file.

GRFOUT   outputs graphics data to a terminal or file.

ARCHON   presides over the calculation. It calls the output routines as well as the routines that do the calculations, and often performs calculations itself.

A.7 Menu Structure

We give here a brief overview of the variables and subroutines that create and execute the menu. The menu is made up of up to 99 individual items, each item is described by eight parameters: MTYPE, MINTS, MREALS, MLOLIM, MHILIM, MSTRNG, MPRMPT, MTAG. The most important parameter is MTYPE which describes the action to be performed. This parameter falls into one of three categories: 1) those that get input from the screen (FLOAT, NUM, BOOLEAN, CHSTR), 2) those that provide flow control within the menu (YESKIP, NOSKIP, MCHOIC, SKIP, QUIT), 3) and those that control the display of information (WAIT, PPRINT, CLRTRM, MTITLE, TITLE). The function of the other parameters depends on the value of MTYPE and are given below. Each of these parameters are FORTRAN arrays dimensioned to 100. The index of the array corresponds to the index of the menu item. The one exception is MSTRNG which (since rarely used) is dimensioned to 10. These parameter arrays are passed via common blocks in the include file MENU.ALL.

There are four routines that are used in running the menu. Subroutine MENU (in file UTIL.FOR) initializes menu parameters that are common to all programs; subroutine INIT (in each example or project) initializes the menu parameters specific to that program. Subroutine ASK (in file UTIL.FOR) executes the menu items by performing the action dictated by MTYPE. Subroutine PRTAGS (in file UTIL.FOR) prints a summary of menu items.

Presented here is a summary of menu parameter functions for each value of MTYPE. Any parameter not listed for a particular type is not used.

MTYPE=FLOAT :   get a floating point number from the screen
  MPRMPT :   menu prompt
   MTAGS :   shorter prompt used in PRTAGS
  MREALS :   default value
  MLOLIM :   minimum allowed value
  MHILIM :   maximum allowed value

MTYPE=NUM   :   get an integer from the screen
  MPRMPT :   menu prompt
   MTAGS :   shorter prompt used in PRTAGS
  MLOLIM :   minimum allowed value
  MHILIM :   maximum allowed value
   MINTS :   default value

MTYPE=BOOLEAN: get a yes/no answer from the screen
  MPRMPT :   menu prompt
   MTAGS :   shorter prompt used in PRTAGS
   MINTS :   default value (0=no, 1=yes)

MTYPE=CHSTR :   get a character string from the screen
  MPRMPT :   menu prompt
   MTAGS :   shorter prompt used in PRTAGS
  MHILIM :   maximum number of characters allowed
   MINTS :   index of MSTRNG for this item
            (since MSTRNG is dimensioned to 10 only)
  MSTRNG :   default value

MTYPE=MCHOIC :   get a menu choice (integer) from screen and branch
  MPRMPT :   menu prompt
   MTAGS :   of form 'nn nn nn ...', controls menu branching; the m'th
            nn value is the menu item branched to when user chooses
            menu item m. Because of the form of the string, menu
            items can only go from 1 to 99.
  MREALS :   the absolute value equals menu choice; if less than zero the
            default value is not changed, if greater than zero default
            is changed (useful when a menu choice is also a parameter
            choice).
  MLOLIM :   minimum allowed value
  MHILIM :   maximum allowed value
   MINTS :   default value

MTYPE=YESKIP: get a yes/no answer from screen, skip on yes
  MPRMPT: menu prompt
  MTAGS :   shorter prompt used in PRTAGS
  MREALS: menu item to skip to on yes
  MINTS :   default value (0=no, 1=yes)

MTYPE=NOSKIP: get a yes/no answer from screen, skip on no
  MPRMPT: menu prompt
  MTAGS :   shorter prompt used in PRTAGS
  MREALS: menu item to skip to on no
  MINTS :   default value (0=no, 1=yes)

MTYPE=SKIP :   skip unconditionally
  MREALS: menu item to skip to

MTYPE=WAIT :   pause and wait for user to press return before
        continuing
  MPRMPT: menu prompt

MTYPE=CLRTRM: clear terminal

MTYPE=QUIT :   quit menu and return to calling routine

MTYPE=PPRINT: print description of a range of menu items by calling
        PRTAGS
  MLOLIM: index of first menu item
  MHILIM: index of last menu item

MTYPE=TITLE :   print information without getting input (in ASK and
        PRTAGS)
  MPRMPT: information to be printed
  MLOLIM: number of blank lines to skip before printing
  MHILIM: number of blank lines to skip after printing

MTYPE=MTITLE: print information without getting input (in ASK only)
  MPRMPT: information to be printed
  MLOLIM: number of blank lines to skip before printing
  MHILIM: number of blank lines to skip after printing

A.8 Default Value Revision

At some point the default values set in the programs will not correspond
to your needs, and changing values at run-time through the menu will
become tedious. When this happens you will want to change the default

values in the code.

Default I/O choices are set in SETUP.FOR (see A.3 above). Physical and numerical defaults are set in subroutine INIT; the values are stored in menu arrays. These default values are in MINTS for integer values and yes/no values, in MREALS for real values, and in MSTRNG for character values. See the section A.7 above for more details. Minimum and maximum allowable values are contained in MHILIM and MLOLIM arrays. Sometimes these limits are set to variables defined by the FORTRAN PARAMETER statement. For example, the number of lattice points for a field cannot be any larger than the declared array size, so the maximum lattice size is equal to the array dimension. You can alter these array dimensions, which are defined by PARAMETER statements located in the PARAM include file.

Appendix B

Programs for the Examples

B.1 Example 1

Algorithm   This program finds the semiclassical approximations to the bound state energies of the Lennard-Jones potential for the input value of $\gamma = (2ma^2V_0/\hbar^2)^{1/2}$. The basic problem is to find, for each integer n, the value of ϵ_n for which Eq. (1.22) is satisfied. After the number of bound states is estimated (end of subroutine PARAM), the energy for each level is found (DO loop 100 in subroutine ARCHON) using a secant search (subroutine SEARCH) to locate the zero of $f = s-(n+\frac{1}{2})\pi$. Subroutine ACTION calculates the action, s, for a given energy by using simple searches to locate the inner and outer turning points and then using Simpson's rule to calculate the integral, with a special treatment for the square-root behavior near the turning points.

The potential is defined by function POT. If you decide to change the potential, be sure that the minimum is normalized to a value of -1 and that POTMIN in subroutine INIT is set equal to the equilibrium X value of the new potential. These requirements ensure that both the energy and turning point searches begin at reasonable values: the energy search at -1 and the turning point search at POTMIN.

Input   The only physical parameter is γ [50.] which is the dimensionless measure of the quantum nature of the problem. The maximum number of bound levels that can be found is fixed by the parameter MAXLVL= 100. If NLEVEL is calculated to be larger than this, you will be prompted for a new value of γ until NLEVEL is small enough (subroutine PCHECK). The numerical parameters are the energy search tolerance [.0005], the turning point search tolerance [.0005], and the number of points used in evaluating the action integral [100]. The graphing parameter NGRF [80] determines the number of points used in drawing the phase-space trajectories (subroutine GRFOUT). The limit on NGRF is set by the parameter MAXGRF= 1000.

Output   After each level is found, the number of the level, the energy, and classical turning points (Xmin and Xmax) are printed. When all the energies are found, the phase-space trajectories $k(x) = \gamma\sqrt{\epsilon_n - v(x)}$, are graphed for each level (subroutine GRFOUT).

```
CCCCCCCCCCCCCCCCCCCCCCCCCCCCCCCCCCCCCCCCCCCCCCCCCCCCCCCCCCCCCCCCCCCCCCCCCC
      PROGRAM EXMPL1
C     Example 1: Bohr-Sommerfeld quantization for bound states of the
C                Lennard-Jones Potential
C  COMPUTATIONAL PHYSICS (FORTRAN VERSION)
C  by Steven E. Koonin and Dawn C. Meredith
C  Copyright 1989, Addison-Wesley Publishing Company
CCCCCCCCCCCCCCCCCCCCCCCCCCCCCCCCCCCCCCCCCCCCCCCCCCCCCCCCCCCCCCCCCCCCCCCCCC
      CALL INIT             !display header screen, setup parameters
5     CONTINUE              !main loop/ execute once for each set of param
        CALL PARAM          !get input from screen
        CALL ARCHON         !search for bound states
      GOTO 5
      END
CCCCCCCCCCCCCCCCCCCCCCCCCCCCCCCCCCCCCCCCCCCCCCCCCCCCCCCCCCCCCCCCCCCCCCCCCC
      SUBROUTINE ARCHON
C finds the bound states of the Lennard-Jones potential
C from the Bohr-Sommerfeld quantization rule
CCCCCCCCCCCCCCCCCCCCCCCCCCCCCCCCCCCCCCCCCCCCCCCCCCCCCCCCCCCCCCCCCCCCCCCCCC
C Global variables:
      INCLUDE 'IO.ALL'
      INCLUDE 'PARAM.E1'
C Local variables:
      REAL S                    !current value of action
      REAL E1                   !current value of energy
      REAL X1,X2                !current turning points
      REAL F1                   !f=action/2 - pi/2 -ilevel*pi
      INTEGER ILEVEL            !current level
      REAL ENERGY(0:MAXLVL)     !energy of bound state
      REAL XIN(0:MAXLVL)        !inner turning point
      REAL XOUT(0:MAXLVL)       !outer turning point
      INTEGER NLINES            !number of lines printed to terminal
      INTEGER SCREEN            !send to terminal
      INTEGER PAPER             !make a hardcopy
      INTEGER FILE              !send to a file
      DATA SCREEN,PAPER,FILE/1,2,3/
CCCCCCCCCCCCCCCCCCCCCCCCCCCCCCCCCCCCCCCCCCCCCCCCCCCCCCCCCCCCCCCCCCCCCCCCCC
C     output summary of parameters
      IF (TTERM) CALL PRMOUT(OUNIT,NLINES)
      IF (TFILE) CALL PRMOUT(TUNIT,NLINES)
      IF (GFILE) CALL PRMOUT(GUNIT,NLINES)
C
C     search for bound states
      E1=-1.                        !begin at the well bottom
      F1=-PI/2                      !the action is zero there
C
```

```
C       find the NLEVEL bound states
        DO 100 ILEVEL=0,NLEVEL-1
C
          CALL SEARCH(ILEVEL,E1,F1,X1,X2) !search for eigenvalue
          ENERGY(ILEVEL)=E1              !store values for this state
          XIN(ILEVEL)=X1
          XOUT(ILEVEL)=X2
C
C         text output
          IF (TTERM) CALL TXTOUT(OUNIT,ILEVEL,E1,X1,X2,NLINES)
          IF (TFILE) CALL TXTOUT(TUNIT,ILEVEL,E1,X1,X2,NLINES)
C
          F1=F1-PI                      !guess to begin search for next level
C
100     CONTINUE
C
        IF (TTERM) CALL PAUSE('to continue...',1)
        IF (TTERM) CALL CLEAR
C
C       graphics output
        IF (GTERM) CALL GRFOUT(SCREEN,ENERGY,XIN,XOUT)
        IF (GFILE) CALL GRFOUT(FILE,ENERGY,XIN,XOUT)
        IF (GHRDCP) CALL GRFOUT(PAPER,ENERGY,XIN,XOUT)
C
        RETURN
        END
CCCCCCCCCCCCCCCCCCCCCCCCCCCCCCCCCCCCCCCCCCCCCCCCCCCCCCCCCCCCCCCCCCCCCCCCCC
        SUBROUTINE SEARCH(N,E1,F1,X1,X2)
C finds the N'th bound state
C E1 is passed in as initial guess for the bound state energy
C    and returned as the true bound state energy with turning points
C    X1 and X2
C F1 is the function which goes to zero at a bound state
C    F1 = action/2-(n+1/2)*pi
CCCCCCCCCCCCCCCCCCCCCCCCCCCCCCCCCCCCCCCCCCCCCCCCCCCCCCCCCCCCCCCCCCCCCCCCCC
C Global variables:
        INCLUDE 'PARAM.E1'
C Input/Output variables:
        INTEGER N                !current level (input)
        REAL E1,E2               !trial energies (I/O)
        REAL F1,F2               !f=action/2-pi*(n+1/2) (I/O)
        REAL S                   !action (output)
        REAL X1,X2               !turning points (output)
C Local variables:
        REAL DE                  !increment in energy search
CCCCCCCCCCCCCCCCCCCCCCCCCCCCCCCCCCCCCCCCCCCCCCCCCCCCCCCCCCCCCCCCCCCCCCCCCC
C       guess the next energy in order to begin search
        E2=E1+ABS(E1)/4.
        DE=2*ETOL
C
C       use secant search to find the bound state
50      IF (ABS(DE) .GT. ETOL) THEN
```

```
          CALL ACTION(E2,X1,X2,S)           !S at new energy
          F2=S-(N+.5)*PI                    !F at new energy
          IF (F1 .NE. F2) THEN              !calculate new DE
             DE=-F2*(E2-E1)/(F2-F1)
          ELSE
             DE=0.
          END IF
C
          E1=E2
          F1=F2                             !roll values
          E2=E1+DE                          !increment energy
          IF (E2 .GE. 0) E2=-ETOL           !keep energy negative
       GOTO 50
       END IF
C
       RETURN
       END
CCCCCCCCCCCCCCCCCCCCCCCCCCCCCCCCCCCCCCCCCCCCCCCCCCCCCCCCCCCCCCCCCCCCCCCC
       SUBROUTINE ACTION(E,X1,X2,S)
C calculates the (action integral)/2 (S) and the classical turning
C points (X1,X2) for a given energy (E)
CCCCCCCCCCCCCCCCCCCCCCCCCCCCCCCCCCCCCCCCCCCCCCCCCCCCCCCCCCCCCCCCCCCCCCCC
C Global variables:
       INCLUDE 'PARAM.E1'
C Input/Output variables:
       REAL E                    !energy (input)
       REAL S                    !action (output)
       REAL X1,X2                !turning points (output)
C Local variables:
       REAL DX                   !increment in turning point search
       REAL H                    !quadrature step size
       REAL SUM                  !sum for integral
       INTEGER IFAC              !coefficient for Simpson's rule
       INTEGER IX                !index on X
       REAL X                    !current X value in sum
       REAL POT                  !potential as a function of X (function)
CCCCCCCCCCCCCCCCCCCCCCCCCCCCCCCCCCCCCCCCCCCCCCCCCCCCCCCCCCCCCCCCCCCCCCCC
C      find inner turning point; begin search at the well bottom
       X1=POTMIN
       DX=.1
50     IF (DX .GT. XTOL) THEN
          X1=X1-DX                !use simple search, going inward
          IF (POT(X1) .GE. E) THEN
             X1=X1+DX
             DX=DX/2
          END IF
       GOTO 50
       END IF
C
C      find the outer turning point; begin search at the well bottom
       X2=POTMIN
       DX=.1
```

```
120   IF (DX .GT. XTOL) THEN
         X2=X2+DX                  !use simple search going outward
         IF (POT(X2) .GE. E) THEN
            X2=X2-DX
            DX=DX/2
         END IF
      GOTO 120
      END IF
C
C     Simpson's rule from X1+H to X2-H
      IF (MOD(NPTS,2) .EQ. 1) NPTS=NPTS+1    !NPTS must be even
      H=(X2-X1)/NPTS                         !step size
      SUM=SQRT(E-POT(X1+H))
      IFAC=2
      DO 200 IX=2,NPTS-2
         X=X1+IX*H
         IF (IFAC .EQ. 2) THEN               !alternate factors
            IFAC=4
         ELSE
            IFAC=2
         END IF
         SUM=SUM+IFAC*SQRT(E-POT(X))
200   CONTINUE
      SUM=SUM+SQRT(E-POT(X2-H))
      SUM=SUM*H/3
C
C     special handling for sqrt behavior of first and last intervals
      SUM=SUM+SQRT(E-POT(X1+H))*2*H/3
      SUM=SUM+SQRT(E-POT(X2-H))*2*H/3
      S=SUM*GAMMA
C
      RETURN
      END
CCCCCCCCCCCCCCCCCCCCCCCCCCCCCCCCCCCCCCCCCCCCCCCCCCCCCCCCCCCCCCCCCCCCCC
      REAL FUNCTION POT(X)
C evaluates the Lennard-Jones potential at X
CCCCCCCCCCCCCCCCCCCCCCCCCCCCCCCCCCCCCCCCCCCCCCCCCCCCCCCCCCCCCCCCCCCCCC
C Passed variables:
      REAL X
CCCCCCCCCCCCCCCCCCCCCCCCCCCCCCCCCCCCCCCCCCCCCCCCCCCCCCCCCCCCCCCCCCCCCC
C If you change the potential, normalize to a minimum of -1
C and change the value of POTMIN in subroutine INIT to the
C new equilibrium position (i.e. the X value at which the force is zero)
C     Lennard-Jones potential in scaled variables
      POT=4*(X**(-12)-X**(-6))
      RETURN
      END
CCCCCCCCCCCCCCCCCCCCCCCCCCCCCCCCCCCCCCCCCCCCCCCCCCCCCCCCCCCCCCCCCCCCCC
      SUBROUTINE INIT
C initializes constants, displays header screen,
C initializes menu arrays for input parameters
CCCCCCCCCCCCCCCCCCCCCCCCCCCCCCCCCCCCCCCCCCCCCCCCCCCCCCCCCCCCCCCCCCCCCC
```

```
C Global variables:
      INCLUDE 'IO.ALL'
      INCLUDE 'MENU.ALL'
      INCLUDE 'PARAM.E1'
C Local parameters:
      CHARACTER*80 DESCRP              !program description
      DIMENSION DESCRP(20)
      INTEGER NHEAD,NTEXT,NGRAPH       !number of lines for each description
CCCCCCCCCCCCCCCCCCCCCCCCCCCCCCCCCCCCCCCCCCCCCCCCCCCCCCCCCCCCCCCCCCCCCCCCCCCCC
C     get environment parameters
      CALL SETUP
C
C     display header screen
      DESCRP(1)= 'EXAMPLE 1'
      DESCRP(2)= 'Bohr-Sommerfeld quantization for bound state'
      DESCRP(3)= 'energies of the 6-12 potential'
      NHEAD=3
C
C     text output description
      DESCRP(4)= 'energy and classical turning points for each state'
      NTEXT=1
C
C     graphics output description
      DESCRP(5)= 'phase space (wavenumber vs. position) portrait'
      DESCRP(6)= 'of classical trajectories'
      NGRAPH=2
C
      CALL HEADER(DESCRP,NHEAD,NTEXT,NGRAPH)
C
C     calculate constants
      PI=4*ATAN(1.0)
      POTMIN=2**(1.0/6.0)
C
C     setup menu arrays, beginning with constant part
      CALL MENU
C
      MTYPE(13)=FLOAT
      MPRMPT(13)= 'Enter gamma=sqrt(2*m*a**2*V/hbar**2) (dimensionless)'
      MTAG(13)= 'Gamma (dimensionless)'
      MLOLIM(13)=1.
      MHILIM(13)=500.
      MREALS(13)=50.
C
      MTYPE(14)=SKIP
      MREALS(14)=35.
C
      MTYPE(38)=FLOAT
      MPRMPT(38)= 'Enter tolerance for energy search (scaled units)'
      MTAG(38)= 'Energy search tolerance (scaled units)'
      MLOLIM(38)=.00001
      MHILIM(38)=.01
      MREALS(38)=.0005
```

```
C
      MTYPE(39)=FLOAT
      MPRMPT(39)=
     +'Enter tolerance for turning point search (scaled units)'
      MTAG(39)= 'Turning point search tolerance (scaled units)'
      MLOLIM(39)=.00001
      MHILIM(39)=.01
      MREALS(39)=.0005
C
      MTYPE(40)=NUM
      MPRMPT(40)= 'Enter number of points for action integral'
      MTAG(40)= 'Number of quadrature points for action integral'
      MLOLIM(40)=20.
      MHILIM(40)=5000.
      MINTS(40)=100
C
      MTYPE(41)=SKIP
      MREALS(41)=60.
C
      MSTRNG(MINTS(75))= 'exmpl1.txt'
C
      MTYPE(76)=SKIP
      MREALS(76)=80.
C
      MSTRNG(MINTS(86))= 'exmpl1.grf'
C
      MTYPE(87)=NUM
      MPRMPT(87)= 'Enter number of points to be used in graphing'
      MTAG(87)= 'Number of graphing points'
      MLOLIM(87)= 10.
      MHILIM(87)= MAXGRF-2
      MINTS(87)= 80
C
      MTYPE(88)=SKIP
      MREALS(88)=90.
C
      RETURN
      END
CCCCCCCCCCCCCCCCCCCCCCCCCCCCCCCCCCCCCCCCCCCCCCCCCCCCCCCCCCCCCCCCCCCCCCCC
      SUBROUTINE PARAM
C gets parameters from screen
C ends program on request
C closes old files
C maps menu variables to program variables
C opens new files
C calculates all derivative parameters
C performs checks on parameters
CCCCCCCCCCCCCCCCCCCCCCCCCCCCCCCCCCCCCCCCCCCCCCCCCCCCCCCCCCCCCCCCCCCCCCCC
C Global variables:
      INCLUDE 'MENU.ALL'
      INCLUDE 'IO.ALL'
      INCLUDE 'PARAM.E1'
```

```
          INCLUDE 'MAP.E1'
C Local variables:
          REAL S                    !current value of action
          REAL E1                   !current value of energy
          REAL X1,X2                !current turning points
C Function:
          LOGICAL LOGCVT            !converts 1 and 0 to true and false
CCCCCCCCCCCCCCCCCCCCCCCCCCCCCCCCCCCCCCCCCCCCCCCCCCCCCCCCCCCCCCCCCCCCCCCCC
C     get input from terminal
          CALL CLEAR
          CALL ASK(1,ISTOP)
C
C     stop program if requested
          IF (MREALS(IMAIN) .EQ. STOP) CALL FINISH
C
C     close files if necessary
          IF (TNAME .NE. MSTRNG(MINTS(ITNAME)))
     +        CALL FLCLOS(TNAME,TUNIT)
          IF (GNAME .NE. MSTRNG(MINTS(IGNAME)))
     +        CALL FLCLOS(GNAME,GUNIT)
C
C     physical and numerical parameters
          GAMMA=MREALS(IGAMMA)
          ETOL=MREALS(IETOL)
          XTOL=MREALS(IXTOL)
          NPTS=MINTS(INPTS)
C
C     text output parameters
          TTERM=LOGCVT(MINTS(ITTERM))
          TFILE=LOGCVT(MINTS(ITFILE))
          TNAME=MSTRNG(MINTS(ITNAME))
C
C     graphics output parameters
          GTERM=LOGCVT(MINTS(IGTERM))
          GHRDCP=LOGCVT(MINTS(IGHRD))
          GFILE=LOGCVT(MINTS(IGFILE))
          GNAME=MSTRNG(MINTS(IGNAME))
          NGRF=MINTS(INGRF)
C
C     open files
          IF (TFILE) CALL FLOPEN(TNAME,TUNIT)
          IF (GFILE) CALL FLOPEN(GNAME,GUNIT)
C     files may have been renamed
          MSTRNG(MINTS(ITNAME))=TNAME
          MSTRNG(MINTS(IGNAME))=GNAME
C
C     calculate total number of levels
          E1=-ETOL
          CALL ACTION(E1,X1,X2,S)
          NLEVEL=INT(S/PI-.5)+1
C     check value of GAMMA
          CALL PCHECK
```

```
C
      CALL CLEAR
C
      RETURN
      END
CCCCCCCCCCCCCCCCCCCCCCCCCCCCCCCCCCCCCCCCCCCCCCCCCCCCCCCCCCCCCCCCCCCCCCCCC
      SUBROUTINE PCHECK
C ensure that the number of states is not greater than the size of
C the data arrays; if so prompt for smaller GAMMA
CCCCCCCCCCCCCCCCCCCCCCCCCCCCCCCCCCCCCCCCCCCCCCCCCCCCCCCCCCCCCCCCCCCCCCCCC
C Global parameters:
      INCLUDE 'PARAM.E1'
      INCLUDE 'MENU.ALL'
      INCLUDE 'IO.ALL'
      INCLUDE 'MAP.E1'
C Local parameters:
      REAL S                    !action
      REAL E                    !small negative energy
      REAL X1,X2                !classical turning points
C Function:
      REAL GETFLT               !returns a floating point variable
CCCCCCCCCCCCCCCCCCCCCCCCCCCCCCCCCCCCCCCCCCCCCCCCCCCCCCCCCCCCCCCCCCCCCCCCC
10    IF ((NLEVEL-1) .GT. MAXLVL) THEN
          WRITE (OUNIT,15) NLEVEL,MAXLVL
          MHILIM(IGAMMA)=GAMMA
          MREALS(IGAMMA) = GETFLT(MREALS(IGAMMA)/2,MLOLIM(IGAMMA),
     +                    MHILIM(IGAMMA), 'Enter a smaller gamma')
          GAMMA=MREALS(IGAMMA)
C
          E=-ETOL
          CALL ACTION(E,X1,X2,S)
          NLEVEL=INT(S/PI+.5)+1
      GOTO 10
      END IF
C
15    FORMAT (' Total number of levels (=',i5,
     +        ') is larger than maximum allowable (=',i3,')')
C
      RETURN
      END
CCCCCCCCCCCCCCCCCCCCCCCCCCCCCCCCCCCCCCCCCCCCCCCCCCCCCCCCCCCCCCCCCCCCCCCCC
      SUBROUTINE PRMOUT(MUNIT,NLINES)
C outputs parameter summary to the specified unit
CCCCCCCCCCCCCCCCCCCCCCCCCCCCCCCCCCCCCCCCCCCCCCCCCCCCCCCCCCCCCCCCCCCCCCCCC
C Global variables:
      INCLUDE 'IO.ALL'
      INCLUDE 'PARAM.E1'
C Input/Output variables:
      INTEGER MUNIT             !unit number for output (input)
      INTEGER NLINES            !number of lines written so far (output)
CCCCCCCCCCCCCCCCCCCCCCCCCCCCCCCCCCCCCCCCCCCCCCCCCCCCCCCCCCCCCCCCCCCCCCCCC
      IF (MUNIT .EQ. OUNIT) CALL CLEAR
```

```
C
        WRITE (MUNIT,2)
        WRITE (MUNIT,4)
        WRITE (MUNIT,6) ETOL,XTOL
        WRITE (MUNIT,8) NPTS
        WRITE (MUNIT,10) GAMMA,NLEVEL
        WRITE (MUNIT,12)
        WRITE (MUNIT,2)
C
        IF (MUNIT .NE. GUNIT) THEN
          WRITE (MUNIT,20)
          WRITE (MUNIT,25)
        END IF
C
        NLINES=7
C
2       FORMAT (' ')
4       FORMAT (' Output from example 1: Bohr Sommerfeld Quantization')
6       FORMAT (' Energy tolerance =', E12.5,
     +         '    position tolerance =', E12.5)
8       FORMAT (' number of quadrature points =', I4)
10      FORMAT (' For gamma =', F8.2,' there are ', I4, ' levels:')
12      FORMAT (' (all quantities are expressed in scaled units)')
20      FORMAT (8X, 'Level', 8X, 'Energy', 12X, 'Xmin', 12X, 'Xmax')
25      FORMAT (8X, '-----', 8X, '------', 12X, '----', 12X ,'----')
C
        RETURN
        END
CCCCCCCCCCCCCCCCCCCCCCCCCCCCCCCCCCCCCCCCCCCCCCCCCCCCCCCCCCCCCCCCCCCCCCC
        SUBROUTINE TXTOUT(MUNIT,ILEVEL,E,X1,X2,NLINES)
C writes results for one state to the requested unit
CCCCCCCCCCCCCCCCCCCCCCCCCCCCCCCCCCCCCCCCCCCCCCCCCCCCCCCCCCCCCCCCCCCCCCC
C Global variables:
        INCLUDE 'IO.ALL'
C Input variables:
        INTEGER MUNIT                  !output unit specifier
        INTEGER ILEVEL                 !current level
        REAL E                         !eigen energy
        REAL X1,X2                     !classical turning points
        INTEGER NLINES                 !number of lines printed so far
CCCCCCCCCCCCCCCCCCCCCCCCCCCCCCCCCCCCCCCCCCCCCCCCCCCCCCCCCCCCCCCCCCCCCCC
C       if screen is full, clear screen and retype headings
        IF ((MOD(NLINES,TRMLIN-6) .EQ. 0)
     +                        .AND. (MUNIT .EQ. OUNIT)) THEN
          CALL PAUSE('to continue...',1)
          CALL CLEAR
          WRITE (MUNIT,20)
          WRITE (MUNIT,25)
        END IF
C
        WRITE (MUNIT,30) ILEVEL,E,X1,X2
C
```

```
C      keep track of printed lines only for terminal output
       IF (MUNIT .EQ. OUNIT) NLINES=NLINES+1
C
20     FORMAT (8X, 'Level', 8X, 'Energy', 12X, 'Xmin', 12X, 'Xmax')
25     FORMAT (8X, '-----', 8X, '------', 12X, '----', 12X ,'----')
30     FORMAT(8X,I4,3(8X,F8.5))
C
       RETURN
       END
CCCCCCCCCCCCCCCCCCCCCCCCCCCCCCCCCCCCCCCCCCCCCCCCCCCCCCCCCCCCCCCCCCCCCCC
       SUBROUTINE GRFOUT(DEVICE,ENERGY,XIN,XOUT)
C outputs phase space portraits of the bound states to the terminal
C and/or a file
CCCCCCCCCCCCCCCCCCCCCCCCCCCCCCCCCCCCCCCCCCCCCCCCCCCCCCCCCCCCCCCCCCCCCCC
C Global parameters:
       INCLUDE 'IO.ALL'
       INCLUDE 'PARAM.E1'
       INCLUDE 'GRFDAT.ALL'
C Input variables:
       INTEGER DEVICE                  !which device is being used?
       REAL ENERGY(0:MAXLVL)           !energy of bound state
       REAL XIN(0:MAXLVL)              !inner turning point
       REAL XOUT(0:MAXLVL)             !outer turning point
C Local parameters:
       INTEGER ILEVEL                  !level index
       REAL H                          !step size for x
       INTEGER IX                      !x index
       REAL E                          !current energy
       REAL K(MAXGRF)                  !current wavenumber
       REAL X(MAXGRF)                  !current position
       CHARACTER*9 CGAMMA,CN           !Gamma,nlevel as a character string
       REAL POT                        !potential (function)
       INTEGER LEN,NLEN                !length of character data
       INTEGER SCREEN                  !send to terminal
       INTEGER PAPER                   !make a hardcopy
       INTEGER FILE                    !send to a file
       DATA SCREEN,PAPER,FILE/1,2,3/
CCCCCCCCCCCCCCCCCCCCCCCCCCCCCCCCCCCCCCCCCCCCCCCCCCCCCCCCCCCCCCCCCCCCCCC
C      messages for the impatient
       IF (DEVICE .NE. SCREEN) WRITE (OUNIT,100)
C
C      calculate parameters for graphing
       IF (DEVICE .NE. FILE) THEN
          NPLOT=1                      !how many plots
          IPLOT=1
C
          YMAX=GAMMA*SQRT(1.+ENERGY(NLEVEL-1))   !limits on data points
          YMIN=-YMAX
          XMIN=XIN(NLEVEL-1)
          XMAX=XOUT(NLEVEL-1)
          Y0VAL=XMIN
          X0VAL=0.
```

```
C
            IF (MOD(NGRF,2) .EQ. 0) NGRF=NGRF+1
            NPOINT=NGRF                          !keep number of points odd
C
            ILINE=1                              !line and symbol styles
            ISYM=1
            IFREQ=0
            NXTICK=5
            NYTICK=5
C
            CALL CONVRT(GAMMA,CGAMMA,LEN) !titles and labels
            CALL ICNVRT(NLEVEL,CN,NLEN)
            INFO=' NLEVEL = '//CN(1:NLEN)
            TITLE='Semiclassically Quantized Trajectories, Gamma='//CGAMMA
            LABEL(1)='scaled position'
            LABEL(2)='scaled wave number'
C
            CALL GTDEV(DEVICE)                   !device nomination
            IF (DEVICE .EQ. SCREEN) CALL GMODE   !change to graphics mode
            CALL LNLNAX                          !draw axes
         END IF
C
C      calculate classical phase space trajectory for each bound state
C      by finding the scaled wavenumber as a function of X and Energy
       DO 50 ILEVEL=0,NLEVEL-1
          E=ENERGY(ILEVEL)
          H=(XOUT(ILEVEL)-XIN(ILEVEL))/((NGRF-1)/2)   !step size
          X(1)=XIN(ILEVEL)
          K(1)=0.
C
          DO 20 IX=1,(NGRF-1)/2
             X(IX+1)=XIN(ILEVEL)+(IX)*H
             K(IX+1)=(E-POT(X(IX+1)))                  !scaled wave number
             IF (K(IX) .LE. 0) THEN
                K(IX)=0.
             ELSE
                K(IX)=GAMMA*SQRT(K(IX))
             END IF
20        CONTINUE
C
          DO 30 IX=(NGRF+1)/2,NGRF-1    !graph is symmetric about x-axis
             X(IX+1)=X(NGRF-IX)
             K(IX+1)=-K(NGRF-IX)
30        CONTINUE
C
C         output results
          IF (DEVICE .EQ. FILE) THEN
             WRITE (GUNIT,75) E
             WRITE (GUNIT,70) (X(IX),K(IX),IX=1,NGRF)
          ELSE
             CALL XYPLOT(X,K)
          END IF
```

```
 50     CONTINUE
 C
 C      end graphing session
        IF (DEVICE .NE. FILE) CALL GPAGE(DEVICE)    !close graphing package
        IF (DEVICE .EQ. SCREEN) CALL TMODE          !switch to text mode
 C
 70     FORMAT (2(5X,E11.3))
 75     FORMAT (/,' Position vs. wave number for Energy =',1PE11.3)
100     FORMAT (/,' Patience, please; output going to a file.')
 C
        RETURN
        END
ccccccccccccccccccccccccccccccccccccccccccccccccccccccccccccccccccccccccc
ccccccccccccccccccccccccccccccccccccccccccccccccccccccccccccccccccccccccc
C param.e1
C
        REAL PI           !pi=3.15159.....
        REAL POTMIN       !x value at equilibrium
        REAL GAMMA        !=sqrt(2*mass*length**2*potential/hbar**2)
        REAL ETOL         !tolerance for energy search
        REAL XTOL         !tolerance for turning point search
        INTEGER NPTS      !number of integration points
        INTEGER NLEVEL    !total number of bounds states (depends on gamma)
C
        INTEGER NGRF      !number of points for graphics output
        INTEGER MAXGRF    !maximum number of graphing points
        INTEGER MAXLVL    !maximum number of quantum levels
C
        PARAMETER (MAXGRF=1000)
        PARAMETER (MAXLVL=100)
C
        COMMON / CONST / PI,POTMIN
        COMMON / PPARAM / GAMMA
        COMMON / NPARAM / ETOL,XTOL,NPTS,NGRF
        COMMON / PCALC / NLEVEL
ccccccccccccccccccccccccccccccccccccccccccccccccccccccccccccccccccccccccc
ccccccccccccccccccccccccccccccccccccccccccccccccccccccccccccccccccccccccc
C map.e1
        INTEGER IGAMMA,IETOL,IXTOL,INPTS,INGRF
        PARAMETER (IGAMMA = 13 )
        PARAMETER (IETOL  = 38 )
        PARAMETER (IXTOL  = 39 )
        PARAMETER (INPTS  = 40 )
        PARAMETER (INGRF  = 87 )
```

B.2 Example 2

Algorithm   The program integrates trajectories in the Hénon-Heiles potential (Eq. 2.35) and finds the corresponding surfaces of section (abbreviated to SOS in the code). Subroutine ARCHON creates the surface of section for one value of the energy by calling subroutine SOS once for each set of initial conditions requested. Subroutine SOS prompts for initial values of y and p_y, calculates p_x from energy conservation (recall $x = 0$ on the surface of section), and executes the time integration (subroutine STEP) until the requested number of surface of section points have been found. Subroutine STEP takes a time step using the fourth-order Runge-Kutta step (subroutine RUNGE); derivatives of the coordinates and momenta are calculated by subroutine EVAL and functions XDERIV and YDERIV. This routine also checks for crossings of the the trajectory with the y-axis (i.e., whenever $x = 0$). If a crossing is found, it temporarily switches to x as the independent variable and integrates back to find the surface of section intersection as described in the text.

The form of the potential may be changed by editing the potential V and its derivatives XDERIV and YDERIV which are located in function V. You will also need to change the graphing limits for the trajectory in subroutine TRJINT and should review the values DY0 and TOLY (subroutine INIT) used in searching. The limits on the energy (also subroutine INIT) should be the smallest and largest values of energy that give rise to bounded motion.

Input   The physical parameter entered from the main menu is the energy [.1]. The initial conditions for y [0.] and p_y [0.] are requested from subroutine SOS so that more than one set can be used in creating a single surface of section. Limits on y depend only on the energy and are calculated by subroutine LIMITS; limits of p_y depend on the energy and y, and are calculated in SOS. The numerical parameters are the time step [.12] and the number of surface of section points [100]. As soon as the requested number of surface of section points are calculated for one set of initial conditions, you will be prompted for the number of additional surface of section points [0]; if you answer none, you will be asked if you want additional initial conditions; if you answer no, the surface of section is complete and you return to the main menu. The maximum surface of section points is fixed by MAXSS= 10,000. The only graphics parameter is the number of points in the trajectory to be plotted [4000]. This is needed to prevent output files from becoming too large and plots from becoming too cluttered.

Output The text output at each surface of section crossing includes the time, kinetic energy, potential energy, total energy, and percent change in energy since the integration began. The graphics output is the trajectory (output as the integration proceeds) and the surface of section points (output when the requested number of points have been found). All surface of section points accumulated since the last exit from the main menu are plotted together, allowing the creation of sections similar to those shown in Figure 2.3.

Because the trajectory is described by a great deal of data, the trajectory is not saved in an array, and the following output procedures have been adopted: 1) the trajectory is never output to a file (only surface of section data are saved to file if requested), 2) if both graphics to the screen and graphics to hardcopy are requested, the trajectory is sent only to the screen, but the surface of section is sent to both the screen and hardcopy, and 3) if both text and graphics are requested to the screen, only the graphics is sent to the screen.

```
CCCCCCCCCCCCCCCCCCCCCCCCCCCCCCCCCCCCCCCCCCCCCCCCCCCCCCCCCCCCCCCCCCCCCCCCC
      PROGRAM EXMPL2
C     Example 2:  Trajectories in the Henon-Heiles potential
C  COMPUTATIONAL PHYSICS (FORTRAN VERSION)
C  by Steven E. Koonin and Dawn C. Meredith
C  Copyright 1989, Addison-Wesley Publishing Company
CCCCCCCCCCCCCCCCCCCCCCCCCCCCCCCCCCCCCCCCCCCCCCCCCCCCCCCCCCCCCCCCCCCCCCCCC
      CALL INIT          !display header screen, setup parameters
5     CONTINUE           !main loop/ execute once for each set of param
      CALL PARAM         !get input from screen
      CALL ARCHON        !calculates surface of section for one energy
      GOTO 5
      END
CCCCCCCCCCCCCCCCCCCCCCCCCCCCCCCCCCCCCCCCCCCCCCCCCCCCCCCCCCCCCCCCCCCCCCCCC
      SUBROUTINE ARCHON
C calculates surface of section (SOS) for one energy
C allows for several sets of initial conditions
CCCCCCCCCCCCCCCCCCCCCCCCCCCCCCCCCCCCCCCCCCCCCCCCCCCCCCCCCCCCCCCCCCCCCCCCC
C Global variables:
      INCLUDE 'IO.ALL'
      INCLUDE 'PARAM.E2'
C Local variables:
      REAL SSY(MAXSS),SSPY(MAXSS)  !surface of section points
      INTEGER NCROSS               !total number of SOS crossings
      INTEGER ANOTHR               !start with another set of init cond?
      INTEGER DEF                  !default number of SOS points
      INTEGER DEFMIN               !minimum number of SOS points
      INTEGER MORE                 !how many more SOS points?
      INTEGER SCREEN               !send to terminal
      INTEGER PAPER                !make a hardcopy
```

```
            INTEGER FILE                !send to a file
C Functions:
            INTEGER YESNO               !yes or no input
            INTEGER GETINT              !integer input
      DATA SCREEN,PAPER,FILE/1,2,3/
CCCCCCCCCCCCCCCCCCCCCCCCCCCCCCCCCCCCCCCCCCCCCCCCCCCCCCCCCCCCCCCCCCCCCCC
      NCROSS=0                          !no surface of section points yet
50    CONTINUE                          !loop over initial conditions
         CALL SOS(NCROSS,SSPY,SSY)      !calculate surface of section
         CALL CLEAR                     !allow for another set of init cond
         ANOTHR=YESNO(0,'Do you want another set of initial conditions?')
         IF (ANOTHR .EQ. 1) THEN
            DEF=MIN(100,MAXSS-NSOS)      !don't exceed storage space
            DEFMIN=MIN(10,MAXSS-NSOS)
            MORE=GETINT(DEF,DEFMIN,MAXSS-NSOS,
     +           'How many surface of section points?')
            NSOS=NSOS+MORE
         GOTO 50
         END IF
C
      IF (GHRDCP) CALL GRFOUT(PAPER,SSY,SSPY,1)   !hrdcpy once per energy
C
      RETURN
      END
CCCCCCCCCCCCCCCCCCCCCCCCCCCCCCCCCCCCCCCCCCCCCCCCCCCCCCCCCCCCCCCCCCCCCCC
      SUBROUTINE SOS(NCROSS,SSPY,SSY)
C finds the surface of section points (SSPY,SSY)
C for a single set of initial conditions;
C NCROSS keeps track of the total number of SOS points found
C for all initial conditions
CCCCCCCCCCCCCCCCCCCCCCCCCCCCCCCCCCCCCCCCCCCCCCCCCCCCCCCCCCCCCCCCCCCCCCC
C Global variables:
      INCLUDE 'IO.ALL'
      INCLUDE 'PARAM.E2'
C Passed variables:
      REAL SSY(MAXSS),SSPY(MAXSS)  !surface of section points (output)
      INTEGER NCROSS               !total number of SOS crossings (I/O)
C Local variables:
      REAL T                       !time
      REAL VAR(4)                  !coordinates and momenta
      REAL OLDVAR(4)               !previous values of VAR
      INTEGER NLINES               !number of lines written to terminal
      INTEGER NBEGIN         !index of first SOS point for current init cond
      LOGICAL CROSS                !have we just crossed a SOS?
      REAL PYMAX                   !limits on PY for this E
      INTEGER MORE                 !how many more SOS points?
      INTEGER IVAR                 !dependent variable index
      REAL TSTOP                   !time at last stop in integration
      INTEGER SCREEN               !send to terminal
      INTEGER PAPER                !make a hardcopy
      INTEGER FILE                 !send to a file
C Functions:
```

```
      INTEGER GETINT,YESNO        !get integer,boolean input from screen
      REAL V                      !potential function
      REAL GETFLT                 !get real input from screen
      DATA SCREEN,PAPER,FILE/1,2,3/
CCCCCCCCCCCCCCCCCCCCCCCCCCCCCCCCCCCCCCCCCCCCCCCCCCCCCCCCCCCCCCCCCCCCCCCCCCC
C     prompt for initial Y and PY; obtain PX from Energy conservation
      WRITE (OUNIT,*) ' '
      WRITE (OUNIT,*) ' Input initial conditions (recall X=0):'
      WRITE (OUNIT,*) ' '
      VAR(1)=0.
      VAR(2)=GETFLT(0.,CMIN,CMAX,' Enter initial value for y')
      PYMAX=SQRT(2*(EINIT-V(VAR(1),VAR(2))))
      VAR(4)=GETFLT(0.,-PYMAX,PYMAX,' Enter initial value for py')
      VAR(3)=SQRT(2*(EINIT-V(VAR(1),VAR(2)))-VAR(4)**2 )
C
C     output summary of parameters
      IF ((TTERM) .AND. (.NOT. GTERM)) CALL PRMOUT(OUNIT,NLINES,VAR)
      IF (TFILE) CALL PRMOUT(TUNIT,NLINES,VAR)
      IF (GFILE) CALL PRMOUT(GUNIT,NLINES,VAR)
C
C     setup initial values
      DO 500 IVAR=1,4
         OLDVAR(IVAR)=VAR(IVAR)
500   CONTINUE
      T=0.
      TSTOP=T
      NBEGIN=NCROSS+1
      IF (TRAJCT) CALL TRJINT(TDEV)      !prepare for graphing traj
C
C     integrate until we get requested number of SOS crossings
10    CONTINUE
         CALL STEP(T,VAR,OLDVAR,NCROSS,SSY,SSPY,NLINES,TSTOP)
      IF (NCROSS .LT. NSOS) GOTO 10
C
      IF (.NOT. GTERM) THEN
         IF (TTERM) CALL PAUSE('to continue...',1)
         IF (TTERM) CALL CLEAR
      ELSE
         CALL GRFOUT(SCREEN,SSY,SSPY,NBEGIN)      !graph SOS points
      END IF
C
C     see if user wishes more SOS points
      CALL CLEAR
      MORE=GETINT(0,0,MAXSS-NSOS,'How many more SOS points?')
      NSOS=NSOS+MORE
      IF (MORE .GT. 0) THEN
         NLINES=0
         IF (TRAJCT) CALL TRJINT(TDEV)
         TSTOP=T
      GOTO 10
      END IF
C
```

```
C       graphics output to a file, once per initial condition
        IF (GFILE) CALL GRFOUT(FILE,SSY,SSPY,NBEGIN)
C
        RETURN
        END
CCCCCCCCCCCCCCCCCCCCCCCCCCCCCCCCCCCCCCCCCCCCCCCCCCCCCCCCCCCCCCCCCCCCCCCC
        SUBROUTINE STEP(T,VAR,OLDVAR,NCROSS,SSY,SSPY,NLINES,TSTOP)
C integrates one time step and checks for SOS crossing
C finds SSY and SSPY on surface of section
CCCCCCCCCCCCCCCCCCCCCCCCCCCCCCCCCCCCCCCCCCCCCCCCCCCCCCCCCCCCCCCCCCCCCCCC
C Global variables:
        INCLUDE 'IO.ALL'
        INCLUDE 'PARAM.E2'
C Passed variables:
        REAL SSY(MAXSS),SSPY(MAXSS)   !surface of section points (output)
        INTEGER NCROSS                !total number of SOS crossings (I/O)
        REAL T                        !time (I/O)
        REAL VAR(4)                   !coordinates and momenta (I/O)
        REAL OLDVAR(4)                !previous values of VAR (I/O)
        INTEGER NLINES                !num of lines written to terminal(I/O)
        REAL TSTOP                    !time at last stop in integ (input)
C Local variables:
        LOGICAL CROSS                 !have we just crossed a SOS?
        REAL E,EPOT,EKIN              !energies
        INTEGER ITIME                 !number of time steps
        INTEGER IVAR                  !dependent function index
C Function:
        REAL V                        !value of the potential
        INTEGER SCREEN                !send to terminal
        INTEGER PAPER                 !make a hardcopy
        INTEGER FILE                  !send to a file
        DATA SCREEN,PAPER,FILE/1,2,3/
CCCCCCCCCCCCCCCCCCCCCCCCCCCCCCCCCCCCCCCCCCCCCCCCCCCCCCCCCCCCCCCCCCCCCCCC
        CROSS=.FALSE.
        T=T+TSTEP
        CALL RUNGE(CROSS,VAR,OLDVAR)                  !take a Runge-Kutta step
C
        EPOT=V(VAR(1),VAR(2))                         !calculate energies
        EKIN=(VAR(3)**2+VAR(4)**2)/2.0
        E=EKIN+EPOT
C
        IF (TRAJCT) THEN              !output trajectories, but only
          ITIME=(T-TSTOP)/TSTEP      !for a limited time
          IF ((ITIME .EQ. NTRJ+1) .AND. (TDEV .EQ. SCREEN)) THEN
              CALL NOTICE            !let user know we're still integ
          ELSE IF (ITIME .LE. NTRJ) THEN
              CALL TRJOUT(VAR,OLDVAR)   !output trajectories
          END IF
        END IF
C
        CROSS=(VAR(1)*OLDVAR(1)) .LT. 0.0            !check for sos crossing
C
```

```
      DO 100 IVAR=1,4                     !update variables
         OLDVAR(IVAR)=VAR(IVAR)
100   CONTINUE
C
      IF (CROSS) THEN                     !find SOS crossing and output results
         NCROSS=NCROSS+1
         CALL RUNGE(CROSS,VAR,OLDVAR)     !find SSY and SSPY
         SSY(NCROSS)=VAR(2)               !store SOS intersections
         SSPY(NCROSS)=VAR(4)
C
         DO 200 IVAR=1,4                  !reset old values
            VAR(IVAR)=OLDVAR(IVAR)
200      CONTINUE
C
         IF ((TDEV .NE. SCREEN) .AND. (TTERM))
     +      CALL TXTOUT(OUNIT,T,E,EPOT,EKIN,NLINES)  !text output
         IF (TFILE) CALL TXTOUT(TUNIT,T,E,EPOT,EKIN,NLINES)
      ENDIF
C
      RETURN
      END
CCCCCCCCCCCCCCCCCCCCCCCCCCCCCCCCCCCCCCCCCCCCCCCCCCCCCCCCCCCCCCCCCCCCCCC
      SUBROUTINE RUNGE(CROSS,VAR,OLDVAR)
C uses 4th order Runge-Kutta algorithm to integrate the
C four coupled ODE's;
C the independent variable is determined by the value of CROSS
CCCCCCCCCCCCCCCCCCCCCCCCCCCCCCCCCCCCCCCCCCCCCCCCCCCCCCCCCCCCCCCCCCCCCCC
C Global variables:
      INCLUDE 'PARAM.E2'
C Passed variables:
      LOGICAL CROSS                       !is this to find sos crossing?(input)
      REAL VAR(4)                         !coordinates and momenta (output)
      REAL OLDVAR(4)                      !previous values of VAR(input)
C Local variables:
      REAL F(4)                           !derivatives
      REAL K1(4),K2(4),K3(4),K4(4)        !increments
      REAL H                              !step size
      INTEGER I                           !dependent variable index
CCCCCCCCCCCCCCCCCCCCCCCCCCCCCCCCCCCCCCCCCCCCCCCCCCCCCCCCCCCCCCCCCCCCCCC
      H=TSTEP
      IF (CROSS) H=-VAR(1)
C
      CALL EVAL(F,VAR,CROSS)
      DO 100 I=1,4
         K1(I)=H*F(I)
         VAR(I)=OLDVAR(I)+K1(I)/2.0
100   CONTINUE
C
      CALL EVAL(F,VAR,CROSS)
      DO 200 I=1,4
         K2(I)=H*F(I)
         VAR(I)=OLDVAR(I)+K2(I)/2.0
```

```
200     CONTINUE
C
        CALL EVAL(F,VAR,CROSS)
        DO 300 I=1,4
           K3(I)=H*F(I)
           VAR(I)=OLDVAR(I)+K3(I)
300     CONTINUE
C
        CALL EVAL(F,VAR,CROSS)
        DO 400 I=1,4
           K4(I)=H*F(I)
           VAR(I)=OLDVAR(I)+(K1(I)+2*K2(I)+2*K3(I)+K4(I))/6
400     CONTINUE
C
        RETURN
        END
CCCCCCCCCCCCCCCCCCCCCCCCCCCCCCCCCCCCCCCCCCCCCCCCCCCCCCCCCCCCCCCCCCCCCCCC
        SUBROUTINE EVAL(F,VAR,CROSS)
C Evaluate derivatives F evaluated at VAR
C The independent variable is determined by CROSS
CCCCCCCCCCCCCCCCCCCCCCCCCCCCCCCCCCCCCCCCCCCCCCCCCCCCCCCCCCCCCCCCCCCCCCCC
C Passed variables:
        REAL VAR(4)               !variables (input)
        REAL F(4)                 !derivatives of variables (output)
        LOGICAL CROSS             !is this for a surface of section?(input)
C Local variables:
        REAL DENOM                !factor to calc SOS
C Functions:
        REAL XDERIV,YDERIV        !x and y derivatives of the potential
CCCCCCCCCCCCCCCCCCCCCCCCCCCCCCCCCCCCCCCCCCCCCCCCCCCCCCCCCCCCCCCCCCCCCCCC
C       to find surface of section, all derivatives are divided by PX
        DENOM=1.0
        IF (CROSS) DENOM=VAR(3)
C
        F(1)=VAR(3)/DENOM
        F(2)=VAR(4)/DENOM
        F(3)=-1.0*XDERIV(VAR(1),VAR(2))/DENOM
        F(4)=-1.0*YDERIV(VAR(1),VAR(2))/DENOM
C
        RETURN
        END
CCCCCCCCCCCCCCCCCCCCCCCCCCCCCCCCCCCCCCCCCCCCCCCCCCCCCCCCCCCCCCCCCCCCCCCC
        REAL FUNCTION V(X,Y)
C Calculates the potential and forces
C
C If you change the potential, you may also need to change DY0 and TOLY
C for Y limit searches, as well as limits for X and Y in TRJINT
CCCCCCCCCCCCCCCCCCCCCCCCCCCCCCCCCCCCCCCCCCCCCCCCCCCCCCCCCCCCCCCCCCCCCCCC
C Passed variables:
        REAL X,Y                  !coordinates
C Functions:
        REAL XDERIV,YDERIV        !x and y derivatives of the potential
```

```
ccccccccccccccccccccccccccccccccccccccccccccccccccccccccccccccccc
      V=(X**2+Y**2)/2+X**2*Y-Y**3/3
      RETURN
C
      ENTRY XDERIV(X,Y)
      XDERIV=X+2*X*Y
      RETURN
C
      ENTRY YDERIV(X,Y)
      YDERIV=Y+X**2-Y**2
      RETURN
C
      END
ccccccccccccccccccccccccccccccccccccccccccccccccccccccccccccccccc
      SUBROUTINE LIMITS
C Find limits on Y from energy conservation using a simple search
C This limit is on the surface of section where X=0.
ccccccccccccccccccccccccccccccccccccccccccccccccccccccccccccccccc
C Global variables:
      INCLUDE 'PARAM.E2'
C Local variables:
      REAL DY                       !step in simple search
      REAL V                        !potential (function)
ccccccccccccccccccccccccccccccccccccccccccccccccccccccccccccccccc
      DY=DY0                        !search for CMAX starting at 0
      CMAX=0.0
10    CONTINUE
        CMAX=CMAX+DY
        IF (V(0.0,CMAX) .GT. EINIT) THEN   !CMAX is where all the
          CMAX=CMAX-DY                     !energy is potntl energy
          DY=DY/2.                         !recall that X=0.0 on SOS
        ENDIF
      IF (DY.GE.TOLY) GOTO 10
C
      DY=DY0                        !search for CMIN starting at 0
      CMIN=0.0
20    CONTINUE
        CMIN=CMIN-DY
        IF ( V(0.0,CMIN) .GT. EINIT) THEN  !CMIN is where all the
          CMIN=CMIN+DY                     !energy is potntl energy
          DY=DY/2.                         !recall that X=0.0 on SOS
        ENDIF
      IF (DY.GE.TOLY) GOTO 20
C
      RETURN
      END
ccccccccccccccccccccccccccccccccccccccccccccccccccccccccccccccccc
      SUBROUTINE INIT
C initializes constants, displays header screen,
C initializes arrays for input parameters
ccccccccccccccccccccccccccccccccccccccccccccccccccccccccccccccccc
C Global variables:
```

```
            INCLUDE 'IO.ALL'
            INCLUDE 'MENU.ALL'
            INCLUDE 'PARAM.E2'
C Local parameters:
            CHARACTER*80 DESCRP           !program description
            DIMENSION DESCRP(20)
            INTEGER NHEAD,NTEXT,NGRAPH    !number of lines for each description
CCCCCCCCCCCCCCCCCCCCCCCCCCCCCCCCCCCCCCCCCCCCCCCCCCCCCCCCCCCCCCCCCCCCCCCC
            CALL SETUP                    !get environment parameters
C
C       display header screen
            DESCRP(1)= 'EXAMPLE 2'
            DESCRP(2)= 'Trajectories in the Henon-Heiles potential'
            NHEAD=2
C
C       text output description
            DESCRP(3)= 'kinetic, potential, and total energy;'
            DESCRP(4)=
          + 'and percent change in energy at surface of section crossing'
            NTEXT=2
C
C       graphics output description
            DESCRP(5)= 'trajectory and surface of section'
            NGRAPH=1
C
            CALL HEADER(DESCRP,NHEAD,NTEXT,NGRAPH)
C
C       calculate constants   (used in search for limits on y)
            DY0=0.1
            TOLY=.0005
C
            CALL MENU                     !constant part of menu
C
            MTYPE(13)=FLOAT
            MPRMPT(13)= 'Enter Energy'
            MTAG(13)= 'Energy'
            MLOLIM(13)=0.0
            MHILIM(13)=1.0/6.
            MREALS(13)=.1
C
            MTYPE(14)=SKIP
            MREALS(14)=35.
C
            MTYPE(38)=FLOAT
            MPRMPT(38)= 'Enter time step'
            MTAG(38)= 'Time step'
            MLOLIM(38)=.00001
            MHILIM(38)=.2
            MREALS(38)=.12
C
            MTYPE(39)=NUM
            MPRMPT(39)= 'Enter number of surface of section points'
```

```
      MTAG(39)= 'Number of surface of section points'
      MLOLIM(39)=10
      MHILIM(39)=MAXSS
      MINTS(39)=100
C
      MTYPE(40)=SKIP
      MREALS(40)=60.
C
      MSTRNG(MINTS(75))= 'exmpl2.txt'
C
      MTYPE(76)=SKIP
      MREALS(76)=80.
C
      MSTRNG(MINTS(86))= 'exmpl2.grf'
C
      MTYPE(87)=NUM
      MPRMPT(87)= 'Enter number of points in trajectory to plot'
      MTAG(87)= 'Number of points in trajectory'
      MLOLIM(87)=0
      MHILIM(87)=10000
      MINTS(87)=4000
C
      MTYPE(88)=SKIP
      MREALS(88)=90.
C
      RETURN
      END
CCCCCCCCCCCCCCCCCCCCCCCCCCCCCCCCCCCCCCCCCCCCCCCCCCCCCCCCCCCCCCCCCCCCCCCC
      SUBROUTINE PARAM
C gets parameters from screen
C ends program on request
C closes old files
C maps menu variables to program variables
C opens new files
C calculates all derivative parameters
CCCCCCCCCCCCCCCCCCCCCCCCCCCCCCCCCCCCCCCCCCCCCCCCCCCCCCCCCCCCCCCCCCCCCCCC
C Global variables:
      INCLUDE 'MENU.ALL'
      INCLUDE 'IO.ALL'
      INCLUDE 'PARAM.E2'
C Local variables:
      INTEGER SCREEN              !send to terminal
      INTEGER PAPER               !make a hardcopy
      INTEGER FILE                !send to a file
C map between menu items and parameters
      INTEGER IE,ITSTEP,INSOS,INTRJ
      PARAMETER (IE = 13 )
      PARAMETER (ITSTEP =38 )
      PARAMETER (INSOS =39 )
      PARAMETER (INTRJ = 87 )
C Function:
      LOGICAL LOGCVT              !converts 1 and 0 to true and false
```

```
          DATA SCREEN,PAPER,FILE/1,2,3/
CCCCCCCCCCCCCCCCCCCCCCCCCCCCCCCCCCCCCCCCCCCCCCCCCCCCCCCCCCCCCCCCCCCCCCCCCCC
C     get input from terminal
      CALL CLEAR
      CALL ASK(1,ISTOP)
C
C     stop program if requested
      IF (MREALS(IMAIN) .EQ. STOP) CALL FINISH
C
.C    close files if necessary
      IF (TNAME .NE. MSTRNG(MINTS(ITNAME)))
     +    CALL FLCLOS(TNAME,TUNIT)
      IF (GNAME .NE. MSTRNG(MINTS(IGNAME)))
     +    CALL FLCLOS(GNAME,GUNIT)
C
C     physical and numerical parameters
      EINIT=MREALS(IE)
      TSTEP=MREALS(ITSTEP)
      NSOS=MINTS(INSOS)
C
C     text output
      TTERM=LOGCVT(MINTS(ITTERM))
      TFILE=LOGCVT(MINTS(ITFILE))
      TNAME=MSTRNG(MINTS(ITNAME))
C
C     graphics output
      GTERM=LOGCVT(MINTS(IGTERM))
      GHRDCP=LOGCVT(MINTS(IGHRD))
      GFILE=LOGCVT(MINTS(IGFILE))
      GNAME=MSTRNG(MINTS(IGNAME))
      NTRJ=MINTS(INTRJ)
C
C     trajectories are output ONLY if graphics are available
C        (it's too much data to output to a file)
C     since traj can only go to one device, these are sent to the screen
      TRAJCT=.FALSE.
      TDEV=0
      IF (GTERM) THEN
         TDEV=SCREEN
         TRAJCT=.TRUE.
      ELSE IF (GHRDCP) THEN
         TDEV=PAPER
         TRAJCT=.TRUE.
      END IF
C
C     open files
      IF (TFILE) CALL FLOPEN(TNAME,TUNIT)
      IF (GFILE) CALL FLOPEN(GNAME,GUNIT)
C     files may have been renamed
      MSTRNG(MINTS(ITNAME))=TNAME
      MSTRNG(MINTS(IGNAME))=GNAME
C
```

```
C       calculate CMIN and CMAX (limits on Y on the surface of section)
        CALL LIMITS
C
        RETURN
        END
CCCCCCCCCCCCCCCCCCCCCCCCCCCCCCCCCCCCCCCCCCCCCCCCCCCCCCCCCCCCCCCCCCCCCCCCC
        SUBROUTINE PRMOUT(MUNIT,NLINES,VAR)
C outputs parameter summary to the specified unit
CCCCCCCCCCCCCCCCCCCCCCCCCCCCCCCCCCCCCCCCCCCCCCCCCCCCCCCCCCCCCCCCCCCCCCCCC
C Global variables:
        INCLUDE 'IO.ALL'
        INCLUDE 'PARAM.E2'
C Passed variables:
        INTEGER MUNIT           !unit number for output (input)
        INTEGER NLINES          !number of lines written so far (I/O)
        REAL VAR(4)             !initial values of coord and momenta (I)
C Local variables:
        INTEGER I               !independent variable index
CCCCCCCCCCCCCCCCCCCCCCCCCCCCCCCCCCCCCCCCCCCCCCCCCCCCCCCCCCCCCCCCCCCCCCCCC
        IF (MUNIT .EQ. OUNIT) CALL CLEAR
C
        WRITE (MUNIT,2)
        WRITE (MUNIT,4)
        WRITE (MUNIT,6) TSTEP
        WRITE (MUNIT,10) EINIT
        WRITE (MUNIT,8) CMIN,CMAX,SQRT(2*EINIT)
        WRITE (MUNIT,12) (VAR(I),I=1,4)
        WRITE (MUNIT,2)
C
C       different header for text and graphics files
        IF (MUNIT .EQ. GUNIT) THEN
          WRITE (MUNIT,20)
          WRITE (MUNIT,25)
        ELSE
          WRITE (MUNIT,30)
          WRITE (MUNIT,35)
          WRITE (MUNIT,40)
          WRITE (MUNIT,2)
        END IF
C
        NLINES=11
C
2       FORMAT (' ')
4       FORMAT (' Output from example 2: Trajectories in the Henon-',
     +          'Heiles Potential')
6       FORMAT (' Time step =', E12.5)
10      FORMAT (' Energy =',F6.3)
8       FORMAT (' Ymin =',F6.3, 5X,' Ymax =',F6.3,5X,'Pymax =',F6.3)
12      FORMAT (' Xinit=',F6.3, 5X,' Yinit=',F6.3,5X,'PXinit=',F6.3,
     +          5X,'PYinit=',F6.3)
20      FORMAT (7X,'Y on SOS',7X, 'PY on SOS')
25      FORMAT (7X,'--------',7X, '---------')
```

```
30      FORMAT (9X,'Time',9X,'Kinetic',4X,'Potential',5X,'Total',
       +          7X,'Percent')
35      FORMAT (23X,'Energy',6x,'Energy',6X,'Energy',7X,'Change')
40      FORMAT (9X,'----',9X,'-------',4X,'---------',6X,'-----',
       +          6X,'-------')
C
        RETURN
        END
CCCCCCCCCCCCCCCCCCCCCCCCCCCCCCCCCCCCCCCCCCCCCCCCCCCCCCCCCCCCCCCCCCCCCCCC
        SUBROUTINE TXTOUT(MUNIT,T,E,EPOT,EKIN,NLINES)
C writes out energy data at each sos crossing
CCCCCCCCCCCCCCCCCCCCCCCCCCCCCCCCCCCCCCCCCCCCCCCCCCCCCCCCCCCCCCCCCCCCCCCC
C Global variables:
        INCLUDE 'IO.ALL'
        INCLUDE 'PARAM.E2'
C Passed variables:
        INTEGER MUNIT              !output unit specifier (I)
        INTEGER NLINES             !number of lines printed to screen (I/O)
        REAL E,EPOT,EKIN           !energies (I)
        REAL T                     !time (I)
CCCCCCCCCCCCCCCCCCCCCCCCCCCCCCCCCCCCCCCCCCCCCCCCCCCCCCCCCCCCCCCCCCCCCCCC
C       if this is a new page, retype headings
        IF ((NLINES .EQ. 0) .AND. (MUNIT .EQ. OUNIT)) THEN
            CALL CLEAR
            WRITE (MUNIT,30)
            WRITE (MUNIT,35)
            WRITE (MUNIT,40)
            WRITE (MUNIT,2)
            NLINES=NLINES+4
C       else if screen is full, clear screen and retype headings
        ELSE IF ((MOD(NLINES,TRMLIN-4) .EQ. 0)
       +                    .AND. (MUNIT .EQ. OUNIT)) THEN
            CALL PAUSE('to continue...',1)
            CALL CLEAR
            WRITE (MUNIT,30)
            WRITE (MUNIT,35)
            WRITE (MUNIT,40)
            WRITE (MUNIT,2)
            NLINES=NLINES+4
        END IF
C
        WRITE (MUNIT,20)T,EKIN,EPOT,E,ABS((E-EINIT)/EINIT)
C
C       keep track of printed lines only for terminal output
        IF (MUNIT .EQ. OUNIT) NLINES=NLINES+1
C
20      FORMAT (5X,1PE12.5,3(5X,0PF7.5),5X,1PE10.3)
2       FORMAT (' ')
30      FORMAT (9X,'Time',9X,'Kinetic',4X,'Potential',5X,'Total',
       +          7X,'Percent')
35      FORMAT (23X,'Energy',6x,'Energy',6X,'Energy',7X,'Change')
40      FORMAT (9X,'----',9X,'-------',4X,'---------',6X,'-----',
```

```
      +           6X,'-------')
C
      RETURN
      END
CCCCCCCCCCCCCCCCCCCCCCCCCCCCCCCCCCCCCCCCCCCCCCCCCCCCCCCCCCCCCCCCCCCCCCC
      SUBROUTINE GRFOUT(DEVICE,SSY,SSPY,NBEGIN)
C outputs surface of section from NBEGIN to NSOS
CCCCCCCCCCCCCCCCCCCCCCCCCCCCCCCCCCCCCCCCCCCCCCCCCCCCCCCCCCCCCCCCCCCCCCC
C Global parameters:
      INCLUDE 'IO.ALL'
      INCLUDE 'PARAM.E2'
      INCLUDE 'GRFDAT.ALL'
C Input variables:
      INTEGER DEVICE                    !which device is being used?
      REAL SSY(MAXSS),SSPY(MAXSS)       !Y and PY values on SOS
      INTEGER NBEGIN                    !beginning SOS for these init cond
C Local parameters:
      INTEGER I                         !indexes SOS points
      INTEGER SCREEN                    !send to terminal
      INTEGER PAPER                     !make a hardcopy
      INTEGER FILE                      !send to a file
      DATA SCREEN,PAPER,FILE/1,2,3/
CCCCCCCCCCCCCCCCCCCCCCCCCCCCCCCCCCCCCCCCCCCCCCCCCCCCCCCCCCCCCCCCCCCCCCC
C     messages for the impatient
      IF (DEVICE .NE. SCREEN) WRITE (OUNIT,100)
C
C     calculate parameters for graphing
      IF (DEVICE .NE. FILE) THEN
C
          NPLOT=2                            !how many plots
          IPLOT=2
C
C         if both screen and paper are used , only SOS are sent to paper
          IF ((GTERM) .AND. (DEVICE .EQ. PAPER)) THEN
              NPLOT=1
              IPLOT=1
              CALL GTDEV(DEVICE)                  !device nomination
          END IF
C
          YMAX=SQRT(2*EINIT)                  !horiz axis is y
          YMIN=-YMAX                          !vert axis is py
          XMAX=CMAX
          XMIN=CMIN
          YOVAL=CMIN
          XOVAL=0.
C
          NPOINT=NSOS
C
          ILINE=5                             !line and symbol styles
          ISYM=5                              !this choice gives unconnected dots
          IFREQ=1
          NXTICK=7
```

```
                NYTICK=7
C
                INFO=' '
                LABEL(1)='Y'
                LABEL(2)='PY'
C
                CALL LNLNAX                           !draw axes
            END IF
C
C       output results
            IF (DEVICE .EQ. FILE) THEN
                WRITE (GUNIT,70) (SSY(I),SSPY(I),I=NBEGIN,NSOS)
            ELSE
                CALL XYPLOT(SSY,SSPY)
            END IF
C
C       end graphing session
            IF (DEVICE .NE. FILE) CALL GPAGE(DEVICE) !close graphing package
            IF (DEVICE .EQ. SCREEN) CALL TMODE       !switch back to text mode
C
70      FORMAT (2(5X,E11.3))
100     FORMAT (/,' Patience, please; output going to a file.')
C
        RETURN
        END
CCCCCCCCCCCCCCCCCCCCCCCCCCCCCCCCCCCCCCCCCCCCCCCCCCCCCCCCCCCCCCCCCCCCCCCCC
        SUBROUTINE TRJINT(DEVICE)
C prepares to graph trajectories
CCCCCCCCCCCCCCCCCCCCCCCCCCCCCCCCCCCCCCCCCCCCCCCCCCCCCCCCCCCCCCCCCCCCCCCCC
C Global parameters:
        INCLUDE 'IO.ALL'
        INCLUDE 'GRFDAT.ALL'
        INCLUDE 'PARAM.E2'
C Input variables:
        INTEGER DEVICE                       !which device is being used?
C Local parameters:
        CHARACTER *9 CE                      !energy as a character string
        INTEGER LEN                          !length of string
        REAL X(4),Y(4)                       !corners of the triangle
        INTEGER SCREEN                       !send to terminal
        INTEGER PAPER                        !make a hardcopy
        INTEGER FILE                         !send to a file
        DATA SCREEN,PAPER,FILE/1,2,3/
CCCCCCCCCCCCCCCCCCCCCCCCCCCCCCCCCCCCCCCCCCCCCCCCCCCCCCCCCCCCCCCCCCCCCCCCC
        NPLOT=2                              !how many plots
        IPLOT=1
C
        YMAX=1.                              !corners of bounding triangle
        YMIN=-.5
        XMAX=SQRT(3.)/2.
        XMIN=-XMAX
        Y0VAL=0.
```

```
      X0VAL=0.
C
      NPOINT=4
C
      ILINE=1                          !line and symbol styles
      ISYM=1
      IFREQ=0
      NXTICK=4
      NYTICK=4
C
      CALL CONVRT(EINIT,CE,LEN)
      TITLE='Henon-Heiles Potential, Energy='//CE
      LABEL(1)='X'
      LABEL(2)='Y'
C
      CALL GTDEV(DEVICE)                       !device nomination
      IF (DEVICE .EQ. SCREEN) CALL GMODE       !change to graphics mode
      CALL LNLNAX                              !draw axes
C
      X(1)=XMAX                                !draw the bounding triangle
      Y(1)=YMIN
      X(2)=XMIN
      Y(2)=YMIN
      X(3)=0.
      Y(3)=YMAX
      X(4)=X(1)
      Y(4)=Y(1)
      CALL XYPLOT(X,Y)
C
      RETURN
      END
CCCCCCCCCCCCCCCCCCCCCCCCCCCCCCCCCCCCCCCCCCCCCCCCCCCCCCCCCCCCCCCCCCCCCCCCC
      SUBROUTINE TRJOUT(VAR,OLDVAR)
C outputs trajectory, one line segment per call
CCCCCCCCCCCCCCCCCCCCCCCCCCCCCCCCCCCCCCCCCCCCCCCCCCCCCCCCCCCCCCCCCCCCCCCCC
C Global parameters:
      INCLUDE 'PARAM.E2'
      INCLUDE 'GRFDAT.ALL'
C Input variables:
      REAL VAR(4)                             !coordinates and momenta
      REAL OLDVAR(4)                          !previous values of VAR
C Local parameters:
      REAL X(2),Y(2)                          !coordinates
CCCCCCCCCCCCCCCCCCCCCCCCCCCCCCCCCCCCCCCCCCCCCCCCCCCCCCCCCCCCCCCCCCCCCCCCC
      NPOINT=2
      X(1)=OLDVAR(1)
      Y(1)=OLDVAR(2)
      X(2)=VAR(1)
      Y(2)=VAR(2)
      CALL XYPLOT(X,Y)
      RETURN
      END
```

```
CCCCCCCCCCCCCCCCCCCCCCCCCCCCCCCCCCCCCCCCCCCCCCCCCCCCCCCCCCCCCCCCCCCCCCCCCC
      SUBROUTINE NOTICE
C let user know we're still computing, even though trajectory isn't
C being plotted
CCCCCCCCCCCCCCCCCCCCCCCCCCCCCCCCCCCCCCCCCCCCCCCCCCCCCCCCCCCCCCCCCCCCCCCCCC
C Global variables:
      INCLUDE 'GRFDAT.ALL'
      INFO=' Still computing, but no longer plotting ...'
      CALL LEGEND
      RETURN
      END
CCCCCCCCCCCCCCCCCCCCCCCCCCCCCCCCCCCCCCCCCCCCCCCCCCCCCCCCCCCCCCCCCCCCCCCCCC
CCCCCCCCCCCCCCCCCCCCCCCCCCCCCCCCCCCCCCCCCCCCCCCCCCCCCCCCCCCCCCCCCCCCCCCCCC
C param.e2
C
      REAL DY0,TOLY         !constants for search to find limits on y
      REAL EINIT            !initial energy
      REAL TSTEP            !time step value
      INTEGER NSOS          !number of surface of section points
      REAL CMIN,CMAX        !limits on Y coordinate for this value of E
C
      LOGICAL TRAJCT        !are trajectories to be plotted
      INTEGER TDEV          !where are trajectories to be plotted
      INTEGER NTRJ          !points in traj to be plotted
C
      INTEGER MAXSS         !maximum number of sos points
      PARAMETER (MAXSS=10000)
C
      COMMON / CONST  / DY0,TOLY
      COMMON / PPARAM / EINIT
      COMMON / NPARAM / TSTEP,NSOS
      COMMON / PCALC  / CMIN,CMAX
      COMMON / TPLOT  / TDEV,TRAJCT,NTRJ
```

B.3 Example 3

Algorithm This program finds the stationary states of the one-dimensional Schroedinger equation for a particle in a box (i.e., the potential goes to infinity at the edges of the lattice) with various shapes for the bottom. The parameter γ, defined in the text, controls the classical nature of the system; the larger γ, the more classical the system.

Three analytical forms of the potential bottom are available (subroutine POTNTL): square well, parabolic, and Lennard-Jones. In addition, the square well can have a square bump, and the parabolic well can be flattened symmetrically about $x = 0$. The potentials are all normalized so that the minimum is at -1, and the maximum (exclusive of the hard walls) is at $+1$. If you code in another potential, be sure to follow this convention. The minimum and maximum values of x are set in subroutine INIT and are different for each potential. Also in subroutine POTNTL, guesses are made for the ground state energy (EBOTTM) and ground state energy spacing (DEBOT). These are based on the analytical values for the square and parabolic wells, and on a parabolic approximation for the Lennard-Jones potential, and are functions of γ. DEBOT is actually one-fifth of the expected spacing, and EBOTTM is two energy steps below the expected ground state. The search will find the ground state energy if it starts with these values. If you code in your own potential, you should also code in appropriate values for EBOTTM and DEBOT.

For each level sought above the starting energy (IF loop 5 in subroutine ARCHON), a search is made to find the energy for which Eq. (3.23) is satisfied. A simple search is used until f changes sign, whereupon the secant method is employed (IF loop 10 in subroutine SEARCH). To find f for a given energy, the Schroedinger equation is integrated leftward and rightward (subroutine NUMERV) using the Numerov algorithm, and the solutions are matched at the leftmost turning point (IF loop 1 in subroutine NUMERV). The search normally stops when f becomes smaller than TOLF, but will also stop if DE becomes smaller than machine precision, or if the number of steps taken exceeds the value of MAXSTP input. This last feature is added to allow graceful recovery from searches that have "gone wild"; this situation is likely to arise, for example, when two levels are nearly degenerate. When a search is ended, you have the option of finding another level or returning to the main menu.

Subroutine DETAIL determines the phase of the wave function based on the procedure given in the text, counts the number of nodes, and detects integration into classically forbidden regions.

Input   The physical parameters are γ [30] and the shape of the potential [Lennard-Jones]. The numerical parameters are the matching tolerance [5.E-05] (when f is smaller than this value, the search ends), the number of lattice points [160] whose upper limit is determined by the parameter MAXLAT= 1000, and the maximum number of search steps allowed [50].

At the beginning of each search you are prompted for a starting energy and energy step. If you have just left the main menu, the default values are EBOTTM and DEBOT from subroutine POTNTL; if you have just found a level, the default spacing is the spacing used at the *beginning* of the most recent search, and the default energy is the sum of the eigenvalue just found and the default spacing. The energy is in units of the potential depth V_0.

Output   As the search proceeds, the program outputs the energy, the energy spacing, f, and the number of nodes. In addition, the last column, labeled "Frbddn", displays a "yes" if the integration has entered a classically forbidden region; otherwise, a "no". When an eigenvalue is found, the eigenvalue, number of nodes, and classical turning points are displayed. At the end of each search, the graphics display the potential, all eigenvalues found since you last left the main menu (indicated on the graph by a line at the appropriate level between the left-most and right-most turning points), and the current eigenfunction. A maximum of MAXLEV= 50 eigenvalues can be stored; if you search for more levels, the earliest eigenvalues will be overwritten.

```
cccccccccccccccccccccccccccccccccccccccccccccccccccccccccccccccccccccc
      PROGRAM EXMPL3
C     Example 3: Bound states in a one-dimensional potential
C  COMPUTATIONAL PHYSICS (FORTRAN VERSION)
C  by Steven E. Koonin and Dawn C. Meredith
C  Copyright 1989, Addison-Wesley Publishing Company
cccccccccccccccccccccccccccccccccccccccccccccccccccccccccccccccccccccc
      CALL INIT              !display header screen, setup parameters
5     CONTINUE               !main loop; execute once for each set of param
         CALL PARAM          !get input parameters
         CALL ARCHON         !solve time-independent Schroedinger equation
      GOTO 5
      END
cccccccccccccccccccccccccccccccccccccccccccccccccccccccccccccccccccccc
      SUBROUTINE ARCHON
C solves the time-independent Schroedinger equation for 1-dimensional
C potential; allows for more than one eigenvalue search
cccccccccccccccccccccccccccccccccccccccccccccccccccccccccccccccccccccc
C Global variables:
```

```
      INCLUDE 'IO.ALL'
      INCLUDE 'PARAM.E3'
C Local variables:
      REAL ENERGY(MAXLEV)                  !array of eigenvalues
      INTEGER LEFTTP(MAXLEV),RGHTTP(MAXLEV)   !array of turning points
      INTEGER NFOUND                       !number of levels found
      REAL PSI(MAXLAT)                     !wave function
      INTEGER ITER                         !continue finding levels?
      INTEGER NODES                        !number of nodes
      REAL E                               !trial eigenenergy
      INTEGER LTP,RTP                      !classical turning points (lattice)
      REAL DE,DESAVE                       !step in energy search
      INTEGER NSTP                         !number of steps taken in search
      INTEGER SCREEN                       !send to terminal
      INTEGER PAPER                        !make a hardcopy
      INTEGER FILE                         !send to a file
C Functions:
      REAL GETFLT                          !obtain real input
      INTEGER YESNO                        !obtain yes or no input
      DATA SCREEN,PAPER,FILE/1,2,3/
CCCCCCCCCCCCCCCCCCCCCCCCCCCCCCCCCCCCCCCCCCCCCCCCCCCCCCCCCCCCCCCCCCCCCCCCC
C     output summary of parameters
      IF (TTERM) CALL PRMOUT(OUNIT)
      IF (TFILE) CALL PRMOUT(TUNIT)
      IF (GFILE) CALL PRMOUT(GUNIT)
C
      ITER=1                               !initialize searching variables
      NFOUND=0
      E=EBOTTM                             !reasonable starting values
      DE=DEBOT
C
5     IF (ITER .EQ . 1) THEN               !loop over energy levels
C
          E=GETFLT(E,EMIN,EMAX,'Enter energy')     !get starting values
          DE=GETFLT(DE,DEMIN,DEMAX,'Enter step in energy search')
          DESAVE=DE                        !save for next level
C
          IF (TTERM) CALL CTITL(OUNIT)!output titles
          IF (TFILE) CALL CTITL(TUNIT)
C
          CALL SEARCH(E,DE,LTP,RTP,NODES,NSTP,PSI) !search for energy
C
C         prompt for new E and DE if too many steps taken in search
          IF (NSTP .GE. MAXSTP) THEN
              WRITE (OUNIT,*) ' '
              WRITE (OUNIT,*) ' Eigenstate not yet found'
              ITER=YESNO(1,
     +            'Do you want to try again with new E or DE?')
          ELSE
C         or output eigenvalue information
              IF (TTERM) CALL EOUT(E,LTP,RTP,NODES,OUNIT)
              IF (TFILE) CALL EOUT(E,LTP,RTP,NODES,TUNIT)
```

```
                IF (GFILE) CALL EOUT(E,LTP,RTP,NODES,GUNIT)
C               save energy and turning points
                IF (NFOUND .EQ. MAXLEV) THEN
                   NFOUND=0
                   WRITE (OUNIT,*) ' Storage capacity for levels exceeded'
                   WRITE (OUNIT,*) ' Old information will be overwritten'
                END IF
                NFOUND=NFOUND+1
                ENERGY(NFOUND)=E
                LEFTTP(NFOUND)=LTP
                RGHTTP(NFOUND)=RTP
C
                IF (TTERM) CALL PAUSE('to continue...',1)
                IF (TTERM) CALL CLEAR
C
                IF (GTERM) THEN              !graphics output
                   CALL GRFOUT(SCREEN,ENERGY,LEFTTP,RGHTTP,NFOUND,PSI)
                   CALL CLEAR
                END IF
                IF (GFILE) CALL GRFOUT(FILE,ENERGY,LEFTTP,RGHTTP,NFOUND,PSI)
                IF (GHRDCP)
       +           CALL GRFOUT(PAPER,ENERGY,LEFTTP,RGHTTP,NFOUND,PSI)
C
C               check if another level is to be found
                WRITE (OUNIT,*) ' '
                WRITE (OUNIT,*) ' '
                ITER=YESNO(1,'Do you want to find another level?')
C
             END IF
C
             E=E+DESAVE                 !reasonable starting values for next level
             DE=DESAVE
C
          GOTO 5
          END IF
C
       RETURN
       END
CCCCCCCCCCCCCCCCCCCCCCCCCCCCCCCCCCCCCCCCCCCCCCCCCCCCCCCCCCCCCCCCCCCCCCCCCCCCC
       SUBROUTINE SEARCH(E,DE,LTP,RTP,NODES,NSTP,PSI)
C search for one eigenvalue beginning with energy E and step DE
CCCCCCCCCCCCCCCCCCCCCCCCCCCCCCCCCCCCCCCCCCCCCCCCCCCCCCCCCCCCCCCCCCCCCCCCCCCCC
C Global variables:
       INCLUDE 'IO.ALL'
       INCLUDE 'PARAM.E3'
C Passed variables:
       INTEGER NODES           !number of nodes (output)
       REAL E                  !trial eigenenergy, true eigenenergy (I/O)
       INTEGER LTP,RTP         !classical turning points (lattice)(output)
       REAL DE                 !step in energy search (input)
       INTEGER NSTP            !number of steps taken in search (output)
       REAL PSI(MAXLAT)        !wave function (output)
```

```
C Local variables:
      LOGICAL SECANT           !are we doing secant search yet?
      REAL F,FOLD,FSTART       !values of the mismatch
      REAL EOLD                !last value of the energy
      INTEGER NCROSS           !classically forbidden regions
      LOGICAL FIRST            !signal first step in search
      INTEGER NLINES           !number of lines sent to screen
C Functions:
      REAL GETFLT              !obtain real input
CCCCCCCCCCCCCCCCCCCCCCCCCCCCCCCCCCCCCCCCCCCCCCCCCCCCCCCCCCCCCCCCCCCCCCCC
      FIRST=.TRUE.             !first call to NUMERV for this level
      SECANT=.FALSE.           !start off with simple search
      F=TOLF                   !dummy value to get into loop
      NLINES=2
      NSTP=0
C
C     search for eigenenergy until F is small enough or too many
C     steps are taken
10    IF ((ABS(F) .GE. TOLF) .AND. (NSTP .LT. MAXSTP)) THEN
         NSTP=NSTP+1
C        integ Schroedinger equation; find discontinuity in derivative
         CALL NUMERV (F,NODES,E,LTP,RTP,NCROSS,FIRST,PSI)
         IF (FIRST .EQV. .TRUE.) THEN
            FSTART=F
            FIRST=.FALSE.
         END IF
         IF (F*FSTART .LE. 0.) SECANT=.TRUE.
         IF ((SECANT .EQV. .TRUE.) .AND. (F .NE. FOLD)) THEN
            DE=-F*(E-EOLD)/(F-FOLD)
         END IF
C
         IF (TTERM) CALL TXTOUT(OUNIT,E,DE,F,NODES,NCROSS,NLINES)
         IF (TFILE) CALL TXTOUT(TUNIT,E,DE,F,NODES,NCROSS,NLINES)
C
         EOLD=E                !update values for next step
         FOLD=F
         E=E+DE
C
C        keep energy greater than VMIN=-1
         IF (E .LT. -1.) THEN
            WRITE (OUNIT,*) ' Energy is less than -1.'
            E=GETFLT(-.99,EMIN,EMAX,' Enter new energy')
            DE=GETFLT(DE,DEMIN,DEMAX,'Enter step in energy search')
         END IF
C
C        end search if DE is smaller than machine precision
         IF ((ABS(DE) .LT. DEMIN) .AND. (ABS(F) .GT. TOLF)) THEN
            WRITE (OUNIT,20)
20          FORMAT('0',21X,'Delta E is too small, search must stop')
            F=TOLF/2           !dummy value to end search
         END IF
C
```

```
            GOTO 10
            END IF
C
            RETURN
            END
CCCCCCCCCCCCCCCCCCCCCCCCCCCCCCCCCCCCCCCCCCCCCCCCCCCCCCCCCCCCCCCCCCCC
            SUBROUTINE NUMERV (F,NODES,E,LTP,RTP,NCROSS,FIRST,PSI)
C subroutine to integrate time independent Schroedinger equation
C with energy=E
CCCCCCCCCCCCCCCCCCCCCCCCCCCCCCCCCCCCCCCCCCCCCCCCCCCCCCCCCCCCCCCCCCCC
C Global variables:
            INCLUDE 'PARAM.E3'
C Passed variables:
            REAL F                       !size of mismatch (output)
            INTEGER NODES                !number of nodes (output)
            REAL E                       !trial eigenvalue (input)
            INTEGER LTP,RTP              !classical turning points (output)
            INTEGER NCROSS               !classically forbidden regions (output)
            LOGICAL FIRST                !signal first step in search (output)
            REAL PSI(MAXLAT)             !wave function (output)
C Local variables:
            INTEGER IMATCH               !matching lattice point
            REAL C                       !useful constant
            INTEGER IX,KX                !X indices
            REAL KI,KIM1,KIP1            !terms in Numerov algorithm
            REAL NORM                    !norm of wave function
            REAL PMMTCH,PPMTCH           !PSI at match point
            LOGICAL FLIP                 !do we need to flip sign of PSI
            Real PSIMAX                  !max value of wave function
CCCCCCCCCCCCCCCCCCCCCCCCCCCCCCCCCCCCCCCCCCCCCCCCCCCCCCCCCCCCCCCCCCCC
      IF (FIRST .EQV. .TRUE.) IMATCH=0     !signals search for IMATCH
      C=(DX*DX/12)*GAMMA*GAMMA             !evaluate constant
C
C     find IMATCH once per energy level by looking for entry
C     into classically forbidden regions
      IX=1
      KIM1=C*(E-V(IX))
1     IF (IMATCH .EQ. 0) THEN
            IX=IX+1
            KI=C*(E-V(IX))        !neg value of K indicates class frbdn region
            IF ((KI*KIM1 .LT. 0) .AND. (KIM1 .GT. 0)) IMATCH=IX
C           if other procedure fails to find IMATCH, set IMATCH=NPTS-10
            IF (IX .EQ. NPTS-10) IMATCH=IX
            KIM1=KI
      GOTO 1
      END IF
C
      PSI(1)=0                          !left hand side bound. cond.
      PSI(2)=9.999999E-10
      KIM1=C*(E-V(1))                   !initial K*K values
      KI=C*(E-V(2))
C
```

```
C      Numerov algorithm; S=0; integrate until enter class. frbdn. region
       DO 10 IX=2,IMATCH
          KIP1=C*(E-V(IX+1))
          PSI(IX+1)=(PSI(IX)*(2.-10.*KI)-PSI(IX-1)*(1+KIM1))/(1+KIP1)
          KIM1=KI                        !roll values of K*K
          KI=KIP1
C
C         if PSI grows too large rescale all previous points
          IF (ABS(PSI(IX+1)) .GT. (1.0E+10)) THEN
             DO 20 KX=1,IX+1
                PSI(KX)=PSI(KX)*9.999999E-06
20           CONTINUE
          END IF
10     CONTINUE
       PMMTCH=PSI(IMATCH)               !save value for normalization
C
       PSI(NPTS)=0                      !rhs boundary conditions
       PSI(NPTS-1)=9.999999E-10
       KIP1=C*(E-V(NPTS))              !initial K*K values
       KI=C*(E-V(NPTS-1))
C
C      Numerov algorithm, S=0;integrate from rhs to IMATCH
       DO 30 IX=NPTS-1,IMATCH+1,-1
          KIM1=C*(E-V(IX-1))
          PSI(IX-1)=(PSI(IX)*(2.-10.*KI)-PSI(IX+1)*(1+KIP1))/(1+KIM1)
          KIP1=KI                       !roll values of K*K
          KI=KIM1
C
C         if PSI grows too large rescale all previous points
          IF (ABS(PSI(IX-1)) .GT. (1.0E+10)) THEN
             DO 40 KX=NPTS-1,IX-1,-1
                PSI(KX)=PSI(KX)*9.999999E-06
40           CONTINUE
          END IF
30     CONTINUE
C
       KIM1=C*(E-V(IMATCH-1))           !finds values needed for log deriv
       PPMTCH=(PSI(IMATCH)*(2-10*KI)-PSI(IMATCH+1)*(1+KIP1))/(1+KIM1)
C
       NORM=PMMTCH/PSI(IMATCH)          !norm PSI right to PSI left
       DO 5 IX=IMATCH,NPTS
          PSI(IX)=PSI(IX)*NORM
5      CONTINUE
       PPMTCH=PPMTCH*NORM
C
C      find nodes, turning points, entry into classically forb regions
C      and determine if overall sign must be flipped
       CALL DETAIL(IMATCH,NODES,E,LTP,RTP,NCROSS,PMMTCH,FLIP,FIRST,PSI)
C
C      find maximum PSI value and flip sign if necessary
       PSIMAX=ABS(PSI(1))
       DO 7 IX=1,NPTS
```

```
              IF (ABS(PSI(IX)) .GT. PSIMAX) PSIMAX=ABS(PSI(IX))
              IF (FLIP .EQV. .TRUE.) PSI(IX)=-PSI(IX)
7      CONTINUE
       IF (FLIP) PPMTCH=-PPMTCH
C
       F=(PSI(IMATCH-1)-PPMTCH)/PSIMAX      !evaluate matching condition
C
       RETURN
       END
CCCCCCCCCCCCCCCCCCCCCCCCCCCCCCCCCCCCCCCCCCCCCCCCCCCCCCCCCCCCCCCCCCCCCCCC
       SUBROUTINE
     +     DETAIL(IMATCH,NODES,E,LTP,RTP,NCROSS,PMMTCH,FLIP,FIRST,PSI)
C subroutine to calculates nodes, turning points, entry
C into classically forbidden regions, and whether or not the
C overall sign of PSI must be flipped to make F continuous function of E
CCCCCCCCCCCCCCCCCCCCCCCCCCCCCCCCCCCCCCCCCCCCCCCCCCCCCCCCCCCCCCCCCCCCCCCC
C Global variables:
       INCLUDE 'PARAM.E3'
C Passed variables:
       INTEGER IMATCH              !matching lattice point (input)
       INTEGER NODES               !number of nodes (output)
       REAL E                      !trial eigenvalue (input)
       INTEGER LTP,RTP             !classical turning points (output)
       INTEGER NCROSS              !classically forbidden regions (output)
       REAL PMMTCH                 !PSI at match point (input)
       LOGICAL FIRST               !signal first step in search(input)
       LOGICAL FLIP                !do we need to flip sign of PSI (output)
       REAL PSI(MAXLAT)            !wave function (input)
C Local variables:
       LOGICAL LSAME               !does PSI left retain its sign?
       INTEGER IX                  !X index
       REAL K,KLAST                !wavenumbers
       INTEGER NLOLD,NLFT          !left nodes
       INTEGER NROLD,NRT           !right nodes
       REAL PSIR,PSIL              !sign of PSI which must be constant
CCCCCCCCCCCCCCCCCCCCCCCCCCCCCCCCCCCCCCCCCCCCCCCCCCCCCCCCCCCCCCCCCCCCCCCC
C      find number of nodes on each side
       NLOLD=NLFT                  !save old values first
       NROLD=NRT
       NLFT=0                      !zero sums
       NRT=0
       DO 8 IX=2,IMATCH-1          !left side
          IF (PSI(IX)*PSI(IX-1) .LT. 0) NLFT=NLFT+1
8      CONTINUE
       IF (PSI(IMATCH-1)*PMMTCH .LT. 0) NLFT=NLFT+1
       DO 10 IX=IMATCH+1,NPTS           !right side
          IF (PSI(IX)*PSI(IX-1) .LT. 0) NRT=NRT+1
10     CONTINUE
C
C      check for change in node number
C      if one side of the wave function has gained a node,
C      that side must maintain the same sign between steps
```

```
C       in order that F be a continuous function of Energy.
C       if no new node has appeared, keep same sign same
C       as last time
        IF ((NLOLD .NE. NLFT) .AND. (NRT .EQ. NROLD)) THEN
            IF (LSAME .EQV. .FALSE.) LSAME=.TRUE.
        ELSE IF ((NROLD .NE. NRT) .AND. (NLOLD .EQ. NLFT)) THEN
            IF (LSAME .EQV. .TRUE.) LSAME=.FALSE.
        END IF
C
        IF (FIRST .EQV. .TRUE.) LSAME=.TRUE.       !initialize variables
        IF (FIRST .EQV. .TRUE.) PSIL=1.
C
C       now, finally, determine if the sign needs to be flipped
        FLIP=.FALSE.
        IF ((LSAME) .AND. (PSIL*PSI(2) .LT. 0.)) FLIP=.TRUE.
        IF (( .NOT. LSAME ) .AND. (PSIR*PSI(NPTS-1) .LT. 0.))
     +      FLIP=.TRUE.
C
C       save values for next step
        PSIL=PSI(2)
        PSIR=PSI(NPTS-1)
        IF (FLIP) PSIL=-PSIL
        IF (FLIP) PSIR=-PSIR
C
        NODES=NLFT+NRT                 !total number of nodes
C
C       find leftmost turning point
        LTP=0
        IX=0
110     IF (LTP .EQ. 0) THEN
            IX=IX+1
            IF ((E-V(IX)) .GE. 0) LTP=IX
        GOTO 110
        END IF
C
C       find rightmost turning point
        RTP=0
        IX=NPTS+1
120     IF (RTP .EQ. 0) THEN
            IX=IX-1
            IF ((E-V(IX)) .GE. 0) RTP=IX
        GOTO 120
        END IF
C
C       do we integrate into classically forbidden regions?
        KLAST=(E-V(LTP))
        NCROSS=0
        DO 90 IX=LTP+1,RTP-1
            K=(E-V(IX))
            IF ((K*KLAST .LE. 0) .AND. (K .LT. 0)) NCROSS=NCROSS+1
            KLAST=K
90      CONTINUE
```

```
C
      RETURN
      END
CCCCCCCCCCCCCCCCCCCCCCCCCCCCCCCCCCCCCCCCCCCCCCCCCCCCCCCCCCCCCCCCCCCCCCCCC
      SUBROUTINE POTNTL
C sets up array for current potential;
C all potentials have a minimum value of -1 and max of 1;
C if you change the potential, also change XMIN and XMAX in INIT,
C and estimate DEBOT and EBOTTM (one fifth of the expected spacing
C at the bottom of the well, and two DEBOT's below the expected ground
C state energy).
CCCCCCCCCCCCCCCCCCCCCCCCCCCCCCCCCCCCCCCCCCCCCCCCCCCCCCCCCCCCCCCCCCCCCCCCC
C Global variables:
      INCLUDE 'PARAM.E3'
C Local variables:
      INTEGER ILEFT,IWID,IRIGHT     !lattice values for width, left
      REAL VSCALE                   !scaling factor for renorm of parab
      INTEGER IX                    !labels lattice points
      REAL VMIN                     !minimum value of potential
      REAL CURVE                    !curvature for smoothing bump
CCCCCCCCCCCCCCCCCCCCCCCCCCCCCCCCCCCCCCCCCCCCCCCCCCCCCCCCCCCCCCCCCCCCCCCCC
C     define limits on X and DX
      IF (POT .EQ. SQUARE) THEN
          VXMIN=XMIN1
          VXMAX=XMAX1
      ELSE IF (POT .EQ. PARAB) THEN
          VXMIN=XMIN2
          VXMAX=XMAX2
      ELSE IF (POT .EQ. LENRD) THEN
          VXMIN=XMIN3
          VXMAX=XMAX3
      END IF
C
      DX=(VXMAX-VXMIN)/(NPTS-1)
C
C     setup X (space coordinate) array
      DO 5 IX=1,NPTS
          X(IX)=VXMIN+(IX-1)*DX
5     CONTINUE
C
C     setup V (potential) array
      IF (POT .EQ. SQUARE) THEN                    !square well
          DO 10 IX=1,NPTS
              V(IX)=-1.
10        CONTINUE
          VMAX=1.
          DEBOT=.2*(3.1415926/(4*GAMMA))**2!energy spacing at well bottom
          IF (BUMP .EQ. 1) THEN              !bump in well bottom
              ILEFT=NINT((LEFT-XMIN1)/DX)+1 !locate bump edges
              IRIGHT=NPTS+NINT((RIGHT-XMAX1)/DX)
              RIGHT=X(IRIGHT)
              LEFT=X(ILEFT)
```

```
              CURVE=-.025/((RIGHT-LEFT)/2.)**2
C             top of bump is curved to avoid discontinuities
              DO 20 IX=ILEFT,IRIGHT
                 V(IX)=-1.+HEIGHT*(1.+CURVE*(X(IX)-(RIGHT+LEFT)/2.)**2)
20            CONTINUE
           END IF
C
        ELSE IF (POT .EQ. PARAB) THEN    !parabolic well
           DO 30 IX=1,NPTS
              V(IX)=-(1.-.5*X(IX)*X(IX))
30         CONTINUE
           VMAX=1.
           DEBOT=.2/(SQRT(2.)*GAMMA)      !energy spacing at well bottom
           IF (FLAT .EQ. 1) THEN
C             the parabolic potential is flattened symmetrically about x=0
              IWID=INT(PWID/DX)
              IF (MOD(IWID,2) .NE. MOD(NPTS,2)) IWID=IWID-1
              ILEFT=(NPTS-IWID)/2+1
              VMIN=V(ILEFT)
              DO 40 IX=ILEFT,ILEFT+IWID-1
                 V(IX)=VMIN
40            CONTINUE
C
              VSCALE=2./(V(1)-VMIN)    !normalize so that min value is -1
              DO 50 IX=1,NPTS
                 V(IX)=VSCALE*(V(IX)-VMIN)-1.
50            CONTINUE
           END IF
C
        ELSE IF (POT .EQ. LENRD) THEN    !Lennard-Jones potential
           DO 60 IX=1,NPTS
              V(IX)=4*(X(IX)**(-12.)-X(IX)**(-6))
60         CONTINUE
           VMAX=V(1)
           DEBOT=1.0692/GAMMA !energy spacing at well bottom (see Ex. 1.8)
C
        END IF
C
C       educated guesses for ground state energy and energy spacing;
C       with these values, it will take three steps for F to change sign
        EBOTTM=-1.+2.5*DEBOT
C
        RETURN
        END
CCCCCCCCCCCCCCCCCCCCCCCCCCCCCCCCCCCCCCCCCCCCCCCCCCCCCCCCCCCCCCCCCCCCCCCCC
        SUBROUTINE INIT
C initializes constants, displays header screen,
C initializes arrays for input parameters
CCCCCCCCCCCCCCCCCCCCCCCCCCCCCCCCCCCCCCCCCCCCCCCCCCCCCCCCCCCCCCCCCCCCCCCCC
C Global variables:
        INCLUDE 'IO.ALL'
        INCLUDE 'MENU.ALL'
```

```
            INCLUDE 'PARAM.E3'
C Local parameters:
            CHARACTER*80 DESCRP          !program description
            DIMENSION DESCRP(22)
            INTEGER NHEAD,NTEXT,NGRF     !number of lines for each description
CCCCCCCCCCCCCCCCCCCCCCCCCCCCCCCCCCCCCCCCCCCCCCCCCCCCCCCCCCCCCCCCCCCCCCCCCCCCCCC
            CALL SETUP                   !get environment parameters
C
C       display header screen
            DESCRP(1)= 'EXAMPLE 3'
            DESCRP(2)= 'Bound states in a one-dimensional potential'
            NHEAD=2
C
C       text output description
            DESCRP(3)= 'during search: energy, delta energy, discontinuity,'
            DESCRP(4)='nodes, and entrance into classically forbidden regions'
            DESCRP(5)= 'after search: energy, classical turning points,'
         +  //' number of nodes'
            NTEXT=3
C
C       graphics output description
            DESCRP(6)= 'potential, eigenvalues, and eigenfunctions'
            NGRF=1
C
C       setup constant values for the potentials
            !all potentials have VMIN=-1, VMAX=1
            XMIN1=-2.
            XMAX1=2.
            XMIN2=-2.
            XMAX2=2.
            XMIN3=(2.*(SQRT(2.)-1.))**(.1666666)
            XMAX3=1.9
C.
C       reasonable limits on energy and energy increment
            EMIN=-.99999
            EMAX=100.
            DEMIN=5.E-06
            DEMAX=1.
C
        CALL HEADER(DESCRP,NHEAD,NTEXT,NGRF)
C
        CALL MENU                        !setup constant part of header
C
        MTYPE(13)=FLOAT
        MPRMPT(13)= 'Enter Gamma=sqrt[2m(a**2)V/hbar**2] (dimensionless)'
        MTAG(13)= 'Gamma=sqrt[2m(a**2)V/hbar**2] (dimensionless)'
        MLOLIM(13)=0.001
        MHILIM(13)=150.
        MREALS(13)=30.0
C
        MTYPE(14)=TITLE
        MPRMPT(14)= 'POTENTIAL FUNCTION MENU'
```

```
      MLOLIM(14)=2.
      MHILIM(14)=1.
C
      MTYPE(15)=MTITLE
      MPRMPT(15)='1) Square-well potential'
      MLOLIM(15)=0.
      MHILIM(15)=0.
C
      MTYPE(16)=MTITLE
      MPRMPT(16)='2) Parabolic-well potential'
      MLOLIM(16)=0.
      MHILIM(16)=0.
C
      MTYPE(17)=MTITLE
      MPRMPT(17)='3) Lennard-Jones potential'
      MLOLIM(17)=0.
      MHILIM(17)=1.
C
      MTYPE(18)=MCHOIC
      MPRMPT(18)= 'Make menu choice and press Return'
      MTAG(18)='19 24 26'
      MLOLIM(18)=1.
      MHILIM(18)=3.
      MINTS(18)=3
      MREALS(18)=3.
C
      MTYPE(19)=NOSKIP
      MPRMPT(19)= 'Do you want a bump in the square well?'
      MTAG(19)= 'Square Well Bump:'
      MREALS(19)=35.
      MINTS(19)=0
C
      MTYPE(20)=FLOAT
      MPRMPT(20)= 'Enter the left hand edge of bump (scaled units)'
      MTAG(20)= 'Left hand edge of bump (scaled units)'
      MLOLIM(20)=XMIN1
      MHILIM(20)=XMAX1
      MREALS(20)=-.1
C
      MTYPE(21)=FLOAT
      MPRMPT(21)= 'Enter the right hand edge of bump (scaled units)'
      MTAG(21)= 'Right hand edge of bump (scaled units)'
      MLOLIM(21)=XMIN1
      MHILIM(21)=XMAX1
      MREALS(21)=.1
C
      MTYPE(22)=FLOAT
      MPRMPT(22)= 'Enter the height of the bump (scaled units)'
      MTAG(22)= 'Height of the bump (scaled units)'
      MLOLIM(22)=0.
      MHILIM(22)=20
      MREALS(22)=1.
```

```
C
          MTYPE(23)=SKIP
          MREALS(23)=35.
C
          MTYPE(24)=NOSKIP
          MPRMPT(24)= 'Do you want a flattened bottom in parabolic-well'
          MTAG(24)= 'Parabolic Flattened area:'
          MREALS(24)=35.
          MINTS(24)=0
C
          MTYPE(25)=FLOAT
          MPRMPT(25)= 'Enter length of the flattened area (scaled units)'
          MTAG(25)= 'Length of the flattened area (scaled units)'
          MLOLIM(25)=0
          MHILIM(25)=XMAX2-XMIN2
          MREALS(25)=.1
C
          MTYPE(26)=SKIP
          MREALS(26)=35.
C
          MTYPE(38)=FLOAT
          MPRMPT(38)= 'Enter the matching tolerance'
          MTAG(38)= 'Matching tolerance'
          MLOLIM(38)=5.E-06
          MHILIM(38)=1.000
          MREALS(38)=.00005
C
          MTYPE(39)=NUM
          MPRMPT(39)= 'Enter number of lattice points'
          MTAG(39)= 'Number of lattice points'
          MLOLIM(39)=50
          MHILIM(39)=MAXLAT
          MINTS(39)=160
C
          MTYPE(40)=NUM
          MPRMPT(40)= 'Enter maximum number of steps in search'
          MTAG(40)= 'Maximum number of search steps'
          MLOLIM(40)=10
          MHILIM(40)=300
          MINTS(40)=50
C
          MTYPE(41)=SKIP
          MREALS(41)=60.
C
          MSTRNG(MINTS(75))= 'exmpl3.txt'
C
          MTYPE(76)=SKIP
          MREALS(76)=80.
C
          MSTRNG(MINTS(86))= 'exmpl3.grf'
C
          MTYPE(87)=SKIP
```

```
      MREALS(87)=90.
C
      RETURN
      END
CCCCCCCCCCCCCCCCCCCCCCCCCCCCCCCCCCCCCCCCCCCCCCCCCCCCCCCCCCCCCCCCCCCCCC
      SUBROUTINE PARAM
C gets parameters from screen
C ends program on request
C closes old files
C maps menu variables to program variables
C opens new files
C calculates all derivative parameters
CCCCCCCCCCCCCCCCCCCCCCCCCCCCCCCCCCCCCCCCCCCCCCCCCCCCCCCCCCCCCCCCCCCCCC
C Global variables:
      INCLUDE 'MENU.ALL'
      INCLUDE 'IO.ALL'
      INCLUDE 'PARAM.E3'
C Local variables:
C map between menu items and parameters
      INTEGER IGAMMA,ITOLR,INPTS,IMXSTP
      INTEGER IPOT,IBUMP,ILEFT,IHEIGH,IRIGHT,IFLAT,IPWID
      PARAMETER (IGAMMA = 13 )
      PARAMETER (IPOT   = 18 )
      PARAMETER (IBUMP  = 19 )
      PARAMETER (ILEFT  = 20 )
      PARAMETER (IRIGHT = 21 )
      PARAMETER (IHEIGH = 22 )
      PARAMETER (IFLAT  = 24 )
      PARAMETER (IPWID  = 25 )
      PARAMETER (ITOLR  = 38 )
      PARAMETER (INPTS  = 39 )
      PARAMETER (IMXSTP = 40 )
C Function:
      LOGICAL LOGCVT              !converts 1 and 0 to true and false
CCCCCCCCCCCCCCCCCCCCCCCCCCCCCCCCCCCCCCCCCCCCCCCCCCCCCCCCCCCCCCCCCCCCCC
C     get input from terminal
      CALL CLEAR
      CALL ASK(1,ISTOP)
      CALL CLEAR
C
C     stop program if requested
      IF (MREALS(IMAIN) .EQ. STOP) CALL FINISH
C
C     close files if necessary
      IF (TNAME .NE. MSTRNG(MINTS(ITNAME)))
     +    CALL FLCLOS(TNAME,TUNIT)
      IF (GNAME .NE. MSTRNG(MINTS(IGNAME)))
     +    CALL FLCLOS(GNAME,GUNIT)
C
C     set new parameter values
C     physical and numerical:
      GAMMA=MREALS(IGAMMA)
```

```
        TOLF=MREALS(ITOLR)
        NPTS=MINTS(INPTS)
        MAXSTP=MINTS(IMXSTP)
C
C       potential parameters
        POT=MREALS(IPOT)
        BUMP=MINTS(IBUMP)
        LEFT=MREALS(ILEFT)
        RIGHT=MREALS(IRIGHT)
        HEIGHT=MREALS(IHEIGH)
        FLAT=MINTS(IFLAT)
        PWID=MREALS(IPWID)
C
        CALL POTNTL    !setup potential array
C
C       text output
        TTERM=LOGCVT(MINTS(ITTERM))
        TFILE=LOGCVT(MINTS(ITFILE))
        TNAME=MSTRNG(MINTS(ITNAME))
C
C       graphics output
        GTERM=LOGCVT(MINTS(IGTERM))
        GHRDCP=LOGCVT(MINTS(IGHRD))
        GFILE=LOGCVT(MINTS(IGFILE))
        GNAME=MSTRNG(MINTS(IGNAME))
C
C       open files
        IF (TFILE) CALL FLOPEN(TNAME,TUNIT)
        IF (GFILE) CALL FLOPEN(GNAME,GUNIT)
        !files may have been renamed
        MSTRNG(MINTS(ITNAME))=TNAME
        MSTRNG(MINTS(IGNAME))=GNAME
C
        RETURN
        END
CCCCCCCCCCCCCCCCCCCCCCCCCCCCCCCCCCCCCCCCCCCCCCCCCCCCCCCCCCCCCCCCCCCCCCC
        SUBROUTINE PRMOUT(MUNIT)
C outputs parameter summary to the specified unit
CCCCCCCCCCCCCCCCCCCCCCCCCCCCCCCCCCCCCCCCCCCCCCCCCCCCCCCCCCCCCCCCCCCCCCC
C Global variables:
        INCLUDE 'IO.ALL'
        INCLUDE 'PARAM.E3'
C Passed variables:
        INTEGER MUNIT              !unit number for output
        REAL VAR(4)                !initial values of coord and momenta
CCCCCCCCCCCCCCCCCCCCCCCCCCCCCCCCCCCCCCCCCCCCCCCCCCCCCCCCCCCCCCCCCCCCCCC
        IF (MUNIT .EQ. OUNIT) CALL CLEAR
C
        WRITE (MUNIT,2)
        WRITE (MUNIT,4)
        WRITE (MUNIT,6) GAMMA
        WRITE (MUNIT,8) TOLF
```

```
      WRITE (MUNIT,12) NPTS
C
      IF (POT .EQ. SQUARE) THEN
         WRITE (MUNIT,14)
         IF (BUMP .EQ. 1) WRITE (MUNIT,16) LEFT,RIGHT,HEIGHT
      ELSE IF (POT .EQ. PARAB) THEN
         WRITE (MUNIT,18)
         IF (FLAT .EQ. 1) WRITE (MUNIT,20) -PWID,PWID
      ELSE IF (POT .EQ. LENRD) THEN
         WRITE (MUNIT,22)
      END IF
C
      WRITE (MUNIT,24)VXMIN,VXMAX,DX
      WRITE (MUNIT,26)
      WRITE (MUNIT,2)
C
2     FORMAT (' ')
4     FORMAT (' Output from Example 3: Time Independent Schroedinger',
     +        ' Equation')
6     FORMAT (' Gamma =',F9.3)
8     FORMAT (' Matching tolerance = ', 1PE10.3)
12    FORMAT (' Number of lattice points =', I5)
14    FORMAT (' Square Well potential')
16    FORMAT (' with bump from ',F6.3,' to ',F6.3,' of height =',F8.3)
18    FORMAT (' Parabolic potential')
20    FORMAT (' with flat bottom from ',F6.3,' to ',F6.3)
22    FORMAT (' Lennard-Jones potential')
24    FORMAT (' Xmin=',F6.3,' Xmax=',F6.3,' Dx=',1PE10.3)
26    FORMAT (' Energies and lengths are in scaled units')
C
      RETURN
      END
CCCCCCCCCCCCCCCCCCCCCCCCCCCCCCCCCCCCCCCCCCCCCCCCCCCCCCCCCCCCCCCCCCCCCCC
      SUBROUTINE CTITL(MUNIT)
C writes title for text output to requested device
CCCCCCCCCCCCCCCCCCCCCCCCCCCCCCCCCCCCCCCCCCCCCCCCCCCCCCCCCCCCCCCCCCCCCCC
C Passed variables:
      INTEGER MUNIT                    !unit to which we are writing
CCCCCCCCCCCCCCCCCCCCCCCCCCCCCCCCCCCCCCCCCCCCCCCCCCCCCCCCCCCCCCCCCCCCCCC
      WRITE (MUNIT,30)
      WRITE (MUNIT,35)
C
30    FORMAT
     +     (10X,'Energy',14X,'De',16X,'F',10x,'Nodes',5x,'Frbddn')
35    FORMAT
     +     (10X,'------',14X,'--',16X,'-',10x,'-----',5X,'------')
C
      RETURN
      END
CCCCCCCCCCCCCCCCCCCCCCCCCCCCCCCCCCCCCCCCCCCCCCCCCCCCCCCCCCCCCCCCCCCCCCC
      SUBROUTINE TXTOUT(MUNIT,E,DE,F,NODES,NCROSS,NLINES)
C writes results for one level to the requested unit
```

```
CCCCCCCCCCCCCCCCCCCCCCCCCCCCCCCCCCCCCCCCCCCCCCCCCCCCCCCCCCCCCCCCCCCCCCCCCC
C Global variables:
      INCLUDE 'IO.ALL'
      INCLUDE 'PARAM.E3'
C Input variables:
      INTEGER MUNIT              !unit to which we are writing
      REAL F                     !values of the mismatch
      INTEGER NODES              !number of nodes
      REAL E                     !trial eigenenergy
      REAL DE                    !step in energy search
      INTEGER NCROSS             !classically forbidden regions
      CHARACTER*3 FORBDN         !have we integrated into forb regions
      INTEGER NLINES             !number of lines sent to screen (I/O)
CCCCCCCCCCCCCCCCCCCCCCCCCCCCCCCCCCCCCCCCCCCCCCCCCCCCCCCCCCCCCCCCCCCCCCCCCC
C     if screen is full, clear screen and retype headings
      IF ((MOD(NLINES,TRMLIN-4) .EQ. 0)
     +                         .AND. (MUNIT .EQ. OUNIT)) THEN
          CALL PAUSE('to continue...',1)
          CALL CLEAR
          CALL CTITL(OUNIT)
          NLINES=2
      END IF
C
      IF (NCROSS .GT. 0) THEN
          FORBDN='Yes'
      ELSE
          FORBDN='No '
      END IF
      WRITE (MUNIT,20) E,DE,F,NODES,FORBDN
      NLINES=NLINES+1
C
20    FORMAT (8X,F10.7,7X,1PE10.3,7X,1PE10.3,8X,I2,8X,A3)
C
      RETURN
      END
CCCCCCCCCCCCCCCCCCCCCCCCCCCCCCCCCCCCCCCCCCCCCCCCCCCCCCCCCCCCCCCCCCCCCCCCCC
      SUBROUTINE EOUT(E,LTP,RTP,NODES,MUNIT)
C output information about eigenstate
CCCCCCCCCCCCCCCCCCCCCCCCCCCCCCCCCCCCCCCCCCCCCCCCCCCCCCCCCCCCCCCCCCCCCCCCCC
C Global variables:
      INCLUDE 'IO.ALL'
      INCLUDE 'PARAM.E3'
C Passed variables:
      INTEGER MUNIT              !unit to which we are writing
      INTEGER NODES              !number of nodes
      REAL E                     !trial eigenenergy
      INTEGER LTP,RTP            !classical turning points (lattice)
      REAL XLTP,XRTP             !classical turning points
CCCCCCCCCCCCCCCCCCCCCCCCCCCCCCCCCCCCCCCCCCCCCCCCCCCCCCCCCCCCCCCCCCCCCCCCCC
      XLTP=VXMIN+(LTP-1)*DX      !turning points
      XRTP=VXMIN+(RTP-1)*DX
      WRITE (MUNIT,500) E, NODES
```

```
        WRITE (MUNIT,510) XLTP,XRTP
C
500     FORMAT ('0',20X,'Eigenvalue = ',F10.7,' with ',I2,' nodes.')
510     FORMAT (16X,'and classical turning points ',F7.4,' and ',F7.4)
C
        RETURN
        END
CCCCCCCCCCCCCCCCCCCCCCCCCCCCCCCCCCCCCCCCCCCCCCCCCCCCCCCCCCCCCCCCCCCCCCCCCC
        SUBROUTINE GRFOUT(DEVICE,ENERGY,LEFTTP,RGHTTP,NFOUND,PSI)
C outputs potential and eigenvalues (plot 1) and eigenfunctions (plot 2)
CCCCCCCCCCCCCCCCCCCCCCCCCCCCCCCCCCCCCCCCCCCCCCCCCCCCCCCCCCCCCCCCCCCCCCCCCC
C Global variables
        INCLUDE 'IO.ALL'
        INCLUDE 'PARAM.E3'
        INCLUDE 'GRFDAT.ALL'
C Input variables:
        INTEGER DEVICE                     !which device is being used?
        REAL ENERGY(MAXLEV)                !array of eigenvalues
        INTEGER LEFTTP(MAXLEV),RGHTTP(MAXLEV)    !array of turning points
        INTEGER NFOUND                     !number of levels found
        REAL PSI(MAXLAT)                   !wave function
C Local variables
        INTEGER IX                         !index of lattice
        INTEGER ILEVEL                     !index of eigenvalue
        REAL XE(2),GRAPHE(2)               !arrays to graph eigenvalue
        CHARACTER*9 CGAMMA,CE              !GAMMA, ENERGY as a character string
        INTEGER LEN                        !length of string
        REAL NORM,MAXPSI                   !norm and max value for PSI
        INTEGER SCREEN                     !send to terminal
        INTEGER PAPER                      !make a hardcopy
        INTEGER FILE                       !send to a file
        DATA SCREEN,PAPER,FILE/1,2,3/
CCCCCCCCCCCCCCCCCCCCCCCCCCCCCCCCCCCCCCCCCCCCCCCCCCCCCCCCCCCCCCCCCCCCCCCCCC
        !messages for the impatient
        IF (DEVICE .NE. SCREEN) WRITE (OUNIT,100)
C
        NORM=0.
        DO 10 IX=1,NPTS                    !normalize wave function
           NORM=NORM+PSI(IX)**2
10      CONTINUE
        NORM=SQRT(NORM*DX)
        MAXPSI=0.
        DO 30 IX=1,NPTS
           PSI(IX)=PSI(IX)/NORM
           IF (ABS(PSI(IX)) .GT. MAXPSI) MAXPSI = ABS(PSI(IX))
30      CONTINUE
C
        IF (DEVICE .NE. FILE) THEN
           NPLOT=2                         !how many plots?
C
C          parameters that are the same for both plots
           XMIN=VXMIN                      !axis parameters
```

```
                YOVAL=VXMIN
                XMAX=VXMAX
                NPOINT=NPTS
                LABEL(1)= 'x (scaled units)'
 C
                ILINE=1                          !line and symbol styles
                ISYM=1
                IFREQ=0
                NXTICK=5
                NYTICK=5
 C
                CALL CONVRT(GAMMA,CGAMMA,LEN)
                TITLE='Solutions to Schroedinger''s Equation, Gamma='//CGAMMA
 C
                CALL GTDEV(DEVICE)                !device nomination
                IF (DEVICE .EQ. SCREEN) CALL GMODE  !change to graphics mode
 C
 C              first plot: potential and eigenvalues
                IPLOT=1
                YMIN=-1.
                YMAX=VMAX
                XOVAL=0.
                INFO=' '
                LABEL(2)= 'potential and eigenvalues (scaled units)'
                CALL LNLNAX
                CALL XYPLOT (X,V)                 !plot potential
 C
                ILINE=3                           !plot eigenvalues
                NPOINT=2
                DO 20 ILEVEL=1,NFOUND
                  XE(1)=VXMIN+(LEFTTP(ILEVEL)-1)*DX
                  XE(2)=VXMIN+(RGHTTP(ILEVEL)-1)*DX
                  GRAPHE(1)=ENERGY(ILEVEL)
                  GRAPHE(2)=ENERGY(ILEVEL)
                  CALL XYPLOT(XE,GRAPHE)
 20             CONTINUE
 C
 C              second plot:  wave function
                IPLOT=2
                ILINE=1
                NPOINT=NPTS
                YMIN=-MAXPSI
                YMAX=+MAXPSI
                XOVAL=0.
                CALL CONVRT(ENERGY(NFOUND),CE,LEN)
                INFO='Eigenvalue = '//CE
                LABEL(2)= 'normalized wave function'
                CALL LNLNAX
                CALL XYPLOT (X,PSI)
 C
        ELSE
 C          output to file
```

```
      WRITE (GUNIT,80)
      WRITE (GUNIT,85)
      WRITE (GUNIT,70)  (X(IX),V(IX),PSI(IX),IX=1,NPTS)
      END IF
C
C   end graphing session
      IF (DEVICE .NE. FILE) CALL GPAGE(DEVICE)   !end graphing package
      IF (DEVICE .EQ. SCREEN) CALL TMODE         !switch to text mode
C
100   FORMAT (/,' Patience, please; output going to a file.')
80    FORMAT ('0',13X,'Position',14X,'Potential',13X,'Wave function')
85    FORMAT (13X,'--------',14X,'---------',13X,'------------')
70    FORMAT (3(11X,1PE12.5))
C
      RETURN
      END
CCCCCCCCCCCCCCCCCCCCCCCCCCCCCCCCCCCCCCCCCCCCCCCCCCCCCCCCCCCCCCCCCCCCCCCC
CCCCCCCCCCCCCCCCCCCCCCCCCCCCCCCCCCCCCCCCCCCCCCCCCCCCCCCCCCCCCCCCCCCCCCCC
C param.e3
C
      REAL GAMMA             !degree of 'quantumness'
      REAL TOLF              !tolerance in search
      INTEGER NPTS           !number of lattice points
      INTEGER MAXSTP         !max number of steps in search
C
      INTEGER MAXLAT         !maximum number of lattice points
      PARAMETER (MAXLAT=1000)
C
      INTEGER SQUARE,PARAB,LENRD  !potential types
      PARAMETER (SQUARE=1)
      PARAMETER (PARAB=2)
      PARAMETER (LENRD=3)
C
      INTEGER POT            !which potential
      INTEGER BUMP,FLAT      !which features do we want
      REAL LEFT,RIGHT,HEIGHT,PWID !potential characteristics
      REAL V(MAXLAT)         !potential array
      REAL X(MAXLAT)         !space coordinate
      REAL VMAX              !max value of potential
      REAL EBOTTM,DEBOT      !guesses for g.s. energy and spacing
C
      REAL EMIN,EMAX,DEMIN,DEMAX  !limits on energy and energy increment
C
      REAL XMIN1,XMAX1,DX1   !limits on each potential
      REAL XMIN2,XMAX2,DX2
      REAL XMIN3,XMAX3,DX3
      REAL VXMIN,VXMAX,DX
C
      INTEGER MAXLEV              !maximum number of levels
      PARAMETER (MAXLEV=50)
C
      COMMON / PPARAM / GAMMA
```

```
COMMON / NPARAM / TOLF,NPTS,MAXSTP
COMMON / POTPRM / POT,BUMP,FLAT,LEFT,RIGHT,HEIGHT,PWID
COMMON / CONST  / XMIN1,XMAX1,XMIN2,XMAX2,XMIN3,XMAX3,
+                 EMIN,EMAX,DEMIN,DEMAX
COMMON / PCALC  / VXMIN,VXMAX,DX,DX1,DX2,DX3,V,VMAX,EBOTTM,DEBOT,X
```

B.4 Example 4

Algorithm   This program calculates the Born and eikonal scattering amplitudes and cross sections (Eqs. (4.21) and (4.29,30), respectively) for an electron incident on a square-well, Gaussian well, or Lenz-Jensen potential, as described in the text. The angles at which the differential cross section are calculated are fixed by the number of quadrature points input (NLEG) and the value of θ_{cut} as defined in the text (beginning of subroutine ARCHON). The differential cross section at each angle (subroutine DIFFCS) is calculated in both the Born and eikonal approximations by subroutines BORN and EIKONL, respectively, with the Bessel function J_0 calculated by function BESSJO, and the potential defined in function V. The total cross section is then calculated by integrating the differential cross section over these NLEG forward and backward angles (DO loops 100 and 200 in subroutine ARCHON). All integrations are done by the same order Gauss-Legendre quadrature for which the weights and abscissae are established in subroutine QUAD.

Input   The physical parameters are the type of potential [Lenz-Jensen], the nuclear charge [4] (Lenz-Jensen only), or well depth [20] (square or Gaussian well only), and the energy of the incident particle (in eV) [20]. The only numerical parameter is the number of quadrature points [20]. The program checks that the number of quadrature points is one of the fourteen allowed values: $2, 3, 4, 5, 6, 8, 10, 12, 16, 20, 24, 32, 40,$ or 48.

Output   For each of the 2*NLEG scattering angles, the text output displays the scattering angle (in degrees), the momentum transfer (in inverse Angstroms), and both the Born and eikonal differential cross sections (in square Angstroms). When all of the differential cross sections have been calculated, the program prints out the total cross section in the Born and eikonal approximations, as well as the total cross section from the optical theorem. The graphics then displays the differential cross section as a function of angle for both the Born and eikonal approximations.

```
CCCCCCCCCCCCCCCCCCCCCCCCCCCCCCCCCCCCCCCCCCCCCCCCCCCCCCCCCCCCCCCCCCCCCCCC
      PROGRAM EXMPL4
C     Example 4: Born and Eikonal Approximations to Quantum Scattering
C  COMPUTATIONAL PHYSICS (FORTRAN VERSION)
C  by Steven E. Koonin and Dawn C. Meredith
C  Copyright 1989, Addison-Wesley Publishing Company
CCCCCCCCCCCCCCCCCCCCCCCCCCCCCCCCCCCCCCCCCCCCCCCCCCCCCCCCCCCCCCCCCCCCCCCC
      CALL INIT         !display header screen, setup parameters
5     CONTINUE          !main loop/ execute once for each set of param
      CALL PARAM        !get input from screen
```

```
            CALL ARCHON        !calculate cross sections
            GOTO 5
            END
CCCCCCCCCCCCCCCCCCCCCCCCCCCCCCCCCCCCCCCCCCCCCCCCCCCCCCCCCCCCCCCCCCCCC
            SUBROUTINE ARCHON
C calculates the Born, eikonal, and optical theorem total cross sections
CCCCCCCCCCCCCCCCCCCCCCCCCCCCCCCCCCCCCCCCCCCCCCCCCCCCCCCCCCCCCCCCCCCCC
C Global variables:
            INCLUDE 'IO.ALL'
            INCLUDE 'PARAM.E4'
C Local variables:
            REAL SINHLF              !sin theta_cut/2
            REAL THTCUT,COSCUT       !theta_cut/2, cos of the same
            REAL SIGBT,SIGBTF,SIGBTB !Born cross sections
            REAL SIGET,SIGETF,SIGETB !eikonal cross sections
            REAL SIGOPT,IMFE0        !optical cross sect, imag part of f(0)
            REAL COSTH               !integration variable
            INTEGER ILEG,LANG        !index quad points, angle
            INTEGER NLINES           !number of lines printed to terminal
            INTEGER SCREEN           !send to terminal
            INTEGER PAPER            !make a hardcopy
            INTEGER FILE             !send to a file
            DATA SCREEN,PAPER,FILE/1,2,3/
CCCCCCCCCCCCCCCCCCCCCCCCCCCCCCCCCCCCCCCCCCCCCCCCCCCCCCCCCCCCCCCCCCCCC
C      output summary of parameters
            IF (TTERM) CALL PRMOUT(OUNIT,NLINES)
            IF (TFILE) CALL PRMOUT(TUNIT,NLINES)
            IF (GFILE) CALL PRMOUT(GUNIT,NLINES)
C
C      find angle at which q*rcut=2*pi to divide integral into
C         forward and backward scattering
            SINHLF=PI/(K*RCUT)
            IF (SINHLF.GT.SQHALF) SINHLF=SQHALF
            THTCUT=2*ATAN(SINHLF/SQRT(1-SINHLF*SINHLF))
            COSCUT=COS(THTCUT)
C
            SIGBT=0.0                        !zero sums
            SIGET=0.0
            IMFE0=0.0
C
C      integrate from theta=0 to theta=theta_cut
C         to find forward angle cross section
            DO 100 ILEG=1,NLEG
                COSTH=COSCUT+(XLEG(ILEG)+1)*(1-COSCUT)/2.0
                LANG=ILEG
                CALL DIFFCS(COSTH,ILEG,LANG,SIGET,SIGBT,IMFE0,NLINES)
100         CONTINUE
            SIGBTF=PI*(1-COSCUT)*SIGBT
            SIGETF=PI*(1-COSCUT)*SIGET
C
            SIGBT=0                          !zero sums
            SIGET=0
```

```
C
C     integrate from theta=theta_cut to theta=pi
C     to find backward angle cross section
      DO 200 ILEG=1,NLEG
         COSTH=-1+(XLEG(ILEG)+1)*(COSCUT+1)/2.0
         LANG=ILEG+NLEG
         CALL DIFFCS(COSTH,ILEG,LANG,SIGET,SIGBT,IMFE0,NLINES)
200   CONTINUE
      SIGBTB=PI*(COSCUT+1)*SIGBT
      SIGETB=PI*(COSCUT+1)*SIGET
C
C     add backward and forward scattering cross sections
      SIGBT=SIGBTF+SIGBTB
      SIGET=SIGETF+SIGETB
      SIGOPT=4*PI/K*IMFE0
C
C     output results
      IF (TTERM) CALL TOTOUT(OUNIT,THTCUT,SIGBT,SIGET,SIGOPT)
      IF (TFILE) CALL TOTOUT(TUNIT,THTCUT,SIGBT,SIGET,SIGOPT)
C
      IF (TTERM) CALL PAUSE('to continue...',1)
      IF (TTERM) CALL CLEAR
C
C     graphics output
      IF (GTERM) CALL GRFOUT(SCREEN)
      IF (GFILE) CALL GRFOUT(FILE)
      IF (GHRDCP) CALL GRFOUT(PAPER)
C
      RETURN
      END
CCCCCCCCCCCCCCCCCCCCCCCCCCCCCCCCCCCCCCCCCCCCCCCCCCCCCCCCCCCCCCCCCCCCCCCCC
      SUBROUTINE DIFFCS(COSTH,ILEG,LANG,SIGET,SIGBT,IMFE0,NLINES)
C calculate the differential cross section at a give angle (COSTH)
C and its contribution to the total cross section
CCCCCCCCCCCCCCCCCCCCCCCCCCCCCCCCCCCCCCCCCCCCCCCCCCCCCCCCCCCCCCCCCCCCCCCCC
C Global Variables:
      INCLUDE 'PARAM.E4'
      INCLUDE 'IO.ALL'
C Passed Variables
      REAL SIGBT              !Born cross sections (output)
      REAL SIGET              !eikonal cross sections (output)
      REAL IMFE0              !imag part of f(0) (output)
      REAL COSTH              !integration variable (input)
      INTEGER ILEG,LANG       !index quad points, angle (input)
      INTEGER NLINES          !number of lines printed to terminal (I/O)
C Local Variables:
      REAL THETA,Q            !angle, momentum transfer
      REAL DEGREE(96)         !angle
      REAL SIGE(96)           !eikonal differential cross section
      REAL SIGB(96)           !born differential cross section
      COMMON/RESULT/DEGREE,SIGE,SIGB  !pass to graphics routine only
CCCCCCCCCCCCCCCCCCCCCCCCCCCCCCCCCCCCCCCCCCCCCCCCCCCCCCCCCCCCCCCCCCCCCCCCC
```

```
          THETA=ATAN(SQRT(1-COSTH*COSTH)/COSTH)      !find the angle
          IF (THETA.LT.0.0) THETA=THETA+PI
          DEGREE(LANG)=THETA*180.0/PI
C
          Q=2*K*SIN(THETA/2.0)                       !momentum transfer at this angle
C
          CALL BORN(Q,LANG,SIGB)                     !calculate differential cross sect
          SIGBT=SIGBT+SIGB(LANG)*WLEG(ILEG)          !and add to total cross sect
          CALL EIKONL(Q,LANG,SIGE,IMFE0)
          SIGET=SIGET+SIGE(LANG)*WLEG(ILEG)
C
          IF (TTERM)                                 !text output
     &    CALL TXTOUT(OUNIT,DEGREE(LANG),Q,SIGB(LANG),SIGE(LANG),NLINES)
          IF (TFILE)
     &    CALL TXTOUT(TUNIT,DEGREE(LANG),Q,SIGB(LANG),SIGE(LANG),NLINES)
C
          RETURN
          END
CCCCCCCCCCCCCCCCCCCCCCCCCCCCCCCCCCCCCCCCCCCCCCCCCCCCCCCCCCCCCCCCCCCCCCCCCCCCC
          SUBROUTINE BORN(Q,LANG,SIGB)
C calculates the differential cross section in the Born approximation
C for one value of the momentum transfer (Q)
CCCCCCCCCCCCCCCCCCCCCCCCCCCCCCCCCCCCCCCCCCCCCCCCCCCCCCCCCCCCCCCCCCCCCCCCCCCCC
C Global Variables:
          INCLUDE 'PARAM.E4'
C Passed Variables
          REAL     Q                    !momentum transfer (input)
          INTEGER LANG                  !index of the angle (input)
          REAL SIGB(96)                 !Born differential cross section (output)
C Local Variables:
          REAL     R                    !radial variable
          REAL     FBORN                !scattering amplitude
          INTEGER J                     !index of quadrature points
C Function:
          REAL V                        !potential
CCCCCCCCCCCCCCCCCCCCCCCCCCCCCCCCCCCCCCCCCCCCCCCCCCCCCCCCCCCCCCCCCCCCCCCCCCCCC
          FBORN=0                       !zero sum
          DO 100 J=1,NLEG
            R=RMAX*(XLEG(J)+1)/2.0                   !scale the variable
            FBORN=FBORN+SIN(Q*R)*V(R)*R*WLEG(J)      !Gaussian quadrature
100       CONTINUE
          FBORN=-RMAX/(Q*HBARM)*FBORN
          SIGB(LANG)=FBORN*FBORN
C
          RETURN
          END
CCCCCCCCCCCCCCCCCCCCCCCCCCCCCCCCCCCCCCCCCCCCCCCCCCCCCCCCCCCCCCCCCCCCCCCCCCCCC
          SUBROUTINE EIKONL(Q,LANG,SIGE,IMFE0)
C calculates differential cross section in the eikonal approximation
C for one value of the momentum transfer Q
CCCCCCCCCCCCCCCCCCCCCCCCCCCCCCCCCCCCCCCCCCCCCCCCCCCCCCCCCCCCCCCCCCCCCCCCCCCCC
C Global Variables:
```

```
      INCLUDE 'PARAM.E4'
C Input/Output Variables
      REAL Q                          !momentum transfer (input)
      INTEGER LANG                    !index of the angle (input)
      REAL SIGE(96)                   !eikonal diff cross section (output)
      REAL IMFE0                      !imag part of f(0) (output)
C Local Variables:
      REAL CHI(96)                    !profile function
      REAL B                          !impact parameter
      COMPLEX FE                      !scattering amplitude
      REAL ZMAX,ZZ                    !integration variable for CHI
      REAL R                          !radius
      REAL J0                         !Bessel function
      INTEGER M,J                     !index for quadrature
      REAL X                          !argument of Bessel function
      COMPLEX SQRTM1                  !square root of minus 1
C Functions:
      REAL BESSJ0                     !zeroth Bessel function eval at X
      REAL V                          !potential
CCCCCCCCCCCCCCCCCCCCCCCCCCCCCCCCCCCCCCCCCCCCCCCCCCCCCCCCCCCCCCCCCCCCCCCCCCC
      SQRTM1=(0.,1.)                  !constant
      FE=(0.,0.)                      !zero the scatt ampl
C
      DO 100 J=1,NLEG                 !integrate over B
      B=RMAX*(XLEG(J)+1)/2.0
C
         IF (LANG .EQ. 1) THEN        !calculate profile function once
            CHI(J)=0.0
            ZMAX=SQRT(RMAX*RMAX-B*B)  !integration variable z
            DO 200 M=1,NLEG
               ZZ=ZMAX*(XLEG(M)+1)/2. !scale variable
               R=SQRT(ZZ*ZZ+B*B)      !calculate radius
               CHI(J)=CHI(J)+V(R)*WLEG(M)          !Gaussian quad
200         CONTINUE
            CHI(J)=-(ZMAX/(2*K*HBARM))*CHI(J)
            IMFE0=IMFE0+B*(SIN(CHI(J)))**2*WLEG(J) !imag f(0)
         ENDIF
C
         X=Q*B
         J0=BESSJ0(X)
         FE=FE+B*J0*WLEG(J)*(EXP(2*SQRTM1*CHI(J))-1.) !Gaussian quad
C
100   CONTINUE
C
      FE=-SQRTM1*K*RMAX*FE/2. !factor of RMAX/2 from change in variables
      SIGE(LANG)=ABS(FE)**2
      IF (LANG .EQ. 1) IMFE0=K*RMAX*IMFE0
C
      RETURN
      END
CCCCCCCCCCCCCCCCCCCCCCCCCCCCCCCCCCCCCCCCCCCCCCCCCCCCCCCCCCCCCCCCCCCCCCCCCCC
      REAL FUNCTION V(R)
```

```
C calculates the potential at fixed R
CCCCCCCCCCCCCCCCCCCCCCCCCCCCCCCCCCCCCCCCCCCCCCCCCCCCCCCCCCCCCCCCCCCCCCCC
C Global variables:
      INCLUDE 'PARAM.E4'
C Input/Output Variables:
      REAL R                        !radius (input)
C Local Variables
      REAL RR                       !radius for Lenz-Jensen (.ge. R1S)
      REAL U                        !temp variable for Lenz-Jensen
CCCCCCCCCCCCCCCCCCCCCCCCCCCCCCCCCCCCCCCCCCCCCCCCCCCCCCCCCCCCCCCCCCCCCCCC
      IF (POT .EQ. SQUARE) THEN     !square well
         V=-VZERO
C
      ELSE IF (POT .EQ. GAUSS) THEN !Gaussian well
         V=-VZERO*EXP(-2*R*R)
C
      ELSE                          !Lenz-Jensen
         RR=R                       !cutoff radius at location of 1S shell
         IF (R .LT. R1S) RR=R1S
         U=4.5397*Z6*SQRT(RR)
         V=-((Z*E2)/RR)*EXP(-U)*(1+U+U*U*(.3344+U*(.0485+.002647*U)))
      ENDIF
C
      RETURN
      END
CCCCCCCCCCCCCCCCCCCCCCCCCCCCCCCCCCCCCCCCCCCCCCCCCCCCCCCCCCCCCCCCCCCCCCCC
      REAL FUNCTION BESSJ0(X)
C calculates the zeroth order Bessel function at X
CCCCCCCCCCCCCCCCCCCCCCCCCCCCCCCCCCCCCCCCCCCCCCCCCCCCCCCCCCCCCCCCCCCCCCCC
C Global Variables:
      INCLUDE 'PARAM.E4'
C Input Variables:
      REAL X                        !argument of J_0
C Local Variables
      REAL Y,Y2,TEMP,TEMP2          !temporary variables to save values
CCCCCCCCCCCCCCCCCCCCCCCCCCCCCCCCCCCCCCCCCCCCCCCCCCCCCCCCCCCCCCCCCCCCCCCC
      Y=X/3.
      Y2=Y*Y
C
      IF ( ABS(X) .LE. 3.) THEN
         TEMP=-.0039444+.00021*Y2
         TEMP=.0444479+Y2*TEMP
         TEMP=-.3163866+Y2*TEMP
C
         TEMP=1.2656208+Y2*TEMP
         TEMP=-2.2499997+Y2*TEMP
         BESSJ0=1.+Y2*TEMP
      ELSE
         Y=1./Y
         TEMP=-7.2805E-04+1.4476E-04*Y
         TEMP=1.37237E-03+TEMP*Y
         TEMP=-9.512E-05+TEMP*Y
```

```
      TEMP=-.0055274+TEMP*Y
      TEMP=-7.7E-07+TEMP*Y
      TEMP=.79788456+TEMP*Y
      TEMP2=-2.9333E-04+1.3558E-04*Y
      TEMP2=-5.4125E-04+TEMP2*Y
      TEMP2=2.62573E-03+TEMP2*Y
      TEMP2=-3.954E-05+TEMP2*Y
      TEMP2=-4.166397E-02+TEMP2*Y
      TEMP2=X-.78539816+TEMP2*Y
      BESSJ0=TEMP*COS(TEMP2)/SQRT(X)
      ENDIF
C
      RETURN
      END
CCCCCCCCCCCCCCCCCCCCCCCCCCCCCCCCCCCCCCCCCCCCCCCCCCCCCCCCCCCCCCCCCCCCCCC
      SUBROUTINE QUAD(NLBAD)
C subroutine to establish Gauss-Legendre abscissae and weights
CCCCCCCCCCCCCCCCCCCCCCCCCCCCCCCCCCCCCCCCCCCCCCCCCCCCCCCCCCCCCCCCCCCCCCC
C Global variables:
      INCLUDE 'PARAM.E4'
C Passed Variables
      LOGICAL NLBAD                    !allowed value of nleg?(output)
C Local Variables:
      INTEGER ILEG                     !index for abscissae and weights
CCCCCCCCCCCCCCCCCCCCCCCCCCCCCCCCCCCCCCCCCCCCCCCCCCCCCCCCCCCCCCCCCCCCCCC
      NLBAD=.FALSE.
C
      IF (NLEG.EQ.2) THEN
         XLEG(1)=.577350269189626
         WLEG(1)=1.                    !nleg=2
      ELSE IF (NLEG.EQ.3) THEN
         XLEG(1)=.774596669241483
         WLEG(1)=.555555555555556      !NLEG=3
         XLEG(2)=0.
         WLEG(2)=.888888888888889
      ELSE IF (NLEG.EQ.4) THEN
         XLEG(1)=.861136311594053
         WLEG(1)=.347854845137454      !NLEG=4
         XLEG(2)=.339981043584856
         WLEG(2)=.652145154862546
      ELSE IF (NLEG.EQ.5) THEN
         XLEG(1)=.906179845938664
         WLEG(1)=.236926885056189      !NLEG=5
         XLEG(2)=.538469310105683
         WLEG(2)=.478628670499366
         XLEG(3)=0.
         WLEG(3)=.568888888888889
      ELSE IF (NLEG.EQ.6) THEN
         XLEG(1)=.932469514203152
         WLEG(1)=.17132449237917       !NLEG=6
         XLEG(2)=.661209386466266
         WLEG(2)=.360761573048139
```

```
      XLEG(3)=.238619186083197
      WLEG(3)=.467913934572691
ELSE IF (NLEG.EQ.8) THEN
      XLEG(1)=.960289856497536
      WLEG(1)=.101228536290376      !NLEG=8
      XLEG(2)=.796666477413627
      WLEG(2)=.222381034453374
      XLEG(3)=.525532409916329
      WLEG(3)=.313706645877887
      XLEG(4)=.18343464249565
      WLEG(4)=.362683783378362
ELSE IF (NLEG.EQ.10) THEN
      XLEG(1)=.973906528517172
      WLEG(1)=.066671344308688      !NLEG=10
      XLEG(2)=.865063366688985
      WLEG(2)=.149451349150581
      XLEG(3)=.679409568299024
      WLEG(3)=.219086362515982
      XLEG(4)=.433395394129247
      WLEG(4)=.269266719309996
      XLEG(5)=.148874338981631
      WLEG(5)=.295524224714753
ELSE IF (NLEG.EQ.12) THEN
      XLEG(1)=.981560634246719
      WLEG(1)=.047175336386512      !NLEG=12
      XLEG(2)=.904117256370475
      WLEG(2)=.106939325995318
      XLEG(3)=.769902674194305
      WLEG(3)=.160078328543346
      XLEG(4)=.587317954286617
      WLEG(4)=.203167426723066
      XLEG(5)=.36783149899818
      WLEG(5)=.233492536538355
      XLEG(6)=.125233408511469
      WLEG(6)=.249147045813403
ELSE IF (NLEG.EQ.16) THEN
      XLEG(1)=.98940093499165
      WLEG(1)=.027152459411754      !NLEG=16
      XLEG(2)=.944575023073233
      WLEG(2)=.062253523938648
      XLEG(3)=.865631202387832
      WLEG(3)=.095158511682493
      XLEG(4)=.755404408355003
      WLEG(4)=.124628971255534
      XLEG(5)=.617876244402644
      WLEG(5)=.149595988816577
      XLEG(6)=.458016777657227
      WLEG(6)=.169156519395003
      XLEG(7)=.281603550779259
      WLEG(7)=.182603415044924
      XLEG(8)=.095012509837637
      WLEG(8)=.189450610455069
```

```
ELSE IF (NLEG.EQ.20) THEN
  XLEG(1)=.993128599185094
  WLEG(1)=.017614007139152        !NLEG=20
  XLEG(2)=.963971927277913
  WLEG(2)=.040601429800386
  XLEG(3)=.912234428251325
  WLEG(3)=.062672048334109
  XLEG(4)=.839116971822218
  WLEG(4)=.083276741576704
  XLEG(5)=.74633190646015
  WLEG(5)=.10193011981724
  XLEG(6)=.636053680726515
  WLEG(6)=.118194531961518
  XLEG(7)=.510867001950827
  WLEG(7)=.131688638449176
  XLEG(8)=.373706088715419
  WLEG(8)=.142096109318382
  XLEG(9)=.227785851141645
  WLEG(9)=.149172986472603
  XLEG(10)=.076526521133497
  WLEG(10)=.152753387130725
ELSE IF (NLEG.EQ.24) THEN
  XLEG(1)=.995187219997021
  WLEG(1)=.012341229799987        !NLEG=24
  XLEG(2)=.974728555971309
  WLEG(2)=.028531388628933
  XLEG(3)=.938274552002732
  WLEG(3)=.044277438817419
  XLEG(4)=.886415527004401
  WLEG(4)=.059298584915436
  XLEG(5)=.820001985973902
  WLEG(5)=.07334648141108
  XLEG(6)=.740124191578554
  WLEG(6)=.086190161531953
  XLEG(7)=.648093651936975
  WLEG(7)=.097618652104113
  XLEG(8)=.545421471388839
  WLEG(8)=.107444270115965
  XLEG(9)=.433793507626045
  WLEG(9)=.115505668053725
  XLEG(10)=.315042679696163
  WLEG(10)=.121670472927803
  XLEG(11)=.191118867473616
  WLEG(11)=.125837456346828
  XLEG(12)=.064056892862605
  WLEG(12)=.127938195346752
ELSE IF (NLEG.EQ.32) THEN
  XLEG(1)=.997263861849481
  WLEG(1)=.00701861000947         !NLEG=32
  XLEG(2)=.985611511545268
  WLEG(2)=.016274394730905
  XLEG(3)=.964762255587506
```

```
     WLEG(3)=.025392065309262
     XLEG(4)=.934906075937739
     WLEG(4)=.034273862913021
     XLEG(5)=.896321155766052
     WLEG(5)=.042835898022226
     XLEG(6)=.849367613732569
     WLEG(6)=.050998059262376
     XLEG(7)=.794483795967942
     WLEG(7)=.058684093478535
     XLEG(8)=.732182118740289
     WLEG(8)=.065822222776361
     XLEG(9)=.663044266930215
     WLEG(9)=.072345794108848
     XLEG(10)=.587715757240762
     WLEG(10)=.07819389578707
     XLEG(11)=.506899908932229
     WLEG(11)=.083311924226946
     XLEG(12)=.421351276130635
     WLEG(12)=.087652093004403
     XLEG(13)=.331868602282127
     WLEG(13)=.091173878695763
     XLEG(14)=.239287362252137
     WLEG(14)=.093844399080804
     XLEG(15)=.144471961582796
     WLEG(15)=.095638720079274
     XLEG(16)=.048307665687738
     WLEG(16)=.096540088514727
 ELSE IF (NLEG.EQ.40) THEN
     XLEG(1)=.998237709710559
     WLEG(1)=.004521277098533          !NLEG=40
     XLEG(2)=.990726238699457
     WLEG(2)=.010498284531152
     XLEG(3)=.977259949983774
     WLEG(3)=.016421058381907
     XLEG(4)=.957916819213791
     WLEG(4)=.022245849194166
     XLEG(5)=.932812808278676
     WLEG(5)=.027937006980023
     XLEG(6)=.902098806968874
     WLEG(6)=.033460195282547
     XLEG(7)=.865959503212259
     WLEG(7)=.038782167974472
     XLEG(8)=.824612230833311
     WLEG(8)=.043870908185673
     XLEG(9)=.778305651426519
     WLEG(9)=.048695807635072
     XLEG(10)=.727318255189927
     WLEG(10)=.053227846983936
     XLEG(11)=.671956684614179
     WLEG(11)=.057439769099391
     XLEG(12)=.61255388966798
     WLEG(12)=.061306242492928
```

```
        XLEG(13)=.549467125095128
        WLEG(13)=.064804013456601
        XLEG(14)=.483075801686178
        WLEG(14)=.067912045815233
        XLEG(15)=.413779204371605
        WLEG(15)=.070611647391286
        XLEG(16)=.341994090825758
        WLEG(16)=.072886582395804
        XLEG(17)=.268152185007253
        WLEG(17)=.074723169057968
        XLEG(18)=.192697580701371
        WLEG(18)=.076110361900626
        XLEG(19)=.116084070675255
        WLEG(19)=.077039818164247
        XLEG(20)=.03877241750605
        WLEG(20)=.077505947978424
ELSE IF (NLEG.EQ.48) THEN
        XLEG(1)=.998771007252426
        WLEG(1)=.003153346052305      !NLEG=48
        XLEG(2)=.99353017226635
        WLEG(2)=.007327553901276
        XLEG(3)=.984124583722826
        WLEG(3)=.011477234579234
        XLEG(4)=.970591592546247
        WLEG(4)=.015579315722943
        XLEG(5)=.95298770316043
        WLEG(5)=.019616160457355
        XLEG(6)=.931386690706554
        WLEG(6)=.023570760839324
        XLEG(7)=.905879136715569
        WLEG(7)=.027426509708356
        XLEG(8)=.876572020274247
        WLEG(8)=.031167227832798
        XLEG(9)=.843588261624393
        WLEG(9)=.03477722256477
        XLEG(10)=.807066204029442
        WLEG(10)=.03824135106583
        XLEG(11)=.76715903251574
        WLEG(11)=.041545082943464
        XLEG(12)=.724034130923814
        WLEG(12)=.044674560856694
        XLEG(13)=.677872379632663
        WLEG(13)=.04761665849249
        XLEG(14)=.628867396776513
        WLEG(14)=.050359035553854
        XLEG(15)=.577224726083972
        WLEG(15)=.052890189485193
        XLEG(16)=.523160974722233
        WLEG(16)=.055199503699984
        XLEG(17)=.466902904750958
        WLEG(17)=.057277292100403
        XLEG(18)=.408686481990716
```

```
                 WLEG(18)=.059114839698395
                 XLEG(19)=.34875588629216
                 WLEG(19)=.060704439165893
                 XLEG(20)=.287362487355455
                 WLEG(20)=.062039423159892
                 XLEG(21)=.224763790394689
                 WLEG(21)=.063114192286254
                 XLEG(22)=.161222356068891
                 WLEG(22)=.063924238584648
                 XLEG(23)=.097004699209462
                 WLEG(23)=.06446616443595
                 XLEG(24)=.032380170962869
                 WLEG(24)=.064737696812683
           ELSE
              NLBAD=.TRUE.        !if NLEG is a disallowed value
           ENDIF
     C
           IF (.NOT. NLBAD) THEN
              DO 100 ILEG=1,NLEG/2              !since the weights and abscissa
                 XLEG(NLEG-ILEG+1)=-XLEG(ILEG) !are even and odd functions
                 WLEG(NLEG-ILEG+1)=WLEG(ILEG)  !functions respectively
     100       CONTINUE
           ENDIF
     C
           RETURN
           END
     CCCCCCCCCCCCCCCCCCCCCCCCCCCCCCCCCCCCCCCCCCCCCCCCCCCCCCCCCCCCCCCCCCCCCCC
           SUBROUTINE INIT
     C initializes constants, displays header screen,
     C initializes arrays for input parameters
     CCCCCCCCCCCCCCCCCCCCCCCCCCCCCCCCCCCCCCCCCCCCCCCCCCCCCCCCCCCCCCCCCCCCCCC
     C Global variables:
           INCLUDE 'IO.ALL'
           INCLUDE 'MENU.ALL'
           INCLUDE 'PARAM.E4'
     C Local parameters:
           CHARACTER*80 DESCRP              !program description
           DIMENSION DESCRP(20)
           INTEGER NHEAD,NTEXT,NGRAPH    !number of lines for each description
     CCCCCCCCCCCCCCCCCCCCCCCCCCCCCCCCCCCCCCCCCCCCCCCCCCCCCCCCCCCCCCCCCCCCCCC
           CALL SETUP                       !get environment parameters
     C
     C     display header screen
           DESCRP(1)= 'EXAMPLE 4'
           DESCRP(2)= 'Born and Eikonal Approximations'
           DESCRP(3)= 'to Quantum Scattering'
           NHEAD=3
     C
     C     text output description
           DESCRP(4)= 'angle, momentum transfer, Born and Eikonal '
         +    //'differential cross sections,'
           DESCRP(5)= 'and Born, Eikonal, and optical theorem '
```

```
     +    //'total cross section'
          NTEXT=2
C
C         graphics output description
          DESCRP(6)= 'differential cross section vs. theta'
          NGRAPH=1
C
          CALL HEADER(DESCRP,NHEAD,NTEXT,NGRAPH)
C
C         set constant values
          PI=3.1415926
          SQHALF=SQRT(0.5)
          E2=14.409                       !electron charge, squared
          RMAX=2.0
          HBARM=7.6359                    !hbar**2/mass electron
C
          CALL MENU                       !setup constant part of menu
C
          MTYPE(13)=MTITLE
          MPRMPT(13)='Potential Function Options:'
          MLOLIM(13)=2
          MHILIM(13)=1
C
          MTYPE(14)=MTITLE
          MPRMPT(14)='1) Lenz-Jensen Potential: electron & neutral atom'
          MLOLIM(14)=0
          MHILIM(14)=0
C
          MTYPE(15)=MTITLE
          MPRMPT(15)='2) Square Well'
          MLOLIM(15)=0
          MHILIM(15)=0
C
          MTYPE(16)=MTITLE
          MPRMPT(16)='3) Gaussian Well'
          MLOLIM(16)=0
          MHILIM(16)=1
C
          MTYPE(17)=MCHOIC
          MPRMPT(17)='Enter Choice'
          MTAG(17)='18 20 20'
          MLOLIM(17)=1
          MHILIM(17)=3
          MINTS(17)=1
          MREALS(17)=1.
C
          MTYPE(18)=NUM
          MPRMPT(18)='Enter charge of the atomic nucleus'
          MTAG(18)='Z'
          MLOLIM(18)=1
          MHILIM(18)=108
          MINTS(18)=4
```

```
C
        MTYPE(19)=SKIP
        MREALS(19)=21
C
        MTYPE(20)=FLOAT
        MPRMPT(20)='Enter Vzero (eV)'
        MTAG(20)='Vzero (eV)'
        MLOLIM(20)=0.0
        MHILIM(20)=5000.
        MREALS(20)=20.0
C
        MTYPE(21)=FLOAT
        MPRMPT(21)='Enter Energy (eV)'
        MTAG(21)='Energy (eV)'
        MLOLIM(21)=0.0
        MHILIM(21)=5.0E+06
        MREALS(21)=20.0
C
        MTYPE(22)=SKIP
        MREALS(22)=35
C
        MTYPE(38)=NUM
        MPRMPT(38)='Number of quadrature points'
        MTAG(38)='number of quadrature points'
        MLOLIM(38)=2
        MHILIM(38)=48
        MINTS(38)=20
C
        MTYPE(39)=SKIP
        MREALS(39)=60
C
        MSTRNG(MINTS(75))= 'exmpl4.txt'
C
        MTYPE(76)=SKIP
        MREALS(76)=80.
C
        MSTRNG(MINTS(86))= 'exmpl4.grf'
C
        MTYPE(87)=SKIP
        MREALS(87)=90.
C
        RETURN
        END
CCCCCCCCCCCCCCCCCCCCCCCCCCCCCCCCCCCCCCCCCCCCCCCCCCCCCCCCCCCCCCCCCCCCCCCC
        SUBROUTINE PARAM
C gets parameters from screen
C ends program on request
C closes old files
C maps menu variables to program variables
C opens new files
C calculates all derivative parameters
CCCCCCCCCCCCCCCCCCCCCCCCCCCCCCCCCCCCCCCCCCCCCCCCCCCCCCCCCCCCCCCCCCCCCCCC
```

```
C Global variables:
      INCLUDE 'MENU.ALL'
      INCLUDE 'IO.ALL'
      INCLUDE 'PARAM.E4'
C Local Variables:
      LOGICAL NLBAD                      !allowed value of nleg?
C map between menu items and parameters
      INTEGER IPOT,IZ,IVZERO,IE,INLEG
      PARAMETER (IPOT   = 17 )
      PARAMETER (IZ     = 18 )
      PARAMETER (IVZERO = 20 )
      PARAMETER (IE     = 21 )
      PARAMETER (INLEG  = 38 )
C Functions:
      INTEGER GETINT                     !gets integer value from screen
      LOGICAL LOGCVT                     !converts 1 and 0 to true and false
CCCCCCCCCCCCCCCCCCCCCCCCCCCCCCCCCCCCCCCCCCCCCCCCCCCCCCCCCCCCCCCCCCCCCCCCCC
C     get input from terminal
      CALL CLEAR
      CALL ASK(1,ISTOP)
C
C     stop program if requested
      IF (MREALS(IMAIN) .EQ. STOP) CALL FINISH
C
C     close files if necessary
      IF (TNAME .NE. MSTRNG(MINTS(ITNAME)))
     +    CALL FLCLOS(TNAME,TUNIT)
      IF (GNAME .NE. MSTRNG(MINTS(IGNAME)))
     +    CALL FLCLOS(GNAME,GUNIT)
C
C     physical and numerical parameters
      POT=MINTS(IPOT)
      Z=MINTS(IZ)
      VZERO=MREALS(IVZERO)
      E=MREALS(IE)
      NLEG=MINTS(INLEG)
C
C     derived constants
      K=SQRT(2*E/HBARM)
      IF (POT .EQ. LENZ) THEN
         RCUT=1.
         Z6=Z**.166667
         R1S=.529/Z
      ELSE
         RCUT=2.
      END IF
C
C     find weights and abscissae for Gauss-Legendre integration
      CALL QUAD(NLBAD)
100   IF (NLBAD) THEN
         WRITE (OUNIT,*) ' Allowed values for quadrature points are ',
     +                   '2 3 4 5 6 8 10 12 16 20 24 32 40 48'
```

```
                NLEG=GETINT(20,2,48,'Enter new number of quadrature points')
                MINTS(INLEG)=NLEG
                CALL QUAD(NLBAD)
            GOTO 100
            END IF
            CALL CLEAR
C
C       text output
            TTERM=LOGCVT(MINTS(ITTERM))
            TFILE=LOGCVT(MINTS(ITFILE))
            TNAME=MSTRNG(MINTS(ITNAME))
C
C       graphics output
            GTERM=LOGCVT(MINTS(IGTERM))
            GHRDCP=LOGCVT(MINTS(IGHRD))
            GFILE=LOGCVT(MINTS(IGFILE))
            GNAME=MSTRNG(MINTS(IGNAME))
C
C       open files
            IF (TFILE) CALL FLOPEN(TNAME,TUNIT)
            IF (GFILE) CALL FLOPEN(GNAME,GUNIT)
            !files may have been renamed
            MSTRNG(MINTS(ITNAME))=TNAME
            MSTRNG(MINTS(IGNAME))=GNAME
C
            RETURN
            END
CCCCCCCCCCCCCCCCCCCCCCCCCCCCCCCCCCCCCCCCCCCCCCCCCCCCCCCCCCCCCCCCCCCCCCCCC
            SUBROUTINE PRMOUT(MUNIT,NLINES)
C outputs parameters to MUNIT
CCCCCCCCCCCCCCCCCCCCCCCCCCCCCCCCCCCCCCCCCCCCCCCCCCCCCCCCCCCCCCCCCCCCCCCCC
C Global variables:
            INCLUDE 'IO.ALL'
            INCLUDE 'PARAM.E4'
C Passed variables:
            INTEGER MUNIT            !unit number for output (input)
            INTEGER NLINES           !number of lines written so far (I/O)
CCCCCCCCCCCCCCCCCCCCCCCCCCCCCCCCCCCCCCCCCCCCCCCCCCCCCCCCCCCCCCCCCCCCCCCCC
            IF (MUNIT .EQ. OUNIT) CALL CLEAR
C
            WRITE (MUNIT,2)
            WRITE (MUNIT,4)
            WRITE (MUNIT,6) E
            WRITE (MUNIT,8) NLEG
C
            IF (POT .EQ. LENZ) THEN
                WRITE (MUNIT,10) Z
            ELSE IF (POT .EQ. SQUARE) THEN
                WRITE (MUNIT,12) VZERO
            ELSE IF (POT .EQ. GAUSS) THEN
                WRITE (MUNIT,14) VZERO
            END IF
```

```
      WRITE (MUNIT,15)
C
C     different header for text and graphics files
      IF (MUNIT .EQ. GUNIT) THEN
         WRITE (MUNIT,2)
      ELSE
         WRITE (MUNIT,2)
         WRITE (MUNIT,16)
         WRITE (MUNIT,17)
         WRITE (MUNIT,18)
      END IF
C
      NLINES=10
C
2     FORMAT (' ')
4     FORMAT (' Output from example 4: Born and eikonal '
     +        'approximations to quantum scattering')
6     FORMAT (' Energy (eV) =', 1PE10.3)
8     FORMAT (' Number of quadrature points = ' , I2)
10    FORMAT (' Lenz Jensen potential with Z = ', I3)
12    FORMAT (' Square well potential with Vzero = ' 1PE10.3)
14    FORMAT (' Gaussian potential with Vzero = ' 1PE10.3)
15    FORMAT (' Theta is in degrees')
16    FORMAT (11X,'Theta',13X,'Q',14X,'Born',14X,'Eikonal')
17    FORMAT (10X,'degrees',11X,'A*-1',12X,'A**2',14X,' A**2 ')
18    FORMAT (10X,'-------',11X,'----',12X,'----',14X,'-------')
C
      RETURN
      END
CCCCCCCCCCCCCCCCCCCCCCCCCCCCCCCCCCCCCCCCCCCCCCCCCCCCCCCCCCCCCCCCCCCCCCC
      SUBROUTINE TXTOUT(MUNIT,DEGREE,Q,SIGB,SIGE,NLINES)
C writes out differential cross section at each angle
CCCCCCCCCCCCCCCCCCCCCCCCCCCCCCCCCCCCCCCCCCCCCCCCCCCCCCCCCCCCCCCCCCCCCCC
C Global variables:
      INCLUDE 'IO.ALL'
      INCLUDE 'PARAM.E4'
C Input variables:
      INTEGER MUNIT            !output unit specifier
      INTEGER NLINES           !number of lines printed to screen (I/O)
      REAL DEGREE,Q            !angle and momentum transfer
      REAL SIGB,SIGE           !Born and eikonal diff cross section
CCCCCCCCCCCCCCCCCCCCCCCCCCCCCCCCCCCCCCCCCCCCCCCCCCCCCCCCCCCCCCCCCCCCCCC
C     if screen is full, clear and type headings again
      IF ((MOD(NLINES,TRMLIN-5) .EQ. 0)
     +                     .AND. (MUNIT .EQ. OUNIT)) THEN
         CALL PAUSE('to continue...',1)
         CALL CLEAR
         WRITE (MUNIT,16)
         WRITE (MUNIT,17)
         WRITE (MUNIT,18)
         WRITE (MUNIT,2)
         NLINES=NLINES+3
```

```
            END IF
C
            WRITE (MUNIT,20) DEGREE,Q,SIGB,SIGE
C
C           keep track of printed lines only for terminal output
            IF (MUNIT .EQ. OUNIT) NLINES=NLINES+1
C
20          FORMAT(9X,F7.2,9X,F7.2,9X,1PE10.3,9X,1PE10.3)
2           FORMAT (' ')
16          FORMAT (11X,'Theta',13X,'Q',14X,'Born',14X,'Eikonal')
17          FORMAT (10X,'degrees',11X,'A*-1',12X,'A**2',14X,'  A**2 ')
18          FORMAT (10X,'-------',11X,'----',12X,'----',14X,'-------')
C
            RETURN
            END
CCCCCCCCCCCCCCCCCCCCCCCCCCCCCCCCCCCCCCCCCCCCCCCCCCCCCCCCCCCCCCCCCCCCCCCC
            SUBROUTINE TOTOUT(MUNIT,THTCUT,SIGBT,SIGET,SIGOPT)
C writes out total cross section to MUNIT
CCCCCCCCCCCCCCCCCCCCCCCCCCCCCCCCCCCCCCCCCCCCCCCCCCCCCCCCCCCCCCCCCCCCCCCC
C Global variables:
            INCLUDE 'IO.ALL'
            INCLUDE 'PARAM.E4'
C Input variables:
            INTEGER MUNIT                !output unit specifier
            REAL THTCUT                  !theta cut
            REAL SIGBT,SIGET,SIGOPT      !total cross section with diff methods
CCCCCCCCCCCCCCCCCCCCCCCCCCCCCCCCCCCCCCCCCCCCCCCCCCCCCCCCCCCCCCCCCCCCCCCC
            WRITE (MUNIT,*) ' '
            WRITE (MUNIT,10) THTCUT*180/PI
            WRITE (MUNIT,20) SIGBT
            WRITE (MUNIT,30) SIGET
            WRITE (MUNIT,40) SIGOPT
C
10          FORMAT ('                        theta cut = ', F7.2,' degrees')
20          FORMAT ('                Born sigma total =',1PE10.3,' Angstroms**2')
30          FORMAT ('             Eikonal sigma total =',1PE10.3,' Angstroms**2')
40          FORMAT (' Optical theorem sigma total =',1PE10.3,' Angstroms**2')
C
            RETURN
            END
CCCCCCCCCCCCCCCCCCCCCCCCCCCCCCCCCCCCCCCCCCCCCCCCCCCCCCCCCCCCCCCCCCCCCCCC
            SUBROUTINE GRFOUT(DEVICE)
C graphs differential cross section vs. angle on a semi-log scale
CCCCCCCCCCCCCCCCCCCCCCCCCCCCCCCCCCCCCCCCCCCCCCCCCCCCCCCCCCCCCCCCCCCCCCCC
C Global variables
            INCLUDE 'IO.ALL'
            INCLUDE 'PARAM.E4'
            INCLUDE 'GRFDAT.ALL'
C Input variables:
            INTEGER DEVICE               !which device is being used?
C Local variables
            INTEGER ILEG                 !indexes angle
```

```
      INTEGER EXPMAX,EXPMIN   !min and max exp for diff cross section
      CHARACTER*9 CE          !E as a character string
      REAL DEGREE(96)         !angle
      REAL SIGE(96)           !eikonal differential cross section
      REAL SIGB(96)           !Born differential cross section
      INTEGER LEN             !string length
      INTEGER SCREEN          !send to terminal
      INTEGER PAPER           !make a hardcopy
      INTEGER FILE            !send to a file
      COMMON/RESULT/DEGREE,SIGE,SIGB
      DATA SCREEN,PAPER,FILE/1,2,3/
CCCCCCCCCCCCCCCCCCCCCCCCCCCCCCCCCCCCCCCCCCCCCCCCCCCCCCCCCCCCCCCCCCCCCCCC
C     messages for the impatient
      IF (DEVICE .NE. SCREEN) WRITE (OUNIT,100)
C
C     calculate parameters for graphing
      IF (DEVICE .NE. FILE) THEN
         NPLOT=1                          !how many plots?
         IPLOT=1
C
         YMAX=0.                          !find limits on data points
         YMIN=SIGB(1)
         DO 20 ILEG=1,2*NLEG
            IF (SIGB(ILEG) .GT. YMAX) YMAX=SIGB(ILEG)
            IF (SIGE(ILEG) .GT. YMAX) YMAX=SIGE(ILEG)
            IF (SIGB(ILEG) .LT. YMIN) YMIN=SIGB(ILEG)
            IF (SIGE(ILEG) .LT. YMIN) YMIN=SIGE(ILEG)
20       CONTINUE
C        find integer limits on exponent
         EXPMAX=INT(LOG10(YMAX))
         IF (YMAX .GT. 1.) EXPMAX =EXPMAX+1
         EXPMIN=INT(LOG10(YMIN))
         IF (YMIN .LT. 1.) EXPMIN=EXPMIN-1
         YMAX=10.**EXPMAX
         YMIN=10.**EXPMIN
C
         XMIN=DEGREE(1)
         XMAX=DEGREE(2*NLEG)
         Y0VAL=XMIN
         X0VAL=YMIN
C
         NPOINT=2*NLEG
C
         ILINE=1                          !line and symbol styles
         ISYM=4
         IFREQ=1
         NXTICK=5
         NYTICK=EXPMAX-EXPMIN
         IF (NYTICK .GT. 8) THEN          !keep number of ticks small
            IF (MOD(NYTICK,2) .EQ. 0) THEN
               NYTICK=NYTICK/2
            ELSE
```

```
                  NYTICK=8
               END IF
            END IF
C
         CALL CONVRT(E,CE,LEN)                        !titles and labels
         INFO = 'Born(X) and Eikonal(O) Approximations'
         LABEL(1)= 'Angle (Degrees)'
         IF (POT .EQ. LENZ) THEN
            TITLE = ' Lenz-Jensen Potential, Energy (eV)='//CE
            LABEL(2)= 'Differential Cross Section (Angstroms**2)'
         ELSE IF (POT .EQ. SQUARE) THEN
            TITLE = 'Square Well Potential, Energy (eV)='//CE
            LABEL(2)= 'Differential Cross Section (Angstroms**2)'
         ELSE IF (POT .EQ. GAUSS) THEN
            TITLE = 'Gaussian Potential, Energy (eV)='//CE
            LABEL(2)= 'Differential Cross Section (Angstroms**2)'
         END IF
C
         CALL GTDEV(DEVICE)                           !device nomination
         IF (DEVICE .EQ. SCREEN) CALL GMODE           !change to graphics mode
         CALL LGLNAX                                  !draw axes
      END IF
C
C     output results
      IF (DEVICE .EQ. FILE) THEN
         WRITE (GUNIT,16)
         WRITE (GUNIT,18)
         WRITE (GUNIT,70) (DEGREE(ILEG),SIGB(ILEG),SIGE(ILEG),
     +        ILEG=1,2*NLEG)
      ELSE
         CALL XYPLOT (DEGREE,SIGB)                    !plot born
         ISYM=1
         CALL XYPLOT (DEGREE,SIGE)                    !plot eikonal
      END IF
C
C     end graphing session
      IF (DEVICE .NE. FILE) CALL GPAGE(DEVICE)   !end graphics package
      IF (DEVICE .EQ. SCREEN) CALL TMODE         !switch to text mode
C
16    FORMAT (10X,'Theta',13X,'Born',14X,'Eikonal')
18    FORMAT (10X,'-----',13X,'----',14X,'-------')
70    FORMAT (9X,F7.2,9X,1PE10.3,9X,1PE10.3)
100   FORMAT (/,' Patience, please; output going to a file.')
C
      RETURN
      END
CCCCCCCCCCCCCCCCCCCCCCCCCCCCCCCCCCCCCCCCCCCCCCCCCCCCCCCCCCCCCCCCCCCCCCCC
CCCCCCCCCCCCCCCCCCCCCCCCCCCCCCCCCCCCCCCCCCCCCCCCCCCCCCCCCCCCCCCCCCCCCCCC
C param.e4
      INTEGER Z                       !charge of nucleus
      REAL E                          !energy of particle
      REAL VZERO                      !depth of potential
```

```
      INTEGER NLEG                    !number of quadrature points
      INTEGER POT                     !which potential
      INTEGER LENZ,SQUARE,GAUSS       !types of potentials
C
      PARAMETER (LENZ=1)
      PARAMETER (SQUARE=2)
      PARAMETER (GAUSS=3)
C
      REAL E2                         !square of charge on electron
      REAL HBARM                      !Planck's constant
      REAL RMAX                       !maximum radius of potential
      REAL SQHALF,PI                  !constants
C
      REAL K                          !wave number
      REAL Z6                         !sixth root of Z
      REAL R1S                        !radius of 1s shell
      REAL RCUT                       !spatial extent of potential
      REAL WLEG,XLEG                  !Gauss-Legendre weights and abscissae
C
      COMMON/PPARAM/Z,E,VZERO,NLEG
      COMMON/NPARAM/POT
      COMMON/CONSTS/E2,HBARM,RMAX,SQHALF,PI
      COMMON/PCALC/Z6,K,RCUT,WLEG(48),XLEG(48),R1S
```

B.5 Example 5

Algorithm   This program fits electron-nucleus elastic scattering cross sections to determine the nuclear charge density using the method described in the text. Once the target nucleus is chosen, the experimental data is read in (from file CA.DAT, NI.DAT, or PB.DAT), and the recoil and Coulomb corrections are made (subroutine NDATA). Before iterations begin, the program calculates the quantities that do not depend on the Fourier expansion coefficients: f_{outer} (Eq. 5.63), part of the profile function $\chi_n(b)$ (Eq. 5.66b), and a factor in the integrand for f_{inner} (subroutine TABLES). All integrals are done by 20-point Gauss-Legendre quadrature. The initial density is taken to be the Fermi function, and the initial Fourier coefficients are obtained from a discrete Fourier transform of that function properly normalized to the total charge of the nucleus (subroutine FERMI). For each iteration (DO loop 100 in subroutine ARCHON), the improved expansion coefficients are found (subroutine MINIMZ) by constructing the matrix A and vector B defined by Eq. (5.49), inverting the matrix (subroutine MATINV), and solving Eq. (5.51). Note that since Eq. (5.47b) is homogeneous, it can be suitably scaled to the same order of magnitude as the rest of the rows of A. At each iteration, the profile function, scattering amplitude, theoretical cross section, χ^2, density, and uncertainties in the density are calculated (subroutine FIT).

Input   The only physical parameter is the nucleus [Ca40]. Once the nuclear data has been read in, the program displays the beam energy, maximum momentum transfer, and estimated nuclear radius. (All energies are in MeV, all lengths in fermi, and all cross-sections in mbarns/steradian.) You are then prompted for the value of the maximum radius R [$1.07A^{1/3}$ fermis], the total number of basis functions to use [QMAX*RMAX/π] (see discussion in text after Eq. 5.59), and the number to start with. The maximum number of sines is fixed by CMAX= 15. After each iteration (except the "zeroth") you are given the chance to increase the number of sines (if the current basis is smaller than the total requested), to continue iterating (if the current basis is the total requested), or to end iterations.

Output   After each iteration (including the "zeroth" iteration when the Fourier coefficients are determined only by the Fermi function), the text output displays the iteration number, χ^2, degrees of freedom, number of sines used, total charge, maximum momentum transfer for this basis, expansion coefficients, and their uncertainty. The graphics then displays both the experimental cross section (with error bars) and the theoretical fit on a semi-log plot. The density and its uncertainty as a function of radius are also plotted. The number of radii at which the density is

calculated is fixed by NGRF= 100. Note that graphics goes to hardcopy
or a file only after the final iteration, but goes to the screen after each
iteration.

```
ccccccccccccccccccccccccccccccccccccccccccccccccccccccccccccccccccc
      PROGRAM EXMPL5
C     Example 5: Determining Nuclear Charge Density
C  COMPUTATIONAL PHYSICS (FORTRAN VERSION)
C  by Steven E. Koonin and Dawn C. Meredith
C  Copyright 1989, Addison-Wesley Publishing Company
ccccccccccccccccccccccccccccccccccccccccccccccccccccccccccccccccccccc
      CALL INIT            !display header screen, setup parameters
5     CONTINUE             !main loop/ execute once for each set of param
      CALL PARAM           !get input from screen
      CALL ARCHON          !calculate nuclear charge density
      GOTO 5
      END
ccccccccccccccccccccccccccccccccccccccccccccccccccccccccccccccccccccc
      SUBROUTINE ARCHON
C calculates nuclear charge density from best fit to electron scattering
C data; the density is expanded in a finite Fourier series
ccccccccccccccccccccccccccccccccccccccccccccccccccccccccccccccccccccc
C Global variables:
      INCLUDE 'PARAM.E5'
      INCLUDE 'IO.ALL'
C Local variables:
      INTEGER ITER                   !number of iterations
      INTEGER INCRS,CONT             !increase nsine? continue iter?
      INTEGER NSINE                  !number of coefficients
      REAL CZERO(CMAX)               !Fourier coefficients
      REAL SIGT(DATMAX)              !total cross section
      COMPLEX FTOTAL(DATMAX)         !total scattering amplitude
      REAL QBASIS                    !max num momentum transfer
      DOUBLE PRECISION A(CMAX+1,CMAX+1) !matrix for inverting
      REAL CHI(NLEG)                 !profile functions
      REAL RHO(NGRF)                 !density for graphing
      REAL DRHO(NGRF)                !error in density
      INTEGER SCREEN                 !send to terminal
      INTEGER PAPER                  !make a hardcopy
      INTEGER FILE                   !send to a file
C Function:
      INTEGER GETINT,YESNO           !get input from screen
      DATA SCREEN,PAPER,FILE/1,2,3/
ccccccccccccccccccccccccccccccccccccccccccccccccccccccccccccccccccccc
C     read in nuclear data and calculate derivative parameters
      CALL NDATA
C     set up arrays that do not depend on Cn's
      CALL TABLES(A)
C     prompt for value of N to begin with
      NSINE=GETINT(NBASIS,1,NBASIS,
     +           'Enter number of sine functions to begin with')
```

```
      CALL CLEAR
C
C     output summary of parameters
      IF (TTERM) CALL PRMOUT(OUNIT)
      IF (TFILE) CALL PRMOUT(TUNIT)
      IF (GFILE) CALL PRMOUT(GUNIT)
C
C     obtain initial Cn's from Fermi distribution
      ITER=0
      QBASIS=NSINE*PI/RMAX                       !largest Q described by basis
      CALL FERMI(NSINE,CZERO)
C     calculate initial fit and display
      CALL FIT(NSINE,CZERO,ITER,SIGT,FTOTAL,QBASIS,CHI,A,RHO,DRHO)
C
100   CONTINUE                                   !loop over iterations
      ITER=ITER+1
C        find changes in CZERO's to minimize the error
         CALL MINIMZ(CZERO,NSINE,SIGT,FTOTAL,QBASIS,CHI,A)
C        calculate quality of fit, errors; display results
         CALL FIT(NSINE,CZERO,ITER,SIGT,FTOTAL,QBASIS,CHI,A,RHO,DRHO)
C
C        various options for continuing
         IF (NSINE .LT. NBASIS) THEN
           INCRS=
     +     YESNO(0,'Do you want to increment number of sines by one?')
           IF (INCRS .EQ. 1) THEN
             NSINE=NSINE+1
             CZERO(NSINE)=0.
             CONT=1
             QBASIS=NSINE*PI/RMAX               !largest Q described by basis
           ELSE
             CONT=YESNO(1,'Do you want to continue iterating?')
           END IF
         ELSE
           CONT=YESNO(1,'Do you want to continue iterating?')
         END IF
C
      IF (CONT .EQ. 1) GOTO 100
C
C     write out only the final version to a file
      IF (GHRDCP) CALL GRFOUT(PAPER,RHO,SIGT,NSINE,DRHO)
      IF (GFILE) CALL GRFOUT(FILE,RHO,SIGT,NSINE,DRHO)
C
      RETURN
      END
CCCCCCCCCCCCCCCCCCCCCCCCCCCCCCCCCCCCCCCCCCCCCCCCCCCCCCCCCCCCCCCCCCCCCCC
      SUBROUTINE NDATA
C read in nuclear data
C define variables that are nucleus-dependent
C take recoil out of data; calculate effective momentum transfer
C prompt for NBASIS and RMAX
CCCCCCCCCCCCCCCCCCCCCCCCCCCCCCCCCCCCCCCCCCCCCCCCCCCCCCCCCCCCCCCCCCCCCCC
```

```
C Global variables:
      INCLUDE 'PARAM.E5'
      INCLUDE 'IO.ALL'
C Local variables:
      INTEGER I,IR               !index of input data, radius
      REAL QLAB,RECOIL           !variables to adjust input data
      REAL MTARGT                !mass of the target in MeV
      INTEGER NZERO              !suggestion for NBASIS
C Function:
      INTEGER GETINT             !get integer input from terminal
      REAL GETFLT                !get real input from terminal
CCCCCCCCCCCCCCCCCCCCCCCCCCCCCCCCCCCCCCCCCCCCCCCCCCCCCCCCCCCCCCCCCCCCCCC
C     open data files
      IF (NUCL .EQ. CA) THEN
          OPEN(UNIT=DUNIT,FILE='CA.DAT',STATUS='OLD')
      ELSE IF (NUCL .EQ. NI) THEN
          OPEN(UNIT=DUNIT,FILE='NI.DAT',STATUS='OLD')
      ELSE IF (NUCL .EQ. PB) THEN
          OPEN(UNIT=DUNIT,FILE='PB.DAT',STATUS='OLD')
      END IF
C
C     read in data description
      READ (DUNIT,5) TARGET
      READ (DUNIT,*) ZTARGT,ATARGT,EBEAM,NPTS
5     FORMAT (2X,A10)
C
C     calculate target parameters
      RZERO=1.07*ATARGT**(1./3.)
      MTARGT=940*ATARGT
      ZA=ZTARGT*ALPHA
      KBEAM=EBEAM/HBARC
      VC1=1.+4./3.*ZA*HBARC/(EBEAM*RZERO)
C
C     read in cross sections; close file
      READ (DUNIT,20) (THETA(I),SIGE(I),DSIGE(I),
     +              THETA(I+1),SIGE(I+1),DSIGE(I+1),I=1,NPTS,2)
20    FORMAT (2(1X,F7.3,1X,E10.3,1X,E10.3))
      CLOSE (UNIT=DUNIT)
C
C     do some preliminary adjustments to the data
      QMAX=0.
      DO 30 I=1,NPTS
C        keep the error small
         IF (DSIGE(I) .GT. SIGE(I)) DSIGE(I)=.8*SIGE(I)
C        angle in radians
         THETA(I)=THETA(I)*PI/180.
C        correction to momentum transfer
         QLAB=2*KBEAM*SIN(THETA(I)/2.)
         QEFF(I)=QLAB*VC1
         IF (QEFF(I) .GT. QMAX) QMAX=QEFF(I)
C        take out recoil
         RECOIL=1.+2.*EBEAM*SIN(THETA(I)/2)**2/MTARGT
```

```
                    SIGE(I)=SIGE(I)*RECOIL
                    DSIGE(I)=DSIGE(I)*RECOIL
      30    CONTINUE
      C
      C     prompt for input of RMAX and NBASIS
            WRITE (OUNIT,40) TARGET
            WRITE (OUNIT,50) EBEAM
            WRITE (OUNIT,60) QMAX
            WRITE (OUNIT,70) RZERO
            WRITE (OUNIT,*) ' '
      C
            RMAX=GETFLT(RZERO+4.,1.,25.,
           +       ' Enter the boundary radius to use (in fermis)')
            NZERO=INT(QMAX*RMAX/PI)
            WRITE (OUNIT,80) NZERO
            NBASIS=GETINT(NZERO,1,15,' Enter NBASIS')
      C
      C     fill array RGRF that is used for charge density with radial values
            DRGRF=RMAX/NGRF
            DO 90 IR=1,NGRF
                RGRF(IR)=IR*DRGRF
      90    CONTINUE
      C
      40    FORMAT (' For the target ',A10)
      50    FORMAT (' the data are at a beam energy of ',F7.3,' MeV')
      60    FORMAT (' the maximum momentum transfer covered is ',
           +        F7.3,' fm**-1')
      70    FORMAT (' the nuclear radius is about',F7.3,' fm')
      80    FORMAT (' to keep QMAX less than Qexperimental, NBASIS'
           +  ' must be less than ',I2)
      C
            RETURN
            END
CCCCCCCCCCCCCCCCCCCCCCCCCCCCCCCCCCCCCCCCCCCCCCCCCCCCCCCCCCCCCCCCCCCCCCCCC
            SUBROUTINE TABLES(A)
C calculates CHI(N), sets up table of B*J_0(QR), calculates FOUTER
C and zeroes the matrix A
C (none of these variables depend on Cn's, and therefore need
C to be done only once per calculation)
CCCCCCCCCCCCCCCCCCCCCCCCCCCCCCCCCCCCCCCCCCCCCCCCCCCCCCCCCCCCCCCCCCCCCCCCC
C Global variables:
            INCLUDE 'PARAM.E5'
            INCLUDE 'IO.ALL'
C Output variables:
            DOUBLE PRECISION A(CMAX+1,CMAX+1) !matrix for inverting
C Local variables:
            REAL B                           !impact parameter/RMAX
            INTEGER N,ILEG,JLEG,I,J          !indices
            REAL Z,X,Y                       !integration variables
            REAL SUM                         !sum for integration
            REAL PHI,BESSJ0,BESSJ1           !real functions
            REAL J0,J1                       !temp values for Fouter
```

```
      COMPLEX MU                    !arguments of the Lommel function
      REAL NU
      COMPLEX FACTOR                !common factor in FOUTER
      COMPLEX LOMMEL                !Lommel function * X**(1-MU)
      COMPLEX TEMP1,TEMP2           !temp storage for Lommel function
ccccccccccccccccccccccccccccccccccccccccccccccccccccccccccccccccccccc
c     function for CHIN
      PHI(Y)=ALOG((1.+SQRT(1.-Y**2))/Y)-SQRT(1.-Y**2)
c     Lommel function * X**(1-MU)
      LOMMEL(X,MU,NU)=1.- ((MU-1.)**2-NU**2)/X**2 +
     +              ((MU-1)**2-NU**2)*((MU-3.)**2-NU**2)/X**4
c
c     zero the matrix A
      DO 2 I=1,CMAX+1
         DO 1 J=1,CMAX+1
            A(I,J)=0.
1        CONTINUE
2     CONTINUE
c
c     CHIN = part of profile function independent of Cn's
      WRITE (OUNIT,*) ' Calculating chi(n)...'
      DO 20 N=1,NBASIS
         DO 10 ILEG=1,NLEG
            B=(1+XLEG(ILEG))/2.        !all B's are scaled by RMAX
            SUM=0.
            DO 5 JLEG=1,NLEG
               Z=B+(1.-B)/2.*(1.+XLEG(JLEG))          !Eq. 5.66b
               SUM=SUM+WLEG(JLEG)*Z*SIN(N*PI*Z)*PHI(B/Z)  !Gaussian quad
5           CONTINUE
            CHIN(ILEG,N)=-4*PI*ALPHA*RMAX**2*SUM*(1.-B)/2.
10       CONTINUE
20    CONTINUE
c
c     part of F_inner integrand independent of Cn's
      WRITE (OUNIT,*) ' Calculating Bessel functions * b ...'
      DO 60 ILEG=1,NLEG
         B=(1+XLEG(ILEG))/2.          !all B's are scaled by RMAX
         DO 50 I=1,NPTS               !Eq. 5.62
            X=QEFF(I)*RMAX*B
            JTABLE(ILEG,I)=BESSJ0(X)*B*WLEG(ILEG)
50       CONTINUE
60    CONTINUE
c
c     calculate F_outer (which only sees a point charge)
      WRITE (OUNIT,*) ' Calculating outer scattering amplitude...'
      WRITE (OUNIT,*) ' '
      FACTOR=-2*SQRTM1*ZA
      DO 80 I=1,NPTS
         X=QEFF(I)*RMAX
         J0=BESSJ0(X)
         J1=BESSJ1(X)                 !Eq. 5.63
         TEMP1=LOMMEL(X,FACTOR,-1.)
```

```
               TEMP2=LOMMEL(X,1.+FACTOR,0.)
               FOUTER(I)=FACTOR*J0*TEMP1+X*J1*TEMP2-X*J1
               FOUTER(I)=SQRTM1*KBEAM*FOUTER(I)/(QEFF(I)**2)
 80      CONTINUE
 C
         RETURN
         END
CCCCCCCCCCCCCCCCCCCCCCCCCCCCCCCCCCCCCCCCCCCCCCCCCCCCCCCCCCCCCCCCCCCCCCCCCCC
         SUBROUTINE FERMI(NSINE,CZERO)
C calculate initial CZERO by finding Fourier coefficients
C for the Fermi Function
CCCCCCCCCCCCCCCCCCCCCCCCCCCCCCCCCCCCCCCCCCCCCCCCCCCCCCCCCCCCCCCCCCCCCCCCCCC
C Global variables:
         INCLUDE 'PARAM.E5'
C Input variables:
         INTEGER NSINE                     !number of coefficients
C Output variables
         REAL CZERO(CMAX)                  !Fourier coefficients
C Local variables:
         INTEGER ILEG,N                    !indices
         REAL RADIUS,R                     !radial step,radius
         REAL FERMIF,RRHO                  !Fermi Function
         REAL SUM                          !sum for normalization
         REAL THICK                        !thickness of Fermi function
         DATA THICK / 2.4 /
CCCCCCCCCCCCCCCCCCCCCCCCCCCCCCCCCCCCCCCCCCCCCCCCCCCCCCCCCCCCCCCCCCCCCCCCCCC
         FERMIF(R)=1./(1.+EXP(4.4*(R-RZERO)/THICK))
 C
         DO 30 N=1,NSINE                   !zero sums
            CZERO(N)=0.
 30      CONTINUE
 C       radial integrals to find C's
         DO 50 ILEG=1,NLEG
            RADIUS=RMAX*(1+XLEG(ILEG))/2
            RRHO=RADIUS*FERMIF(RADIUS)     !Fourier transform of Eq. 5.57
            DO 40 N=1,NSINE                !Gaussian quad
               CZERO(N)=CZERO(N)+RRHO*SIN(N*PI*RADIUS/RMAX)*WLEG(ILEG)
 40         CONTINUE
 50      CONTINUE
 C
 C       normalize the C's to charge ZTARGT (Eq. 5.58)
         SUM=0.
         DO 70 N=1,NSINE
               IF (MOD(N,2) .EQ. 1) SUM=SUM+CZERO(N)/N
               IF (MOD(N,2) .EQ. 0) SUM=SUM-CZERO(N)/N
 70      CONTINUE
         SUM=ZTARGT/(4*RMAX**2*SUM)
         DO 80 N=1,NBASIS
            CZERO(N)=CZERO(N)*SUM
 80      CONTINUE
 C
         RETURN
```

```
      END
CCCCCCCCCCCCCCCCCCCCCCCCCCCCCCCCCCCCCCCCCCCCCCCCCCCCCCCCCCCCCCCCCCCCCCC
      SUBROUTINE FIT(NSINE,CZERO,ITER,SIGT,FTOTAL,QBASIS,CHI,A,RHO,DRHO)
C calculates sigma (given CZERO), chi squared, the charge density,
C and error in charge density; displays the results
CCCCCCCCCCCCCCCCCCCCCCCCCCCCCCCCCCCCCCCCCCCCCCCCCCCCCCCCCCCCCCCCCCCCCCC
C Global variables:
      INCLUDE 'PARAM.E5'
      INCLUDE 'IO.ALL'
C Input variables:
      REAL CZERO(CMAX)                    !Fourier coefficients
      INTEGER NSINE                       !number of coefficients
      INTEGER ITER                        !number of iterations
C Output variables:
      REAL SIGT(DATMAX)                   !total cross section
      COMPLEX FTOTAL(DATMAX)              !total scattering amplitude
      REAL QBASIS                         !max num momentum transfer
      REAL RHO(NGRF)                      !charge density
      REAL DRHO(NGRF)                     !error in density
      DOUBLE PRECISION A(CMAX+1,CMAX+1)   !matrix for inverting
C Local variables:
      REAL CHI(NLEG)                      !profile functions
      COMPLEX FINNER                      !inner scattering amplitude
      INTEGER ILEG,IQ,N,IR,M              !indices
      REAL CHISQ                          !goodness of fit
      INTEGER NDOF                        !degrees of freedom in fit
      REAL SINES(CMAX)                    !table of sines
      REAL SUM,FAC                        !temp variable for DRHO
      REAL Z                              !total nuclear charge
      INTEGER SCREEN                      !send to terminal
      INTEGER PAPER                       !make a hardcopy
      INTEGER FILE                        !send to a file
      DATA SCREEN,PAPER,FILE/1,2,3/
CCCCCCCCCCCCCCCCCCCCCCCCCCCCCCCCCCCCCCCCCCCCCCCCCCCCCCCCCCCCCCCCCCCCCCC
C     calculate total profile function at each b (Eq. 5.66a)
      DO 20 ILEG=1,NLEG
         CHI(ILEG)=-ZA*LOG((1+XLEG(ILEG))/2)
         DO 10 N=1,NSINE
            CHI(ILEG)=CHI(ILEG)+CZERO(N)*CHIN(ILEG,N)
10       CONTINUE
20    CONTINUE
C
C     calculate fit sigma, chisquare
      CHISQ=0
      NDOF=0
      DO 50 IQ=1,NPTS                        !loop over data points
C        calculate the inner and total scattering amplitude
         FINNER=0.
         DO 40 ILEG=1,NLEG                    !Eq. 5.62
            FINNER=FINNER+JTABLE(ILEG,IQ)*(EXP(2*SQRTM1*CHI(ILEG))-1.)
40       CONTINUE
         FINNER=-SQRTM1*KBEAM*RMAX**2*FINNER/2.
```

```
C          scattering amplitude
           FTOTAL(IQ)=(FINNER+FOUTER(IQ))
C
C          sigma in mbarnes/sr, with momentum transfer corr   (Eq. 5.60)
           SIGT(IQ)=10*COS(THETA(IQ)/2)**2*VC1**2*ABS(FTOTAL(IQ))**2
           IF (QEFF(IQ) .LE. QBASIS) THEN
              CHISQ=CHISQ+((SIGE(IQ)-SIGT(IQ))/DSIGE(IQ))**2   !Eq. 5.42
              NDOF=NDOF+1
           END IF
50         CONTINUE
           NDOF=NDOF-NSINE                                     !degrees of freedom
C
C       calculate the density and its error
           Z=0.
           DO 70 IR=1,NGRF
              RHO(IR)=0.                                       !zero sums
              DRHO(IR)=0.
              DO 60 N=1,NSINE
                 SINES(N)=SIN(N*PI*RGRF(IR)/RMAX)   !common term
                 RHO(IR)=RHO(IR)+CZERO(N)*SINES(N)  !Eq. 5.57; density*radius
                 SUM=0.
                 DO 80 M=1,N                        !sums to find err in dens
                    FAC=2.
                    IF (M .EQ. N) FAC=1.
                    SUM=SUM+FAC*A(N,M)*SINES(M)     !Eq. 5.55, sum over M
80               CONTINUE
                 DRHO(IR)=DRHO(IR)+SUM*SINES(N)     !Eq. 5.55, sum over N
60            CONTINUE
              DRHO(IR)=SQRT(ABS(DRHO(IR)))/RGRF(IR) !take out radius from RHO
              Z=Z+RHO(IR)*RGRF(IR)                  !integ dens for tot charge
              RHO(IR)=RHO(IR)/RGRF(IR)              !now RHO is just density
70         CONTINUE
           Z=Z*4*PI*RMAX/NGRF
C
        IF (TTERM)
     +  CALL TXTOUT(OUNIT,ITER,CHISQ,NDOF,NSINE,QBASIS,CZERO,A,Z)
        IF (TFILE)
     +  CALL TXTOUT(TUNIT,ITER,CHISQ,NDOF,NSINE,QBASIS,CZERO,A,Z)
        IF (GTERM) THEN
           CALL GRFOUT(SCREEN,RHO,SIGT,NSINE,DRHO)
           CALL CLEAR
        END IF
C
        RETURN
        END
CCCCCCCCCCCCCCCCCCCCCCCCCCCCCCCCCCCCCCCCCCCCCCCCCCCCCCCCCCCCCCCCCCCCCCC
        SUBROUTINE MINIMZ(CZERO,NSINE,SIGT,FTOTAL,QBASIS,CHI,A)
C finds the corrections to CZERO to minimize the fit
CCCCCCCCCCCCCCCCCCCCCCCCCCCCCCCCCCCCCCCCCCCCCCCCCCCCCCCCCCCCCCCCCCCCCCC
C Global variables:
        INCLUDE 'IO.ALL'
        INCLUDE 'PARAM.E5'
```

```
C Input variables:
      REAL CZERO(CMAX)                      !Fourier coefficients (I/O)
      INTEGER NSINE                         !number of coefficients
      REAL SIGT(DATMAX)                     !total cross section
      REAL CHI(NLEG)                        !profile functions
      COMPLEX FTOTAL(DATMAX)                !total scattering amplitude
      REAL QBASIS                           !max num momentum transfer
      DOUBLE PRECISION A(CMAX+1,CMAX+1)     !matrix for inverting (I/O)
C Local variables:
      REAL BVEC(CMAX+1)                     !lhs of linear equation
      INTEGER NDIM                          !dimension of matrix
      COMPLEX DFDC                          !diff of F w.r.t. CZERO
      REAL W(CMAX)                          !diff of SIGT w.r.t. CZERO
      INTEGER IQ,N,ILEG,M                   !indices
      REAL SUM                              !temp value for matrix mulplct
CCCCCCCCCCCCCCCCCCCCCCCCCCCCCCCCCCCCCCCCCCCCCCCCCCCCCCCCCCCCCCCCCCCCCCCCCCCC
      WRITE (OUNIT,*) ' Calculating and inverting A...'
      NDIM=NSINE+1
C
C     zero the vector B and matrix A
      DO 10 N=1,NSINE
        BVEC(N)=0
        DO 20 M=1,N
          A(N,M)=0.
20      CONTINUE
10    CONTINUE
C
C     calculate the matrix A and vector B
      DO 70 IQ=1,NPTS
        IF (QEFF(IQ) .LT. QBASIS) THEN   !leave off Q's not desc by basis
          DO 40 N=1,NSINE                !first find W terms
            DFDC=0
            DO 30 ILEG=1,NLEG
              DFDC=DFDC+                     !terms in W (Eq. 5.67b)
     +          JTABLE(ILEG,IQ)*EXP(2*SQRTM1*CHI(ILEG))*CHIN(ILEG,N)
30          CONTINUE
            DFDC=DFDC*KBEAM*RMAX**2
            W(N)=20.*VC1**4*COS(THETA(IQ)/2)**2     !Eq. 5.67a
     +        *REAL(CONJG(FTOTAL(IQ))*DFDC)         !W in mbarnes/sr
40        CONTINUE
C
          DO 60 N=1,NSINE                          !calculate matrix A
            DO 50 M=1,N                            !Eq. 5.48
              A(N,M)=A(N,M)+DBLE(W(N)*W(M)/DSIGE(IQ)**2)
50          CONTINUE                               !and vector B(Eq 5.48)
            BVEC(N)=BVEC(N)+(SIGE(IQ)-SIGT(IQ))*W(N)/DSIGE(IQ)**2
60        CONTINUE
        END IF
70    CONTINUE
C
C     include dZ/dCn terms, and fill out A using symmetry
      DO 100 N=1,NSINE                             !Eq 5.49a, 5.58
```

```
          IF (MOD(N,2) .EQ. 1) A(N,NDIM)=-A(1,1)/N   !scale by A(1,1)
          IF (MOD(N,2) .EQ. 0) A(N,NDIM)=A(1,1)/N
          A(NDIM,N)=A(N,NDIM)
          DO 90 M=1,N              !fill out A matrix using symmetry
             A(M,N)=A(N,M)
90        CONTINUE
100    CONTINUE
       A(NSINE+1,NSINE+1)=0.
C
       CALL MATINV(A,NDIM)         !invert the matrix
C
       DO 120 N=1,NSINE            !multiply A**-1 * B to solve for
          SUM=0.                   !corrections to Czero
          DO 110 M=1,NSINE
             SUM=SUM+A(N,M)*BVEC(M)
110       CONTINUE
          CZERO(N)=CZERO(N)+SUM
120    CONTINUE
C
       RETURN
       END
CCCCCCCCCCCCCCCCCCCCCCCCCCCCCCCCCCCCCCCCCCCCCCCCCCCCCCCCCCCCCCCCCCCCCCCCC
       SUBROUTINE MATINV(A,NDIM)
C inverts the matrix A of dimension NDIM using Gauss-Jordan elimination;
C A is replaced by its inverse
CCCCCCCCCCCCCCCCCCCCCCCCCCCCCCCCCCCCCCCCCCCCCCCCCCCCCCCCCCCCCCCCCCCCCCCCC
C Global variables:
       INCLUDE 'IO.ALL'
       INCLUDE 'PARAM.E5'
C Input/output variables:
       DOUBLE PRECISION A(CMAX+1,CMAX+1) !matrix for inverting
       INTEGER NDIM                      !dimension of matrix
C Local variables:
       DOUBLE PRECISION P(CMAX+1),Q(CMAX+1) !temporary storage
       LOGICAL ZEROED(CMAX+1)           !keeps track of who's zeroed
       INTEGER I,J,K                    !indices
       INTEGER KBIG                     !index of row being zeroed
       DOUBLE PRECISION PIVOT           !largest diag element in A
       REAL BIG                         !largest diag elem of non zeroed rows
CCCCCCCCCCCCCCCCCCCCCCCCCCCCCCCCCCCCCCCCCCCCCCCCCCCCCCCCCCCCCCCCCCCCCCCCC
C      which rows are zeroed? none so far
       DO 10 I=1,NDIM
          ZEROED(I)=.FALSE.
10     CONTINUE
C
C      loop over all rows
       DO 100 I=1,NDIM
          KBIG=0                        !search for largest diag element
          BIG=0                         !in rows not yet zeroed
          DO 20 J=1,NDIM
             IF (.NOT. ZEROED(J)) THEN
                IF (ABS(A(J,J)) .GT. BIG) THEN
```

```
                 BIG=ABS(A(J,J))
                 KBIG=J
              END IF
           END IF
20      CONTINUE
C
C       store the largest diagonal element
        IF (I .EQ. 1) PIVOT=A(KBIG,KBIG)
C
C       if all diagonals are zero, then the matrix is singular
        IF (KBIG .EQ. 0) THEN
           WRITE (OUNIT,*) ' Matrix is singular'
           RETURN
        END IF
C       matrix is ill conditioned if size of elements varies greatly
        IF (ABS(A(KBIG,KBIG)/PIVOT) .LT. 1.E-14) THEN
           WRITE (OUNIT,*) ' Matrix is ill conditioned'
           RETURN
        END IF
C
C       begin zeroing row KBIG
        ZEROED(KBIG)=.TRUE.
        Q(KBIG)=1/A(KBIG,KBIG)
        P(KBIG)=1.
        A(KBIG,KBIG)=0.
C
C       elements above the diagonal
        IF (KBIG .GT. 1) THEN
           DO 30 J=1,KBIG-1
              P(J)=A(J,KBIG)
              Q(J)=A(J,KBIG)*Q(KBIG)
              IF (.NOT. ZEROED(J)) Q(J)=-Q(J)
              A(J,KBIG)=0.
30         CONTINUE
        END IF
C
C       elements to the right of the diagonal
        IF (KBIG .LT. NDIM) THEN
           DO 40 J=KBIG+1,NDIM
              P(J)=A(KBIG,J)
              Q(J)=-A(KBIG,J)*Q(KBIG)
              IF (ZEROED(J)) P(J)=-P(J)
              A(KBIG,J)=0.
40         CONTINUE
        END IF
C
C       transform all of A
        DO 60 J=1,NDIM
           DO 50 K=J,NDIM
              A(J,K)=A(J,K)+P(J)*Q(K)
50         CONTINUE
60      CONTINUE
```

```
100     CONTINUE
C
C       symmetrize A**-1
        DO 110 J=2,NDIM
           DO 120 K=1,J-1
              A(J,K)=A(K,J)
120        CONTINUE
110     CONTINUE
C
        RETURN
        END
CCCCCCCCCCCCCCCCCCCCCCCCCCCCCCCCCCCCCCCCCCCCCCCCCCCCCCCCCCCCCCCCCCCCCCC
        SUBROUTINE INIT
C initializes constants, displays header screen,
C initializes arrays for input parameters
CCCCCCCCCCCCCCCCCCCCCCCCCCCCCCCCCCCCCCCCCCCCCCCCCCCCCCCCCCCCCCCCCCCCCCC
C Global variables:
        INCLUDE 'IO.ALL'
        INCLUDE 'MENU.ALL'
        INCLUDE 'PARAM.E5'
C Local parameters:
        CHARACTER*80 DESCRP             !program description
        DIMENSION DESCRP(20)
        INTEGER NHEAD,NTEXT,NGRAPH      !number of lines for each description
CCCCCCCCCCCCCCCCCCCCCCCCCCCCCCCCCCCCCCCCCCCCCCCCCCCCCCCCCCCCCCCCCCCCCCC
C       get environment parameters
        CALL SETUP
C
C       display header screen
        DESCRP(1)= 'EXAMPLE 5'
        DESCRP(2)= 'Determining Nuclear Charge Densities'
        NHEAD=2
C
C       text output description
        DESCRP(3)= 'Density parameters and quality of fit'
        NTEXT=1
C
C       graphics output description
        DESCRP(4)= 'Charge density and scattering cross section'
        NGRAPH=1
C
        CALL HEADER(DESCRP,NHEAD,NTEXT,NGRAPH)
C
C       calculate constants
        PI=4*ATAN(1.0)
        ALPHA=1./137.036
        HBARC=197.329
        SQRTM1=(0.,1.)
C       get data for Gauss-Legendre integration
        CALL GAUSS
C
C       setup menu arrays, beginning with constant part
```

```
      CALL MENU
C
      MPRMPT(4)='2) (not used)'
C
      MTYPE(13)=MTITLE
      MPRMPT(13)='Choice of nuclei:'
      MLOLIM(13)=2
      MHILIM(13)=1
C
      MTYPE(14)=MTITLE
      MPRMPT(14)='1) Calcium 40 '
      MLOLIM(14)=0
      MHILIM(14)=0
C
      MTYPE(15)=MTITLE
      MPRMPT(15)='2) Nickel 58 '
      MLOLIM(15)=0
      MHILIM(15)=0
C
      MTYPE(16)=MTITLE
      MPRMPT(16)='3) Lead 208 '
      MLOLIM(16)=0
      MHILIM(16)=1
C
      MTYPE(17)=MCHOIC
      MPRMPT(17)='Enter your choice'
      MTAG(17)='18 18 18'
      MLOLIM(17)=1
      MHILIM(17)=3
      MINTS(17)=1
      MREALS(17)=1.
C
      MTYPE(18)=SKIP
      MREALS(18)=35
C
      MTYPE(36)=SKIP
      MREALS(36)=60.
C
      MSTRNG(MINTS(75))= 'exmpl5.txt'
C
      MTYPE(76)=SKIP
      MREALS(76)=80.
C
      MSTRNG(MINTS(86))= 'exmpl5.grf'
C
      MTYPE(87)=SKIP
      MREALS(87)=90.
      RETURN
      END
CCCCCCCCCCCCCCCCCCCCCCCCCCCCCCCCCCCCCCCCCCCCCCCCCCCCCCCCCCCCCCCCCCCCCCC
      SUBROUTINE PARAM
C gets parameters from screen
```

```
C ends program on request
C closes old files
C maps menu variables to program variables
C opens new files
C calculates all derivative parameters
C performs checks on parameters
CCCCCCCCCCCCCCCCCCCCCCCCCCCCCCCCCCCCCCCCCCCCCCCCCCCCCCCCCCCCCCCCCCCC
C Global variables:
      INCLUDE 'MENU.ALL'
      INCLUDE 'IO.ALL'
      INCLUDE 'PARAM.E5'
C Local variables:
      INTEGER INUCL                    !map menu arrays to parameters
      PARAMETER (INUCL  = 17 )
C Function:
      LOGICAL LOGCVT                   !converts 1 to true, others to false
CCCCCCCCCCCCCCCCCCCCCCCCCCCCCCCCCCCCCCCCCCCCCCCCCCCCCCCCCCCCCCCCCCCC
C     get input from terminal
      CALL CLEAR
      CALL ASK(1,ISTOP)
C
C     stop program if requested
      IF (MREALS(IMAIN) .EQ. STOP) CALL FINISH
C
C     close files if necessary
      IF (TNAME .NE. MSTRNG(MINTS(ITNAME)))
     +     CALL FLCLOS(TNAME,TUNIT)
      IF (GNAME .NE. MSTRNG(MINTS(IGNAME)))
     +     CALL FLCLOS(GNAME,GUNIT)
C
C     set new parameter values
C     physical and numerical
      NUCL=MINTS(INUCL)
C
C     text output
      TTERM=LOGCVT(MINTS(ITTERM))
      TFILE=LOGCVT(MINTS(ITFILE))
      TNAME=MSTRNG(MINTS(ITNAME))
C
C     graphics output
      GTERM=LOGCVT(MINTS(IGTERM))
      GHRDCP=LOGCVT(MINTS(IGHRD))
      GFILE=LOGCVT(MINTS(IGFILE))
      GNAME=MSTRNG(MINTS(IGNAME))
C
C     open files
      IF (TFILE) CALL FLOPEN(TNAME,TUNIT)
      IF (GFILE) CALL FLOPEN(GNAME,GUNIT)
      !files may have been renamed
      MSTRNG(MINTS(ITNAME))=TNAME
      MSTRNG(MINTS(IGNAME))=GNAME
C
```

```
      RETURN
      END
CCCCCCCCCCCCCCCCCCCCCCCCCCCCCCCCCCCCCCCCCCCCCCCCCCCCCCCCCCCCCCCCCCCCCCCC
      SUBROUTINE GAUSS
C establish Gauss-Legendre weights and abscissae for 20 points
CCCCCCCCCCCCCCCCCCCCCCCCCCCCCCCCCCCCCCCCCCCCCCCCCCCCCCCCCCCCCCCCCCCCCCCC
C Global variables:
      INCLUDE 'PARAM.E5'
C Local variables:
      INTEGER ILEG                    !index for weights and abscissae
CCCCCCCCCCCCCCCCCCCCCCCCCCCCCCCCCCCCCCCCCCCCCCCCCCCCCCCCCCCCCCCCCCCCCCCC
      XLEG(1)=.993128599185094
      WLEG(1)=.017614007139152
      XLEG(2)=.963971927277913
      WLEG(2)=.040601429800386
      XLEG(3)=.912234428251325
      WLEG(3)=.062672048334109
      XLEG(4)=.839116971822218
      WLEG(4)=.083276741576704
      XLEG(5)=.74633190646015
      WLEG(5)=.10193011981724
      XLEG(6)=.636053680726515
      WLEG(6)=.118194531961518
      XLEG(7)=.510867001950827
      WLEG(7)=.131688638449176
      XLEG(8)=.373706088715419
      WLEG(8)=.142096109318382
      XLEG(9)=.227785851141645
      WLEG(9)=.149172986472603
      XLEG(10)=.076526521133497
      WLEG(10)=.152753387130725
C
      DO 10 ILEG=1,NLEG/2       !weights and abscissae are even and odd
         XLEG(21-ILEG)=-XLEG(ILEG)
         WLEG(21-ILEG)=WLEG(ILEG)
10    CONTINUE
C
      RETURN
      END
CCCCCCCCCCCCCCCCCCCCCCCCCCCCCCCCCCCCCCCCCCCCCCCCCCCCCCCCCCCCCCCCCCCCCCCC
      REAL FUNCTION BESSJ0(X)
C calculates zeroth order Bessel function at X
CCCCCCCCCCCCCCCCCCCCCCCCCCCCCCCCCCCCCCCCCCCCCCCCCCCCCCCCCCCCCCCCCCCCCCCC
C Passed variables:
      REAL X
C Local variables:
      REAL TEMP,TEMP2,Y2,Y
CCCCCCCCCCCCCCCCCCCCCCCCCCCCCCCCCCCCCCCCCCCCCCCCCCCCCCCCCCCCCCCCCCCCCCCC
      Y=X/3.
      IF(ABS(X) .LT. 3.0) THEN
         Y2=Y*Y
         TEMP=-.0039444+.00021*Y2
```

```
            TEMP=.0444479+Y2*TEMP
            TEMP=-.3163866+Y2*TEMP
            TEMP=1.2656208+Y2*TEMP
            TEMP=-2.2499997+Y2*TEMP
            BESSJ0=1+Y2*TEMP
      ELSE
            Y=1/Y
            TEMP=-7.2805E-04+1.4476E-04*Y
            TEMP=1.37237E-03+TEMP*Y
            TEMP=-9.512E-05+TEMP*Y
            TEMP=-.0055274+TEMP*Y
            TEMP=-7.7E-07+TEMP*Y
            TEMP=.79788456+TEMP*Y
            TEMP2=-2.9333E-04+1.3558E-04*Y
            TEMP2=-5.4125E-04+TEMP2*Y
            TEMP2=2.62573E-03+TEMP2*Y
            TEMP2=-3.954E-05+TEMP2*Y
            TEMP2=-4.166397E-02+TEMP2*Y
            TEMP2=X-.78539816+TEMP2*Y
            BESSJ0=TEMP*COS(TEMP2)/SQRT(X)
      END IF
      RETURN
      END
CCCCCCCCCCCCCCCCCCCCCCCCCCCCCCCCCCCCCCCCCCCCCCCCCCCCCCCCCCCCCCCCCCCCCCCC
      REAL FUNCTION BESSJ1(X)
C calculates first order Bessel function at X
CCCCCCCCCCCCCCCCCCCCCCCCCCCCCCCCCCCCCCCCCCCCCCCCCCCCCCCCCCCCCCCCCCCCCCCC
C Passed variables:
      REAL X
C Local variables:
      REAL TEMP,TEMP2,Y2,Y
CCCCCCCCCCCCCCCCCCCCCCCCCCCCCCCCCCCCCCCCCCCCCCCCCCCCCCCCCCCCCCCCCCCCCCCC
      Y=X/3.
      IF(ABS(X) .LT. 3.0) THEN
            Y2=Y*Y
            TEMP=-3.1761E-04+1.109E-05*Y2
            TEMP=4.43319E-03+Y2*TEMP
            TEMP=-3.954289E-02+Y2*TEMP
            TEMP=.21093573+Y2*TEMP
            TEMP=-.56249985+Y2*TEMP
            BESSJ1=X*(.5+Y2*TEMP)
      ELSE
            Y=1/Y
            TEMP=1.13653E-03-2.0033E-04*Y
            TEMP=-2.49511E-03+TEMP*Y
            TEMP=1.7105E-04+TEMP*Y
            TEMP=1.659667E-02+TEMP*Y
            TEMP=1.56E-06+TEMP*Y
            TEMP=.79788456+TEMP*Y
            TEMP2=7.9824E-04-2.9166E-04*Y
            TEMP2=7.4348E-04+TEMP2*Y
            TEMP2=-6.37879E-03+TEMP2*Y
```

```
        TEMP2=.0000565+TEMP2*Y
        TEMP2=.12499612+TEMP2*Y
        TEMP2=X-2.35619449+TEMP2*Y
        BESSJ1=TEMP*COS(TEMP2)/SQRT(X)
      END IF
      RETURN
      END
CCCCCCCCCCCCCCCCCCCCCCCCCCCCCCCCCCCCCCCCCCCCCCCCCCCCCCCCCCCCCCCCCCCCC
      SUBROUTINE PRMOUT(MUNIT)
C outputs parameter summary to the specified unit
CCCCCCCCCCCCCCCCCCCCCCCCCCCCCCCCCCCCCCCCCCCCCCCCCCCCCCCCCCCCCCCCCCCCC
C Global variables:
      INCLUDE 'IO.ALL'
      INCLUDE 'PARAM.E5'
C Passed variables:
      INTEGER MUNIT            !unit number for output
CCCCCCCCCCCCCCCCCCCCCCCCCCCCCCCCCCCCCCCCCCCCCCCCCCCCCCCCCCCCCCCCCCCCC
      IF (MUNIT .EQ. OUNIT) CALL CLEAR
C
      WRITE (MUNIT,2)
      WRITE (MUNIT,4)
      WRITE (MUNIT,6) TARGET
      WRITE (MUNIT,8) EBEAM
      WRITE (MUNIT,10) RMAX
      WRITE (MUNIT,12) NBASIS
      WRITE (MUNIT,14) QMAX
      WRITE (MUNIT,16) NBASIS*PI/RMAX
      WRITE (MUNIT,2)
C
2     FORMAT (' ')
4     FORMAT (' Output from example 5: Nuclear Charge Densities')
6     FORMAT (' For the target ',A10)
8     FORMAT (' the data are at a beam energy of ',F7.3,' MeV')
10    FORMAT (' the radial cutoff for the charge density=',F7.3,' fm')
12    FORMAT (' the number of sines used=', I2)
14    FORMAT (' the maximum experimental momentum transfer='
     +        ,F7.3,' fm**-1')
16    FORMAT (' the maximum numerical momentum transfer=',F7.3,'fm**-1')
C
      RETURN
      END
CCCCCCCCCCCCCCCCCCCCCCCCCCCCCCCCCCCCCCCCCCCCCCCCCCCCCCCCCCCCCCCCCCCCC
      SUBROUTINE TXTOUT(MUNIT,ITER,CHISQ,NDOF,NSINE,QBASIS,CZERO,A,Z)
C outputs the charge density parameters and goodness of fit to MUNIT
CCCCCCCCCCCCCCCCCCCCCCCCCCCCCCCCCCCCCCCCCCCCCCCCCCCCCCCCCCCCCCCCCCCCC
C Global variables:
      INCLUDE 'IO.ALL'
      INCLUDE 'PARAM.E5'
C Input variables:
      INTEGER MUNIT            !unit to write to
      REAL CZERO(CMAX)         !Fourier coefficients
      INTEGER NSINE            !number of coefficients
```

```
            INTEGER ITER                          !number of iterations
            REAL CHISQ                            !goodness of fit
            INTEGER NDOF                          !degrees of freedom in fit
            DOUBLE PRECISION A(CMAX+1,CMAX+1)     !matrix we're inverting
            REAL QBASIS                           !max num momentum transfer
            REAL Z                                !total charge
      C Local variables:
            INTEGER N                             !index of CZERO
            REAL SIGMA                            !error of CZERO
      CCCCCCCCCCCCCCCCCCCCCCCCCCCCCCCCCCCCCCCCCCCCCCCCCCCCCCCCCCCCCCCCCCCCCCCC
            WRITE (MUNIT,*) ' '
            WRITE (MUNIT,*) ' '
            WRITE (MUNIT,10) ITER
            WRITE (MUNIT,15) CHISQ,NDOF
            WRITE (MUNIT,20) NSINE,Z
            WRITE (MUNIT,25) QBASIS
            WRITE (MUNIT,*) ' '
            WRITE (MUNIT,30)
      C
            DO 50 N=1,NSINE
               SIGMA=0.
               IF (A(N,N) .GT. 0) SIGMA=SQRT(A(N,N))
               WRITE (MUNIT,40) N,CZERO(N),SIGMA
      50    CONTINUE
      C
            IF (MUNIT .EQ. OUNIT) CALL PAUSE('to continue...',1)
      C
      10    FORMAT (' Iteration ', I4)
      15    FORMAT (' Chi**2 = ',1PE15.8,' for ',I4,' degrees of freedom')
      20    FORMAT (' Number of sines = ',I2, '  total charge =',F10.4)
      25    FORMAT (' Maximum momentum transfer for this basis = ',1PE15.8,
           +         ' fm**-1')
      30    FORMAT (' The expansion coefficients of the charge density '
           +         'and their errors are:')
      40    FORMAT (' C(',I2,') = ',1PE15.8,' +- ',1PE15.8)
      C
            RETURN
            END
      CCCCCCCCCCCCCCCCCCCCCCCCCCCCCCCCCCCCCCCCCCCCCCCCCCCCCCCCCCCCCCCCCCCCCCCC
            SUBROUTINE GRFOUT(DEVICE,RHO,SIGT,NSINE,DRHO)
      C outputs differential cross section vs. momentum transfer
      C and nuclear charge density (with error bars) vs. radius
      CCCCCCCCCCCCCCCCCCCCCCCCCCCCCCCCCCCCCCCCCCCCCCCCCCCCCCCCCCCCCCCCCCCCCCCC
      C Global variables
            INCLUDE 'IO.ALL'
            INCLUDE 'PARAM.E5'
            INCLUDE 'GRFDAT.ALL'
      C Input variables:
            INTEGER DEVICE          !which device is being used?
            REAL SIGT(DATMAX)       !total cross section
            REAL RHO(NGRF)          !density for graphing
            REAL DRHO(NGRF)         !error in density
```

```
      INTEGER NSINE         !number of sines
C Local variables
      INTEGER IDATA,IR      !indexes data, density
      INTEGER EXPMAX,EXPMIN !min and max exp for diff cross section
      CHARACTER*9 CEBEAM    !EBEAM as a character string
      INTEGER LENGTH        !length of character strings
      CHARACTER*9 CSINE     !number of sines as a string
      INTEGER SINLEN
      REAL X(2),Y(2)        !arrays for error bars
      INTEGER SCREEN             !send to terminal
      INTEGER PAPER             !make a hardcopy
      INTEGER FILE             !send to a file
      DATA SCREEN,PAPER,FILE/1,2,3/
CCCCCCCCCCCCCCCCCCCCCCCCCCCCCCCCCCCCCCCCCCCCCCCCCCCCCCCCCCCCCCCCCCCC
C     messages for the impatient
      IF (DEVICE .NE. SCREEN) WRITE (OUNIT,100)
C
C     calculate parameters for graphing the cross sections
      IF (DEVICE .NE. FILE) THEN
         NPLOT=1                        !how many plots?
         IPLOT=1
C
         YMAX=0.                        !find limits on data points
         YMIN=SIGT(1)
         DO 20 IDATA=1,NPTS
            IF (SIGT(IDATA) .GT. YMAX) YMAX=SIGT(IDATA)
            IF (SIGE(IDATA) .GT. YMAX) YMAX=SIGE(IDATA)
            IF (SIGT(IDATA) .LT. YMIN) YMIN=SIGT(IDATA)
            IF (SIGE(IDATA) .LT. YMIN) YMIN=SIGE(IDATA)
20       CONTINUE
C        find integer limits on exponent
         EXPMAX=INT(LOG10(YMAX))
         IF (YMAX .GT. 1.) EXPMAX =EXPMAX+1
         EXPMIN=INT(LOG10(YMIN))
         IF (YMIN .LT. 1.) EXPMIN=EXPMIN-1
         YMAX=10.**EXPMAX
         YMIN=10.**EXPMIN
C
         XMIN=QEFF(1)                   !more limits
         XMAX=QEFF(NPTS)
         Y0VAL=XMIN
         X0VAL=YMIN
C
         NPOINT=NPTS
C
         ILINE=1                        !line and symbol styles
         ISYM=1
         IFREQ=1
         NXTICK=5
         NYTICK=EXPMAX-EXPMIN
         IF (NYTICK .GT. 8) THEN        !keep number of ticks small
            IF (MOD(NYTICK,2) .EQ. 0) THEN
```

```
                    NYTICK=NYTICK/2
                 ELSE
                    NYTICK=8
                 END IF
              END IF
C
              CALL CONVRT(EBEAM,CEBEAM,LENGTH)                !titles and labels
              INFO = 'Calculated(X) and Experimental(O)'
              LABEL(1)= 'Qeffective (fm**-1)'
              LABEL(2)= 'Differential Cross Section (mBarnes/sr)'
              TITLE = ' Electron Scattering on '//TARGET
       +             //' at '//CEBEAM(1:LENGTH)//' MeV'
C
              CALL GTDEV(DEVICE)                              !device nomination
              IF (DEVICE .EQ. SCREEN) CALL GMODE             !change to graphics mode
              CALL LGLNAX                                     !draw axes
           END IF
C
C      output cross sections vs. angle
       IF (DEVICE .EQ. FILE) THEN
           WRITE (GUNIT,200)
           WRITE (GUNIT,210)
           WRITE (GUNIT,220) (QEFF(IDATA),SIGT(IDATA),IDATA=1,NPTS)
       ELSE
           CALL XYPLOT(QEFF,SIGE)                            !plot experimental data
           NPOINT=2
           IFREQ=0
           DO 80 IDATA=1,NPTS                                !with error bars
              X(1)=QEFF(IDATA)
              X(2)=QEFF(IDATA)
              Y(1)=SIGE(IDATA)+DSIGE(IDATA)
              Y(2)=SIGE(IDATA)-DSIGE(IDATA)
              CALL XYPLOT(X,Y)
80         CONTINUE
           NPOINT=NPTS
           IFREQ=1
           ISYM=4
           CALL XYPLOT (QEFF,SIGT)                           !plot calculated sigma
           CALL GPAGE(DEVICE)
       END IF
C
C      calculate parameters for charge density
       IF (DEVICE .NE. FILE) THEN
           YMAX=0.                                           !find limits on data points
           YMIN=0.
           DO 120 IR=1,NGRF
              IF (RHO(IR) .GT. YMAX) YMAX=RHO(IR)
120        CONTINUE
           XMIN=0.
           XMAX=RMAX
           Y0VAL=XMIN
           X0VAL=YMIN
```

```
C
         NPOINT=NGRF
C
         ILINE=1                              !line and symbol styles
         ISYM=4
         IFREQ=1
         NXTICK=5
         NYTICK=5
         IFREQ=0
C
         CALL ICNVRT(NSINE,CSINE,SINLEN)    !titles and labels
         INFO=' '
         LABEL(1)= 'Radius (fermis)'
         LABEL(2)= 'Nuclear charge density (fermi**-3)'
         TITLE = TARGET//' using '//CSINE(1:SINLEN)//' sine functions'
C
         CALL LNLNAX                          !draw axes
      END IF
C
C     output charge density and its error
      IF (DEVICE .EQ. FILE) THEN
         WRITE (GUNIT,*) ' '
         WRITE (GUNIT,300)
         WRITE (GUNIT,310)
         WRITE (GUNIT,320) (RGRF(IR),RHO(IR),DRHO(IR),IR=1,NGRF)
      ELSE
         CALL XYPLOT (RGRF,RHO)                !charge density
         NPOINT=2
         IFREQ=0
         DO 180 IR=1,NGRF                      !with error bars
            X(1)=RGRF(IR)
            X(2)=X(1)
            Y(1)=RHO(IR)-DRHO(IR)
            Y(2)=Y(1)+2*DRHO(IR)
            CALL XYPLOT(X,Y)
180      CONTINUE
      END IF
C
C     end graphing session
      IF (DEVICE .NE. FILE) CALL GPAGE(DEVICE)  !end graphics package
      IF (DEVICE .EQ. SCREEN) CALL TMODE        !switch to text mode
C
100   FORMAT (/,' Patience, please; output going to a file.')
200   FORMAT (10X,'Qeff (1/fermi)',9X,'Sigma (mBarnes/sr)')
210   FORMAT (10X,'--------------',9X,'------------------')
220   FORMAT (2(10X,1PE15.8))
300   FORMAT (12X,'R (fermi)',8X,'Charge density (fm**-3)',10X,'Error')
310   FORMAT (12X,'---------',8X,'-----------------------',10X,'-----')
320   FORMAT (3(9X,1PE15.8))
C
      RETURN
      END
```

```
CCCCCCCCCCCCCCCCCCCCCCCCCCCCCCCCCCCCCCCCCCCCCCCCCCCCCCCCCCCCCCCCCCCCCCCCCCC
CCCCCCCCCCCCCCCCCCCCCCCCCCCCCCCCCCCCCCCCCCCCCCCCCCCCCCCCCCCCCCCCCCCCCCCCCCC
C param.p5
C
      REAL PI                         !3.14159
      COMPLEX SQRTM1                  !square root of minus 1
      REAL ALPHA                      !fine structure constant
      REAL HBARC                      !hbar times c (MeV-fermi)
      INTEGER NLEG                    !number of points for GL integ
      PARAMETER (NLEG=20)
      REAL WLEG(NLEG),XLEG(NLEG)      !weights and abscissae
C
      INTEGER CA,NI,PB                !flags for each nucleus
      PARAMETER (CA=1)
      PARAMETER (NI=2)
      PARAMETER (PB=3)
C
      REAL RMAX                       !radius at which density =0
      INTEGER NUCL                    !nucleus of choice
      INTEGER NBASIS                  !number of sin waves to include
C
      INTEGER DATMAX                  !maximum number of data points
      PARAMETER (DATMAX=100)
      INTEGER NGRF                    !number of points for graphing
      PARAMETER (NGRF=100)
C
      REAL THETA(DATMAX),SIGE(DATMAX),DSIGE(DATMAX)!experimental data
      REAL QEFF(DATMAX)
      REAL ZTARGT,ATARGT              !charge, number of nucleons
      INTEGER NPTS                    !number of data points
      REAL EBEAM,KBEAM                !beam energy and wavenumber
      REAL ZA                         !charge * fine struct const
      REAL QMAX                       !maximum momentum transfer
      REAL RZERO                      !model radius for nucleus
      CHARACTER*10 TARGET
      REAL VC1                        !correction to momentum transfer
      REAL DRGRF,RGRF(NGRF)           !radial values for graphing
C
      INTEGER CMAX                    !maximum number of Cn's
      PARAMETER (CMAX=15)
      REAL CHIN(NLEG,CMAX)            !chi(n) array
      REAL JTABLE(NLEG,DATMAX)        !Bessel functions * b
      COMPLEX FOUTER(DATMAX)          !outer part of scatt amplitude
C
      COMMON/CONST/PI,ALPHA,HBARC,WLEG,XLEG,SQRTM1
      COMMON/PPARM/RMAX,NUCL,NBASIS
      COMMON/PCALC/ZTARGT,ATARGT,NPTS,EBEAM,KBEAM,ZA,QMAX,RZERO,VC1
      COMMON/NUCDAT/THETA,SIGE,DSIGE,QEFF
      COMMON/TABLE/CHIN,JTABLE,FOUTER
      COMMON/GRF/DRGRF,RGRF
      COMMON/CHARC/TARGET
```

B.6 Example 6

Algorithm   This program solves Laplace's equation in two dimensions on a uniform rectangular lattice by Gauss-Seidel iteration. The program currently allows the input of Dirichlet boundary conditions only. However, the array BNDCN can be used to indicate other types of boundary conditions (i.e., periodic and Neumann). Once all parameters are input, the relaxation iterations proceed with Eq. (6.17) being solved successively at each lattice point (subroutine RELAX). If requested, these relaxation sweeps will be restricted to a sublattice. The total energy per area is also calculated in RELAX. Note that the lattice spacing in both x and y is set equal to one. Every NFREQ'th iteration, you are given option to continue iterating or to return to the main menu.

Input   The input procedure is a bit different for this program, as well as for Project 6. For these two codes, you may find that during the iteration procedure you want to keep the field values and all parameters, except one or two, the same. For example, you may wish to alter just the relaxation parameter, change a boundary condition, or change the frequency at which the field is displayed. Such a procedure is straightforward, but only as long as the lattice dimensions remain the same. Therefore, the menu is broken into two pieces: the first level obtains the lattice dimensions; the second level (or main menu) obtains all other parameters. As long as you do not enter the first level menu, *all parameters and field values are unchanged unless you specifically request otherwise.*

The first level prompts you for the lattice parameters NX [20] and NY [20]. The maximum lattice sizes are fixed by two considerations: first, the parameters NXMAX= 79 and NYMAX= 79 which determine the array sizes; second, the size of your terminal as defined by TRMWID and TRMLIN in subroutine SETUP. This last restriction is imposed so that the fields can be displayed on the screen, using characters to indicate field strength (subroutine DISPLY). If you will be using a graphics package, you can remove the terminal size restrictions on NX and NY in subroutine LATTIC. If it is not your first time through this menu, you will also be given the option of ending program execution [no], and (if that answer is *no*) resetting input parameters to their default values. Note that all parameters that depend on lattice dimensions (i.e., the field values, interior boundary conditions, and sublattice parameters) are automatically reset every time you enter this first level menu. The main menu allows for input of all remaining parameters. The physical parameters are the value of the field on the upper [0.], lower [25.], left [25.], and right [0.] boundaries, which are fixed for Dirichlet boundary conditions. The limits on field values are given by FMIN= 0 and FMAX= 25, and are imposed so that the scale for character display of the field can be fixed (see Output below). You can also fix Dirichlet boundary conditions on the interior of the lattice. Two geometries are allowed (subroutine BNDRY): rectangles defined by the lower left and upper right corners; and "45 degree" lines defined by one corner, the quadrant, and the length (i.e., quadrant one produces a line at 45°, quadrant two at 135°, three at 225°, and four at 315°, all with respect to the positive x-axis). Note that x and y locations are input as integers to indicate the corresponding lattice point. A maximum of BCMAX= 10 interior boundary conditions are permitted. The menu also allows for the display (subroutine DSPLBC or GRFBC) and review (subroutine REVIEW) of interior boundary conditions. REVIEW allows for deletion of interior boundary conditions or alteration of their fixed field values.

The numerical parameters are the relaxation parameter [1.], the option to relax a sublattice [no] specified by the lower left and upper right corners, and the initial field value. Initial field values at non-boundary points can: 1) be read in from a file [exmpl6.in] created earlier by this program; 2) be set to a specified value [12.5]; or 3) remain unchanged. Option 2 is the default if the calculation is just beginning; otherwise, option 3 is the default. The two graphics parameters are the number of contours for plotting [10] (used only if you have a graphics package), and the frequency at which the field is displayed [10]. This is also the frequency at which you are asked if you would like to continue iterations. The maximum number of contours is fixed by MAXLEV= 20.

Output   After each relaxation sweep, the text output displays the iteration number (iteration count is zeroed on return from the main menu), energy, change in energy from the last iteration, and the maximum change in the field. Every NFREQ'th iteration, the program will display the field. Note that the field will be displayed as contours (if you have requested graphics sent to the screen, see subroutine GRFOUT) or as characters (if you have requested text to the screen, see subroutine DISPLY). In the character display, small letters indicate boundary points and capital letters indicate non-boundary points. Currently the program sets the maximum field value to 25 and the minimum to 0, so that the letter A corresponds to a field value of 0 and the letter Z to a field value of 25. For the contour plots, the boundary points are indicated by dashed lines; the appearance of the contour lines will depend on your graphics package. Because display of the field can slow down calculations considerably, you may want to set NREQ to a large number.

Again, to avoid voluminous output, the field is not written out to a
file or sent to a hardcopy device each time it is displayed on the screen.
Instead, when you request to stop iterations, you will be asked if you want
the field values to be written to the requested output devices. Note that
the field may be written out in three ways: as character data (if text is
requested to a file), as numerical data (if graphics is requested a file), or
as a contour plot (if graphics is requested to a hardcopy device).

```
ccccccccccccccccccccccccccccccccccccccccccccccccccccccccccccccccccccccc
      PROGRAM EXMPL6
C     Example 6: Solving Laplace's equation in two dimensions
C  COMPUTATIONAL PHYSICS (FORTRAN VERSION)
C  by Steven E. Koonin and Dawn C. Meredith
C  Copyright 1989, Addison-Wesley Publishing Company
ccccccccccccccccccccccccccccccccccccccccccccccccccccccccccccccccccccccc
      CALL INIT              !display header screen, setup parameters
5     CONTINUE               !main loop/ execute once for each set of param
      CALL LATTIC            !get lattice size, allow for ending
      CALL ARCHON            !get detailed bound cond and relax lattice
      GOTO 5
      END
ccccccccccccccccccccccccccccccccccccccccccccccccccccccccccccccccccccccc
      SUBROUTINE ARCHON
C subroutine to get parameters, relax the lattice, and output results
ccccccccccccccccccccccccccccccccccccccccccccccccccccccccccccccccccccccc
C Global variables:
      INCLUDE 'PARAM.E6'
      INCLUDE 'IO.ALL'
C Local variables:
      REAL P(NXMAX,NYMAX)         !field values
      REAL PGRF(NXMAX,NYMAX)      !field values for graphing
      INTEGER XMIN,XMAX,YMIN,YMAX !corners of relaxing lattice
      REAL DELMAX,OLDMAX          !keep track of changes in P
      REAL ENERGY,DELE            !current and delta energy
      INTEGER NITER               !number of iterations
      LOGICAL ITER                !continue iterating?
      LOGICAL PRM                 !print out parameters?
      LOGICAL END                 !end this run?
      INTEGER IX,IY               !lattice indices
      INTEGER CHOICE              !write out field now?
      INTEGER NLINES              !number of lines sent to terminal
      INTEGER SCREEN              !send to terminal
      INTEGER PAPER               !make a hardcopy
      INTEGER FILE                !send to a file
C Functions:
      REAL GETFLT                 !user input function
      INTEGER GETINT,YESNO        !user input functions
      LOGICAL LOGCVT              !convert 0,1 to false,true
```

```
          DATA SCREEN,PAPER,FILE/1,2,3/
CCCCCCCCCCCCCCCCCCCCCCCCCCCCCCCCCCCCCCCCCCCCCCCCCCCCCCCCCCCCCCCCCCCCCCCC
          END=.FALSE.                    !initial values
          DO 5 IX=1,NX
            DO 5 IY=1,NY
              P(IX,IY)=0
5         CONTINUE
C
200       CONTINUE                       !allow for many runs with same lattice
              CALL PARAM(P,END)          !get new parameters
              IF (END) RETURN            !start fresh or end altogether
C
          IF (SUBLAT) THEN
            XMIN=NXLL                    !set limits for relaxation
            XMAX=NXUR
            YMIN=NYLL
            YMAX=NYUR
          ELSE
            XMIN=1
            YMIN=1
            XMAX=NX
            YMAX=NY
          END IF
C
          IF (TFILE) CALL PRMOUT(TUNIT,NLINES)
          PRM=.TRUE.
          NITER=0
C
99        CONTINUE                       !begin iterations
              IF ((TTERM) .AND. (PRM)) CALL PRMOUT(OUNIT,NLINES)
              NITER=NITER+1
C                                        !relax lattice
          CALL RELAX(P,XMIN,XMAX,YMIN,YMAX,ENERGY,DELE,DELMAX,OLDMAX)
              IF (NITER .EQ. 1) THEN     !no changes if this is the first time
                DELE=0
                OLDMAX=0
              END IF
C
          IF. (TFILE) CALL TXTOUT(TUNIT,NITER,ENERGY,DELE,DELMAX,NLINES)
          IF (TTERM) CALL TXTOUT(OUNIT,NITER,ENERGY,DELE,DELMAX,NLINES)
C
          ITER=.FALSE.
          PRM=.TRUE.
          IF (MOD(NITER,NFREQ) .NE. 0) THEN
              ITER=.TRUE.                !continue iterating without asking
              PRM=.FALSE.                !don't print out header
          ELSE                           !display fields every NFREQ iteration
              IF (GTERM) THEN
                  CALL PAUSE('to see the field ...',1)
                  CALL CLEAR
                  CALL GRFOUT(SCREEN,P,PGRF,NX,NY,ENERGY)
              ELSE IF (TTERM) THEN
```

```
          CALL PAUSE('to see the field ...',1)
          CALL CLEAR
          CALL DISPLY(P,OUNIT)
       END IF
       IF (SUBLAT) THEN    !and provide options for continuing
          SUBLAT=LOGCVT(YESNO(1,'Continue relaxing sublattice?'))
          IF (SUBLAT) ITER=.TRUE.
       ELSE
          ITER=LOGCVT(YESNO(1,'Continue relaxing lattice?'))
       END IF
     END IF
C
     IF (ITER) THEN        !continue iterating
       GOTO 99
C
     ELSE
C        prompt for writing out of field values
       IF ((GFILE) .OR. (GHRDCP) .OR. (TFILE)) THEN
          CHOICE=YESNO(1,'Do you want to write out field now?')
          IF (CHOICE .EQ. 1) THEN
             IF (GFILE) CALL WRTOUT(P)     !write out field if requested
             IF (GHRDCP) CALL GRFOUT(PAPER,P,PGRF,NX,NY,ENERGY)
             IF (TFILE) CALL DISPLY(P,TUNIT)
          END IF
       END IF
       GOTO 200         !display menu
     END IF
C
     END
CCCCCCCCCCCCCCCCCCCCCCCCCCCCCCCCCCCCCCCCCCCCCCCCCCCCCCCCCCCCCCCCCCCCCC
     SUBROUTINE RELAX(P,XMIN,XMAX,YMIN,YMAX,ENERGY,DELE,DELMAX,OLDMAX)
C relaxes the lattice (specified by XMIN, etc), calculates the new
C energy (ENERGY) and maximum change in the field (DELMAX)
CCCCCCCCCCCCCCCCCCCCCCCCCCCCCCCCCCCCCCCCCCCCCCCCCCCCCCCCCCCCCCCCCCCCCC
C Global variables:
     INCLUDE 'PARAM.E6'
     INCLUDE 'IO.ALL'
C Input/output variables:
     REAL P(NXMAX,NYMAX)               !field values (I/O)
     INTEGER XMIN,XMAX,YMIN,YMAX       !corners of relaxing lattice(input)
     REAL DELMAX,OLDMAX                !keep track of change in P (output)
     REAL ENERGY,DELE                  !current and delta energy (output)
C Local variables:
     INTEGER IX,IY                     !lattice indices
     REAL A,B,C,D                      !values of P at neighboring points
     REAL PNEW,POLD,DELP               !temp value of P at current point
     REAL OLDE                         !old energy
CCCCCCCCCCCCCCCCCCCCCCCCCCCCCCCCCCCCCCCCCCCCCCCCCCCCCCCCCCCCCCCCCCCCCC
     OLDMAX=DELMAX                     !roll values
     OLDE=ENERGY
     DELMAX=0.                         !initialize values
     ENERGY=0.
```

```
C
      DO 200 IY=YMIN,YMAX
         DO 100 IX=XMIN,XMAX
            A=0.
            B=0.
            C=0.
            D=0.
            IF (IX .LT. NX) A=P(IX+1,IY)    !field values at neighboring
            IF (IX .GT. 1)  B=P(IX-1,IY)    !lattice points
            IF (IY .LT. NY) C=P(IX,IY+1)
            IF (IY .GT. 1)  D=P(IX,IY-1)
C
            POLD=P(IX,IY)
            IF (BNDCND(IX,IY) .EQ. NON) THEN
               PNEW=(1.-OMEGA)*POLD+OMEGA/4*(A+B+C+D)       !relax
               P(IX,IY)=PNEW
               IF (POLD .NE. 0.) THEN
                  DELP=ABS((POLD-PNEW)/POLD)
               ELSE
                  DELP=ABS(PNEW)
               END IF
               IF (DELP .GT. DELMAX) DELMAX=DELP            !find max change
            ELSE
               PNEW=POLD                        !don't relax DIRCHLT bound cond
            END IF
C
            IF (IX .GT. 1) ENERGY=ENERGY+(PNEW-B)**2
            IF (IY .GT. 1) ENERGY=ENERGY+(PNEW-D)**2
C
100      CONTINUE
200   CONTINUE
      ENERGY=ENERGY/2/(NX-1)/(NY-1)
      DELE=OLDE-ENERGY
C
      RETURN
      END
CCCCCCCCCCCCCCCCCCCCCCCCCCCCCCCCCCCCCCCCCCCCCCCCCCCCCCCCCCCCCCCCCCCCCCCC
      SUBROUTINE INIT
C initializes constants, displays header screen,
C initializes menu arrays for input parameters
CCCCCCCCCCCCCCCCCCCCCCCCCCCCCCCCCCCCCCCCCCCCCCCCCCCCCCCCCCCCCCCCCCCCCCCC
C Global variables:
      INCLUDE 'IO.ALL'
      INCLUDE 'MENU.ALL'
      INCLUDE 'PARAM.E6'
C Local parameters:
      CHARACTER*80 DESCRP                !program description
      DIMENSION DESCRP(20)
      INTEGER NHEAD,NTEXT,NGRAPH         !number of lines for each description
CCCCCCCCCCCCCCCCCCCCCCCCCCCCCCCCCCCCCCCCCCCCCCCCCCCCCCCCCCCCCCCCCCCCCCCC
C     get environment parameters
      CALL SETUP
```

```
C
C       display header screen
        DESCRP(1)= 'EXAMPLE 6'
        DESCRP(2)= 'Solving Laplace''s Equation in Two Dimensions'
        NHEAD=2
C
C       text output description
        DESCRP(3)= 'iteration, energy, change in energy,'
        DESCRP(4)= 'and the maximum change in field'
        NTEXT=2
C
C       graphics output description
        DESCRP(5)= 'field values'
        NGRAPH=1
C
        CALL HEADER(DESCRP,NHEAD,NTEXT,NGRAPH)
C
        FIRST=.TRUE.              !is this the first time through the menu?
C
C       set up constant part of menu
        CALL MENU
C
C       item 9 has a different use in this program
        MTYPE(9)=MTITLE
        MPRMPT(9)='7) Change lattice size or end altogether'
        MLOLIM(9)=0
        MHILIM(9)=1
C
        MTYPE(13)=FLOAT
        MPRMPT(13)= 'Enter fixed value of the field for UPPER boundary'
        MTAG(13)= 'Field on UPPER boundary'
        MLOLIM(13)=FMIN
        MHILIM(13)=FMAX
        MREALS(13)=FMIN
C
        MTYPE(14)=FLOAT
        MPRMPT(14)= 'Enter fixed value of the field for LOWER boundary'
        MTAG(14)= 'Field on LOWER boundary'
        MLOLIM(14)=FMIN
        MHILIM(14)=FMAX
        MREALS(14)=FMAX
C
        MTYPE(15)=FLOAT
        MPRMPT(15)= 'Enter fixed value of the field for LEFT boundary'
        MTAG(15)= 'Field on LEFT boundary'
        MLOLIM(15)=FMIN
        MHILIM(15)=FMAX
        MREALS(15)=FMAX
C
        MTYPE(16)=FLOAT
        MPRMPT(16)= 'Enter fixed value of the field for RIGHT boundary'
        MTAG(16)= 'Field on RIGHT boundary'
```

```
               MLOLIM(16)=FMIN
               MHILIM(16)=FMAX
               MREALS(16)=FMIN
      C
               MTYPE(17)=NOSKIP
               MPRMPT(17)=
      +        'Do you wish to enter/review/display interior bound cond?'
               MTAG(17)='Enter/review/display interior boundary conditions'
               MINTS(17)=0
               MREALS(17)=35
      C
               MTYPE(18)=QUIT
      C
      C        This section of the menu (from 25-33) is called from BNDRY
               MTYPE(25)=MTITLE
               MPRMPT(25)='Interior Boundary Conditions Menu'
               MLOLIM(25)=2
               MHILIM(25)=1
      C
               MTYPE(26)=MTITLE
               MPRMPT(26)='1) input rectangular boundary conditions'
               MLOLIM(26)=0
               MHILIM(26)=0
      C
               MTYPE(27)=MTITLE
               MPRMPT(27)='2) input 45 degree line boundary conditions'
               MLOLIM(27)=0
               MHILIM(27)=0
      C
               MTYPE(28)=MTITLE
               MPRMPT(28)='3) review boundary conditions '
               MLOLIM(28)=0
               MHILIM(28)=0
      C
               MTYPE(29)=MTITLE
               MPRMPT(29)='4) display boundary conditions '
               MLOLIM(29)=0
               MHILIM(29)=0
      C
               MTYPE(30)=MTITLE
               MPRMPT(30)='5) return to main menu'
               MLOLIM(30)=0
               MHILIM(30)=0
      C
               MTYPE(31)=MCHOIC
               MPRMPT(31)='Enter Choice'
               MTAG(31)='32 32 32 32 32 32'
               MLOLIM(31)=1
               MHILIM(31)=5
               MINTS(31)=5
               MREALS(31)=-5
      C
```

```
    MTYPE(38)=FLOAT
    MPRMPT(38)= 'Enter relaxation parameter'
    MTAG(38)= 'Relaxation parameter'
    MLOLIM(38)=0.
    MHILIM(38)=2.
    MREALS(38)=1.
C
    MTYPE(39)=NOSKIP
    MPRMPT(39)='Do you want to specify a sublattice to relax first?'
    MTAG(39)='Relax a sublattice first?'
    MINTS(39)=0
    MREALS(39)=44.
C
    MTYPE(40)=NUM
    MPRMPT(40)='Enter lower left X value for sublattice'
    MTAG(40)='Lower left X sublattice value'
    MLOLIM(40)=1
    MHILIM(40)=NXMAX
    MINTS(40)=1
C
    MTYPE(41)=NUM
    MPRMPT(41)='Enter lower left Y value for sublattice'
    MTAG(41)='Lower left Y sublattice value'
    MLOLIM(41)=1
    MHILIM(41)=NYMAX
    MINTS(41)=1
C
    MTYPE(42)=NUM
    MPRMPT(42)='Enter upper right X value for sublattice'
    MTAG(42)='Upper right X sublattice value'
    MLOLIM(42)=1
    MHILIM(42)=NXMAX
    MINTS(42)=1
C
    MTYPE(43)=NUM
    MPRMPT(43)='Enter upper right Y value for sublattice'
    MTAG(43)='Upper right Y sublattice value'
    MLOLIM(43)=1
    MHILIM(43)=NXMAX
    MINTS(43)=1
C
    MTYPE(44)=MTITLE
    MPRMPT(44)='Initial Field Value Menu'
    MLOLIM(44)=1
    MHILIM(44)=1
C
    MTYPE(45)=MTITLE
    MPRMPT(45)='1) Read in initial values from a file'
    MLOLIM(45)=0
    MHILIM(45)=0
C
    MTYPE(46)=MTITLE
```

```
            MPRMPT(46)='2) Set field at all lattice points to one value'
            MLOLIM(46)=0
            MHILIM(46)=0
C
            MTYPE(47)=MTITLE
            MPRMPT(47)='3) Leave field values unchanged'
            MLOLIM(47)=0
            MHILIM(47)=0
C
            MTYPE(48)=MCHOIC
            MPRMPT(48)='Enter Choice'
            MTAG(48)='49 51 52'
            MLOLIM(48)=1
            MHILIM(48)=3
            MINTS(48)=2
            MREALS(48)=2
C
            MTYPE(49)=CHSTR
            MPRMPT(49)= 'Enter name of data file'
            MTAG(49)= 'File with initial values for the field'
            MHILIM(49)=12.
            MINTS(49)=3.
            MSTRNG(MINTS(49))= 'exmpl6.in'
C
            MTYPE(50)=SKIP
            MREALS(50)=60.
C
            MTYPE(51)=FLOAT
            MPRMPT(51)= 'Enter initial value for the field'
            MTAG(51)= 'Initial Field Value'
            MLOLIM(51)=FMIN
            MHILIM(51)=FMAX
            MREALS(51)=(FMAX-FMIN)/2
C
            MTYPE(52)=SKIP
            MREALS(52)=60.
C
            MSTRNG(MINTS(75))= 'exmpl6.txt'
C
            MTYPE(76)=SKIP
            MREALS(76)=80.
C
            MSTRNG(MINTS(86))= 'exmpl6.grf'
C
            MTYPE(87)=NUM
            MPRMPT(87)= 'Enter the display frequency for the field'
            MTAG(87)= 'Field display frequency'
            MLOLIM(87)= 1.
            MHILIM(87)= 100.
            MINTS(87)= 10
C
            MTYPE(88)=NUM
```

```
      MPRMPT(88)= 'Enter number of contour levels'
      MTAG(88)= 'Number of contour levels'
      MLOLIM(88)= 1.
      MHILIM(88)= MAXLEV
      MINTS(88)= 10
C
      MTYPE(89)=SKIP
      MREALS(89)=90.
C
      RETURN
      END
CCCCCCCCCCCCCCCCCCCCCCCCCCCCCCCCCCCCCCCCCCCCCCCCCCCCCCCCCCCCCCCCCCCCCCC
      SUBROUTINE LATTIC
C gets lattice size from screen and calculates best way to display the
C field as ascii characters based on lattice size and terminal size;
C resets all boundary conditions and default menu values
CCCCCCCCCCCCCCCCCCCCCCCCCCCCCCCCCCCCCCCCCCCCCCCCCCCCCCCCCCCCCCCCCCCCCCC
C Global variables:
      INCLUDE 'PARAM.E6'
      INCLUDE 'IO.ALL'
      INCLUDE 'MENU.ALL'
      INCLUDE 'MAP.E6'
C Local variables:
      INTEGER END                   !end program
      INTEGER IX,IY,IBC             !lattice indices, BC index
      INTEGER NXHI,NYHI,NXLO,NYLO   !limits on lattice size
      INTEGER NXDEF,NYDEF           !default lattice sizes
      LOGICAL RESET                 !reset parameters?
C Functions:
      INTEGER YESNO,GETINT          !user input functions
      LOGICAL LOGCVT                !change 1,0 to true and false
CCCCCCCCCCCCCCCCCCCCCCCCCCCCCCCCCCCCCCCCCCCCCCCCCCCCCCCCCCCCCCCCCCCCCCC
C     allow user to end the program
      CALL CLEAR
      IF (.NOT. FIRST) THEN
          END=YESNO(0,' Do you want to end the program?')
          IF (END .EQ. 1) CALL FINISH
      ELSE
C         the lattice size is determined by array size and terminal size;
C         if you're using graphics, terminal size won't matter,
C         set NXHI=NXMAX and NYHI=NYMAX
          NXHI=MIN(TRMWID-2,NXMAX)
          NYHI=MIN(TRMLIN-3,NYMAX)
          NXDEF=MIN(NXHI,20)
          NYDEF=MIN(NYHI,20)
          NXLO=MIN(5,NXDEF)
          NYLO=MIN(5,NYDEF)
      END IF
C
C     get lattice parameters from the terminal
      NX=GETINT(NXDEF,NXLO,NXHI,' Enter number of X lattice points')
      NY=GETINT(NYDEF,NYLO,NYHI,' Enter number of Y lattice points')
```

```
            NXDEF=NX
            NYDEF=NY
  C
  C       calculate parameters for best looking display
          IF (2*NX .LE. TRMWID) THEN
                XSKIP=.TRUE.               !skip spaces in x
                XCNTR=(TRMWID-2*NX)/2      !how to center display
          ELSE
                XSKIP=.FALSE.
                XCNTR=(TRMWID-NX)/2
          END IF
          IF (XCNTR .LT. 1) XCNTR=1
  C
          IF (2*NY .LE. TRMLIN-3) THEN
                YSKIP=.TRUE.               !skip lines in y
                YCNTR=(TRMLIN-2*NY)/2-2    !how to center display
          ELSE
                YSKIP=.FALSE.
                YCNTR=(TRMLIN-NY)/2-2
          END IF
          IF (YCNTR .LT. 0) YCNTR=0
  C
  C       set all bound cond parameters to zero
          NBC=0
          DO 10 IBC=1,BCMAX
              XLL(IBC)=0
              YLL(IBC)=0
              XUR(IBC)=0
              YUR(IBC)=0
              BCGEOM(IBC)=0
              BCVAL(IBC)=0.
  10      CONTINUE
          DO 20,IX=1,NXMAX
              DO 30 IY=1,NYMAX
                  BNDCND(IX,IY)=0
  30          CONTINUE
  20      CONTINUE
  C
  C       set up limits on sublattice
          MHILIM(INXLL)=NX
          MHILIM(INYLL)=NY
          MHILIM(INXUR)=NX
          MHILIM(INYUR)=NY
  C
  C       allow for resetting of defaults
          IF (FIRST) THEN
              FIRST=.FALSE.
          ELSE
              RESET=LOGCVT(YESNO(0,' Do you want to reset default values?'))
              IF (RESET) THEN
                  MREALS(13)=FMIN        !upper bound. cond.
                  MREALS(14)=FMAX        !lower bound. cond.
```

```
         MREALS(15)=FMAX      !left  bound. cond.
         MREALS(16)=FMIN      !right bound. cond.
         MREALS(38)=1.        !omega
         MINTS(39)=0          !sublattice?
         MINTS(40)=1          !lower left  x sublattice
         MINTS(41)=1          !upper right x sublattice
         MINTS(42)=1          !lower left  y sublattice
         MINTS(43)=1          !upper right y sublattice
         MINTS(48)=2          !PTYPE
         MREALS(48)=2
         MSTRNG(MINTS(49))= 'exmpl6.in'
         MREALS(51)=(FMAX-FMIN)/2 !initial P value
         MINTS(73)=TXTTRM
         MSTRNG(MINTS(75))= 'exmpl6.txt'
         MINTS(74)=TXTFIL
         MINTS(83)=GRFTRM
         MINTS(84)=GRFHRD
         MINTS(85)=GRFFIL
         MSTRNG(MINTS(86))= 'exmpl6.grf'
         MINTS(87)= 10        !graphing frequency
         MINTS(88)= 10        !number of contour lines
        END IF
       END IF
C
       RETURN
       END
CCCCCCCCCCCCCCCCCCCCCCCCCCCCCCCCCCCCCCCCCCCCCCCCCCCCCCCCCCCCCCCCCCCCCCCCCCC
       SUBROUTINE PARAM(P,END)
C gets parameters from screen
C ends program on request
C closes old files
C maps menu variables to program variables
C opens new files
C calculates all derivative parameters
C performs checks on sublattice parameters
C set the field to its initial values (P)
C and controls ending of program (END)
CCCCCCCCCCCCCCCCCCCCCCCCCCCCCCCCCCCCCCCCCCCCCCCCCCCCCCCCCCCCCCCCCCCCCCCCCCC
C Global variables:
       INCLUDE 'MENU.ALL'
       INCLUDE 'IO.ALL'
       INCLUDE 'PARAM.E6'
       INCLUDE 'MAP.E6'
C Input/output variables:
       REAL P(NXMAX,NYMAX)    !field values
       LOGICAL END            !end program?
C Local variables:
       INTEGER TYPE,GEOM      !type and geom of bound cond
       INTEGER IX,IY          !lattice indices
C Functions:
       LOGICAL LOGCVT         !converts 1 and 0 to true and false
       INTEGER GETINT         !get integer from screen
```

```
cccccccccccccccccccccccccccccccccccccccccccccccccccccccccccccccccccccc
c     get input from terminal
10    CONTINUE
      MINTS(INTERR)=0          !reset value
c     get all parameters except interior bound cond
      CALL CLEAR
      CALL ASK(1,ISTOP)
c
c     start fresh or end altogether, if so requested
      IF (MREALS(IMAIN) .EQ. STOP)  THEN
         END=.TRUE.
         RETURN
      END IF
c
c     set basic boundary conditions before doing interior bc
      UPPER=MREALS(IUPPER)
      LOWER=MREALS(ILOWER)
      LEFT=MREALS(ILEFT)
      RIGHT=MREALS(IRIGHT)
      GEOM=RECTGL
      TYPE=DRCHLT
      CALL SETBC(P,GEOM,TYPE,UPPER,1,NX,NY,NY)
      CALL SETBC(P,GEOM,TYPE,LOWER,1,NX,1,1)
      CALL SETBC(P,GEOM,TYPE,LEFT,1,1,1,NY)
      CALL SETBC(P,GEOM,TYPE,RIGHT,NX,NX,1,NY)
c     need to know if graphics are available
      GTERM=LOGCVT(MINTS(IGTERM))
c
c     allow for input of interior boundary conditions
      IF (MINTS(INTERR) .EQ. 1) CALL BNDRY(P)
      IF (MINTS(INTERR) .EQ. 1) GOTO 10
c
c     close files if necessary
      IF (TNAME .NE. MSTRNG(MINTS(ITNAME)))
     +     CALL FLCLOS(TNAME,TUNIT)
c
c     physical and numerical parameters
      OMEGA=MREALS(IOMEGA)
      SUBLAT=LOGCVT(MINTS(ISUB))
      NXLL=MINTS(INXLL)
      NYLL=MINTS(INYLL)
      NXUR=MINTS(INXUR)
      NYUR=MINTS(INYUR)
      PTYPE=ABS(MREALS(IPTYPE))
      PFILE=MSTRNG(MINTS(IPFILE))
      PINIT=MREALS(IPINIT)
c
c     text output
      TTERM=LOGCVT(MINTS(ITTERM))
      TFILE=LOGCVT(MINTS(ITFILE))
      TNAME=MSTRNG(MINTS(ITNAME))
c
```

```
C      graphics output
       GHRDCP=LOGCVT(MINTS(IGHRD))
       GFILE=LOGCVT(MINTS(IGFILE))
       GNAME=MSTRNG(MINTS(IGNAME))
       NFREQ=MINTS(INFREQ)
       NLEV=MINTS(INLEV)
C
C      open files
       IF (TFILE) CALL FLOPEN(TNAME,TUNIT)
       !files may have been renamed
       MSTRNG(MINTS(ITNAME))=TNAME
C
C      check sublattice parameters
39     IF ((NXLL .GT. NXUR) .OR. (NYLL .GT. NYUR)) THEN
          WRITE (OUNIT,40)
          CALL ASK(INXLL,INYUR)
          NXLL=MINTS(INXLL)
          NYLL=MINTS(INYLL)
          NXUR=MINTS(INXUR)
          NYUR=MINTS(INYUR)
          GOTO 39
       END IF
40     FORMAT (' Sublattice parameters must have upper right values '
      +          'greater than lower left')
       CALL CLEAR
C
C      set P to its initial value
       IF (PTYPE .EQ. 1) CALL READIN(P)            !read data in
C      sometimes in READIN PTYPE is changed
       IF (PTYPE .EQ. 2) THEN
          DO 60 IX=1,NX
             DO 70 IY=1,NY              !set all non bound cond points to PINIT
                IF (BNDCND(IX,IY) .EQ. 0) P(IX,IY)=PINIT
70           CONTINUE
60        CONTINUE
       ELSE IF (PTYPE .EQ. 3) THEN               !keep P the same
          CONTINUE
       END IF
C
C      reset PTYPE menu parameters so that P remains unchanged during
C      runs unless the user explicitly requests otherwise
       MINTS(IPTYPE)=3
       MREALS(IPTYPE)=3
C
       RETURN
       END
CCCCCCCCCCCCCCCCCCCCCCCCCCCCCCCCCCCCCCCCCCCCCCCCCCCCCCCCCCCCCCCCCCCCCCCCC
       SUBROUTINE BNDRY(P)
C obtains interior boundary conditions from user (P)
CCCCCCCCCCCCCCCCCCCCCCCCCCCCCCCCCCCCCCCCCCCCCCCCCCCCCCCCCCCCCCCCCCCCCCCCC
C Global variables:
       INCLUDE 'PARAM.E6'
```

```
            INCLUDE 'MAP.E6'
            INCLUDE 'MENU.ALL'
            INCLUDE 'IO.ALL'
C Input/output variables:
            REAL P(NXMAX,NYMAX)                  !field values
C Local variables:
            INTEGER TYPE                         !type of bound cond
            INTEGER XLDIST,XRDIST                !dist of diag to left and right bc
            INTEGER YUDIST,YLDIST                !dist of diag to upper and lower bc
            INTEGER UPLIM                        !upper limit to diag length
C Functions:
            INTEGER GETINT                       !get integer from screen
            REAL GETFLT                          !get float from screen
CCCCCCCCCCCCCCCCCCCCCCCCCCCCCCCCCCCCCCCCCCCCCCCCCCCCCCCCCCCCCCCCCCCCCCCC
            TYPE=DRCHLT                          !currently only dirichlet bc implemented
C
1           CALL CLEAR
            CALL ASK(IBCMNU,IBNDRY)                          !display bound cond menu
C
C       get detailed geometrical information for rectangular b.c.
            IF (MREALS(IBNDRY) .EQ. -1)  THEN
              IF (NBC .LT. BCMAX) THEN
                NBC=NBC+1                            !update number of b.c.
                BCGEOM(NBC)=RECTGL                   !set geometry
                XLL(NBC)= GETINT(2,2,NX-1,' Enter lower left X value')
                YLL(NBC)= GETINT(2,2,NY-1,' Enter lower left Y value')
                XUR(NBC)=
     +          GETINT(XLL(NBC),XLL(NBC),NX-1,' Enter upper right X value')
                YUR(NBC)=
     +          GETINT(YLL(NBC),YLL(NBC),NY-1,' Enter upper right Y value')
                BCVAL(NBC)=GETFLT(FMIN,FMIN,FMAX,' Enter field value ')
                CALL SETBC(P,BCGEOM(NBC),TYPE,BCVAL(NBC),XLL(NBC),XUR(NBC),
     +                     YLL(NBC),YUR(NBC))
              ELSE                                 !too many b.c.'s
                WRITE (OUNIT,30)                   !display warning message
                WRITE (OUNIT,40)
                CALL PAUSE('to continue...',1)
              END IF
C
C       get detailed geometrical information for diagonal b.c.
            ELSE IF (MREALS(IBNDRY) .EQ. -2) THEN
              IF (NBC .LT. BCMAX) THEN
                NBC=NBC+1                            !update number of  b.c.'s
                BCGEOM(NBC)=DIAGNL                   !set geometry
                XLL(NBC)= GETINT(2,2,NX-1,' Enter first X value ')
                YLL(NBC)= GETINT(2,2,NY-1,' Enter first Y value ')
                XUR(NBC)= GETINT(1,1,4,'Enter quadrant: 1, 2, 3, or 4')
C               make sure that diagonal stays within the lattice
                XRDIST=NX-1-XLL(NBC)
                XLDIST=XLL(NBC)-2
                YUDIST=NY-1-YLL(NBC)
                YLDIST=YLL(NBC)-2
```

```
      IF (XUR(NBC) .EQ. 1) THEN
          UPLIM=MIN(XRDIST,YUDIST)
      ELSE IF (XUR(NBC) .EQ. 2) THEN
          UPLIM=MIN(XLDIST,YUDIST)
      ELSE IF (XUR(NBC) .EQ. 3) THEN
          UPLIM=MIN(XLDIST,YLDIST)
      ELSE IF (XUR(NBC) .EQ. 4) THEN
          UPLIM=MIN(XRDIST,YLDIST)
      END IF
      YUR(NBC)= GETINT(UPLIM,0,UPLIM,
     +        'Enter length of 45 degree line (max is default)')
      BCVAL(NBC)=GETFLT(FMIN,FMIN,FMAX,' Enter field value ')
      CALL SETBC(P,BCGEOM(NBC),TYPE,BCVAL(NBC),XLL(NBC),XUR(NBC),
     +           YLL(NBC),YUR(NBC))
      ELSE                          !too many b.c.'s
      WRITE (OUNIT,30)              !display warning message
      WRITE (OUNIT,40)
      CALL PAUSE('to continue...',1)
C
      END IF
C
      ELSE IF (MREALS(IBNDRY) .EQ. -3) THEN    !review bc
      IF (NBC .EQ. 0) THEN
          WRITE(OUNIT,2)                        !nothing to review
          CALL PAUSE('to continue...',1)
      ELSE
          CALL REVIEW(P)
      END IF
C
      ELSE IF (MREALS(IBNDRY) .EQ. -4) THEN    !display bc
      IF (NBC .EQ. 0) THEN
          WRITE(OUNIT,2)                        !nothing to display
          CALL PAUSE('to continue...',1)
      ELSE
          IF (GTERM) THEN                       !graphics
              CALL GRFBC
          ELSE
              CALL DSPLBC(P)                    !no graphics
          END IF
      END IF
C
      ELSE IF (MREALS(IBNDRY) .EQ. -5) THEN    !go back to Main menu
      RETURN
      END IF
C
      GOTO 1
C
30    FORMAT ( ' You have entered the maximum number of boundary'
     +         ' conditions allowed')
40    FORMAT ( ' You can add more only if you delete others first'
     +         ' using the REVIEW option')
2     FORMAT (' No interior boundary conditions have been entered')
```

```
C
      END
CCCCCCCCCCCCCCCCCCCCCCCCCCCCCCCCCCCCCCCCCCCCCCCCCCCCCCCCCCCCCCCCCCCCCCCCC
      SUBROUTINE SETBC(P,GEOM,TYPE,PVALUE,X1,X2,Y1,Y2)
C given an interior boundary condition whose geometry is given by GEOM,
C X1, X2, Y1, Y2; set P array elements to PVALUE and set BNCCND to TYPE
CCCCCCCCCCCCCCCCCCCCCCCCCCCCCCCCCCCCCCCCCCCCCCCCCCCCCCCCCCCCCCCCCCCCCCCCC
C Global variables:
      INCLUDE 'PARAM.E6'
C Input variables:
      REAL P(NXMAX,NYMAX)              !field values (I/O)
      INTEGER GEOM                    !geometry type of bound cond
      INTEGER TYPE                    !type of bound cond
      REAL PVALUE                     !value of P on bound
      INTEGER X1,X2,Y1,Y2             !specify location of bound cond
C Local variables:
      INTEGER IX,IY,I                 !lattice indices
      INTEGER XSLOPE,YSLOPE           !slopes for diagonal bc
CCCCCCCCCCCCCCCCCCCCCCCCCCCCCCCCCCCCCCCCCCCCCCCCCCCCCCCCCCCCCCCCCCCCCCCCC
      IF (GEOM .EQ. RECTGL) THEN
         DO 10 IX=X1,X2
            P(IX,Y1)=PVALUE           !lower side
            P(IX,Y2)=PVALUE           !upper side
            BNDCND(IX,Y1)=TYPE
            BNDCND(IX,Y2)=TYPE
10       CONTINUE
         DO 20 IY=Y1,Y2
            P(X1,IY)=PVALUE           !left side
            P(X2,IY)=PVALUE           !right side
            BNDCND(X1,IY)=TYPE
            BNDCND(X2,IY)=TYPE
20       CONTINUE
C
C     for diagonal, Y2 is length and X2 is angle
      ELSE IF (GEOM .EQ. DIAGNL) THEN
         IF ((X2 .EQ. 1) .OR. (X2 .EQ. 2)) YSLOPE=1
         IF ((X2 .EQ. 3) .OR. (X2 .EQ. 4)) YSLOPE=-1
         IF ((X2 .EQ. 1) .OR. (X2 .EQ. 4)) XSLOPE=1
         IF ((X2 .EQ. 2) .OR. (X2 .EQ. 3)) XSLOPE=-1
         DO 30 I=0,Y2
            IY=Y1+I*YSLOPE
            IX=X1+I*XSLOPE
            P(IX,IY)=PVALUE
C display boundary conditions (P) as small letters
C all non-bound conditions are displayed as '-', regardless of value
C so user may clearly see boundary conditions; always written to OUNIT
            BNDCND(IX,IY)=TYPE
30       CONTINUE
      END IF
C
      RETURN
      END
```

```
ccccccccccccccccccccccccccccccccccccccccccccccccccccccccccccccccccccccc
      SUBROUTINE DSPLBC(P)
C display boundary conditions (P) as small letters
C all non-bound conditions are displayed as '-', regardless of value
C so user may clearly see boundary conditions; always written to OUNIT
ccccccccccccccccccccccccccccccccccccccccccccccccccccccccccccccccccccccc
C Global variables:
      INCLUDE 'PARAM.E6'
      INCLUDE 'IO.ALL'
C Input variables:
      REAL P(NXMAX,NYMAX)             !field values
C Local variables:
      INTEGER IX,IY                   !lattice indices
      INTEGER PTEMP                   !field at current lattice site
      CHARACTER*1 FIELD(NXMAX)        !field as character data
      CHARACTER*80 BLNK               !blanks for centering in X
      CHARACTER*1 ASKII,NEGASK(0:25)  !charac data for display
      DATA BLNK /' '/
      DATA ASKII/'-'/
      DATA NEGASK/'a','b','c','d','e','f','g','h','i','j','k','l','m',
     +            'n','o','p','q','r','s','t','u','v','w','x','y','z'/
ccccccccccccccccccccccccccccccccccccccccccccccccccccccccccccccccccccccc
      CALL CLEAR
      DO 20 IY=1,YCNTR                !center output
         WRITE (OUNIT,*) ' '
20    CONTINUE
C
      DO 100 IY=NY,1,-1
         DO 50 IX=1,NX
            IF (BNDCND(IX,IY) .EQ. 0) THEN  !set non-b.c. to '-'
               FIELD(IX)=ASKII
            ELSE
               PTEMP=NINT(P(IX,IY))         !change b.c. to ascii values
               IF (PTEMP .GT. FMAX) PTEMP=FMAX
               IF (PTEMP .LT. FMIN) PTEMP=FMIN
               FIELD(IX)=NEGASK(PTEMP)
            END IF
50       CONTINUE
         IF (XSKIP) THEN
            WRITE (OUNIT,10) BLNK(1:XCNTR),(FIELD(IX),IX=1,NX)
         ELSE
            WRITE (OUNIT,15) BLNK(1:XCNTR),(FIELD(IX),IX=1,NX)
         END IF
         IF (YSKIP) WRITE (OUNIT,*) ' '
10       FORMAT (1X,A,100(A1,1X))
15       FORMAT (1X,A,100(A1))
100   CONTINUE
      CALL PAUSE(' to go back to menu...',1)
C
      RETURN
      END
ccccccccccccccccccccccccccccccccccccccccccccccccccccccccccccccccccccccc
```

```
          SUBROUTINE GRFBC
C display boundary conditions as dashed lines so user may inspect
C boundary conditions; always written to OUNIT using graphics package
CCCCCCCCCCCCCCCCCCCCCCCCCCCCCCCCCCCCCCCCCCCCCCCCCCCCCCCCCCCCCCCCCCCCCCCC
C Global variables:
          INCLUDE 'IO.ALL'                    '
          INCLUDE 'PARAM.E6'
          INCLUDE 'GRFDAT.ALL'
C Local variables:
          REAL BCX(5),BCY(5)            !corners of boundary conditions
          INTEGER I,IX,IY               !level index, lattice indices
          INTEGER IBC                   !index on boundary conditions
          REAL XSLOPE, YSLOPE           !slope of diagonal BC
          INTEGER SCREEN                !send to terminal
          INTEGER PAPER                 !make a hardcopy
          INTEGER FILE                  !send to a file
          DATA SCREEN,PAPER,FILE/1,2,3/
CCCCCCCCCCCCCCCCCCCCCCCCCCCCCCCCCCCCCCCCCCCCCCCCCCCCCCCCCCCCCCCCCCCCCCCC
          NPLOT=1                              !how many plots
          IPLOT=1
C
          ILINE=2
          YMAX=NY                              !axes data
          YMIN=1.
          XMIN=1.
          XMAX=NX
          YOVAL=XMIN
          XOVAL=YMIN
          NXTICK=5
          NYTICK=5
C
          LABEL(1)='NX'                        !descriptions
          LABEL(2)='NY'
          TITLE='Boundary conditions'
C
          CALL GTDEV(SCREEN)                   !device nomination
          CALL GMODE                           !change to graphics mode
          CALL LNLNAX                          !draw axes
C
C         display interior boundary conditions as dashed lines
          DO 200 IBC=1,NBC                     !loop over b.c.
            IF (BCGEOM(IBC) .EQ. DIAGNL) THEN
              NPOINT=2
              IF ((XUR(IBC) .EQ. 1) .OR. (XUR(IBC) .EQ. 2)) YSLOPE=1
              IF ((XUR(IBC) .EQ. 3) .OR. (XUR(IBC) .EQ. 4)) YSLOPE=-1
              IF ((XUR(IBC) .EQ. 1) .OR. (XUR(IBC) .EQ. 4)) XSLOPE=1
              IF ((XUR(IBC) .EQ. 2) .OR. (XUR(IBC) .EQ. 3)) XSLOPE=-1
              BCX(1)=XLL(IBC)
              BCX(2)=XLL(IBC)+XSLOPE*YUR(IBC)
              BCY(1)=YLL(IBC)
              BCY(2)=YLL(IBC)+YSLOPE*YUR(IBC)
            ELSE IF (BCGEOM(IBC) .EQ. RECTGL) THEN
```

```
            NPOINT=5
            BCX(1)=XLL(IBC)
            BCX(2)=BCX(1)
            BCX(3)=XUR(IBC)
            BCX(4)=BCX(3)
            BCY(1)=YLL(IBC)
            BCY(2)=YUR(IBC)
            BCY(3)=BCY(2)
            BCY(4)=BCY(1)
            BCX(5)=BCX(1)
            BCY(5)=BCY(1)
         END IF
         CALL XYPLOT(BCX,BCY)           !plot boundaries
200   CONTINUE
C
C     end graphing session
      CALL GPAGE(SCREEN)    !end graphics package
      CALL TMODE            !switch to text mode
C
      RETURN
      END
CCCCCCCCCCCCCCCCCCCCCCCCCCCCCCCCCCCCCCCCCCCCCCCCCCCCCCCCCCCCCCCCCCCCCCCCC
      SUBROUTINE REVIEW(P)
C routine to allow user to review interior boundary conditions;
C they can 1) leave as is , 2) delete, 3) alter field value
CCCCCCCCCCCCCCCCCCCCCCCCCCCCCCCCCCCCCCCCCCCCCCCCCCCCCCCCCCCCCCCCCCCCCCCCC
C Global variables:
      INCLUDE 'PARAM.E6'
      INCLUDE 'IO.ALL'
C Input/output variables:
      REAL P(NXMAX,NYMAX)            !field values
C Local variables:
      INTEGER IBC,JBC               !index on bound cond
      INTEGER GETINT,CHOICE         !get user choice
      LOGICAL DELETE(BCMAX)         !delete this bound cond?
      INTEGER NDEL                  !number deleted
      INTEGER TYPE                  !type of boundary condition
      REAL GETFLT                   !get float from user
CCCCCCCCCCCCCCCCCCCCCCCCCCCCCCCCCCCCCCCCCCCCCCCCCCCCCCCCCCCCCCCCCCCCCCCCC
      NDEL=0
C
      DO 100 IBC=1,NBC
         DELETE(IBC)=.FALSE.        !initialize
C
C        write out description of this bound cond
         WRITE (OUNIT,5) IBC
         IF (BCGEOM(IBC) .EQ. DIAGNL) THEN
             WRITE (OUNIT,10) XLL(IBC),YLL(IBC),XUR(IBC),YUR(IBC)
         ELSE IF (BCGEOM(IBC) .EQ. RECTGL) THEN
             WRITE (OUNIT,15) XLL(IBC),YLL(IBC),XUR(IBC),YUR(IBC)
         END IF
         WRITE (OUNIT,20) BCVAL(IBC)
```

```
C
            CHOICE=GETINT(1,1,3,
     +      'Do you want to 1)keep it 2)delete it 3) change field value?')
C
            IF (CHOICE .EQ. 1) THEN          !do nothing
               CONTINUE
            ELSE IF (CHOICE .EQ. 2) THEN     !delete it
               NDEL=NDEL+1
               DELETE(IBC)=.TRUE.
               TYPE= NON                     !releases b.c.
               CALL SETBC(P,BCGEOM(IBC),TYPE,BCVAL(IBC),XLL(IBC),XUR(IBC),
     +                   YLL(IBC),YUR(IBC))
            ELSE IF (CHOICE .EQ. 3) THEN   !get new field value
               BCVAL(IBC)=GETFLT(FMIN,FMIN,FMAX,' Enter field value ')
            END IF
100    CONTINUE
C
C   get rid of spaces left in bound cond set when one or more is deleted
            IF (NDEL .NE. 0) THEN
               IBC=0                         !JBC labels old set of bound cond
               DO 200 JBC=1,NBC              !IBC labels new set of bound cond
                  IF (DELETE(JBC)) THEN
                     CONTINUE
                  ELSE
                     IBC=IBC+1               !next empty place
                     XLL(IBC)=XLL(JBC)       !move up parameter values to
                     YLL(IBC)=YLL(JBC)       !next empty place
                     XUR(IBC)=XUR(JBC)
                     YUR(IBC)=YUR(JBC)
                     BCGEOM(IBC)=BCGEOM(JBC)
                     BCVAL(IBC)=BCVAL(JBC)
                  END IF
200       CONTINUE
               NBC=NBC-NDEL                  !update number of b.c.'s
            END IF
C
            DO 300 IBC=1,NBC
C          reset boundary conditions in case some of deleted
C          bound cond wrote over others
C          also adjusts those for which the field value was changed
               TYPE=DRCHLT
               CALL SETBC(P,BCGEOM(IBC),TYPE,BCVAL(IBC),XLL(IBC),XUR(IBC),
     +                   YLL(IBC),YUR(IBC))
300       CONTINUE
            RETURN
C
5      FORMAT (' Boundary Condition Number ',I2)
10     FORMAT (' Diagonal with first coordinates=(',I2,',',I2,
     +         '),  quadrant =',I2,', and length='I2)
15     FORMAT (' Rectangle with lower left at (',I2,',',I2,
     +         ') and upper right at (',I2,',',I2,')')
20     FORMAT (' The field is fixed at ',1PE12.5)
```

```
C
      END
CCCCCCCCCCCCCCCCCCCCCCCCCCCCCCCCCCCCCCCCCCCCCCCCCCCCCCCCCCCCCCCCCC
      SUBROUTINE READIN(P)
C read in field values and boundary conditions;
C reads in only files written by subroutine WRTOUT, and files written
C with same lattice size
CCCCCCCCCCCCCCCCCCCCCCCCCCCCCCCCCCCCCCCCCCCCCCCCCCCCCCCCCCCCCCCCCC
C Global variables:
      INCLUDE 'IO.ALL'
      INCLUDE 'PARAM.E6'
      INCLUDE 'MENU.ALL'
      INCLUDE 'MAP.E6'
C Output variables:
      REAL P(NXMAX,NYMAX)            !field values
C Local variables:
      INTEGER IX,IY                 !lattice indices
      CHARACTER*80 JUNK             !first line if of no interest
      LOGICAL SUCESS                !did we find a file to open?
      INTEGER CHOICE                !user answer
      INTEGER MX,MY,MOMEGA          !parameters from PFILE
C Function:
      INTEGER YESNO                 !get yesno input from user
      CHARACTER*40 CHARAC           !returns character input
CCCCCCCCCCCCCCCCCCCCCCCCCCCCCCCCCCCCCCCCCCCCCCCCCCCCCCCCCCCCCCCCCC
10    CALL FLOPN2(PFILE,DUNIT,SUCESS)     !open the file for input
      MSTRNG(MINTS(IPFILE))=PFILE          !file may have been renamed
C
      IF (.NOT. SUCESS) THEN
         CALL REASK                 !prompt again for init P
         RETURN                     !if no file was found
      ELSE
C
         READ (DUNIT,5) JUNK        !skip over title
5        FORMAT (A)
C
C        lattice sizes must match; if they don't, allow for other options
         READ (DUNIT,*) MX,MY,MOMEGA
         IF ((MX .NE. NX) .OR. (MY .NE. NY)) THEN
            CALL FLCLOS(PFILE,DUNIT)        !close it up
            WRITE (OUNIT,15) MX,MY
15          FORMAT (' Input file has does not have the correct'
     +             ' lattice size, it is ',I2,' by ',I2)
            CHOICE=YESNO(1,' Do you want to try another file?')
            IF (CHOICE .EQ. 0) THEN
                 CALL REASK                 !prompt again for init P
                 RETURN
            ELSE
               PFILE=CHARAC(PFILE,12,'Enter another filename')
               MSTRNG(MINTS(IPFILE))=PFILE
               GOTO 10         !try to open this one
            END IF
```

```
            END IF
C
C          if we've gotten this far, we've opened a file with data
C          from a lattice of the same size; finally read in field
            DO 100 IX=1,NX
               READ (DUNIT,110) (P(IX,IY),IY=1,NY)
100         CONTINUE
110         FORMAT (5(2X,1PE14.7))
            CALL FLCLOS(PFILE,DUNIT)                !close file
         END IF
C
         RETURN
         END
CCCCCCCCCCCCCCCCCCCCCCCCCCCCCCCCCCCCCCCCCCCCCCCCCCCCCCCCCCCCCCCCCCCCCCC
         SUBROUTINE REASK
C prompt again for initial P configuration;
C called only if reading in from a file is requested and failed
CCCCCCCCCCCCCCCCCCCCCCCCCCCCCCCCCCCCCCCCCCCCCCCCCCCCCCCCCCCCCCCCCCCCCCC
C Global variables:
         INCLUDE 'MENU.ALL'
         INCLUDE 'MAP.E6'
         INCLUDE 'PARAM.E6'
CCCCCCCCCCCCCCCCCCCCCCCCCCCCCCCCCCCCCCCCCCCCCCCCCCCCCCCCCCCCCCCCCCCCCCC
C          redisplay initial Field Value Menu, but disallow choice 1
         MPRMPT(45)='1) (not allowed)'
         MLOLIM(48)=2
         MINTS(48)=3
         MREALS(48)=3
         CALL ASK(44,52)
C
         PTYPE=ABS(MREALS(IPTYPE))                !set parameter choices
         PINIT=MREALS(IPINIT)
C
         MPRMPT(45)='1) Read in initial values from a file'   !reset menu
         MLOLIM(48)=1
C
         RETURN
         END
CCCCCCCCCCCCCCCCCCCCCCCCCCCCCCCCCCCCCCCCCCCCCCCCCCCCCCCCCCCCCCCCCCCCCCC
         SUBROUTINE PRMOUT(MUNIT,NLINES)
C write out parameter summary to MUNIT
CCCCCCCCCCCCCCCCCCCCCCCCCCCCCCCCCCCCCCCCCCCCCCCCCCCCCCCCCCCCCCCCCCCCCCC
C Global variables:
         INCLUDE 'IO.ALL'
         INCLUDE 'PARAM.E6'
C Input variables:
         INTEGER MUNIT                           !fortran unit number
         INTEGER NLINES                          !number of lines sent to terminal
CCCCCCCCCCCCCCCCCCCCCCCCCCCCCCCCCCCCCCCCCCCCCCCCCCCCCCCCCCCCCCCCCCCCCCC
         IF (MUNIT .EQ. OUNIT) CALL CLEAR
C
         WRITE (MUNIT,5)
```

```
      IF (MUNIT .EQ. GUNIT) THEN
          WRITE (MUNIT,*) NX,NY,OMEGA
      ELSE
          WRITE (MUNIT,10) NX,NY,OMEGA
          WRITE (MUNIT,12) UPPER,LOWER,LEFT,RIGHT
          IF (SUBLAT) WRITE (MUNIT,15) NXLL,NYLL,NXUR,NYUR
          WRITE (MUNIT,*) ' '
          WRITE (MUNIT,30)
          WRITE (MUNIT,40)
      END IF
      NLINES=4
      IF (SUBLAT) NLINES=NLINES+1
C
5     FORMAT (' Output from example 6: Solutions to Laplace''s Equation'
     +        ' in two dimensions')
10    FORMAT (' NX =',I3,5X,' NY =',I3,5X,'  Omega = ',F6.3)
12    FORMAT (' field values on upper, lower, left, and right are ',
     +        4(2X,F5.2))
15    FORMAT (' Sublattice defined by  (',I3,',',I3,') and (',
     +        I3,',',I3,')')
30    FORMAT (8X,'Iter',11X,'Energy',13X,'Delta E',12X,'max Delta P')
40    FORMAT (8X,'----',11X,'------',13X,'-------',12X,'-----------')
C
      RETURN
      END
cccccccccccccccccccccccccccccccccccccccccccccccccccccccccccccccccccccc
      SUBROUTINE TXTOUT(MUNIT,NITER,ENERGY,DELE,DELMAX,NLINES)
C output text results to MUNIT
cccccccccccccccccccccccccccccccccccccccccccccccccccccccccccccccccccccc
C Global variables:
      INCLUDE 'IO.ALL'
C Input variables:
      INTEGER NITER              !number of iterations
      REAL DELMAX,OLDMAX         !keep track of changes in P
      REAL ENERGY,DELE           !current and delta energy
      INTEGER MUNIT              !fortran unit number
      INTEGER NLINES             !number of lines sent to terminal(I/O)
cccccccccccccccccccccccccccccccccccccccccccccccccccccccccccccccccccccc
C     if screen is full, clear screen and retype headings
      IF ((MOD(NLINES,TRMLIN-6) .EQ. 0)
     +                    .AND. (MUNIT .EQ. OUNIT)) THEN
          CALL PAUSE('to continue...',1)
          CALL CLEAR
          WRITE (MUNIT,30)
          WRITE (MUNIT,40)
      END IF
      IF (MUNIT .EQ. OUNIT) NLINES=NLINES+1
C
      WRITE (MUNIT,50) NITER,ENERGY,DELE,DELMAX
50    FORMAT (8X,I4,3(8X,1PE12.5))
30    FORMAT (8X,'Iter',11X,'Energy',13X,'Delta E',12X,'max Delta P')
40    FORMAT (8X,'----',11X,'------',13X,'-------',12X,'-----------')
```

```
      C
            RETURN
            END
      CCCCCCCCCCCCCCCCCCCCCCCCCCCCCCCCCCCCCCCCCCCCCCCCCCCCCCCCCCCCCCCCCCCCCCC
            SUBROUTINE DISPLY(P,MUNIT)
      C display field (P) as ascii characters: bound cond are small letters,
      C non bound cond are in capitals; output sent to MUNIT
      CCCCCCCCCCCCCCCCCCCCCCCCCCCCCCCCCCCCCCCCCCCCCCCCCCCCCCCCCCCCCCCCCCCCCCC
      C Global variables:
            INCLUDE 'PARAM.E6'
            INCLUDE 'IO.ALL'
      C Input variables:
            REAL P(NXMAX,NYMAX)                    !field values
            INTEGER MUNIT                          !unit we're writing to
      C Local variables:
            INTEGER IX,IY                          !lattice indices
            INTEGER PTEMP                          !field at current lattice site
            CHARACTER*1 FIELD(NXMAX)               !field as character data
            CHARACTER*80 BLNK                      !blanks for centering in X
            CHARACTER*1 ASKII(0:25),NEGASK(0:25)   !charac data for display
            DATA BLNK /' '/
            DATA ASKII/'A','B','C','D','E','F','G','H','I','J','K','L','M',
           +           'N','O','P','Q','R','S','T','U','V','W','X','Y','Z'/
            DATA NEGASK/'a','b','c','d','e','f','g','h','i','j','k','l','m',
           +            'n','o','p','q','r','s','t','u','v','w','x','y','z'/
      CCCCCCCCCCCCCCCCCCCCCCCCCCCCCCCCCCCCCCCCCCCCCCCCCCCCCCCCCCCCCCCCCCCCCCC
            IF (MUNIT .EQ. OUNIT) THEN
               DO 20 IY=1,YCNTR                    !center output
                  WRITE (OUNIT,*) ' '
       20      CONTINUE
            ELSE
               WRITE (MUNIT,*) ' Field values are:'
            END IF
      C
            DO 100 IY=NY,1,-1
               DO 50 IX=1,NX
                  PTEMP=NINT(P(IX,IY))             !convert field value to ascii
                  IF (PTEMP .GT. FMAX) PTEMP=FMAX  !keep things in bounds
                  IF (PTEMP .LT. FMIN) PTEMP=FMIN
                  IF (BNDCND(IX,IY) .EQ. 0) FIELD(IX)=ASKII(PTEMP)
                  IF (BNDCND(IX,IY) .NE. 0) FIELD(IX)=NEGASK(PTEMP)
       50      CONTINUE
      C        write out a line at a time (no centering done for TUNIT)
               IF (MUNIT .EQ. TUNIT) THEN
                  WRITE (TUNIT,16) (FIELD(IX),IX=1,NX)
               ELSE
                  IF (XSKIP) THEN
                     WRITE (OUNIT,10) BLNK(1:XCNTR),(FIELD(IX),IX=1,NX)
                  ELSE
                     WRITE (OUNIT,15) BLNK(1:XCNTR),(FIELD(IX),IX=1,NX)
                  END IF
                  IF (YSKIP) WRITE (OUNIT,*) ' '
```

```
          END IF
10        FORMAT (1X,A,100(A1,1X))
15        FORMAT (1X,A,100(A1))
16        FORMAT (1X,100(A1))
100   CONTINUE
C
      RETURN
      END
CCCCCCCCCCCCCCCCCCCCCCCCCCCCCCCCCCCCCCCCCCCCCCCCCCCCCCCCCCCCCCCCCCCCCCCCC
      SUBROUTINE GRFOUT(DEVICE,P,PGRF,MX,MY,E)
C display contours of the field P
C     the field values must be in an array that is exactly NX by NY;
C     this can be accomplished with implicit dimensioning which
C     requires that PGRF and its dimensions be passed to this routine
CCCCCCCCCCCCCCCCCCCCCCCCCCCCCCCCCCCCCCCCCCCCCCCCCCCCCCCCCCCCCCCCCCCCCCCCC
C Global variables:
      INCLUDE 'IO.ALL'
      INCLUDE 'PARAM.E6'
      INCLUDE 'GRFDAT.ALL'
C Input variables:
      INTEGER DEVICE              !which device
      REAL P(NXMAX,NYMAX)         !field values
      INTEGER MX,MY               !NX and NY in disguise
      REAL PGRF(MX,MY)            !field values for graphing
      REAL E                      !energy
C Local variables:
      REAL BCX(5),BCY(5)          !corners of boundary conditions
      INTEGER I,IX,IY             !level index, lattice indices
      REAL PMAX,PMIN              !largest and smallest field value
      INTEGER IBC                 !index on boundary conditions
      REAL XSLOPE, YSLOPE         !slope of diagonal BC
      CHARACTER*9 CMIN,CMAX,CE     !data as characters
      INTEGER LMIN,LMAX,ELEN      !string lengths
      INTEGER SCREEN              !send to terminal
      INTEGER PAPER               !make a hardcopy
      INTEGER FILE                !send to a file
      DATA SCREEN,PAPER,FILE/1,2,3/
CCCCCCCCCCCCCCCCCCCCCCCCCCCCCCCCCCCCCCCCCCCCCCCCCCCCCCCCCCCCCCCCCCCCCCCCC
      PMAX=0                      !set scale
      PMIN=P(1,1)
      DO 10 IX=1,MX
         DO 20 IY=1,MY
            IF (P(IX,IY) .GT. PMAX) PMAX=P(IX,IY)
            IF (P(IX,IY) .LT. PMIN) PMIN=P(IX,IY)
            PGRF(IX,IY)=P(IX,IY)      !load field into PGRF
20       CONTINUE
10    CONTINUE
C
C     messages for the impatient
      IF (DEVICE .NE. SCREEN) WRITE (OUNIT,100)
C
C     calculate parameters for graphing
```

```
            NPLOT=1                            !how many plots
            IPLOT=1
C
            YMAX=MY                            !axes data
            YMIN=1.
            XMIN=1.
            XMAX=MX
            YOVAL=XMIN
            XOVAL=YMIN
            NXTICK=5
            NYTICK=5
C
            LABEL(1)='NX'                      !descriptions
            LABEL(2)='NY'
            CALL CONVRT(PMIN,CMIN,LMIN)
            CALL CONVRT(PMAX,CMAX,LMAX)
            CALL CONVRT(E,CE,ELEN)
            INFO='Pmin ='//CMIN(1:LMIN)//' Pmax='//
           +            CMAX(1:LMAX)//' E='//CE(1:ELEN)
            TITLE='Solutions to Laplace''s Equation'
C
            CALL GTDEV(DEVICE)                 !device nomination
            IF (DEVICE .EQ. SCREEN) CALL GMODE !change to graphics mode
            CALL LNLNAX                        !draw axes
C
C           display interior boundary conditions as dashed lines
            DO 200 IBC=1,NBC                   !loop over b.c.
              IF (BCGEOM(IBC) .EQ. DIAGNL) THEN
                NPOINT=2
                IF ((XUR(IBC) .EQ. 1) .OR. (XUR(IBC) .EQ. 2)) YSLOPE=1
                IF ((XUR(IBC) .EQ. 3) .OR. (XUR(IBC) .EQ. 4)) YSLOPE=-1
                IF ((XUR(IBC) .EQ. 1) .OR. (XUR(IBC) .EQ. 4)) XSLOPE=1
                IF ((XUR(IBC) .EQ. 2) .OR. (XUR(IBC) .EQ. 3)) XSLOPE=-1
                BCX(1)=XLL(IBC)
                BCX(2)=XLL(IBC)+XSLOPE*YUR(IBC)
                BCY(1)=YLL(IBC)
                BCY(2)=YLL(IBC)+YSLOPE*YUR(IBC)
              ELSE IF (BCGEOM(IBC) .EQ. RECTGL) THEN
                NPOINT=5
                BCX(1)=XLL(IBC)
                BCX(2)=BCX(1)
                BCX(3)=XUR(IBC)
                BCX(4)=BCX(3)
                BCY(1)=YLL(IBC)
                BCY(2)=YUR(IBC)
                BCY(3)=BCY(2)
                BCY(4)=BCY(1)
                BCX(5)=BCX(1)
                BCY(5)=BCY(1)
              END IF
              ILINE=2
              CALL XYPLOT(BCX,BCY)             !plot boundaries
```

```
200    CONTINUE
C
       ILINE=1
       CALL CONTOR(PGRF,MX,MY,PMIN,PMAX,NLEV)
C
C      end graphing session
       IF (DEVICE .NE. FILE) CALL GPAGE(DEVICE)    !end graphics package
       IF (DEVICE .EQ. SCREEN) CALL TMODE          !switch to text mode
C
100    FORMAT (/,' Patience, please; output going to a file.')
C
       RETURN
       END
CCCCCCCCCCCCCCCCCCCCCCCCCCCCCCCCCCCCCCCCCCCCCCCCCCCCCCCCCCCCCCCCCCCCCC
       SUBROUTINE WRTOUT(P)
C write out field (P) and boundary conditions to GUNIT for reading back
C in as initial conditions or for graphing with an external package
CCCCCCCCCCCCCCCCCCCCCCCCCCCCCCCCCCCCCCCCCCCCCCCCCCCCCCCCCCCCCCCCCCCCCC
C Global variables:
       INCLUDE 'IO.ALL'
       INCLUDE 'MENU.ALL'
       INCLUDE 'PARAM.E6'
C Input variables:
       REAL P(NXMAX,NYMAX)             !field values
C Local variables:
       INTEGER IX,IY                   !lattice indices
       INTEGER NLINES                  !lines written in PRMOUT
CCCCCCCCCCCCCCCCCCCCCCCCCCCCCCCCCCCCCCCCCCCCCCCCCCCCCCCCCCCCCCCCCCCCCC
       CALL FLOPEN(GNAME,GUNIT)            !open file
       MSTRNG(MINTS(IGNAME))=GNAME
       CALL PRMOUT(GUNIT,NLINES)          !write out header
C
       DO 100 IX=1,NX                     !write out field
          WRITE (GUNIT,10) (P(IX,IY),IY=1,NY)
100    CONTINUE
10     FORMAT (5(2X,1PE14.7))
C
       DO 200 IX=1,NX                     !write out bound cond
          WRITE (GUNIT,20) (BNDCND(IX,IY),IY=1,NY)
200    CONTINUE
20     FORMAT (40(1X,I1))
       CALL FLCLOS(GNAME,GUNIT)          !close it up
C
       RETURN
       END
CCCCCCCCCCCCCCCCCCCCCCCCCCCCCCCCCCCCCCCCCCCCCCCCCCCCCCCCCCCCCCCCCCCCCC
CCCCCCCCCCCCCCCCCCCCCCCCCCCCCCCCCCCCCCCCCCCCCCCCCCCCCCCCCCCCCCCCCCCCCC
C param.e6
C
       INTEGER NXMAX,NYMAX             !maximum lattice size
       PARAMETER (NXMAX=79,NYMAX=79)
       REAL FMIN,FMAX                  !maximum values of the field
```

```
          PARAMETER (FMIN=0.,FMAX=25.)
          INTEGER BCMAX                       !maximum number of bound cond
          PARAMETER (BCMAX=10)
          INTEGER RECTGL,DIAGNL               !shapes of bound cond
          PARAMETER (RECTGL=1,DIAGNL=2)
          INTEGER DRCHLT,NON                  !type of bound cond
          PARAMETER (DRCHLT=1,NON=0)
          INTEGER MAXLEV                       !maximum number of contour levels
          PARAMETER (MAXLEV=20)
   C
          LOGICAL FIRST                       !first time through menu?
          INTEGER NX,NY                       !lattice size
          REAL UPPER,LOWER,LEFT,RIGHT         !field values on lattice edge
          REAL OMEGA                          !relaxation parameter
          INTEGER PTYPE                       !choice for initial field
          CHARACTER*12 PFILE                  !file to input init data
          REAL PINIT                          !initial guess
          LOGICAL SUBLAT                      !relax sublattice first?
          INTEGER NXLL,NYLL,NXUR,NYUR         !sublattice parameters
   C
          INTEGER NFREQ                       !frequency of display
          INTEGER NLEV                        !number of contour levels
   C
          INTEGER NBC                         !number of bound cond
          INTEGER XLL(BCMAX),YLL(BCMAX),XUR(BCMAX),YUR(BCMAX)
                                              !coordinates for bound cond
          INTEGER BCGEOM(BCMAX)               !geometry of bound cond
          REAL BCVAL(BCMAX)                   !value of bound cond
          INTEGER BNDCND(NXMAX,NYMAX)         !flag to indicate bound cond
   C
          LOGICAL XSKIP,YSKIP                 !skip spaces or lines in display
          INTEGER XCNTR,YCNTR                 !how to center display
   C
          COMMON / FLAG   / FIRST
          COMMON / PPARAM / NX,NY,UPPER,LOWER,LEFT,RIGHT
          COMMON / NPARAM / OMEGA,PINIT,SUBLAT,NXLL,NYLL,NXUR,NYUR,PTYPE
          COMMON / BCPRM  / NBC,XLL,YLL,XUR,YUR,BCGEOM,BCVAL,BNDCND
          COMMON / GPARAM / NLEV,NFREQ,XSKIP,YSKIP,XCNTR,YCNTR
          COMMON / ASCII  / PFILE
   CCCCCCCCCCCCCCCCCCCCCCCCCCCCCCCCCCCCCCCCCCCCCCCCCCCCCCCCCCCCCCCCCCCCCCCCCCCC
   CCCCCCCCCCCCCCCCCCCCCCCCCCCCCCCCCCCCCCCCCCCCCCCCCCCCCCCCCCCCCCCCCCCCCCCCCCCC
   C map.e6
          INTEGER IUPPER,ILOWER,ILEFT,IRIGHT,IOMEGA,IBCMNU,IBNDRY
          INTEGER ISUB,INXLL,INYLL,INXUR,INYUR,IPTYPE,IPFILE,IPINIT
          INTEGER INFREQ,INLEV,INTERR
          PARAMETER (IUPPER = 13 )
          PARAMETER (ILOWER = 14 )
          PARAMETER (ILEFT  = 15 )
          PARAMETER (IRIGHT = 16 )
          PARAMETER (INTERR = 17 )
          PARAMETER (IBCMNU = 25 )
          PARAMETER (IBNDRY = 31 )
```

```
PARAMETER (IOMEGA = 38 )
PARAMETER (ISUB   = 39 )
PARAMETER (INXLL  = 40 )
PARAMETER (INYLL  = 41 )
PARAMETER (INXUR  = 42 )
PARAMETER (INYUR  = 43 )
PARAMETER (IPTYPE = 48 )
PARAMETER (IPFILE = 49 )
PARAMETER (IPINIT = 51 )
PARAMETER (INFREQ = 87 )
PARAMETER (INLEV  = 88 )
```

B.7 Example 7

Algorithm   This program solves the time-dependent Schroedinger equation for a particle moving in a one dimensional box (i.e., the potential is infinite at the edges of the lattice: $x = -1$ and $x = +1$), allowing for various potential obstacles in the bottom of the box. After the potential and initial wavepacket shape have been defined, the time evolution proceeds (loop 15 in subroutine ARCHON). Here, the basic task is to find the new wave function at each time step by solving Eqs. (7.31,7.33) in subroutine EVOLVE using the matrix inversion scheme (7.12–7.16). For this purpose the α and γ coefficients are computed once for each value of DT before the time evolution begins (subroutine TRDIAG). At each iteration the energy and normalization are computed in subroutines ENERGY and NORMLZ, respectively. Every NSTEP'th iteration you have the option to continue iterations, change DT or NSTEP, or return to the main menu.

One important physical parameter is the wavenumber k_0 which describes the oscillations of the initial wavepacket. For an initial Gaussian

$$\phi(x) = \exp(ik_0 x)\exp(-(x - x_0)^2/(2\sigma^2)) \quad .$$

This number fixes the energy and the momentum of the particle. (The energy and momentum of a *free* Gaussian packet are $k_0^2 + 3/(4\sigma^2)$ and k_0, where $\hbar = 2m = 1$. For wide Gaussians packets, the k_0^2 dominates the energy.) KO is not entered directly, but is fixed by entering the number of cycles on the lattice, where number of cycles on the lattice equals $k_0 \times$ length of lattice$/2\pi$. The maximum number of cycles allowed on the lattice is NPTS/4. This restriction ensures that one cycle is described by at least 4 lattice points, and therefore the oscillating part of the wavepacket can be described on the lattice.

From the discussion above, it is clear that the appropriate scale for both the energy of the particle and the height of the potential is given by k_0^2. In addition, the size of the time step is conveniently scaled by k_0^{-2}. To see this, note that the error term scales as $(H\Delta t)^3$. For accurate time evolution we must have (taking $H \approx k_0^2$) $\Delta t < 1/k_0^2$, or in scaled units $\Delta t < 1$.

Input   The physical parameters include the potential [square well] and the initial wavepacket shape [Gaussian]. Various analytical forms for the potential are possible (square-well, Gaussian barrier, potential step, or parabolic well), as defined in subroutine POTNTL. You must also specify the potential's center [.5], half-width at one-twentieth of the maximum [.08], and height (in units of KO**2) [.3], as well as the x value [.5] that divides left and right (used to estimate transmission coefficients). The two analytic choices for the initial wavepacket are a Gaussian or Lorentzian (subroutine INTPHI). The parameters that define the packets are the center [-.5], width [.25], and number of cycles on the lattice [10.] ($=\text{KO}/\pi$). The only numerical parameter entered in the main menu is the number of lattice points [200]. The maximum number of lattice points is fixed by MAXLAT= 1000. The other numerical parameters are the scaled time step [.2], and the number of time steps until you can end the iterations [40]; these are entered after you leave the main menu. The one graphics parameter is NFREQ (also entered after the main menu). This determines the frequency at which "movie" frames are displayed (see Output below).

Output At each time step the text output displays the unscaled time; the energy (in units of KO**2); left, right, and total probabilities; and left, right, and total $\langle x \rangle$. The graphics routine plots both $\phi^2(x)$ and $V(x)$ on the same set of axis. The scale for $V(x)$ goes between VMAX and VMIN. The scale for $\phi^2(x)$ is the same for the entire run (subroutine GRFOUT). This has the disadvantage that the scale could occasionally be too small to display the entire wave function, but a fixed scale makes apparent relative amplitudes of $\phi^2(x)$ at different times.

If both text and graphics are requested to the terminal, the plot of $\phi^2(x)$ and $V(x)$ will be displayed only every NSTEP'th time step. If only graphics is requested to the screen, you will obtain output that resembles movies: every NFREQ'th time step, $\phi^2(x)$ will be plotted, with four plots on one page to allow comparison of $\phi^2(x)$ at different times. The program ensures that NSTEP is an integer multiple of 4*NFREQ (subroutine ARCHON). Graphics output to a file or a hardcopy device is done every NSTEP'th iteration. For the hardcopy device only one plot, not four, is printed per page.

```
CCCCCCCCCCCCCCCCCCCCCCCCCCCCCCCCCCCCCCCCCCCCCCCCCCCCCCCCCCCCCCCCCCCCCCCC
      PROGRAM EXMPL7
C     Example 7: The time-dependent Schroedinger Equation
C  COMPUTATIONAL PHYSICS (FORTRAN VERSION)
C  by Steven E. Koonin and Dawn C. Meredith
C  Copyright 1989, Addison-Wesley Publishing Company
CCCCCCCCCCCCCCCCCCCCCCCCCCCCCCCCCCCCCCCCCCCCCCCCCCCCCCCCCCCCCCCCCCCCCCCC
      CALL INIT          !display header screen, setup parameters
5     CONTINUE           !main loop/ execute once for each set of param
      CALL PARAM         !get input from screen
      CALL ARCHON        !calculate time evolution of a wavepacket
      GOTO 5
      END
CCCCCCCCCCCCCCCCCCCCCCCCCCCCCCCCCCCCCCCCCCCCCCCCCCCCCCCCCCCCCCCCCCCCCCCC
```

```
      SUBROUTINE ARCHON
C calculates the time evolution of a one-dimensional wavepacket
C according to the time-dependent Schroedinger equation
CCCCCCCCCCCCCCCCCCCCCCCCCCCCCCCCCCCCCCCCCCCCCCCCCCCCCCCCCCCCCCCCCCCCCCC
C Global variables:
      INCLUDE 'PARAM.E7'
      INCLUDE 'IO.ALL'
C Local variables:
      COMPLEX PHI(0:MAXLAT)        !wavepacket
      REAL PHI2(0:MAXLAT)          !wavepacket squared
      REAL TIME                    !time
      REAL TPROB,LPROB,RPROB       !normalization of the wavepacket
      REAL LX,RX,TX                !average position
      REAL E                       !energy
      REAL DT                      !time step
      INTEGER IT                   !time index
      REAL DTSCAL                  !scaled time step
      REAL DTMIN,DTMAX             !min,max reasonable size for DTSCAL
      INTEGER NLINES               !number of lines printed to terminal
      INTEGER SCREEN               !send to terminal
      INTEGER PAPER                !make a hardcopy
      INTEGER FILE                 !send to a file
      INTEGER NSTEP                !number of time steps to take
      LOGICAL MORE,NEWDT           !options for continuing
      INTEGER FRAME                !which frame of the movie is it?
      INTEGER NFREQ                !graphing frequency (movies only)
      REAL DIFF                    !temp variables for movies
C Functions:
      REAL GETFLT                  !get floating point number from screen
      INTEGER YESNO                !get yes/no answer from screen
      LOGICAL LOGCVT               !change from 1,0 to true and false
      INTEGER GETINT               !get integer data from screen
      DATA SCREEN,PAPER,FILE/1,2,3/
CCCCCCCCCCCCCCCCCCCCCCCCCCCCCCCCCCCCCCCCCCCCCCCCCCCCCCCCCCCCCCCCCCCCCCC
C     output summary of parameters
      IF (TTERM) CALL PRMOUT(OUNIT,NLINES)
      IF (TFILE) CALL PRMOUT(TUNIT,NLINES)
      IF (GFILE) CALL PRMOUT(GUNIT,NLINES)
C
      CALL INTPHI(PHI,PHI2)        !setup initial PHI array
      TIME=0.                      !initialize time
C
      NSTEP=40                     !def num of time steps until next prompt
      NFREQ=10                     !def graphing freq (movies only)
      DTSCAL=.2                    !def scaled time step
      DTMAX=100.                   !max scaled time step
      DTMIN=DX**2*K0**2/25.        !minimum scaled time step
C
10    CONTINUE                     !loop over different DT and/or NSTEP
      DTSCAL=GETFLT(DTSCAL,-DTMAX,DTMAX,
     +     'Enter time step (units of K0**-2)')
      IF (ABS(DTSCAL) .LT. DTMIN)
```

```
   +      DTSCAL=SIGN(DTMIN,DTSCAL)      !don't let it get too small
          DT=DTSCAL/K0**2               !physical time
          NSTEP=GETINT(NSTEP,1,1000,'Enter number of time steps')
          NLINES=NLINES+4
C
          IF (MOVIES) THEN
             NFREQ=GETINT(NFREQ,1,1000,'Enter graphing frequency')
             !make sure that total num of frames is divisible by 4
             DIFF=MOD(NSTEP,4*NFREQ)
             IF (DIFF .NE. 0) NSTEP=NSTEP+4*NFREQ-DIFF
          END IF
C
          CALL TRDIAG(DT)               !calculate GAMMA for this DT
          IF (TFILE) CALL TITLES(TUNIT,NLINES,DT)
C
15        CONTINUE                      !loop over sets of NSTEP time steps
          IF (TTERM) CALL TITLES(OUNIT,NLINES,DT)
C
          DO 20 IT=1,NSTEP             !time evolution
             TIME=TIME+DT
             CALL EVOLVE(PHI,PHI2,DT)   !take a time step
             CALL NORMLZ(PHI2,LPROB,RPROB,TPROB,LX,RX,TX) !calc norm
             CALL ENERGY(PHI,PHI2,E)    !calc energy
             !output
             IF ((MOVIES) .AND. (MOD(IT,NFREQ) .EQ. 0)) THEN
                FRAME=MOD(IT/NFREQ,4)
                IF (FRAME .EQ. 1) THEN
                   CALL GRFOUT(SCREEN,PHI2,TIME,TPROB,TX,E)
                ELSE
                   CALL GRFSEC(SCREEN,PHI2,TIME,TPROB,TX,FRAME)
                END IF
             END IF
             IF (TTERM) CALL
   +          TXTOUT(OUNIT,E,TIME,LPROB,RPROB,TPROB,LX,RX,TX,NLINES)
             IF (TFILE) CALL
   +          TXTOUT(TUNIT,E,TIME,LPROB,RPROB,TPROB,LX,RX,TX,NLINES)
20        CONTINUE
C
C         graphics output
          IF (MOVIES) THEN
             CALL TMODE                 !switch back to text mode
          ELSE IF (GTERM) THEN
C            print out graphics now if text was being sent to screen
             CALL PAUSE('to see the wave packet ...',1)
             CALL GRFOUT(SCREEN,PHI2,TIME,TPROB,TX,E)
          END IF
          IF (GFILE) CALL GRFOUT(FILE,PHI2,TIME,TPROB,TX,E)
          IF (GHRDCP) CALL GRFOUT(PAPER,PHI2,TIME,TPROB,TX,E)
C
          MORE=LOGCVT(YESNO(1,'Continue iterating?'))
          IF (MORE) THEN
             NLINES=0
```

```
            IF (TTERM) CALL CLEAR
            GOTO 15
        END IF
C
        NEWDT=LOGCVT(YESNO(1,'Change time step and continue?'))
        IF (NEWDT) THEN
            NLINES=0
            IF (TTERM) CALL CLEAR
            GOTO 10
        END IF
C
        RETURN
        END
CCCCCCCCCCCCCCCCCCCCCCCCCCCCCCCCCCCCCCCCCCCCCCCCCCCCCCCCCCCCCCCCCCCCCCCC
        SUBROUTINE EVOLVE(PHI,PHI2,DT)
C take one time step using the implicit algorithm
C (Eq. 7.30-7.33 and 7.11-7.16)
CCCCCCCCCCCCCCCCCCCCCCCCCCCCCCCCCCCCCCCCCCCCCCCCCCCCCCCCCCCCCCCCCCCCCCCC
C Global variables:
        INCLUDE 'PARAM.E7'
C Input/output variables:
        COMPLEX PHI(0:MAXLAT)       !wavepacket (I/O)
        REAL PHI2(0:MAXLAT)         !wavepacket squared (I/O)
        REAL DT                     !time step (input)
C Local variables:
        COMPLEX CONST               !term in matrix inversion
        COMPLEX CHI                 !part of wave function
        COMPLEX BETA(0:MAXLAT)      !term in matrix inversion
        INTEGER IX                  !lattice index
CCCCCCCCCCCCCCCCCCCCCCCCCCCCCCCCCCCCCCCCCCCCCCCCCCCCCCCCCCCCCCCCCCCCCCCC
        CONST=4*SQRTM1*DX*DX/DT
        BETA(NPTS-1)=0.             !initial conditions for BETA
        DO 10 IX=NPTS-2,0,-1        !backward recursion for BETA
            BETA(IX)=GAMMA(IX+1)*(BETA(IX+1)-CONST*PHI(IX+1))
10      CONTINUE
C
        CHI=(0.,0.)                 !boundary conditions
        DO 20 IX=1,NPTS-1           !forward recursion for CHI and PHI
            CHI=GAMMA(IX)*CHI+BETA(IX-1)   !CHI at this lattice point
            PHI(IX)=CHI-PHI(IX)            !PHI at new time
            PHI2(IX)=ABS(PHI(IX))**2       !PHI2 at new time
20      CONTINUE
C
        RETURN
        END
CCCCCCCCCCCCCCCCCCCCCCCCCCCCCCCCCCCCCCCCCCCCCCCCCCCCCCCCCCCCCCCCCCCCCCCC
        SUBROUTINE TRDIAG(DT)
C calculate GAMMA for the inversion of the tridiagonal matrix
C (Eq. 7.30-7.33 and 7.11-7.16)
CCCCCCCCCCCCCCCCCCCCCCCCCCCCCCCCCCCCCCCCCCCCCCCCCCCCCCCCCCCCCCCCCCCCCCCC
C Global variables:
        INCLUDE 'PARAM.E7'
```

```
C Input variables:
      REAL DT                          !time step
C Local variables:
      INTEGER IX                       !lattice index
      COMPLEX AZERO,CONST1             !terms in matrix inversion
      REAL CONST2                      !useful constant
CCCCCCCCCCCCCCCCCCCCCCCCCCCCCCCCCCCCCCCCCCCCCCCCCCCCCCCCCCCCCCCCCCCCCCCC
      CONST1=-2.+2*SQRTM1*DX**2/DT
      CONST2=DX*DX*K0*K0
      GAMMA(NPTS)=0.                   !initial conditions for GAMMA
      DO 20 IX=NPTS-1,0,-1             !backward recursion
         AZERO=CONST1-CONST2*V(IX)     !use GAMMA(IX)=ALPHA(IX-1)
         GAMMA(IX)=-1./(AZERO+GAMMA(IX+1))
20    CONTINUE
C
      RETURN
      END
CCCCCCCCCCCCCCCCCCCCCCCCCCCCCCCCCCCCCCCCCCCCCCCCCCCCCCCCCCCCCCCCCCCCCCCC
      SUBROUTINE NORMLZ(PHI2,LPROB,RPROB,TPROB,LX,RX,TX)
C given PHI2, finds left, right, and total probability of the wavepacket
C as well as the average X value (left, right, and total)
CCCCCCCCCCCCCCCCCCCCCCCCCCCCCCCCCCCCCCCCCCCCCCCCCCCCCCCCCCCCCCCCCCCCCCCC
C Global variables:
      INCLUDE 'PARAM.E7'
C Input variable:
      REAL PHI2(0:MAXLAT)              !wavepacket squared
C Output variables:
      REAL LPROB,RPROB,TPROB           !left, right, and total probability
      REAL LX,RX,TX                    !left, right, and total average position
C Local variables:
      INTEGER IX                       !lattice index
CCCCCCCCCCCCCCCCCCCCCCCCCCCCCCCCCCCCCCCCCCCCCCCCCCCCCCCCCCCCCCCCCCCCCCCC
      LPROB=0.                         !zero sums
      LX=0.
      DO 10 IX=1,IMID-1                !integrate with trapezoidal rule
         LPROB=LPROB+PHI2(IX)
         LX=LX+X(IX)*PHI2(IX)
10    CONTINUE
      LPROB=LPROB+PHI2(IMID)/2         !middle point is shared between
      LX=LX+XMID*PHI2(IMID)/2          !  left and right
C
      RPROB=PHI2(IMID)/2               !middle point contribution
      RX=XMID*PHI2(IMID)/2
      DO 20 IX=IMID+1,NPTS-1           !integrate with trapezoidal rule
         RPROB=RPROB+PHI2(IX)
         RX=RX+X(IX)*PHI2(IX)
20    CONTINUE
C
      TPROB=LPROB+RPROB                !total probability
      IF (LPROB .NE. 0.) LX=LX/LPROB   !normalize LX,RX
      IF (RPROB .NE. 0.) RX=RX/RPROB
      TX=LX*LPROB+RX*RPROB             !total <x> is a weighted sum
```

```
      C
            RETURN
            END
CCCCCCCCCCCCCCCCCCCCCCCCCCCCCCCCCCCCCCCCCCCCCCCCCCCCCCCCCCCCCCCCCCCCC
            SUBROUTINE ENERGY(PHI,PHI2,E)
      C calculate the scaled energy E (units of K0**2) given PHI and PHI2
CCCCCCCCCCCCCCCCCCCCCCCCCCCCCCCCCCCCCCCCCCCCCCCCCCCCCCCCCCCCCCCCCCCCC
      C Global variables:
            INCLUDE 'PARAM.E7'
      C Input/Output variables:
            COMPLEX PHI(0:MAXLAT)       !wavepacket (input)
            REAL PHI2(0:MAXLAT)         !wavepacket squared (input)
            REAL E                      !total scaled energy (output)
      C Local variables:
            INTEGER IX                  !lattice index
            REAL PE                     !potential energy
            COMPLEX KE                  !kinetic energy
CCCCCCCCCCCCCCCCCCCCCCCCCCCCCCCCCCCCCCCCCCCCCCCCCCCCCCCCCCCCCCCCCCCCC
            PE=0.                       !initialize sums
            KE=0.
            DO 10 IX=1,NPTS-1           !integrate using trapezoidal rule
               PE=PE+V(IX)*PHI2(IX)
               KE=KE+CONJG(PHI(IX))*(PHI(IX-1)-2*PHI(IX)+PHI(IX+1))
      10    CONTINUE
            KE=-KE/DX**2/K0**2
            E=PE+REAL(KE)               !energy scale = K0**2
            RETURN
            END
CCCCCCCCCCCCCCCCCCCCCCCCCCCCCCCCCCCCCCCCCCCCCCCCCCCCCCCCCCCCCCCCCCCCC
            SUBROUTINE INTPHI(PHI,PHI2)
      C creates initial PHI array from input parameters
CCCCCCCCCCCCCCCCCCCCCCCCCCCCCCCCCCCCCCCCCCCCCCCCCCCCCCCCCCCCCCCCCCCCC
      C Global variables:
            INCLUDE 'PARAM.E7'
      C Output variables:
            COMPLEX PHI(0:MAXLAT)       !wavepacket
            REAL PHI2(0:MAXLAT)         !wavepacket squared
      C Local variables:
            INTEGER IX                  !lattice index
            REAL NORM,LPROB,RPROB,SQRTN !normalization of the wavepacket
            REAL LX,RX,TX               !average position
CCCCCCCCCCCCCCCCCCCCCCCCCCCCCCCCCCCCCCCCCCCCCCCCCCCCCCCCCCCCCCCCCCCCC
            IF (PACKET .EQ. LORNTZ) THEN           !Lorentzian
               DO 10 IX=1,NPTS-1
                  PHI(IX)=CEXP(SQRTM1*K0*X(IX))/(SIGMA**2+(X(IX)-W0)**2)
                  PHI2(IX)=CABS(PHI(IX))**2
      10       CONTINUE
      C
            ELSE IF (PACKET .EQ. GAUSS) THEN       !Gaussian
               DO 20 IX=1,NPTS-1
                  PHI(IX)=CEXP(SQRTM1*K0*X(IX))*EXP(-(X(IX)-W0)**2/2/SIGMA**2)
                  PHI2(IX)=CABS(PHI(IX))**2
```

```
20        CONTINUE
          END IF
C
          PHI(0)=0.                    !potential is infinite at ends of lattice
          PHI(NPTS)=0.                 ! so PHI is zero there
C
          CALL NORMLZ(PHI2,LPROB,RPROB,NORM,LX,RX,TX) !normalize wavepacket
          SQRTN=SQRT(NORM)
          DO 30 IX=0,NPTS
             PHI(IX)=PHI(IX)/SQRTN
             PHI2(IX)=PHI2(IX)/NORM
30        CONTINUE
C
          RETURN
          END
CCCCCCCCCCCCCCCCCCCCCCCCCCCCCCCCCCCCCCCCCCCCCCCCCCCCCCCCCCCCCCCCCCCCCCCCC
          SUBROUTINE POTNTL
C fills the potential array
CCCCCCCCCCCCCCCCCCCCCCCCCCCCCCCCCCCCCCCCCCCCCCCCCCCCCCCCCCCCCCCCCCCCCCCCC
C Global variables:
          INCLUDE 'PARAM.E7'
C Local variables:
          INTEGER IX                   !lattice index
CCCCCCCCCCCCCCCCCCCCCCCCCCCCCCCCCCCCCCCCCCCCCCCCCCCCCCCCCCCCCCCCCCCCCCCCC
          DO 5 IX=0,NPTS               !array of X values
             X(IX)=VXMIN+IX*DX
5         CONTINUE
C
          IF (POT .EQ. SQUARE) THEN          !Square bump/well
             DO 10 IX=0,NPTS
                IF ((X(IX) .GE. (X0-A)) .AND. (X(IX) .LE. (X0+A))) THEN
                   V(IX)=V0
                ELSE
                   V(IX)=0.
                END IF
10           CONTINUE
C
C         AGAUSS is fixed so that V(X0+A)=V(X0-A)=EPS*V0 (see INIT)
          ELSE IF (POT .EQ. GAUSS) THEN   !Gaussian bump/well
             DO 20 IX=0,NPTS
                V(IX)=V0*EXP(-AGAUSS*(X(IX)-X0)**2/A**2)
20           CONTINUE
C
          ELSE IF (POT .EQ. PARAB) THEN   !Parabola
             DO 30 IX=0,NPTS
                V(IX)=V0*((X(IX)-X0)**2/A**2)
30           CONTINUE
C
C         ASTEP is fixed so that V(X0+A)=1.-EPS*V0 and
C                                V(X0-A)=EPS*V0 (see INIT)
          ELSE IF (POT .EQ. STEP) THEN     !Smooth step function
             DO 40 IX=0,NPTS
```

```
                V(IX)=V0/2*(2/PI*ATAN(ASTEP*(X(IX)-X0)/A)+1.)
40        CONTINUE
        END IF
C
        VMAX=V(0)                              !find VMAX and VMIN to give
        VMIN=V(0)                              !  a scale for graphics
        DO 50 IX=1,NPTS
           IF (V(IX) .GT. VMAX) VMAX=V(IX)
           IF (V(IX) .LT. VMIN) VMIN=V(IX)
50        CONTINUE
        IF (VMAX .EQ. VMIN) VMAX=VMIN+1        !need a nonzero difference
C
        RETURN
        END
CCCCCCCCCCCCCCCCCCCCCCCCCCCCCCCCCCCCCCCCCCCCCCCCCCCCCCCCCCCCCCCCCCCCCCCCC
        SUBROUTINE INIT
C initializes constants, displays header screen,
C initializes menu arrays for input parameters
CCCCCCCCCCCCCCCCCCCCCCCCCCCCCCCCCCCCCCCCCCCCCCCCCCCCCCCCCCCCCCCCCCCCCCCCC
C Global variables:
        INCLUDE 'IO.ALL'
        INCLUDE 'MENU.ALL'
        INCLUDE 'PARAM.E7'
C Local parameters:
        CHARACTER*80 DESCRP                    !program description
        DIMENSION DESCRP(20)
        INTEGER NHEAD,NTEXT,NGRAPH             !number of lines for each description
        REAL EPS                               !small number to set width of potntls
CCCCCCCCCCCCCCCCCCCCCCCCCCCCCCCCCCCCCCCCCCCCCCCCCCCCCCCCCCCCCCCCCCCCCCCCC
C       get environment parameters
        CALL SETUP
C
C       display header screen
        DESCRP(1)= 'EXAMPLE 7'
        DESCRP(2)= 'Solution of time-dependent Schroedinger Equation'
        DESCRP(3)= 'for a wavepacket in a one-dimensional potential'
        NHEAD=3
C
C       text output description
        DESCRP(4)= 'time, energy, probability, and <x>'
        NTEXT=1
C
C       graphics output description
        DESCRP(5)= 'potential and |wavepacket|**2 vs. x'
        NGRAPH=1
C
        CALL HEADER(DESCRP,NHEAD,NTEXT,NGRAPH)
C
C       define constants
        PI=4.0*ATAN(1.0)
        SQRTM1=(0,1)                           !square root of -1
C
```

```
C        if you change VXMIN and/or VXMAX, change prompts below
C        for X0, W0, and CYCLE=K0*(VXMAX-VXMIN)/2/PI
         VXMIN=-1.                    !physical limits on lattice
         VXMAX=1.
C
         EPS=1./20.                   !find constants so that at X=X0+A
         AGAUSS=LOG(1./EPS)           ! or X=X0-A, V(X)=EPS*V0
         ASTEP=TAN(PI*(1./2.-EPS))    !(doesn't apply to parabola)
C
C        setup menu arrays, beginning with constant part
         CALL MENU
C
         MTYPE(13)=TITLE
         MPRMPT(13)='POTENTIAL FUNCTION MENU'
         MLOLIM(13)=2.
         MHILIM(13)=1.
C
         MTYPE(14)=MTITLE
         MPRMPT(14)=
        +'1) Square-well: V=V0 for X0-A < X < X0+A'
         MLOLIM(14)=0.
         MHILIM(14)=0.
C
         MTYPE(15)=MTITLE
         MPRMPT(15)=
        +'2) Gaussian: V(X)=V0*(EXP-(AGAUSS*((X-X0)/A)**2))'
         MLOLIM(15)=0.
         MHILIM(15)=0.
C
         MTYPE(16)=MTITLE
         MPRMPT(16)=
        +'3) Parabolic: V(X)=V0*(X-X0)**2/A**2'
         MLOLIM(16)=0.
         MHILIM(16)=0.
C
         MTYPE(17)=MTITLE
         MPRMPT(17)=
        +'4) Smooth step: V(X)=V0*(2/PI*ATN(ASTEP*(X-X0)/A)+1)/2'
         MLOLIM(17)=0.
         MHILIM(17)=1.
C
         MTYPE(18)=MCHOIC
         MPRMPT(18)='Make a menu choice and press return'
         MTAG(18)='19 19 19 19'
         MLOLIM(18)=1.
         MHILIM(18)=4.
         MINTS(18)=1
         MREALS(18)=1.
C
         MTYPE(19)=FLOAT
         MPRMPT(19)='Enter X0 (center of potential  -1 < X0 < 1 )'
         MTAG(19)='Center of potential (X0)'
```

```
      MLOLIM(19)=VXMIN
      MHILIM(19)=VXMAX
      MREALS(19)=VXMIN+3*(VXMAX-VXMIN)/4
C
      MTYPE(20)=FLOAT
      MPRMPT(20)='Enter A (half width of one-twentieth max)'
      MTAG(20)='Half-width of potential (A)'
      MLOLIM(20)=10*(VXMAX-VXMIN)/MAXLAT
      MHILIM(20)=(VXMAX-VXMIN)*10.
      MREALS(20)=(VXMAX-VXMIN)/25.
C
      MTYPE(21)=FLOAT
      MPRMPT(21)='Enter V0 (height of potential in units of K0**2)'
      MTAG(21)='Height of potential V0 (units of K0**2)'
      MLOLIM(21)=-100.
      MHILIM(21)=100.
      MREALS(21)=.3
C
      MTYPE(22)=FLOAT
      MPRMPT(22)=
     + 'Enter XMID which separates right from left (-1 < XMID < 1)'
      MTAG(22)='middle X value'
      MLOLIM(22)=VXMIN
      MHILIM(22)=VXMAX
      MREALS(22)=VXMIN+3*(VXMAX-VXMIN)/4
C
      MTYPE(23)=TITLE
      MPRMPT(23)='WAVEPACKET MENU'
      MLOLIM(23)=2.
      MHILIM(23)=1.
C
      MTYPE(24)=MTITLE
      MPRMPT(24)=
     +'1) Lorentzian:  PHI(X)=EXP(I*K0*X)/(SIGMA**2+(X-W0)**2)'
      MLOLIM(24)=0.
      MHILIM(24)=0.
C
      MTYPE(25)=MTITLE
      MPRMPT(25)=
     +'2) Gaussian: PHI(X)=EXP(I*K0*X)*EXP(-(X-W0)**2/2/SIGMA**2)'
      MLOLIM(25)=0.
      MHILIM(25)=1.
C
      MTYPE(26)=MCHOIC
      MPRMPT(26)='Make a menu choice and press return'
      MTAG(26)='27 27'
      MLOLIM(26)=1.
      MHILIM(26)=2.
      MINTS(26)=2
      MREALS(26)=2.
C
      MTYPE(27)=FLOAT
```

```
      MPRMPT(27)='Enter W0 (center of packet -1 < W0 < 1 )'
      MTAG(27)='Center of packet (W0)'
      MLOLIM(27)=VXMIN
      MHILIM(27)=VXMAX
      MREALS(27)=VXMIN+(VXMAX-VXMIN)/3
C
      MTYPE(28)=FLOAT
      MPRMPT(28)='Enter SIGMA (width of packet)'
      MTAG(28)='Width of packet (SIGMA)'
      MLOLIM(28)=10*(VXMAX-VXMIN)/MAXLAT
      MHILIM(28)=10*(VXMAX-VXMIN)
      MREALS(28)=.125*(VXMAX-VXMIN)
C
      MTYPE(29)=FLOAT
      MPRMPT(29)='Enter number of cycles on lattice = K0/PI '
      MTAG(29)='Number of cycles = K0/PI'
      MLOLIM(29)=-REAL(MAXLAT)/4.
      MHILIM(29)= REAL(MAXLAT)/4.
      MREALS(29)= 10.
C
      MTYPE(30)=SKIP
      MREALS(30)=35.
C
      MTYPE(38)=NUM
      MPRMPT(38)= 'Enter number of lattice points'
      MTAG(38)= 'Number of lattice points'
      MLOLIM(38)=20.
      MHILIM(38)=MAXLAT
      MINTS(38)=200
C
      MTYPE(39)=SKIP
      MREALS(39)=60.
C
      MSTRNG(MINTS(75))= 'exmpl7.txt'
C
      MTYPE(76)=SKIP
      MREALS(76)=80.
C
      MSTRNG(MINTS(86))= 'exmpl7.grf'
C
      MTYPE(87)=SKIP
      MREALS(87)=90.
C
      RETURN
      END
CCCCCCCCCCCCCCCCCCCCCCCCCCCCCCCCCCCCCCCCCCCCCCCCCCCCCCCCCCCCCCCCCCCCCCCC
      SUBROUTINE PARAM
C gets parameters from screen
C ends program on request
C closes old files
C maps menu variables to program variables
C opens new files
```

```
C calculates all derivative parameters
C performs checks on parameters
CCCCCCCCCCCCCCCCCCCCCCCCCCCCCCCCCCCCCCCCCCCCCCCCCCCCCCCCCCCCCCCCCCCC
C Global variables:
      INCLUDE 'MENU.ALL'
      INCLUDE 'IO.ALL'
      INCLUDE 'PARAM.E7'
C Local variables:
      REAL CYCLES                  !number of cycles on lattice
      REAL MAXCYC                  !maximum number of cycles for given NPTS
C     map between menu items and parameters
      INTEGER IPOT,IX0,IA,IV0,IXMID,IPACK,ISIGMA,IW0,ICYCLE,INPTS
      PARAMETER (IPOT   = 18)
      PARAMETER (IX0    = 19)
      PARAMETER (IA     = 20)
      PARAMETER (IV0    = 21)
      PARAMETER (IXMID  = 22)
      PARAMETER (IPACK  = 26)
      PARAMETER (IW0    = 27)
      PARAMETER (ISIGMA = 28)
      PARAMETER (ICYCLE = 29)
      PARAMETER (INPTS  = 38)
C Functions:
      LOGICAL LOGCVT               !converts 1 and 0 to true and false
      REAL GETFLT                  !get floating point number from screen
CCCCCCCCCCCCCCCCCCCCCCCCCCCCCCCCCCCCCCCCCCCCCCCCCCCCCCCCCCCCCCCCCCCC
C     get input from terminal
      CALL CLEAR
      CALL ASK(1,ISTOP)
C
C     stop program if requested
      IF (MREALS(IMAIN) .EQ. STOP) CALL FINISH
C
C     close files if necessary
      IF (TNAME .NE. MSTRNG(MINTS(ITNAME)))
     +     CALL FLCLOS(TNAME,TUNIT)
      IF (GNAME .NE. MSTRNG(MINTS(IGNAME)))
     +     CALL FLCLOS(GNAME,GUNIT)
C
C     set new parameter values
C     physical and numerical
      POT=MINTS(IPOT)
      X0=MREALS(IX0)
      A=MREALS(IA)
      V0=MREALS(IV0)
      XMID=MREALS(IXMID)
      PACKET=MINTS(IPACK)
      SIGMA=MREALS(ISIGMA)
      W0=MREALS(IW0)
      CYCLES=MREALS(ICYCLE)
      NPTS=MINTS(INPTS)
C
```

```
C       text output
        TTERM=LOGCVT(MINTS(ITTERM))
        TFILE=LOGCVT(MINTS(ITFILE))
        TNAME=MSTRNG(MINTS(ITNAME))
C
C       graphics output
        GTERM=LOGCVT(MINTS(IGTERM))
        GHRDCP=LOGCVT(MINTS(IGHRD))
        GFILE=LOGCVT(MINTS(IGFILE))
        GNAME=MSTRNG(MINTS(IGNAME))
C
C       open files
        IF (TFILE) CALL FLOPEN(TNAME,TUNIT)
        IF (GFILE) CALL FLOPEN(GNAME,GUNIT)
        !files may have been renamed
        MSTRNG(MINTS(ITNAME))=TNAME
        MSTRNG(MINTS(IGNAME))=GNAME
C
C       derivative parameters
        MOVIES=.FALSE.
        IF ((GTERM) .AND. (.NOT. TTERM)) MOVIES=.TRUE.
        DX=(VXMAX-VXMIN)/NPTS            !spatial step
        IMID=NINT((XMID-VXMIN)/DX)       !nearest lattice index to XMID
        XMID=VXMIN+IMID*DX               !force XMID to be a lattice point
        MREALS(IXMID)=XMID               !change default accordingly
C
C       check CYCLES with respect to lattice size
        MAXCYC=REAL(NPTS)/4.             !max number of cycles for NPTS
        IF (CYCLES .GT. MAXCYC) THEN
           WRITE (OUNIT,*) 'Number of cycles is too large for NPTS'
           CYCLES=GETFLT(MAXCYC/10.,-MAXCYC,MAXCYC,
     +        'Enter number of cycles = K0/PI')
           MREALS(ICYCLE)=CYCLES
        END IF
        K0=CYCLES*2*PI/(VXMAX-VXMIN)
        CALL CLEAR
C
        CALL POTNTL                      !setup potential array
        IF ((GTERM) .OR. (GHRDCP)) CALL GRFINT !initialize graphing param
C
        RETURN
        END
CCCCCCCCCCCCCCCCCCCCCCCCCCCCCCCCCCCCCCCCCCCCCCCCCCCCCCCCCCCCCCCCCCCCCCCC
        SUBROUTINE PRMOUT(MUNIT,NLINES)
C outputs parameter summary to MUNIT
CCCCCCCCCCCCCCCCCCCCCCCCCCCCCCCCCCCCCCCCCCCCCCCCCCCCCCCCCCCCCCCCCCCCCCCC
C Global variables:
        INCLUDE 'IO.ALL'
        INCLUDE 'PARAM.E7'
C Passed variables:
        INTEGER MUNIT                    !unit number for output (input)
        INTEGER NLINES                   !number of lines written so far (output)
```

```
CCCCCCCCCCCCCCCCCCCCCCCCCCCCCCCCCCCCCCCCCCCCCCCCCCCCCCCCCCCCCCCCCCCCCCCCCCC
        IF (MUNIT .EQ. OUNIT) CALL CLEAR
C
        WRITE (MUNIT,2)
        WRITE (MUNIT,4)
        WRITE (MUNIT,24) XMID,NPTS,DX
        WRITE (MUNIT,2)
        IF (POT .EQ. SQUARE) THEN
          WRITE (MUNIT,6)
        ELSE IF (POT .EQ. GAUSS) THEN
          WRITE (MUNIT,8)
        ELSE IF (POT .EQ. PARAB) THEN
          WRITE (MUNIT,10)
        ELSE IF (POT .EQ. STEP) THEN
          WRITE (MUNIT,12)
        END IF
        WRITE (MUNIT,14) X0,A,V0
        WRITE (MUNIT,2)
        IF (PACKET .EQ. LORNTZ) THEN
          WRITE (MUNIT,16)
        ELSE IF (PACKET .EQ. GAUSS) THEN
          WRITE (MUNIT,18)
        END IF
        WRITE (MUNIT,20) W0,SIGMA
        WRITE (MUNIT,22) K0,K0**2
        WRITE (MUNIT,2)
C
        NLINES=11
C
2       FORMAT (' ')
4       FORMAT
      + (' Output from example 7: Time-dependent Schroedinger Equation')
6       FORMAT (' Square-barrier/well: V=V0 for X0-A < X < X0+A')
8       FORMAT
      + (' Gaussian barrier/well: V(X)=V0*(EXP-(AGAUSS*((X-X0)/A)**2)))')
10      FORMAT (' Parabolic well: V(X)=V0*(X-X0)**2/A**2')
12      FORMAT (' Smooth step: V(X)=V0*(2/PI*ATN(ASTEP*(X-X0)/A)+1)/2')
14      FORMAT (' X0 = ',1PE12.5,5X,' A = ',1PE12.5,5X,
      +            ' V0 (units K0**2) = ',1PE12.5)
16      FORMAT (' Lorentzian wavepacket:  ',
      +          'PHI(X)=EXP(I*K0*X)/(SIGMA**2+(X-W0)**2)')
18      FORMAT (' Gaussian wavepacket: ',
      +          'PHI(X)=EXP(I*K0*X)*EXP(-(X-W0)**2/2/SIGMA**2)')
20      FORMAT (' W0 = ',1PE12.5,5X,' SIGMA = ',1PE12.5)
22      FORMAT (' K0 = ',1PE12.5,5X,' K0**2 (energy scale) = ',1PE12.5)
24      FORMAT (' XMIDDLE = ',1PE12.5,5X,' NPTS = ',I5,5X,
      +            ' space step = ',1PE12.5)
26      FORMAT (18X,'energy',9X,'probability',19X,'<x>')
28      FORMAT (7X,'time',7X,'K0**2 ',5X,'left',3X,'right',3X,'total',6X,
      +          'left',4X,'right',4X,'total')
30      FORMAT (7X,'----',7X,'------',5X,'----',3X,'-----',3X,'-----',6X,
      +          '----',4X,'-----',4X,'-----')
```

```
C
      RETURN
      END
CCCCCCCCCCCCCCCCCCCCCCCCCCCCCCCCCCCCCCCCCCCCCCCCCCCCCCCCCCCCCCCCCCCCCCC
      SUBROUTINE TITLES(MUNIT,NLINES,DT)
C write out time step and column titles to MUNIT
CCCCCCCCCCCCCCCCCCCCCCCCCCCCCCCCCCCCCCCCCCCCCCCCCCCCCCCCCCCCCCCCCCCCCCC
C Global variables:
      INCLUDE 'IO.ALL'
C Passed variables:
      INTEGER MUNIT           !unit number for output (input)
      INTEGER NLINES          !number of lines written so far (output)
.     REAL DT                 !time step (input)
CCCCCCCCCCCCCCCCCCCCCCCCCCCCCCCCCCCCCCCCCCCCCCCCCCCCCCCCCCCCCCCCCCCCCCC
      WRITE (MUNIT,*)  ' '
      WRITE (MUNIT,20) DT
      WRITE (MUNIT,26)
      WRITE (MUNIT,28)
      WRITE (MUNIT,30)
C
      IF (MUNIT .EQ. OUNIT) NLINES=NLINES+5
C
20    FORMAT (' time step = ',1PE15.8)
26    FORMAT (18X,'energy',9X,'probability',19X,'<x>')
28    FORMAT (7X,'time',7X,'K0**2 ',5X,'left',3X,'right',3X,'total',6X,
     +         'left',4X,'right',4X,'total')
30    FORMAT (7X,'----',7X,'------',5X,'----',3X,'-----',3X,'-----',6X,
     +         '----',4X,'-----',4X,'-----')
C
      RETURN
      END
CCCCCCCCCCCCCCCCCCCCCCCCCCCCCCCCCCCCCCCCCCCCCCCCCCCCCCCCCCCCCCCCCCCCCCC
      SUBROUTINE TXTOUT(MUNIT,E,TIME,LPROB,RPROB,TPROB,LX,RX,TX,NLINES)
C writes results for one state to MUNIT
CCCCCCCCCCCCCCCCCCCCCCCCCCCCCCCCCCCCCCCCCCCCCCCCCCCCCCCCCCCCCCCCCCCCCCC
C Global variables:
      INCLUDE 'IO.ALL'
C Input variables:
      INTEGER MUNIT           !output unit specifier
      REAL TIME               !time
      REAL TPROB,LPROB,RPROB  !normalization of the wavepacket
      REAL LX,RX,TX           !average position
      REAL E                  !energy
      INTEGER NLINES          !number of lines printed so far (I/O)
CCCCCCCCCCCCCCCCCCCCCCCCCCCCCCCCCCCCCCCCCCCCCCCCCCCCCCCCCCCCCCCCCCCCCCC
C     if screen is full, clear screen and retype headings
      IF ((MOD(NLINES,TRMLIN-4) .EQ. 0)
     +                        .AND. (MUNIT .EQ. OUNIT)) THEN
          CALL PAUSE('to continue...',1)
          CALL CLEAR
          WRITE (MUNIT,26)
          WRITE (MUNIT,28)
```

```
              WRITE (MUNIT,30)
              NLINES=NLINES+3
           END IF
C
           WRITE (MUNIT,40) TIME,E,LPROB,RPROB,TPROB,LX,RX,TX
C          keep track of printed lines only for terminal output
           IF (MUNIT .EQ. OUNIT) NLINES=NLINES+1
C
26         FORMAT (18X,'energy',9X,'probability',19X,'<x>')
28         FORMAT (7X,'time',7X,'K0**2 ',5X,'left',3X,'right',3X,'total',6X,
     +            'left',4X,'right',4X,'total')
30         FORMAT (7X,'----',7X,'------',5X,'----',3X,'-----',3X,'-----',6X,
     +            '----',4X,'-----',4X,'-----')
40         FORMAT (4X,1PE10.3,4X,0PF6.3,4X,F5.3,2(3X,F5.3),
     +     4X,F6.3,2(3X,F6.3))
C
           RETURN
           END
CCCCCCCCCCCCCCCCCCCCCCCCCCCCCCCCCCCCCCCCCCCCCCCCCCCCCCCCCCCCCCCCCCCCCCCCC
           SUBROUTINE GRFINT
C initialize all graphing variables which are the same for one run
CCCCCCCCCCCCCCCCCCCCCCCCCCCCCCCCCCCCCCCCCCCCCCCCCCCCCCCCCCCCCCCCCCCCCCCCC
C Global variables
           INCLUDE 'PARAM.E7'
           INCLUDE 'GRFDAT.ALL'
CCCCCCCCCCCCCCCCCCCCCCCCCCCCCCCCCCCCCCCCCCCCCCCCCCCCCCCCCCCCCCCCCCCCCCCCC
           YMAX=VMAX                         !find limits on data points
           YMIN=VMIN
           X0VAL=VMIN
           XMIN=VXMIN
           XMAX=VXMAX
           Y0VAL=VXMIN
C
           NPOINT=NPTS+1
C
           ISYM=1                            !symbols and ticks
           IFREQ=0
           NXTICK=5
           NYTICK=5
           ILINE=1
C
           LABEL(1)='scaled X'              !titles and labels
           LABEL(2)='potential, PHI2'
C
           RETURN
           END
CCCCCCCCCCCCCCCCCCCCCCCCCCCCCCCCCCCCCCCCCCCCCCCCCCCCCCCCCCCCCCCCCCCCCCCCC
           SUBROUTINE GRFOUT(DEVICE,PHI2,TIME,TPROB,TX,E)
C outputs wavepacket and potential to DEVICE
CCCCCCCCCCCCCCCCCCCCCCCCCCCCCCCCCCCCCCCCCCCCCCCCCCCCCCCCCCCCCCCCCCCCCCCCC
C Global variables
           INCLUDE 'IO.ALL'
```

```
      INCLUDE 'PARAM.E7'
      INCLUDE 'GRFDAT.ALL'
C Input variables:
      INTEGER DEVICE                 !which device is being used?
      REAL PHI2(0:MAXLAT)            !wavepacket squared
      REAL TIME                      !time
      REAL TPROB                     !normalization of the wavepacket
      REAL TX                        !average position
      REAL E                         !energy
C Local variables
      REAL PHIGRF(0:MAXLAT)          !scaled wavepacket squared
      REAL VSCALE                    !scaling factor for PHI2
      INTEGER IX                     !lattice index
      CHARACTER *9 CE,CSIG           !Energy, SIGMA as character data
      CHARACTER*9 CPROB,CTX,CTIME    !data as characters
      INTEGER PLEN,TXLEN,TLEN,ELEN,SLEN !length of strings
      INTEGER LENTRU                 !true length of character data
      INTEGER SCREEN                 !send to terminal
      INTEGER PAPER                  !make a hardcopy
      INTEGER FILE                   !send to a file
      DATA SCREEN,PAPER,FILE/1,2,3/
CCCCCCCCCCCCCCCCCCCCCCCCCCCCCCCCCCCCCCCCCCCCCCCCCCCCCCCCCCCCCCCCCCCCCCCCCC
C     messages for the impatient
      IF (DEVICE .NE. SCREEN) WRITE (OUNIT,100)
C
C     calculate parameters for graphing
      IF (DEVICE .NE. FILE) THEN
C
         IF (.NOT. MOVIES) THEN
            NPLOT=1                  !how many plots?
         ELSE
            NPLOT=4                  !for movies, there are
         END IF                      ! 4 plots per page
         IPLOT=1
C
         CALL CONVRT(TPROB,CPROB,PLEN)     !create legend
         CALL CONVRT(TX,CTX,TXLEN)
         CALL CONVRT(TIME,CTIME,TLEN)
         INFO='t='//CTIME(1:TLEN)//'  norm='//
     +      CPROB(1:PLEN)//'  <X>='//CTX(1:TXLEN)
C
         CALL CONVRT(E,CE,ELEN)            !description of  data
         CALL CONVRT(SIGMA,CSIG,SLEN)
         IF (PACKET .EQ. LORNTZ) THEN
            TITLE='Lorentzian packet,'
         ELSE IF (PACKET .EQ. GAUSS) THEN
            TITLE='Gaussian packet,'
         END IF
         TITLE=TITLE(1:LENTRU(TITLE))//'  width='//CSIG(1:SLEN)
     +      //', E ='//CE(1:ELEN)
C
         CALL GTDEV(DEVICE)               !device nomination
```

```
                    IF (DEVICE .EQ. SCREEN) CALL GMODE    !change to graphics mode
                    CALL LNLNAX                           !draw axes
                 END IF
C
C        The largest PHI2(IX) can ever become is 1 (all probability at
C        one lattice point).  The following scale assumes that the maximum
C        value is 10/NPTS (prob equally shared by one-tenth of the points)
         VSCALE=(VMAX-VMIN)*NPTS/10.   !scale PHI2 so that it fits on Vscale
         DO 10 IX=0,NPTS
            PHIGRF(IX)=VMIN+PHI2(IX)*VSCALE
            IF(PHIGRF(IX) .GT. VMAX) PHIGRF(IX)=VMAX     !clip high values
10       CONTINUE
C
C        output results
         IF (DEVICE .EQ. FILE) THEN
            WRITE (GUNIT,60) TIME
            WRITE (GUNIT,70) (X(IX),V(IX),PHI2(IX),IX=1,NPTS)
         ELSE
              CALL XYPLOT (X,V)
              CALL XYPLOT (X,PHIGRF)
         END IF
C
C        end graphing session
         IF (.NOT. MOVIES) THEN
            IF (DEVICE .NE. FILE) CALL GPAGE(DEVICE)     !end graphics package
            IF (DEVICE .EQ. SCREEN) CALL TMODE           !switch to text mode
         END IF
C
60       FORMAT (' X,V, PHI2 at time = ',1PE15.8)
70       FORMAT (3(5X,E15.8))
100      FORMAT (/,' Patience, please; output going to a file.')
C
         RETURN
         END
CCCCCCCCCCCCCCCCCCCCCCCCCCCCCCCCCCCCCCCCCCCCCCCCCCCCCCCCCCCCCCCCCCCCCCCCCCCCC
         SUBROUTINE GRFSEC(DEVICE,PHI2,TIME,TPROB,TX,FRAME)
C outputs wavepacket and potential for movies, frames 2-4
CCCCCCCCCCCCCCCCCCCCCCCCCCCCCCCCCCCCCCCCCCCCCCCCCCCCCCCCCCCCCCCCCCCCCCCCCCCCC
C Global variables
         INCLUDE 'IO.ALL'
         INCLUDE 'PARAM.E7'
         INCLUDE 'GRFDAT.ALL'
C Input variables:
         INTEGER DEVICE                     !which device is being used?
         REAL PHI2(0:MAXLAT)                !wavepacket squared
         REAL TIME                          !time
         REAL TPROB                         !normalization of the wavepacket
         REAL TX                            !average position
         INTEGER FRAME                      !which frame of the movie is it?
C Local variables
         REAL PHIGRF(0:MAXLAT)              !scaled wavepacket squared
         REAL VSCALE                        !scaling factor for PHI2
```

```
        INTEGER IX                   !lattice index
        CHARACTER*9 CPROB,CTX,CTIME  !data as characters
        INTEGER PLEN,TXLEN,TLEN      !length of strings
        INTEGER SCREEN               !send to terminal
        INTEGER PAPER                !make a hardcopy
        INTEGER FILE                 !send to a file
        DATA SCREEN,PAPER,FILE/1,2,3/
ccccccccccccccccccccccccccccccccccccccccccccccccccccccccccccccccccccccccc
        IPLOT=FRAME                  !associate plots with frames
        IF (IPLOT .EQ. 0) IPLOT=4
C
        CALL CONVRT(TPROB,CPROB,PLEN)!create legend
        CALL CONVRT(TX,CTX,TXLEN)
        CALL CONVRT(TIME,CTIME,TLEN)
        INFO='t='//CTIME(1:TLEN)//'  norm='//
     +      CPROB(1:PLEN)//'  <X>='//CTX(1:TXLEN)
        CALL LNLNAX                  !draw axes
C
        VSCALE=(VMAX-VMIN)*NPTS/10.  !scale PHI2 so that it fits on Vscale
        DO 10 IX=0,NPTS
           PHIGRF(IX)=VMIN+PHI2(IX)*VSCALE
           IF(PHIGRF(IX) .GT. VMAX) PHIGRF(IX)=VMAX
10      CONTINUE
C
        CALL XYPLOT (X,V)            !plot data
        CALL XYPLOT (X,PHIGRF)
C
        IF (IPLOT .EQ. 4) CALL GPAGE(DEVICE)  !clear this graphics page
C
        RETURN
        END
ccccccccccccccccccccccccccccccccccccccccccccccccccccccccccccccccccccccccc
ccccccccccccccccccccccccccccccccccccccccccccccccccccccccccccccccccccccccc
C param.e7
C
        REAL PI              !pi=3.15159.....
        REAL VXMIN,VXMAX     !physical limits on lattice
        COMPLEX SQRTM1       !square root of -1
        REAL ASTEP,AGAUSS    !factors in the potentials
        INTEGER POT          !which potential
        REAL X0              !center of potential
        REAL A               !width of potential
        REAL V0              !height of potential
        INTEGER PACKET       !which wavepacket
        REAL SIGMA           !packet width
        REAL W0              !packet initial position
        REAL K0              !packet wavenumber
        REAL XMID            !division between left and right
        INTEGER IMID         !index of division
        REAL DX              !spatial step
        INTEGER NPTS         !number of lattice points
        REAL VMAX,VMIN       !limits on value of potential
```

```
         LOGICAL MOVIES      !when only graphics to screen is requested
C
         INTEGER MAXLAT      !maximum number of lattice points
         PARAMETER (MAXLAT=1000)
         REAL X(0:MAXLAT)    !array of X values
         REAL V(0:MAXLAT)    !array of potential values
         COMPLEX GAMMA(0:MAXLAT) !terms in matrix inversion
C
         INTEGER SQUARE,GAUSS,PARAB,STEP !types of potentials
         INTEGER LORNTZ                  !types of packets
         PARAMETER (SQUARE=1,GAUSS=2,PARAB=3,STEP=4,LORNTZ=1)
C
         COMMON / CONST / PI,VXMIN,VXMAX,SQRTM1,ASTEP,AGAUSS
         COMMON / PPARAM / POT,X0,A,V0,PACKET,SIGMA,W0,XMID
         COMMON / NPARAM / NPTS
         COMMON / PCALC / DX,X,V,GAMMA,IMID,VMAX,VMIN,MOVIES,K0
```

B.8 Example 8

Algorithm This program simulates the two-dimensional Ising model using the algorithm of Metropolis *et al.* Before calculations begin, the acceptance ratio *r* is calculated for the 10 different cases (subroutine PARAM), and the initial configuration is chosen at random (DO loop 5 in subroutine ARCHON). After the thermalization sweeps are performed (loop 10), data taking for the groups begins (loop 20). The two basic subroutines are METROP, which performs a Metropolis sweep of the lattice, and SUM, which calculates the energy and magnetization for the configuration of the lattice every NFREQ'th sweep. During the data taking, both group and total sums for the energy and magnetization are accumulated (subroutine AVERAG) so that effects of sweep-to-sweep correlations can be monitored, as discussed immediately above Exercise 8.6. After the requested number of groups have been calculated, you are prompted for the number of additional groups to be done [10.].

Input The physical parameters are the magnetic field [0.] and interaction strength [.3], both in units of *kT*. The numerical parameters are the number of lattice points in *x* [20] and *y* [20], the random number seed [54767], the number of thermalization sweeps [20], the sampling frequency [5], the group size [10], and the number of groups [10]. The maximum number of *x* points is fixed by MAXX= 79; and *y* points, by MAXY= 20. These limits are to ensure that the character display of the configuration will fit on the typical screen which is 24 lines long and 80 characters wide. If your terminal is not 24 × 80, or if you won't be displaying the configuration, you can adjust these parameters. The text output parameter allows you to chose the short version of the output for which the sample sums are not displayed. Please see the note in C.7 concerning the random number seed and compiler storage of integers.

Output Every NFREQ'th sweep, the text output displays the group index, sample index, group size, acceptance ratio, energy, and magnetization. (All thermodynamic quantities are per spin.) If the short output is requested, this output is not printed. Every NFREQ*NSIZE'th sweep (when a group is completed), the text output displays the group index, the group averages for energy, magnetization, susceptibility, and specific heat, as well as the uncertainty in energy and magnetization. Also, at this time the grand averages (including all groups calculated so far) are displayed with two uncertainties for energy and magnetization, as discussed in the text.

392   B. Programs for the Examples

If you request graphics to the screen or file, every **NFREQ**'th sweep the configuration is displayed or printed using characters: spin up is indicated by an **X**, and spin down is indicated by a blank. No more sophisticated graphics output is available since it could tell you no more than this.

```
CCCCCCCCCCCCCCCCCCCCCCCCCCCCCCCCCCCCCCCCCCCCCCCCCCCCCCCCCCCCCCCCCCCCCCCC
      PROGRAM EXMPL8
C    Example 8: Monte Carlo simulation of the two-dimensional Ising Model
C    COMPUTATIONAL PHYSICS (FORTRAN VERSION)
C    by Steven E. Koonin and Dawn C. Meredith
C    Copyright 1989, Addison-Wesley Publishing Company
CCCCCCCCCCCCCCCCCCCCCCCCCCCCCCCCCCCCCCCCCCCCCCCCCCCCCCCCCCCCCCCCCCCCCCCC
      CALL INIT              !display header screen, setup parameters
5     CONTINUE               !main loop/ execute once for each set of param
      CALL PARAM             !get input from screen
      CALL ARCHON            !calculate the thermodynamic quantities
      GOTO 5
      END
CCCCCCCCCCCCCCCCCCCCCCCCCCCCCCCCCCCCCCCCCCCCCCCCCCCCCCCCCCCCCCCCCCCCCCCC
      SUBROUTINE ARCHON
C calculates the thermodynamic quantities:  energy, magnetization
C susceptibility, and specific heat at constant field and interaction
C strength
CCCCCCCCCCCCCCCCCCCCCCCCCCCCCCCCCCCCCCCCCCCCCCCCCCCCCCCCCCCCCCCCCCCCCCCC
C Global variables:
      INCLUDE 'PARAM.E8'
      INCLUDE 'IO.ALL'
C Local variables:
      INTEGER SPIN(MAXX,MAXY)     !spin configuration
C     all of the thermodynamic quant have 2 indices
C     (not all array elements are used, e.g. CHI(sweep,value))
C     first index is the level: sweep, group, or total
C     second index is the value: quantity, quant**2, or sigma**2
      REAL MAG(3,3)               !magnetization
      REAL ENERGY(3,3)            !energy
      REAL CB(3,3)                !specific heat
      REAL CHI(3,3)               !susceptibility
      INTEGER ITHERM              !thermalization index
      INTEGER ITER                !iteration index
      REAL ACCPT                  !acceptance ratio
      INTEGER IX,IY               !horiz and vert indices
      INTEGER NLINES              !number of lines printed to terminal
      INTEGER MORE,IGRP           !how many more groups, group index
      INTEGER ISWEEP              !sweep index
      INTEGER SWEEP,GROUP,TOTAL   !which level of calculation
      INTEGER VALUE,SQUARE,SIGSQ  !which quantity
C Functions:
      REAL GETFLT                 !get floating point number from screen
      INTEGER YESNO               !get yes/no answer from screen
      LOGICAL LOGCVT              !change from 1,0 to true and false
      INTEGER GETINT              !get integer data from screen
```

```
      REAL RANNOS                    !generates a random number
      DATA SWEEP,GROUP,TOTAL/1,2,3/
      DATA VALUE,SQUARE,SIGSQ/1,2,3/
cccccccccccccccccccccccccccccccccccccccccccccccccccccccccccccccccccc
C     output summary of parameters
      IF (TTERM) CALL PRMOUT(OUNIT,NLINES)
      IF (TFILE) CALL PRMOUT(TUNIT,NLINES)
      IF (GFILE) CALL PRMOUT(GUNIT,NLINES)
C
      DO 5 IX=1,NX               !random initial spin configuration
        DO 6 IY=1,NY
          IF (RANNOS(DSEED) .GT. .5) THEN
            SPIN(IX,IY)=1
          ELSE
            SPIN (IX,IY)=-1
          END IF
6       CONTINUE
5     CONTINUE
C
      DO 10 ITHERM=1,NTHERM      !thermalize init config
        CALL METROP(SPIN,ACCPT)
        IF (TTERM) WRITE (OUNIT,7) ITHERM,ACCPT
        IF (TFILE) WRITE (TUNIT,7) ITHERM,ACCPT
7       FORMAT (5X,' thermalization sweep ',I4,
     +          ', acceptance ratio =',F5.3)
10    CONTINUE
      IF (TTERM) CALL PAUSE('to begin summing...',1)
C
      CALL ZERO(TOTAL,MAG,ENERGY,CHI,CB)  !zero total averages
      MORE=NGROUP
C
15    CONTINUE                              !allow for more groups
        DO 20 IGRP=NGROUP-MORE+1,NGROUP    !loop over groups
          CALL ZERO(GROUP,MAG,ENERGY,CHI,CB) !zero group averages
C
          IF ((TTERM) .AND. (.NOT. GTERM) .AND. (.NOT. TERSE))
     +          CALL TITLES(OUNIT,NLINES)
          IF ((TFILE) .AND. (.NOT. TERSE)) CALL TITLES(TUNIT,NLINES)
C
          DO 30 ITER=1,NFREQ*NSIZE
            CALL METROP(SPIN,ACCPT)        !make a sweep of the lattice
C
            IF (MOD(ITER,NFREQ) .EQ. 0) THEN !include in averages
              ISWEEP=ITER/NFREQ                !which sweep is it
              CALL ZERO(SWEEP,MAG,ENERGY,CHI,CB) !zero sweep totals
              CALL SUM(SPIN,MAG,ENERGY)!sweep totals, add to group
C
              IF (GTERM) THEN            !display data
                CALL DISPLY(OUNIT,SPIN,MAG,
     +                      ENERGY,ACCPT,IGRP,ISWEEP)
              ELSE IF ((TTERM) .AND. (.NOT. TERSE)) THEN
                CALL
```

```
      +                       SWPOUT(OUNIT,MAG,ENERGY,ACCPT,IGRP,ISWEEP,NLINES)
                          END IF
                          IF ((TFILE) .AND. (.NOT. TERSE)) CALL
      +                       SWPOUT(TUNIT,MAG,ENERGY,ACCPT,IGRP,ISWEEP,NLINES)
                          IF ((GFILE) .OR. (GHRDCP)) CALL
      +                       DISPLY(GUNIT,SPIN,MAG,ENERGY,ACCPT,IGRP,ISWEEP)
      C
                       END IF
      30           CONTINUE
                   CALL AVERAG(MAG,ENERGY,CHI,CB,IGRP) !calc total averages
      20       CONTINUE
      C
             MORE=GETINT(10,0,1000,'How many more groups?')
             IF (MORE .GT. 0) THEN
                 NGROUP=NGROUP+MORE
                 NLINES=0
                 IF ((TTERM) .AND. (.NOT. TERSE))CALL CLEAR
                 GOTO 15
             END IF
      C
             RETURN
             END
CCCCCCCCCCCCCCCCCCCCCCCCCCCCCCCCCCCCCCCCCCCCCCCCCCCCCCCCCCCCCCCCCCCCCCCC
             SUBROUTINE METROP(SPIN,ACCPT)
C make one sweep of the lattice using the Metropolis algorithm
C to generate a new configuration
CCCCCCCCCCCCCCCCCCCCCCCCCCCCCCCCCCCCCCCCCCCCCCCCCCCCCCCCCCCCCCCCCCCCCCCC
C Global variables:
             INCLUDE 'PARAM.E8'
C Input/Output variables:
             INTEGER SPIN(MAXX,MAXY)     !spin configuration (I/O)
             REAL ACCPT                  !acceptance ratio (output)
C Local variables:
             INTEGER IX,IY               !horiz and vert indices
             INTEGER IXM1,IXP1,IYM1,IYP1 !indices of nearest neighbors
             INTEGER NNSUM               !sum of nearest neighbors
C Function:
             REAL RANNOS                 !generates a random number
CCCCCCCCCCCCCCCCCCCCCCCCCCCCCCCCCCCCCCCCCCCCCCCCCCCCCCCCCCCCCCCCCCCCCCCC
             ACCPT=0.                         !zero acceptance ratio
             DO 10 IX=1,NX
                 IXP1=IX+1                         !nearest neighbors
                 IF (IX .EQ. NX) IXP1=1            !with periodic b.c.
                 IXM1=IX-1
                 IF (IX .EQ. 1) IXM1=NX
      C
                 DO 20 IY=1,NY
                     IYP1=IY+1                         !nearest neighbors
                     IF (IY .EQ. NY) IYP1=1            !with periodic b.c.
                     IYM1=IY-1
                     IF (IY .EQ. 1) IYM1=NY
      C                                                   !term to weight new config
```

```
         NNSUM=SPIN(IX,IYP1)+SPIN(IX,IYM1)+SPIN(IXP1,IY)+SPIN(IXM1,IY)
C
         IF (RANNOS(DSEED) .LT. RATIO(NNSUM,SPIN(IX,IY))) THEN
            SPIN(IX,IY)=-SPIN(IX,IY)     !flip the spin
            ACCPT=ACCPT+1                !update accept count
         END IF
C
20       CONTINUE
10       CONTINUE
         ACCPT=ACCPT/NSPIN               !make it a ratio
C
         RETURN
         END
CCCCCCCCCCCCCCCCCCCCCCCCCCCCCCCCCCCCCCCCCCCCCCCCCCCCCCCCCCCCCCCCCCCCCCCC
         SUBROUTINE SUM(SPIN,MAG,ENERGY)
C calculate magnetization and energy for this sweep
C add these values to the group averages
CCCCCCCCCCCCCCCCCCCCCCCCCCCCCCCCCCCCCCCCCCCCCCCCCCCCCCCCCCCCCCCCCCCCCCCC
C Global variables:
         INCLUDE 'PARAM.E8'
C Input/output variables:
         INTEGER SPIN(MAXX,MAXY)     !spin configuration (input)
C        all of the thermodynamic quant have 2 indices
C        (not all array elements are used, e.g. CHI(sweep,value))
C        first index is the level: sweep, group, or total
C        second index is the quantity: value, square, or sigma**2
         REAL MAG(3,3)               !magnetization (I/O)
         REAL ENERGY(3,3)            !energy (I/O)
C Local variables:
         INTEGER PAIRS               !interaction sum
         INTEGER SWEEP,GROUP,TOTAL   !which level of calculation
         INTEGER VALUE,SQUARE,SIGSQ  !which quantity
         INTEGER IX,IY               !horiz and vert indices
         INTEGER IXM1,IYM1           !neighbor indices
         DATA SWEEP,GROUP,TOTAL/1,2,3/
         DATA VALUE,SQUARE,SIGSQ/1,2,3/
CCCCCCCCCCCCCCCCCCCCCCCCCCCCCCCCCCCCCCCCCCCCCCCCCCCCCCCCCCCCCCCCCCCCCCCC
         PAIRS=0                     !zero pair sum
         DO 10 IY=1,NY
            IYM1=IY-1
            IF (IY .EQ. 1) IYM1=NY   !neighbor just below
            DO 20 IX=1,NX            !periodic b.c.
               IXM1=IX-1
               IF (IX .EQ. 1) IXM1=NX   !neighbor to the left
                                        !periodic b.c.
C
C              this method of summing pairs does not count twice
               PAIRS=PAIRS+SPIN(IX,IY)*(SPIN(IX,IYM1)+SPIN(IXM1,IY))
C              magnetization is the sum of the spins  (Eq. 8.21a)
               MAG(SWEEP,VALUE)=MAG(SWEEP,VALUE)+SPIN(IX,IY)
C
20          CONTINUE
10       CONTINUE
```

```
C
      MAG(SWEEP,SQUARE)=MAG(SWEEP,VALUE)**2
      ENERGY(SWEEP,VALUE)=-J*PAIRS-B*MAG(SWEEP,VALUE)      !Eq 8.18
      ENERGY(SWEEP,SQUARE)=ENERGY(SWEEP,VALUE)**2
C
C     add sweep contributions to group sums
      MAG(GROUP,VALUE)=MAG(GROUP,VALUE)+MAG(SWEEP,VALUE)
      MAG(GROUP,SQUARE)=MAG(GROUP,SQUARE)+MAG(SWEEP,SQUARE)
      ENERGY(GROUP,VALUE)=ENERGY(GROUP,VALUE)+ENERGY(SWEEP,VALUE)
      ENERGY(GROUP,SQUARE)=ENERGY(GROUP,SQUARE)+ENERGY(SWEEP,SQUARE)
C
      RETURN
      END
CCCCCCCCCCCCCCCCCCCCCCCCCCCCCCCCCCCCCCCCCCCCCCCCCCCCCCCCCCCCCCCCCCCCCCC
      SUBROUTINE AVERAG(MAG,ENERGY,CHI,CB,IGROUP)
C find group averages from group sums and add these to total averages;
C calculate uncertainties and display results
CCCCCCCCCCCCCCCCCCCCCCCCCCCCCCCCCCCCCCCCCCCCCCCCCCCCCCCCCCCCCCCCCCCCCCC
C Global variables:
      INCLUDE 'PARAM.E8'
      INCLUDE 'IO.ALL'
C Input/Output variables:
C     all of the thermodynamic quant have 2 indices
C     (not all array elements are used, e.g. CHI(sweep,value))
C     first index is the level: sweep, group, or total
C     second index is the value: quantity, quant**2, or sigma**2
      REAL MAG(3,3)                    !magnetization
      REAL ENERGY(3,3)                 !energy
      REAL CB(3,3)                     !specific heat
      REAL CHI(3,3)                    !susceptibility
      INTEGER IGROUP                   !group index (input)
C Local variables:
      REAL M,MSIG1,MSIG2               !magnetization and uncertainties
      REAL E,ESIG1,ESIG2               !energy and uncertainties
      REAL SUS,SUSSIG                  !susceptibility and uncertainty
      REAL C,CSIG                      !specific heat and uncertainty
      INTEGER IQUANT                   !index the quantity
      INTEGER SWEEP,GROUP,TOTAL        !which level of calculation
      INTEGER VALUE,SQUARE,SIGSQ       !which quantity
      DATA SWEEP,GROUP,TOTAL/1,2,3/
      DATA VALUE,SQUARE,SIGSQ/1,2,3/
CCCCCCCCCCCCCCCCCCCCCCCCCCCCCCCCCCCCCCCCCCCCCCCCCCCCCCCCCCCCCCCCCCCCCCC
C     calculate group averages and uncertainties from group sums
      DO 10 IQUANT=VALUE,SQUARE
         MAG(GROUP,IQUANT)=MAG(GROUP,IQUANT)/NSIZE
         ENERGY(GROUP,IQUANT)=ENERGY(GROUP,IQUANT)/NSIZE
10    CONTINUE
      CHI(GROUP,VALUE)=MAG(GROUP,SQUARE)-MAG(GROUP,VALUE)**2
      MAG(GROUP,SIGSQ)=CHI(GROUP,VALUE)/NSIZE
      IF (MAG(GROUP,SIGSQ) .LT. 0.) MAG(GROUP,SIGSQ)=0.
      CB(GROUP,VALUE)=ENERGY(GROUP,SQUARE)-ENERGY(GROUP,VALUE)**2
      ENERGY(GROUP,SIGSQ)=CB(GROUP,VALUE)/NSIZE
```

```
      IF (ENERGY(GROUP,SIGSQ) .LT. 0.) ENERGY(GROUP,SIGSQ)=0.
      CHI(GROUP,SQUARE)=CHI(GROUP,VALUE)**2
      CB(GROUP,SQUARE)=CB(GROUP,VALUE)**2
C
C     add group averages to total sums
      DO 20 IQUANT=VALUE,SIGSQ
         MAG(TOTAL,IQUANT)=MAG(TOTAL,IQUANT)+MAG(GROUP,IQUANT)
         ENERGY(TOTAL,IQUANT)=ENERGY(TOTAL,IQUANT)+ENERGY(GROUP,IQUANT)
         CHI(TOTAL,IQUANT)=CHI(TOTAL,IQUANT)+CHI(GROUP,IQUANT)
         CB(TOTAL,IQUANT)=CB(TOTAL,IQUANT)+CB(GROUP,IQUANT)
20    CONTINUE
C
C     find total averages using total sums accumulated so far
      M=MAG(TOTAL,VALUE)/IGROUP
      MSIG1=(MAG(TOTAL,SQUARE)/IGROUP-M**2)/IGROUP/NSIZE
      IF (MSIG1 .LT. 0) MSIG1=0.
      MSIG1=SQRT(MSIG1)
      MSIG2=SQRT(MAG(TOTAL,SIGSQ))/IGROUP
C
      E=ENERGY(TOTAL,VALUE)/IGROUP
      ESIG1=(ENERGY(TOTAL,SQUARE)/IGROUP-E**2)/IGROUP/NSIZE
      IF (ESIG1 .LT. 0) ESIG1=0.
      ESIG1=SQRT(ESIG1)
      ESIG2=SQRT(ENERGY(TOTAL,SIGSQ))/IGROUP
C
      SUS=CHI(TOTAL,VALUE)/IGROUP
      SUSSIG=(CHI(TOTAL,SQUARE)/IGROUP-SUS**2)/IGROUP
      IF (SUSSIG .LT. 0.) SUSSIG=0.
      SUSSIG=SQRT(SUSSIG)
C
      C=CB(TOTAL,VALUE)/IGROUP
      CSIG=(CB(TOTAL,SQUARE)/IGROUP-C**2)/IGROUP
      IF (CSIG .LT. 0.) CSIG=0.
      CSIG=SQRT(CSIG)
C
C     write out summary
      IF (TTERM) CALL TXTOUT(MAG,ENERGY,CB,CHI,E,ESIG1,ESIG2,
     +    M,MSIG1,MSIG2,SUS,SUSSIG,C,CSIG,IGROUP,OUNIT)
      IF (TFILE) CALL TXTOUT(MAG,ENERGY,CB,CHI,E,ESIG1,ESIG2,
     +    M,MSIG1,MSIG2,SUS,SUSSIG,C,CSIG,IGROUP,TUNIT)
C
      RETURN
      END
CCCCCCCCCCCCCCCCCCCCCCCCCCCCCCCCCCCCCCCCCCCCCCCCCCCCCCCCCCCCCCCCCCCCCCCC
      SUBROUTINE ZERO(ILEVEL,MAG,ENERGY,CHI,CB)
C zero sums for ILEVEL thermodynamic values
CCCCCCCCCCCCCCCCCCCCCCCCCCCCCCCCCCCCCCCCCCCCCCCCCCCCCCCCCCCCCCCCCCCCCCCC
C Input/Output variables:
      INTEGER ILEVEL            !which level to zero  (input)
C     all of the thermodynamic quant have 2 indices
C     (not all array elements are used, e.g. CHI(sweep,value))
C     first index is the level: sweep, group, or total
```

```
C       second index is the value: quantity, quant**2, or sigma**2
        REAL MAG(3,3)              !magnetization (output)
        REAL ENERGY(3,3)          !energy (output)
        REAL CB(3,3)              !specific heat (output)
        REAL CHI(3,3)             !susceptibility (output)
C Local variable:
        INTEGER IQUANT            !which quantity
CCCCCCCCCCCCCCCCCCCCCCCCCCCCCCCCCCCCCCCCCCCCCCCCCCCCCCCCCCCCCCCCCCCCCCCC
        DO 10 IQUANT=1,3
           MAG(ILEVEL,IQUANT)=0.
           ENERGY(ILEVEL,IQUANT)=0.
           CHI(ILEVEL,IQUANT)=0.
           CB(ILEVEL,IQUANT)=0.
10      CONTINUE
        RETURN
        END
CCCCCCCCCCCCCCCCCCCCCCCCCCCCCCCCCCCCCCCCCCCCCCCCCCCCCCCCCCCCCCCCCCCCCCCC
        SUBROUTINE INIT
C initializes constants, displays header screen,
C initializes menu arrays for input parameters
CCCCCCCCCCCCCCCCCCCCCCCCCCCCCCCCCCCCCCCCCCCCCCCCCCCCCCCCCCCCCCCCCCCCCCCC
C Global variables:
        INCLUDE 'IO.ALL'
        INCLUDE 'MENU.ALL'
        INCLUDE 'PARAM.E8'
C Local parameters:
        CHARACTER*80 DESCRP              !program description
        DIMENSION DESCRP(20)
        INTEGER NHEAD,NTEXT,NGRAPH       !number of lines for each description
CCCCCCCCCCCCCCCCCCCCCCCCCCCCCCCCCCCCCCCCCCCCCCCCCCCCCCCCCCCCCCCCCCCCCCCC
C       get environment parameters
        CALL SETUP
C
C       display header screen
        DESCRP(1)= 'EXAMPLE 8'
        DESCRP(2)= 'Monte Carlo simulation of the 2-D Ising Model'
        DESCRP(3)= 'using the Metropolis algorithm'
        NHEAD=3
C
C       text output description
        DESCRP(4)= 'acceptance rate, energy, magnetization, specific heat'
        DESCRP(5)= 'and susceptibility (all are values per spin)'
        NTEXT=2
C
C       graphics output description
        DESCRP(6)= 'spin configuration (blank=-1; X=+1)'
        NGRAPH=1
C
        CALL HEADER(DESCRP,NHEAD,NTEXT,NGRAPH)
C
C       setup menu arrays, beginning with constant part
        CALL MENU
```

```
C
      MTYPE(13)=FLOAT
      MPRMPT(13)='Enter value for magnetic field (units of kT)'
      MTAG(13)='Magnetic field (units of kT)'
      MLOLIM(13)=-20.
      MHILIM(13)=20.
      MREALS(13)=0.
C
      MTYPE(14)=FLOAT
      MPRMPT(14)='Enter value for interaction strength (units of kT)'
      MTAG(14)='interaction strength (units of kT)'
      MLOLIM(14)=-20.
      MHILIM(14)=20.
      MREALS(14)=.3
C
      MTYPE(15)=SKIP
      MREALS(15)=35.
C
      MTYPE(38)=NUM
      MPRMPT(38)= 'Enter number of X lattice points'
      MTAG(38)= 'Number of X lattice points'
      MLOLIM(38)=2.
      MHILIM(38)=MAXX
      MINTS(38)=20
C
      MTYPE(39)=NUM
      MPRMPT(39)= 'Enter number of Y lattice points'
      MTAG(39)= 'Number of Y lattice points'
      MLOLIM(39)=2.
      MHILIM(39)=MAXY
      MINTS(39)=20
C
      MTYPE(40)=NUM
      MPRMPT(40)= 'Integer random number seed for init fluctuations'
      MTAG(40)= 'Random number seed'
      MLOLIM(40)=1000.
      MHILIM(40)=99999.
      MINTS(40)=54767
C
      MTYPE(41)=NUM
      MPRMPT(41)= 'Number of thermalization sweeps'
      MTAG(41)= 'Thermalization sweeps'
      MLOLIM(41)=0
      MHILIM(41)=1000
      MINTS(41)=20
C
      MTYPE(42)=NUM
      MPRMPT(42)=  'Enter sampling frequency (to avoid correlations)'
      MTAG(42)= 'Sampling frequency'
      MLOLIM(42)=1
      MHILIM(42)=100
      MINTS(42)=5
```

```
      C
            MTYPE(43)=NUM
            MPRMPT(43)= 'Number of samples in a group'
            MTAG(43)= 'Group size'
            MLOLIM(43)=1
            MHILIM(43)=1000
            MINTS(43)=10
      C
            MTYPE(44)=NUM
            MPRMPT(44)= 'Enter number of groups'
            MTAG(44)= 'Number of groups'
            MLOLIM(44)=1
            MHILIM(44)=1000
            MINTS(44)=10
      C
            MTYPE(45)=SKIP
            MREALS(45)=60.
      C
            MSTRNG(MINTS(75))= 'exmpl8.txt'
      C
            MTYPE(76)=BOOLEN
            MPRMPT(76)='Do you want the short version of the output?'
            MTAG(76)='Short version of output'
            MINTS(76)=0
      C
            MTYPE(77)=SKIP
            MREALS(77)=80.
      C
            MSTRNG(MINTS(86))= 'exmpl8.grf'
      C
            MTYPE(87)=SKIP
            MREALS(87)=90.
      C
            RETURN
            END
CCCCCCCCCCCCCCCCCCCCCCCCCCCCCCCCCCCCCCCCCCCCCCCCCCCCCCCCCCCCCCCCCCCCCCCCC
            SUBROUTINE PARAM
C gets parameters from screen
C ends program on request
C closes old files
C maps menu variables to program variables
C opens new files
C calculates all derivative parameters
C performs checks on parameters
CCCCCCCCCCCCCCCCCCCCCCCCCCCCCCCCCCCCCCCCCCCCCCCCCCCCCCCCCCCCCCCCCCCCCCCCC
C Global variables:
            INCLUDE 'MENU.ALL'
            INCLUDE 'IO.ALL'
            INCLUDE 'PARAM.E8'
C Local variables:
            INTEGER IF                 !possible values for sum of neighb. spins
C map between menu indices and parameters
```

```
      INTEGER IB,IJ,INX,INY,IDSEED,INTHRM,INFREQ,INSIZE,INGRP,ITERSE
      PARAMETER (IB    = 13)
      PARAMETER (IJ    = 14)
      PARAMETER (INX   = 38)
      PARAMETER (INY   = 39)
      PARAMETER (IDSEED = 40)
      PARAMETER (INTHRM = 41)
      PARAMETER (INFREQ = 42)
      PARAMETER (INSIZE = 43)
      PARAMETER (INGRP  = 44)
      PARAMETER (ITERSE = 76)
C Functions:
      LOGICAL LOGCVT            !converts 1 and 0 to true and false
CCCCCCCCCCCCCCCCCCCCCCCCCCCCCCCCCCCCCCCCCCCCCCCCCCCCCCCCCCCCCCCCCCCCCCC
C     get input from terminal
      CALL CLEAR
      CALL ASK(1,ISTOP)
C
C     stop program if requested
      IF (MREALS(IMAIN) .EQ. STOP) CALL FINISH
C
C     close files if necessary
      IF (TNAME .NE. MSTRNG(MINTS(ITNAME)))
     +    CALL FLCLOS(TNAME,TUNIT)
      IF (GNAME .NE. MSTRNG(MINTS(IGNAME)))
     +    CALL FLCLOS(GNAME,GUNIT)
C
C     set new parameter values
C     physical and numerical
      B=MREALS(IB)
      J=MREALS(IJ)
      NX=MINTS(INX)
      NY=MINTS(INY)
      DSEED=DBLE(MINTS(IDSEED))
      NTHERM=MINTS(INTHRM)
      NFREQ=MINTS(INFREQ)
      NSIZE=MINTS(INSIZE)
      NGROUP=MINTS(INGRP)
C
C     text output
      TTERM=LOGCVT(MINTS(ITTERM))
      TFILE=LOGCVT(MINTS(ITFILE))
      TNAME=MSTRNG(MINTS(ITNAME))
      TERSE=LOGCVT(MINTS(ITERSE))
C
C     graphics output
      GTERM=LOGCVT(MINTS(IGTERM))
      GHRDCP=LOGCVT(MINTS(IGHRD))
      GFILE=LOGCVT(MINTS(IGFILE))
      GNAME=MSTRNG(MINTS(IGNAME))
C
C     open files
```

```
        IF (TFILE) CALL FLOPEN(TNAME,TUNIT)
        IF (GFILE) CALL FLOPEN(GNAME,GUNIT)
        !files may have been renamed
        MSTRNG(MINTS(ITNAME))=TNAME
        MSTRNG(MINTS(IGNAME))=GNAME
C
        CALL CLEAR
C
C       calculate derivative parameters
        NSPIN=NX*NY
        DO 10 IF=-4,4,2    !ratio of prob.; not all matrix elem are used
           RATIO(IF,-1)=EXP(2*(J*IF+B))
           RATIO(IF,1) =1./RATIO(IF,-1)
10      CONTINUE
C
C       calculate parameters for best looking text display
        IF (2*NX .LE. TRMWID) THEN
           XSKIP=.TRUE.                !skip spaces in x
           XCNTR=(TRMWID-2*NX)/2       !how to center display
        ELSE
           XSKIP=.FALSE.
           XCNTR=(TRMWID-NX)/2
        END IF
        IF (XCNTR .LT. 1) XCNTR=1
        IF (2*NY .LE. TRMLIN-5) THEN
           YSKIP=.TRUE.                !skip lines in y
           YCNTR=(TRMLIN-2*NY)/2-3     !how to center display
        ELSE
           YSKIP=.FALSE.
           YCNTR=(TRMLIN-NY)/2-3
        END IF
        IF (YCNTR .LT. 0) YCNTR=0
C
        RETURN
        END
CCCCCCCCCCCCCCCCCCCCCCCCCCCCCCCCCCCCCCCCCCCCCCCCCCCCCCCCCCCCCCCCCCCCCCC
        SUBROUTINE PRMOUT(MUNIT,NLINES)
C write out parameter summary to MUNIT
CCCCCCCCCCCCCCCCCCCCCCCCCCCCCCCCCCCCCCCCCCCCCCCCCCCCCCCCCCCCCCCCCCCCCCC
C Global variables:
        INCLUDE 'IO.ALL'
        INCLUDE 'PARAM.E8'
C Input variables:
        INTEGER MUNIT                  !fortran unit number
        INTEGER NLINES                 !number of lines sent to terminal
CCCCCCCCCCCCCCCCCCCCCCCCCCCCCCCCCCCCCCCCCCCCCCCCCCCCCCCCCCCCCCCCCCCCCCC
        IF (MUNIT .EQ. OUNIT) CALL CLEAR
C
        WRITE (MUNIT,5)
        WRITE (MUNIT,6)
        WRITE (MUNIT,7) B
        WRITE (MUNIT,8) J
```

```
      WRITE (MUNIT,10) NX,NY
      WRITE (MUNIT,15) NTHERM
      WRITE (MUNIT,20) NFREQ,NSIZE
      WRITE (MUNIT,*) ' '
      NLINES=8
C
5     FORMAT (' Output from example 8:')
6     FORMAT (' Monte Carlo Simulation of the'
     +        ' 2-D Ising Model using the Metropolis Algorithm')
7     FORMAT (' Magnetic field (units of kT) =',1PE12.5)
8     FORMAT (' Interaction strength (units of kT) =',1PE12.5)
10    FORMAT (' NX =',I3,5X,' NY =',I3)
15    FORMAT (' number of thermalization sweeps =',I4)
20    FORMAT (' sweep frequency = ',I4,' group size =',I4)
C
      RETURN
      END
CCCCCCCCCCCCCCCCCCCCCCCCCCCCCCCCCCCCCCCCCCCCCCCCCCCCCCCCCCCCCCCCCCCCCCC
      SUBROUTINE DISPLY(MUNIT,SPIN,MAG,ENERGY,ACCPT,IGROUP,ISWEEP)
C display spin configuration (spin=-1 is a blank, spin=1 is an X)
C and write out data for this sweep
CCCCCCCCCCCCCCCCCCCCCCCCCCCCCCCCCCCCCCCCCCCCCCCCCCCCCCCCCCCCCCCCCCCCCCC
C Global variables:
      INCLUDE 'PARAM.E8'
      INCLUDE 'IO.ALL'
C Input variables:
      INTEGER SPIN(MAXX,MAXY)    !spin configuration
C     all of the thermodynamic quant have 2 indices
C     (not all array elements are used, e.g. CHI(sweep,value))
C     first index is the level: sweep, group, or total
C     second index is the value: quantity, quant**2, or sigma**2
      REAL MAG(3,3)              !magnetization
      REAL ENERGY(3,3)          !energy
      REAL ACCPT                 !acceptance ratio
      INTEGER MUNIT              !unit we're writing to
      INTEGER ISWEEP,IGROUP      !sweep and group index
C Local variables:
      INTEGER SWEEP,GROUP,TOTAL  !which level of calculation
      INTEGER VALUE,SQUARE,SIGSQ !which quantity
      INTEGER IX,IY              !lattice indices
      CHARACTER*1 CSPIN(MAXX)    !spin as character data
      CHARACTER*80 BLNK          !blanks for centering in X
      DATA BLNK /' '/
      DATA SWEEP,GROUP,TOTAL/1,2,3/
      DATA VALUE,SQUARE,SIGSQ/1,2,3/
CCCCCCCCCCCCCCCCCCCCCCCCCCCCCCCCCCCCCCCCCCCCCCCCCCCCCCCCCCCCCCCCCCCCCCC
      IF (MUNIT .EQ. OUNIT) THEN
         CALL CLEAR
         DO 20 IY=1,YCNTR              !center output
            WRITE (OUNIT,*) ' '
20       CONTINUE
      END IF
```

```
         WRITE (MUNIT,11) IGROUP,ISWEEP,NSIZE,ACCPT,
       +           ENERGY(SWEEP,VALUE)/NSPIN,MAG(SWEEP,VALUE)/NSPIN
11     FORMAT (' group ',I3,', sweep ',I3,' out of ',I3,5X,
       +          ' accpt =',F5.3,' Energy = ',F7.3,'  Mag =',F6.3)
C
         DO 100 IY=NY,1,-1                      !change +-1 to X and blank
           DO 50 IX=1,NX
             IF (SPIN(IX,IY) .EQ. 1) THEN
               CSPIN(IX)='X'
             ELSE
               CSPIN(IX)=' '
             END IF
50         CONTINUE
C          write out a line at a time (no centering done for TUNIT)
           IF (MUNIT .EQ. TUNIT) THEN
                     WRITE (TUNIT,16) (CSPIN(IX),IX=1,NX)
           ELSE
             IF (XSKIP) THEN
               WRITE (MUNIT,10) BLNK(1:XCNTR),(CSPIN(IX),IX=1,NX)
             ELSE
               WRITE (MUNIT,15) BLNK(1:XCNTR),(CSPIN(IX),IX=1,NX)
             END IF
             IF (YSKIP) WRITE (MUNIT,*) ' '
           END IF
10         FORMAT (1X,A,100(A1,1X))
15         FORMAT (1X,A,100(A1))
16         FORMAT (1X,100(A1))
100      CONTINUE
         IF (MUNIT .EQ. OUNIT) CALL PAUSE('to continue...',0)
C
         RETURN
         END
CCCCCCCCCCCCCCCCCCCCCCCCCCCCCCCCCCCCCCCCCCCCCCCCCCCCCCCCCCCCCCCCCCCCCCCC
         SUBROUTINE SWPOUT(MUNIT,MAG,ENERGY,ACCPT,IGROUP,ISWEEP,NLINES)
C and write out data for this sweep
CCCCCCCCCCCCCCCCCCCCCCCCCCCCCCCCCCCCCCCCCCCCCCCCCCCCCCCCCCCCCCCCCCCCCCCC
C Global variables:
         INCLUDE 'PARAM.E8'
         INCLUDE 'IO.ALL'
C Input variables:
C     all of the thermodynamic quant have 2 indices
C     (not all array elements are used, e.g. CHI(sweep,value))
C     first index is the level: sweep, group, or total
C     second index is the value: quantity, quant**2, or sigma**2
         REAL MAG(3,3)                 !magnetization
         REAL ENERGY(3,3)              !energy
         REAL ACCPT                    !acceptance ratio
         INTEGER MUNIT                 !unit we're writing to
         INTEGER ISWEEP,IGROUP         !sweep and group index
         INTEGER NLINES                !lines written to terminal (I/O)
C Local variables:
         INTEGER SWEEP,GROUP,TOTAL     !which level of calculation
```

```
      INTEGER VALUE,SQUARE,SIGSQ !which quantity
      DATA SWEEP,GROUP,TOTAL/1,2,3/
      DATA VALUE,SQUARE,SIGSQ/1,2,3/
ccccccccccccccccccccccccccccccccccccccccccccccccccccccccccccccccccccc
      WRITE (MUNIT,11) IGROUP,ISWEEP,NSIZE,ACCPT,
     +          ENERGY(SWEEP,VALUE)/NSPIN,MAG(SWEEP,VALUE)/NSPIN
11    FORMAT (7X,I3,7X,I3,'/',I3,7X,F5.3,7X,F9.5,7X,F9.3)
      IF (MUNIT .EQ. OUNIT) NLINES=NLINES+1
      RETURN
      END
ccccccccccccccccccccccccccccccccccccccccccccccccccccccccccccccccccccc
      SUBROUTINE TXTOUT(MAG,ENERGY,CB,CHI,E,ESIG1,ESIG2,
     +    M,MSIG1,MSIG2,SUS,SUSSIG,C,CSIG,IGROUP,MUNIT)
C write out averages and uncertainties to MUNIT
ccccccccccccccccccccccccccccccccccccccccccccccccccccccccccccccccccccc
C Global variables:
      INCLUDE 'PARAM.E8'
      INCLUDE 'IO.ALL'
C Input variables:
C     all of the thermodynamic quant have 2 indices
C     (not all array elements are used, e.g. CHI(sweep,value))
C     first index is the level: sweep, group, or total
C     second index is the value: quantity, quant**2, or sigma**2
      REAL MAG(3,3)            !magnetization
      REAL ENERGY(3,3)         !energy
      REAL CB(3,3)             !specific heat
      REAL CHI(3,3)            !susceptibility
      INTEGER IGROUP           !group index
      REAL M,MSIG1,MSIG2       !magnetization and uncertainties
      REAL E,ESIG1,ESIG2       !energy and uncertainties
      REAL SUS,SUSSIG          !susceptibility and uncertainty
      REAL C,CSIG              !specific heat and uncertainty
      INTEGER MUNIT            !which unit number
C Local variables:
      INTEGER SWEEP,GROUP,TOTAL   !which level of calculation
      INTEGER VALUE,SQUARE,SIGSQ  !which quantity
      DATA SWEEP,GROUP,TOTAL/1,2,3/
      DATA VALUE,SQUARE,SIGSQ/1,2,3/
ccccccccccccccccccccccccccccccccccccccccccccccccccccccccccccccccccccc
      WRITE (MUNIT,30) IGROUP,NGROUP
      WRITE (MUNIT,32)
      WRITE (MUNIT,33)
      WRITE (MUNIT,35)
     + ENERGY(GROUP,VALUE)/NSPIN,SQRT(ENERGY(GROUP,SIGSQ))/NSPIN,
     + MAG(GROUP,VALUE)/NSPIN,SQRT(MAG(GROUP,SIGSQ))/NSPIN,
     + CHI(GROUP,VALUE)/NSPIN,CB(GROUP,VALUE)/NSPIN
      WRITE (MUNIT,40) E/NSPIN,ESIG1/NSPIN,ESIG2/NSPIN,
     + M/NSPIN,MSIG1/NSPIN,MSIG2/NSPIN,
     + SUS/NSPIN,SUSSIG/NSPIN,C/NSPIN,CSIG/NSPIN
      WRITE (MUNIT,*) ' '
      IF ((MUNIT .EQ. OUNIT) .AND. (.NOT. TERSE))
     +      CALL PAUSE(' to continue...',1)
```

```
C
30    FORMAT (' Group ',I3,' (out of ',I4,') averages')
32    FORMAT (14X,'Energy',13X,'Magnetization',5X,'Susceptibility',
     +         2X,'Specific Heat')
33    FORMAT (14X,'------',13X,'-------------',5X,'---------------',
     +         2X,'-------------')
35    FORMAT ('  group ',2(1X,F7.3,'+-',F5.3,6X),
     +          2(2X,F6.3,7X))
40    FORMAT ('  total ',2(1X,F7.3,'+-',F5.3,'/',F5.3),
     +          2(1X,F6.3,'+-',F6.3))
C
      RETURN
      END
CCCCCCCCCCCCCCCCCCCCCCCCCCCCCCCCCCCCCCCCCCCCCCCCCCCCCCCCCCCCCCCCCCCCCCC
      SUBROUTINE TITLES(MUNIT,NLINES)
C write out text data titles
CCCCCCCCCCCCCCCCCCCCCCCCCCCCCCCCCCCCCCCCCCCCCCCCCCCCCCCCCCCCCCCCCCCCCCC
C Global variables:
      INCLUDE 'IO.ALL'
C Input/Output variables:
      INTEGER MUNIT                    !output unit (input)
      INTEGER NLINES                   !number of lines written (I/O)
CCCCCCCCCCCCCCCCCCCCCCCCCCCCCCCCCCCCCCCCCCCCCCCCCCCCCCCCCCCCCCCCCCCCCCC
      IF (MUNIT .EQ. OUNIT) CALL CLEAR
C
      WRITE (MUNIT,10)
      WRITE (MUNIT,11)
10    FORMAT (6X,'Group',3X,'Sweep/Out of',5X,'Accpt',9X,'Energy',
     +        6X,'Magnetization')
11    FORMAT (6X,'-----',3X,'-------------',5X,'-----',9X,'------',
     +        6X,'-------------')
C
      IF (MUNIT .EQ. OUNIT) NLINES=NLINES+2
C
      RETURN
      END
CCCCCCCCCCCCCCCCCCCCCCCCCCCCCCCCCCCCCCCCCCCCCCCCCCCCCCCCCCCCCCCCCCCCCCC
CCCCCCCCCCCCCCCCCCCCCCCCCCCCCCCCCCCCCCCCCCCCCCCCCCCCCCCCCCCCCCCCCCCCCCC
C param.e8
C
      REAL B                     !magnetic field strength
      REAL J                     !interaction strength
      INTEGER NX,NY              !horiz and vert number of lattice points
      DOUBLE PRECISION DSEED     !random number seed
      INTEGER NTHERM             !number of thermalization sweeps
      INTEGER NFREQ              !freq of sweeps to avoid correlations
      INTEGER NSIZE              !size of groups
      INTEGER NGROUP             !number of groups
      INTEGER XCNTR,YCNTR        !data for centering display
      LOGICAL XSKIP,YSKIP        !data for centering display
      REAL RATIO(-4:4,-1:1)      !acceptance ratio matrix
      INTEGER NSPIN              !total number of spins
```

```
      LOGICAL TERSE          !terse output
C
      INTEGER MAXX,MAXY       !maximum horiz and vert dimensions
      PARAMETER (MAXX=79)     !these are set assuming that you
      PARAMETER (MAXY=20)     !have no graphics and the length of
                             !your terminal=24
C
      COMMON / PPARAM / B,J
      COMMON / NPARAM / NX,NY,DSEED,NTHERM,NFREQ,NSIZE,NGROUP
      COMMON / GPARAM / XCNTR,YCNTR,XSKIP,YSKIP,TERSE
      COMMON / PCALC / RATIO,NSPIN
```

Programs for
the Projects

C.1 Project 1

Algorithm This program calculates, for a given energy, the deflection function for scattering from the Lennard-Jones potential at NB impact parameters covering the range from BMIN to BMAX. The basic task is to evaluate Eq. (I.8) for each impact parameter requested (DO loop 25 in subroutine ARCHON). This is done in subroutine THETA, where first a simple search is used to locate RMIN, the turning point. Then the integrals are evaluated by the rectangle rule, with a change of variables to regulate the singularities. You may alter the potential by changing FNV(R) at the beginning of subroutine THETA. If you do, don't forget to find the new value for RMAX such that $V(r_{max}) \approx 5 \times 10^{-3} V_0$.

Input The physical parameters are the Energy [1.0], the minimum and maximum impact parameters [.1,2.4], and the number of impact parameters for which the quadrature is performed [20]. The maximum number of impact parameters is fixed by the FORTRAN parameter MAX= 100. Energy is in units of V_0, lengths are in units of a. The program checks that BMIN < BMAX < RMAX in subroutine PCHECK. The numerical parameters include the tolerance for the search on the turning point [.0001], the radius beyond which the potential is negligible [2.5], and the number of quadrature points [40].

Output For each impact parameter the text output displays the impact parameter, B; the radius of closest approach, RMIN; and the scattering angle in degrees, ANGLE. After all values of ANGLE have been calculated, the deflection (scattering) angle is graphed as a function of impact parameter.

```
cccccccccccccccccccccccccccccccccccccccccccccccccccccccccccccccccccccccc
      PROGRAM PROJ1
C     Project 1: Scattering by a central 6-12 potential
C  COMPUTATIONAL PHYSICS (FORTRAN VERSION)
C  by Steven E. Koonin and Dawn C. Meredith
C  Copyright 1989, Addison-Wesley Publishing Company
cccccccccccccccccccccccccccccccccccccccccccccccccccccccccccccccccccccccc
      CALL INIT             !display header screen, setup parameters
5     CONTINUE              !main loop; execute once for each set of param
         CALL PARAM         !get input parameters
         CALL ARCHON        !calculate scattering angles
      GOTO 5
      END
cccccccccccccccccccccccccccccccccccccccccccccccccccccccccccccccccccccccc
      SUBROUTINE ARCHON
C calculate scattering angles as a function of the impact parameter
cccccccccccccccccccccccccccccccccccccccccccccccccccccccccccccccccccccccc
C Global variables
      INCLUDE 'PARAM.P1'
      INCLUDE 'IO.ALL'
C Local variables:
      REAL RMIN             !radius of closest approach
      REAL ANGLE(0:MAX)     !final scattering angle
      REAL IMPACT(0:MAX)    !impact parameter
      REAL INT1, INT2       !values of the two integrals
      INTEGER IB            !index of impact param
      INTEGER NLINES        !number of lines printed so far
      INTEGER SCREEN        !send to terminal
      INTEGER PAPER         !make a hardcopy
      INTEGER FILE          !send to a file
      DATA SCREEN,PAPER,FILE/1,2,3/
cccccccccccccccccccccccccccccccccccccccccccccccccccccccccccccccccccccccc
C     output summary of parameters
      IF (TTERM) CALL PRMOUT(OUNIT,NLINES)
      IF (TFILE) CALL PRMOUT(TUNIT,NLINES)
      IF (GFILE) CALL PRMOUT(GUNIT,NLINES)
C
      DO 25 IB=0,(NB-1)                          !loop over impact param
C
         IMPACT(IB)=BMIN+(IB*DB)                 !current value of B
         CALL THETA(IMPACT(IB),INT1,INT2,RMIN)   !calculate THETA for each B
         ANGLE(IB)=(2*IMPACT(IB)*(INT1-INT2)*180)/PI  !convert to degrees
C
C        text output
         IF (TTERM) CALL TXTOUT(OUNIT,NLINES,IMPACT(IB),RMIN,ANGLE(IB))
         IF (TFILE) CALL TXTOUT(TUNIT,NLINES,IMPACT(IB),RMIN,ANGLE(IB))
C
25    CONTINUE
C
      IF (TTERM) CALL PAUSE('to continue...',1)
      IF (TTERM) CALL CLEAR
C
```

```
      IF (GTERM) CALL GRFOUT(SCREEN,ANGLE,IMPACT)   !graphics output
      IF (GFILE) CALL GRFOUT(FILE,ANGLE,IMPACT)
      IF (GHRDCP) CALL GRFOUT(PAPER,ANGLE,IMPACT)
C
      RETURN
      END
CCCCCCCCCCCCCCCCCCCCCCCCCCCCCCCCCCCCCCCCCCCCCCCCCCCCCCCCCCCCCCCCCCCCCCC
      SUBROUTINE THETA(B,INT1,INT2,RMIN)
C  searches to find RMIN (closest approach)
C  integrates FNI1 from B    to RMAX to find INT1
C  integrates FNI2 from RMIN to RMAX to find INT2
CCCCCCCCCCCCCCCCCCCCCCCCCCCCCCCCCCCCCCCCCCCCCCCCCCCCCCCCCCCCCCCCCCCCCCC
C Global variables:
      INCLUDE 'PARAM.P1'
C Input/Output variables:
      REAL B                     !impact parameter (input)
      REAL INT1, INT2            !values for the two integrals (output)
      REAL RMIN                  !radius of closes approach (output)
C Local variables:
      REAL R                     !current radius
      REAL DR                    !current search step size
      REAL U                     !U=SQRT(R-B) or SQRT(R-RMIN)
      REAL UMAX                  !UMAX=SQRT(RMAX-B) or SQRT(RMAX-RMIN)
      REAL H                     !radial step for quadrature
      REAL SUM                   !sum for quadrature
      INTEGER IU                 !index on U
      REAL FNV                   !scattering potential (function)
      REAL FNI1,FNI2             !first and second integrands (function)
CCCCCCCCCCCCCCCCCCCCCCCCCCCCCCCCCCCCCCCCCCCCCCCCCCCCCCCCCCCCCCCCCCCCCCC
C     define functions
      FNV(R)      = 4*(R**(-12)-R**(-6))
      FNI1(B,R)   = 1/R**2/SQRT(1-(B/R)**2)
      FNI2(B,R,E) = 1/R**2/SQRT(1-(B/R)**2-FNV(R)/E)
C
      RMIN=RMAX                  !inward search for turning point
      DR = 0.2
1     IF (DR.GT.TOLR) THEN
         RMIN=RMIN-DR                  !simple search
         IF ((1-((B/RMIN)**2)-(FNV(RMIN)/E)).LT.0.0) THEN
            RMIN=RMIN +DR              !RMIN is where FNI2 is infinite
            DR=DR/2                    !i.e., the denominator=0
         END IF
         GO TO 1
      END IF
C
      SUM=0                      !first integral by rectangle rule
      UMAX=SQRT(RMAX-B)          !change variable to U=SQRT(R-B)
      H=UMAX/NPTS                !to remove singularity at R=B
      DO 10 IU=1,NPTS
         U=H*(IU-0.5)
         R=U**2+B               !change back to R to eval integrand
         SUM=SUM+U*FNI1(B,R)
```

```
      10 CONTINUE
         INT1=2*H*SUM
C
         SUM=0                            !second integral by rectangle rule
         UMAX=SQRT(RMAX-RMIN)             !change of variable to
         H=UMAX/NPTS                      !to remove singularity at R=RMIN
         DO 22 IU=1,NPTS
            U=H*(IU-0.5)
            R=U**2+RMIN                   !change back to R to eval integrand
            SUM=SUM+U*FNI2(B,R,E)
      22 CONTINUE
         INT2=2*H*SUM
C
         RETURN
         END
CCCCCCCCCCCCCCCCCCCCCCCCCCCCCCCCCCCCCCCCCCCCCCCCCCCCCCCCCCCCCCCCCCCCCCCCC
         SUBROUTINE INIT
C initializes constants, displays header screen,
C initializes arrays for input parameters
CCCCCCCCCCCCCCCCCCCCCCCCCCCCCCCCCCCCCCCCCCCCCCCCCCCCCCCCCCCCCCCCCCCCCCCCC
C Global variables:
         INCLUDE 'IO.ALL'
         INCLUDE 'MENU.ALL'
         INCLUDE 'PARAM.P1'
C Local parameters:
         CHARACTER*80 DESCRP              !program description
         DIMENSION DESCRP(22)
         INTEGER NHEAD,NTEXT,NGRF         !number of lines for each description
CCCCCCCCCCCCCCCCCCCCCCCCCCCCCCCCCCCCCCCCCCCCCCCCCCCCCCCCCCCCCCCCCCCCCCCCC
         CALL SETUP                       !get environment parameters
C
         !display header screen
         DESCRP(1)= 'PROJECT 1'
         DESCRP(2)= 'Scattering by a 6-12 potential'
         NHEAD=2
C
C        text output description
         DESCRP(3)= 'impact parameter, radius of closest approach, '
         DESCRP(4)= 'and the angle of deflection'
         NTEXT=2
C
C        graphics output description
         DESCRP(5)= 'deflection function'
      +  //' (scattering angle vs. impact parameter)'
         NGRF=1
C
         CALL HEADER(DESCRP,NHEAD,NTEXT,NGRF)
C
C        calculate constants
         PI=4*ATAN(1.0)
C
         CALL MENU                        !setup constant part of menu
```

```
C
      MTYPE(13)=FLOAT
      MPRMPT(13)= 'Enter incident energy E (scaled units)'
      MTAG(13)= 'Incident energy (scaled units)'
      MLOLIM(13)=.01
      MHILIM(13)=1000.
      MREALS(13)=1.0
C
      MTYPE(14)=FLOAT
      MPRMPT(14)= 'Enter the minimum impact parameter (scaled units)'
      MTAG(14)= 'Minimum impact parameter Bmin (scaled units)'
      MLOLIM(14)=.0
      MHILIM(14)=2.5
      MREALS(14)=0.1
C
      MTYPE(15)=FLOAT
      MPRMPT(15)= 'Enter the maximum impact parameter (scaled units)'
      MTAG(15)= 'Maximum impact parameter Bmax (scaled units)'
      MLOLIM(15)=.0
      MHILIM(15)=7.
      MREALS(15)=2.4
C
      MTYPE(16)=NUM
      MPRMPT(16)= 'Enter number of values for impact parameter'
      MTAG(16)= 'Number of values for impact parameter'
      MLOLIM(16)=1.
      MHILIM(16)=MAX
      MINTS(16)=20
C
      MTYPE(17)=SKIP
      MREALS(17)=35.
C
      MTYPE(38)=FLOAT
      MPRMPT(38)= 'Enter tolerance for turning point search'
      MTAG(38)= 'Turning point search tolerance'
      MLOLIM(38)=.000005
      MHILIM(38)=.1
      MREALS(38)=.0001
C
      MTYPE(39)=FLOAT
      MPRMPT(39)= 'Enter radius at which V(r)<<E (scaled units)'
      MTAG(39)= 'Maximum radius RMAX (scaled units)'
      MLOLIM(39)=0.
      MHILIM(39)=10.0
      MREALS(39)=2.5
C
      MTYPE(40)=NUM
      MPRMPT(40)= 'Enter number of quadrature points'
      MTAG(40)= 'Number of quadrature points'
      MLOLIM(40)=20.
      MHILIM(40)=300.
      MINTS(40)=40
```

```
C
      MTYPE(41)=SKIP
      MREALS(41)=60.
C
      MSTRNG(MINTS(75))= 'proj1.txt'
C
      MTYPE(76)=SKIP
      MREALS(76)=80.
C
      MSTRNG(MINTS(86))= 'proj1.grf'
C
      MTYPE(87)=SKIP
      MREALS(87)=90.
C
      RETURN
      END
CCCCCCCCCCCCCCCCCCCCCCCCCCCCCCCCCCCCCCCCCCCCCCCCCCCCCCCCCCCCCCCCCCCCCCCC
      SUBROUTINE PARAM
C gets parameters from screen
C ends program on request
C closes old files
C maps menu variables to program variables
C opens new files
C calculates all derivative parameters
CCCCCCCCCCCCCCCCCCCCCCCCCCCCCCCCCCCCCCCCCCCCCCCCCCCCCCCCCCCCCCCCCCCCCCCC
C Global variables:
      INCLUDE 'MENU.ALL'
      INCLUDE 'IO.ALL'
      INCLUDE 'PARAM.P1'
      INCLUDE 'MAP.P1'
C Function:
      LOGICAL LOGCVT              !converts 1 and 0 to true and false
CCCCCCCCCCCCCCCCCCCCCCCCCCCCCCCCCCCCCCCCCCCCCCCCCCCCCCCCCCCCCCCCCCCCCCCC
C     get input from terminal
      CALL CLEAR
      CALL ASK(1,ISTOP)
C
C     stop program if requested
      IF (MREALS(IMAIN) .EQ. STOP) CALL FINISH
C
C     close files if necessary
      IF (TNAME .NE. MSTRNG(MINTS(ITNAME)))
     +     CALL FLCLOS(TNAME,TUNIT)
      IF (GNAME .NE. MSTRNG(MINTS(IGNAME)))
     +     CALL FLCLOS(GNAME,GUNIT)
C
C     physical and numerical parameters
      E=MREALS(IE)
      BMIN=MREALS(IBMIN)
      BMAX=MREALS(IBMAX)
      NB=MINTS(INB)
      RMAX=MREALS(IRMAX)
```

```
         TOLR=MREALS(ITOLR)
         NPTS=MINTS(INPTS)
C
C        text output
         TTERM=LOGCVT(MINTS(ITTERM))
         TFILE=LOGCVT(MINTS(ITFILE))
         TNAME=MSTRNG(MINTS(ITNAME))
C
C        graphics output
         GTERM=LOGCVT(MINTS(IGTERM))
         GHRDCP=LOGCVT(MINTS(IGHRD))
         GFILE=LOGCVT(MINTS(IGFILE))
         GNAME=MSTRNG(MINTS(IGNAME))
C
C        open files
         IF (TFILE) CALL FLOPEN(TNAME,TUNIT)
         IF (GFILE) CALL FLOPEN(GNAME,GUNIT)
C        files may have been renamed
         MSTRNG(MINTS(ITNAME))=TNAME
         MSTRNG(MINTS(IGNAME))=GNAME
C
         CALL PCHECK              !check 0<BMIN<BMAX<=RMAX
         DB=(BMAX-BMIN)/(NB-1)    !calculate step in B
         CALL CLEAR
C
         RETURN
         END
CCCCCCCCCCCCCCCCCCCCCCCCCCCCCCCCCCCCCCCCCCCCCCCCCCCCCCCCCCCCCCCCCCCCCCC
         SUBROUTINE PCHECK
C to ensure that 0<BMIN<BMAX<=RMAX
CCCCCCCCCCCCCCCCCCCCCCCCCCCCCCCCCCCCCCCCCCCCCCCCCCCCCCCCCCCCCCCCCCCCCCC
C Global variables:
         INCLUDE 'MENU.ALL'
         INCLUDE 'IO.ALL'
         INCLUDE 'PARAM.P1'
         INCLUDE 'MAP.P1'
C Function:
         REAL GETFLT                    !gets floating point number
CCCCCCCCCCCCCCCCCCCCCCCCCCCCCCCCCCCCCCCCCCCCCCCCCCCCCCCCCCCCCCCCCCCCCCC
40       IF (BMAX .LE. BMIN) THEN
            WRITE (OUNIT,10) BMAX,BMIN
            WRITE (OUNIT,20)
            MREALS(IBMIN)=GETFLT(MREALS(IBMIN),MLOLIM(IBMIN),MHILIM(IBMIN),
     +                 'Enter BMIN')
            MREALS(IBMAX)=GETFLT(MREALS(IBMAX),BMIN,MHILIM(IBMAX),
     +                 'Enter BMAX')
            BMAX=MREALS(IBMAX)
            BMIN=MREALS(IBMIN)
         GO TO 40
         END IF
C
50       IF (RMAX .LE. BMAX) THEN
```

```
            WRITE (OUNIT,30) RMAX,BMAX
            WRITE (OUNIT,20)
            !prompt for new values
            MREALS(IBMAX)=GETFLT(MREALS(IBMAX),BMIN,MHILIM(IBMAX),
     +                 'Enter BMAX')
            BMAX=MREALS(IBMAX)
            MREALS(IRMAX)=GETFLT(MREALS(IRMAX),MLOLIM(IRMAX),MHILIM(IRMAX),
     +                 'Enter RMAX')
            RMAX=MREALS(IRMAX)
        GO TO 50
        END IF
C
10      FORMAT (/,' BMAX (=',F6.3,')',' is less than BMIN (=',
     +        F6.3,'),')
20      FORMAT ('  but it should be the other way round; '
     +          ' reenter parameter values')
30      FORMAT (/,' RMAX (=',F6.3,')',' is less than BMAX (=',F6.3,')')
C
        RETURN
        END
CCCCCCCCCCCCCCCCCCCCCCCCCCCCCCCCCCCCCCCCCCCCCCCCCCCCCCCCCCCCCCCCCCCCCCCC
        SUBROUTINE PRMOUT(MUNIT,NLINES)
C outputs parameter summary to the specified unit
CCCCCCCCCCCCCCCCCCCCCCCCCCCCCCCCCCCCCCCCCCCCCCCCCCCCCCCCCCCCCCCCCCCCCCCC
C Global variables
        INCLUDE 'IO.ALL'
        INCLUDE 'PARAM.P1'
C Passed variables
        INTEGER MUNIT               !unit number for output (input)
        INTEGER NLINES              !number of lines written so far (output)
CCCCCCCCCCCCCCCCCCCCCCCCCCCCCCCCCCCCCCCCCCCCCCCCCCCCCCCCCCCCCCCCCCCCCCCC
        IF (MUNIT .EQ. OUNIT) CALL CLEAR
C
        WRITE (MUNIT,19)
        WRITE (MUNIT,21)
        WRITE (MUNIT,27)    E
        WRITE (MUNIT,23)    RMAX,NPTS
        WRITE (MUNIT,25)    TOLR
        WRITE (MUNIT,29)    BMIN,BMAX,NB
        WRITE (MUNIT,30)
        WRITE (MUNIT,19)
C
C       different data headers for graphics and text output
        IF (MUNIT .NE. GUNIT) THEN
           WRITE (MUNIT,31)
           WRITE (MUNIT,32)
           WRITE (MUNIT,33)
        ELSE
           WRITE (MUNIT,35)
        END IF
C
        NLINES = 11
```

```
C
19    FORMAT  (' ')
21    FORMAT  (' Output from project 1:',
      +          '  Scattering by a central 6-12 potential ')
23    FORMAT  (' Rmax=',F6.3,5X,'number of quadrature points =',I4)
25    FORMAT  (' Turning point tolerance = ',E12.5)
27    FORMAT  (' Incident energy = 'F8.3)
29    FORMAT  (' Bmin = ',F6.3,5X,'Bmax = ',F6.3,
      +          5X,'number of B values = ',I3)
30    FORMAT  (' Lengths are in scaled units, angles in degrees')
31    FORMAT  (15X,'Impact',14X,'Closest',16X,'Scattering')
32    FORMAT  (13X,'Parameter',13X,'Approach'17X,'Angle')
33    FORMAT  (13X,'---------',13X,'--------'17X,'-----')
35    FORMAT  (2X,'Impact parameter',2X,'Scattering Angle')
C
      RETURN
      END
CCCCCCCCCCCCCCCCCCCCCCCCCCCCCCCCCCCCCCCCCCCCCCCCCCCCCCCCCCCCCCCCCCCCCCCCCC
      SUBROUTINE TXTOUT(MUNIT,NLINES,B,RMIN,ANGLE)
C writes results for one impact parameter to the requested unit
CCCCCCCCCCCCCCCCCCCCCCCCCCCCCCCCCCCCCCCCCCCCCCCCCCCCCCCCCCCCCCCCCCCCCCCCCC
C Global variables:
      INCLUDE 'IO.ALL'
C Passed variables:
      INTEGER MUNIT              !output unit specifier (input)
      INTEGER NLINES             !number of lines printed so far (I/O)
      REAL B                     !impact parameter (input)
      REAL RMIN                  !radius of closest approach (input)
      REAL ANGLE                 !scattering angle(input)
CCCCCCCCCCCCCCCCCCCCCCCCCCCCCCCCCCCCCCCCCCCCCCCCCCCCCCCCCCCCCCCCCCCCCCCCCC
C     if screen is full, clear screen and retype headings
      IF ((MOD(NLINES,TRMLIN-4) .EQ. 0)
      +          .AND. (MUNIT .EQ. OUNIT)) THEN
          CALL PAUSE('to continue...',1)
          CALL CLEAR
          WRITE (MUNIT,31)
          WRITE (MUNIT,32)
          WRITE (MUNIT,33)
          NLINES=3
      END IF
C
      WRITE (MUNIT,35) B,RMIN,ANGLE
C
C     keep track of printed lines only for terminal output
      IF (MUNIT .EQ. OUNIT) NLINES=NLINES+1
C
31    FORMAT (15X,'Impact',14X,'Closest',16X,'Scattering')
32    FORMAT (13X,'Parameter',13X,'Approach'17X,'Angle')
33    FORMAT (13X,'---------',13X,'--------'17X,'-----')
35    FORMAT (15X,F6.3,15X,F6.3,15X,F9.3)
C
      RETURN
```

```
        END
CCCCCCCCCCCCCCCCCCCCCCCCCCCCCCCCCCCCCCCCCCCCCCCCCCCCCCCCCCCCCCCCCCCCCCCC
        SUBROUTINE GRFOUT(DEVICE,ANGLE,IMPACT)
C graphs scattering angle vs. impact parameter
CCCCCCCCCCCCCCCCCCCCCCCCCCCCCCCCCCCCCCCCCCCCCCCCCCCCCCCCCCCCCCCCCCCCCCCC
C Global variables
        INCLUDE 'IO.ALL'
        INCLUDE 'PARAM.P1'
        INCLUDE 'GRFDAT.ALL'
C Input variables:
        INTEGER DEVICE                  !which device is being used?
        REAL ANGLE(0:MAX)               !final scattering angle
        REAL IMPACT(0:MAX)              !impact parameter
C Local variables
        INTEGER IB                      !index on impact parameter
        CHARACTER*9 CE,CR               !energy, RMAX as a character string
        INTEGER LEN,RLEN                !length of string
        INTEGER SCREEN                  !send to terminal
        INTEGER PAPER                   !make a hardcopy
        INTEGER FILE                    !send to a file
        DATA SCREEN,PAPER,FILE/1,2,3/
CCCCCCCCCCCCCCCCCCCCCCCCCCCCCCCCCCCCCCCCCCCCCCCCCCCCCCCCCCCCCCCCCCCCCCCC
C       messages for the impatient
        IF (DEVICE .NE. SCREEN) WRITE (OUNIT,100)
C
C       calculate parameters for graphing
        IF (DEVICE .NE. FILE) THEN
C
            NPLOT=1                             !how many plots?
            IPLOT=1
C
            YMAX=0.                             !find limits on data points
            YMIN=0.
            DO 20 IB=0,NB-1
                IF (ANGLE(IB) .GT. YMAX) YMAX=ANGLE(IB)
                IF (ANGLE(IB) .LT. YMIN) YMIN=ANGLE(IB)
20          CONTINUE
            XMIN=BMIN
            XMAX=BMAX
            Y0VAL=BMIN
            X0VAL=0.
C
            NPOINT=NB
C
            ILINE=1                             !line and symbol styles
            ISYM=1
            IFREQ=1
            NXTICK=5
            NYTICK=5
C
            CALL CONVRT(E,CE,LEN)               !titles and labels
            CALL CONVRT(RMAX,CR,RLEN)
```

```
      INFO='RMAX = '//CR(1:RLEN)
      TITLE = 'Scattering from 6-12 potential, Energy='//CE
      LABEL(1)= 'Impact parameter (scaled units)'
      LABEL(2)= 'Scattering angle (degrees)'
C
      CALL GTDEV(DEVICE)                    !device nomination
      IF (DEVICE .EQ. SCREEN) CALL GMODE    !change to graphics mode
      CALL LNLNAX                           !draw axes
      END IF
C
C     output results
      IF (DEVICE .EQ. FILE) THEN
         WRITE (GUNIT,70) (IMPACT(IB),ANGLE(IB),IB=1,NB)
      ELSE
         CALL XYPLOT (IMPACT,ANGLE)
      END IF
C
C     end graphing session
      IF (DEVICE .NE. FILE) CALL GPAGE(DEVICE)  !close graphing package
      IF (DEVICE .EQ. SCREEN) CALL TMODE        !switch to text mode
C
70    FORMAT (2(5X,E11.3))
100   FORMAT (/,'Patience, please; output going to a file.')
C
      RETURN
      END
CCCCCCCCCCCCCCCCCCCCCCCCCCCCCCCCCCCCCCCCCCCCCCCCCCCCCCCCCCCCCCCCCCCCCCCC
CCCCCCCCCCCCCCCCCCCCCCCCCCCCCCCCCCCCCCCCCCCCCCCCCCCCCCCCCCCCCCCCCCCCCCCC
C param.p1
C
      REAL PI          ! pi=3.14159........
      REAL E           ! incident energy in units of Vzero
      REAL BMIN        ! minimum impact parameter
      REAL BMAX        ! maximum impact parameter
      REAL RMAX        ! radius at which V<<E
      REAL TOLR        ! tolerance for turning point search
      INTEGER NB       ! number of steps in B
      INTEGER NPTS     ! number of quadrature points
      REAL DB          ! size of steps in B
      INTEGER MAX      !max number of B values
C
      PARAMETER (MAX=100)
C
      COMMON / CONST / PI
      COMMON / PPARAM / E,BMIN,BMAX
      COMMON / NPARAM / NB,RMAX,TOLR,NPTS
      COMMON / PCALC / DB
CCCCCCCCCCCCCCCCCCCCCCCCCCCCCCCCCCCCCCCCCCCCCCCCCCCCCCCCCCCCCCCCCCCCCCCC
CCCCCCCCCCCCCCCCCCCCCCCCCCCCCCCCCCCCCCCCCCCCCCCCCCCCCCCCCCCCCCCCCCCCCCCC
C map.p1
      INTEGER IE,IBMIN,IBMAX,INB,IRMAX,INPTS,ITOLR
      PARAMETER (IE =13)
```

```
PARAMETER (IBMIN =14)
PARAMETER (IBMAX =15)
PARAMETER (INB =16)
PARAMETER (ITOLR =38)
PARAMETER (IRMAX =39)
PARAMETER (INPTS =40)
```

C.2 Project 2

Algorithm   This program constructs a series of white dwarf models for a specified electron fraction YE with central densities ranging in equal logarithmic steps between the values of RHO1 and RHO2 specified. For each model (DO loop 20 in subroutine ARCHON), the dimensionless equations (II.18) for the mass and density are integrated by the fourth order Runge-Kutta algorithm (subroutine RUNGE). An empirically scaled radial step DR is used for each model, and initial conditions for the integration are determined by a Taylor expansion of the differential equations about $r = 0$ (beginning of the same DO loop). Integration proceeds until the density has fallen below 10^3 gm cm^{-3}. In addition, the contributions to the internal and gravitational energies are calculated at each radial step (subroutine INTGRT). The radial derivatives of the mass and density are defined in function routine DERS; the derivative of pressure with respect to density is defined in function routine GAMMA; and the integrand for the kinetic energy is defined in function routine EPSRHO.

Input   The physical parameters include the electron fraction [1.], the central density for the first model [1.0E+05], the central density for the last model [1.0E+11], and the number of models to calculate [4.] The programs checks that the initial density is smaller than the final density. There are no numerical parameters: the step size is a fixed function of the central density.

Output   As each model is calculated, the radial step, central density, number of steps, total radius, total mass, kinetic energy, gravitational energy, and total energy for the model are displayed. Densities are in grams per cubic centimeter, energy is in ergs, distances in centimeters, and mass in grams. After all of the models are integrated, the graphics displays two plots. The first is the total mass of the models *vs.* the central density; the second is the density *vs.* radius for each model.

```
ccccccccccccccccccccccccccccccccccccccccccccccccccccccccccccccccccccccc
      PROGRAM PROJ2
C     Project 2: The structure of white dwarf stars
C     COMPUTATIONAL PHYSICS (FORTRAN VERSION)
C     by Steven E. Koonin and Dawn C. Meredith
C     Copyright 1989, Addison-Wesley Publishing Company
ccccccccccccccccccccccccccccccccccccccccccccccccccccccccccccccccccccccc
      CALL INIT            !display header screen, setup parameters
5     CONTINUE             !main loop; execute once for each set of param
      CALL PARAM           !get input parameters
      CALL ARCHON          !calculate density, mass and radius of stars
```

```
          GOTO 5
          END
CCCCCCCCCCCCCCCCCCCCCCCCCCCCCCCCCCCCCCCCCCCCCCCCCCCCCCCCCCCCCCCCCCCCCCCCCCCC
          SUBROUTINE ARCHON
C calculate the mass(r), density(r) for a range of central densities
CCCCCCCCCCCCCCCCCCCCCCCCCCCCCCCCCCCCCCCCCCCCCCCCCCCCCCCCCCCCCCCCCCCCCCCCCCCC
C Global variables:
          INCLUDE 'IO.ALL'
          INCLUDE 'PARAM.P2'
C Local variables:
          REAL MSTOR(MAXMOD)              !mass of star
          REAL RADIUS(MAXMOD)            !radius of star
          REAL STEP(MAXMOD)              !step size
          REAL RMIN(MAXMOD)              !starting radius
          INTEGER NSTEP(MAXMOD)          !number of radial steps
          REAL RHOSTR(MAXMOD,0:MAXSTP)   !density(r) of star
          INTEGER IMODEL                 !index of current model
          INTEGER NLINES                 !number of lines printed out
          REAL M,RHO                     !current mass and density
          REAL R                         !current radius
          REAL DR                        !radial step
          REAL EGRAV                     !gravitational energy of star
          REAL EKINT                     !kinetic energy of star
          REAL EXPONT                    !temporary exponent
          REAL RHOCEN                    !central density for this model
          INTEGER IR                     !number of radial steps
          INTEGER DEVICE                 !current graphing device
          INTEGER SCREEN                 !send to terminal
          INTEGER PAPER                  !make a hardcopy
          INTEGER FILE                   !send to a file
C Functions:
          REAL GAMMA                     !dPressure/dDensity
          DATA SCREEN,PAPER,FILE/1,2,3/
CCCCCCCCCCCCCCCCCCCCCCCCCCCCCCCCCCCCCCCCCCCCCCCCCCCCCCCCCCCCCCCCCCCCCCCCCCCC
C      output summary of parameters
       IF (TTERM) CALL PRMOUT(OUNIT,NLINES)
       IF (TFILE) CALL PRMOUT(TUNIT,NLINES)
       IF (GFILE) CALL PRMOUT(GUNIT,NLINES)
C
       DO 20 IMODEL=1,NMODEL                      !loop over central dens
C
C         calculate central density
          EXPONT=FLOAT(IMODEL-1)/FLOAT(NMODEL-1)  !spacing between dens
          RHOCEN=(RHO1/RHO0)*((RHO2/RHO1)**EXPONT) !equal on log scale
C
          DR=((3.0*0.001/RHOCEN)**(1.0/3.0))/3.0  !radial step
          R=DR/10.                                !begin at finite radius
          RMIN(IMODEL)=R*R0/10.E+08               !starting radius
          STEP(IMODEL)=DR*R0/10.E+08              !step size
C
C         use Taylor expansion to obtain initial conditions
          M=(R**3.0)*(RHOCEN/3.0)                 !initial mass
```

```
C       initial density
        RHO=RHOCEN*(1-RHOCEN*R**2.0/6.0/GAMMA(RHOCEN**(1./3.)))
        RHOSTR(IMODEL,0)=RHO*RHO0
C
C       integrate equations and find energies, mass, and radius of star
        CALL INTGRT(M,RHO,R,DR,EGRAV,EKINT,IMODEL,IR,RHOSTR)
C
        MSTOR(IMODEL)=M*M0/10.E+33           !save values for graph
        RADIUS(IMODEL)=R*R0/10.E+08
        NSTEP(IMODEL)=IR
C
        IF (TTERM)                           !text output
     +      CALL TXTOUT(OUNIT,NLINES,R,M,DR,RHOCEN,EKINT,EGRAV,IR)
        IF (TFILE)
     +      CALL TXTOUT(TUNIT,NLINES,R,M,DR,RHOCEN,EKINT,EGRAV,IR)
C
20      CONTINUE
C
        IF (TTERM) CALL PAUSE('to continue...',1)
        IF (TTERM) CALL CLEAR
C
C       graphics output
        IF (GTERM) CALL GRFOUT(SCREEN,MSTOR,RADIUS,STEP,RMIN,NSTEP,RHOSTR)
        IF (GFILE) CALL GRFOUT(FILE,MSTOR,RADIUS,STEP,RMIN,NSTEP,RHOSTR)
        IF (GHRDCP) CALL GRFOUT(PAPER,MSTOR,RADIUS,STEP,RMIN,NSTEP,RHOSTR)
C
        RETURN
        END
CCCCCCCCCCCCCCCCCCCCCCCCCCCCCCCCCCCCCCCCCCCCCCCCCCCCCCCCCCCCCCCCCCCCCCCCC
        SUBROUTINE   INTGRT(M,RHO,R,DR,EGRAV,EKINT,IMODEL,IR,RHOSTR)
C integrates mass and density beginning with M and RHO at radius R;
C calculates gravitational and kinetic energies
CCCCCCCCCCCCCCCCCCCCCCCCCCCCCCCCCCCCCCCCCCCCCCCCCCCCCCCCCCCCCCCCCCCCCCCCC
C Global variables:
        INCLUDE 'PARAM.P2'
C Passed variables:
        REAL RHOSTR(MAXMOD,0:MAXSTP) !density(r) of star (output)
        INTEGER IMODEL              !index of current model (input)
        REAL M,RHO                  !current mass and density (I/O)
        REAL R                      !current radius (I/O)
        REAL DR                     !radial step (input)
        REAL EGRAV                  !gravitational energy of star (output)
        REAL EKINT                  !kinetic energy of star (output)
        INTEGER IR                  !current lattice point (output)
C Function:
        REAL EPSRHO                 !function in energy density
CCCCCCCCCCCCCCCCCCCCCCCCCCCCCCCCCCCCCCCCCCCCCCCCCCCCCCCCCCCCCCCCCCCCCCCCC
        EGRAV=0.0                   !zero sums
        EKINT=0.0
        IR=0
10      IF ((IR .LT. MAXSTP) .AND. (RHO .GT. RHOCUT)) THEN
          IR=IR+1                   !update radial index
```

```
            CALL RUNGE (M, RHO, R, DR)              !take a Runge-Kutta step
            IF (RHO .LT. RHOCUT) RHO=RHOCUT        !avoid small densities
            RHOSTR(IMODEL,IR)=RHO*RHO0             !save values for graphing
            EGRAV=EGRAV+M*RHO*R                     !contribution to energy integ.
            EKINT=EKINT+R**2*EPSRHO(RHO**(1./3.))
         GOTO 10
         END IF
         EGRAV=-EGRAV*DR*E0                         !unscaled energies
         EKINT=EKINT*DR*E0
         RETURN
         END
CCCCCCCCCCCCCCCCCCCCCCCCCCCCCCCCCCCCCCCCCCCCCCCCCCCCCCCCCCCCCCCCCCCCCCCCCCC
         SUBROUTINE RUNGE (M, RHO, R, DR)
C take a Runge-Kutta step to integrate mass and density
CCCCCCCCCCCCCCCCCCCCCCCCCCCCCCCCCCCCCCCCCCCCCCCCCCCCCCCCCCCCCCCCCCCCCCCCCCC
C Passed variables:
         REAL M, RHO                                !current mass and density (I/O)
         REAL R                                     !current radius (input)
         REAL DR                                    !radial step (input)
C Local variables:
         REAL DMDR, DRHODR                          !radial derivatives of mass and dens
         REAL K1M, K1RHO                            !intermediate increments of mass and
         REAL K2M, K2RHO                            !                        density
         REAL K3M, K3RHO
         REAL K4M, K4RHO
CCCCCCCCCCCCCCCCCCCCCCCCCCCCCCCCCCCCCCCCCCCCCCCCCCCCCCCCCCCCCCCCCCCCCCCCCCC
         CALL DERS (M, RHO, R, DMDR, DRHODR)  !calculate k1's
         K1M=DR*DMDR
         K1RHO=DR*DRHODR
C
         R=R+DR/2.0                                 !calculate k2's
         M=M+.5*K1M
         RHO=RHO+.5*K1RHO
         CALL DERS (M, RHO, R, DMDR, DRHODR)
         K2M=DR*DMDR
         K2RHO=DR*DRHODR
C
         M=M+.5* (K2M-K1M)                          !calculate k3's
         RHO=RHO+.5* (K2RHO-K1RHO)
         CALL DERS (M, RHO, R, DMDR, DRHODR)
         K3M=DR*DMDR
         K3RHO=DR*DRHODR
C
         R=R+DR/2.0                                 !calculate k4's
         M=M+K3M-.5*K2M
         RHO=RHO+K3RHO-.5*K2RHO
         CALL DERS (M, RHO, R, DMDR, DRHODR)
         K4M=DR*DMDR
         K4RHO=DR*DRHODR
C
C        values of mass and density at new radius
         M=M-K3M+ (K1M+2.0*K2M+2.0*K3M+K4M) /6.0
```

```
       RHO=RHO-K3RHO+(K1RHO+2.0*K2RHO+2.0*K3RHO+K4RHO)/6.
C
       RETURN
       END
CCCCCCCCCCCCCCCCCCCCCCCCCCCCCCCCCCCCCCCCCCCCCCCCCCCCCCCCCCCCCCCCCCCCCCCC
       SUBROUTINE DERS(M,RHO,R,DMDR,DRHODR)
C calculate the derivatives of mass and density with respect to radius
CCCCCCCCCCCCCCCCCCCCCCCCCCCCCCCCCCCCCCCCCCCCCCCCCCCCCCCCCCCCCCCCCCCCCCCC
C Passed variables:
       REAL M,RHO             !current mass and density (input)
       REAL R                 !current radius (input)
       REAL DMDR,DRHODR       !radial derivatives of mass and dens (output)
C Functions
       REAL GAMMA             !dPressure/dDensity
CCCCCCCCCCCCCCCCCCCCCCCCCCCCCCCCCCCCCCCCCCCCCCCCCCCCCCCCCCCCCCCCCCCCCCCC
       IF (RHO .GT. 0.0) THEN    !avoid division by zero
          DRHODR=-M*RHO/((R**2)*GAMMA(RHO**(1.0/3.0)))
          DMDR=(R**2)*RHO
       END IF
       RETURN
       END
CCCCCCCCCCCCCCCCCCCCCCCCCCCCCCCCCCCCCCCCCCCCCCCCCCCCCCCCCCCCCCCCCCCCCCCC
       REAL FUNCTION GAMMA(X)
C derivative of pressure with respect to density
CCCCCCCCCCCCCCCCCCCCCCCCCCCCCCCCCCCCCCCCCCCCCCCCCCCCCCCCCCCCCCCCCCCCCCCC
C Passed variables:
       REAL X                 !cubed root of density
       GAMMA=X*X/3.0/SQRT(X*X+1)
       END
CCCCCCCCCCCCCCCCCCCCCCCCCCCCCCCCCCCCCCCCCCCCCCCCCCCCCCCCCCCCCCCCCCCCCCCC
       REAL FUNCTION EPSRHO(X)
C function needed in the calculation of the energy
CCCCCCCCCCCCCCCCCCCCCCCCCCCCCCCCCCCCCCCCCCCCCCCCCCCCCCCCCCCCCCCCCCCCCCCC
C Passed variables:
       REAL X                 !cubed root of density
C Local variables:
       REAL PART1,PART2            !terms in the function
CCCCCCCCCCCCCCCCCCCCCCCCCCCCCCCCCCCCCCCCCCCCCCCCCCCCCCCCCCCCCCCCCCCCCCCC
       PART1=X*(1.0+2.0*X**2)*SQRT(1.0+X**2)
       PART2=LOG(X+SQRT(1.0+X**2))
       EPSRHO=0.375*(PART1-PART2)
       END
CCCCCCCCCCCCCCCCCCCCCCCCCCCCCCCCCCCCCCCCCCCCCCCCCCCCCCCCCCCCCCCCCCCCCCCC
       SUBROUTINE INIT
C initializes constants, displays header screen,
C initializes arrays for input parameters
CCCCCCCCCCCCCCCCCCCCCCCCCCCCCCCCCCCCCCCCCCCCCCCCCCCCCCCCCCCCCCCCCCCCCCCC
C Global variables:
       INCLUDE 'IO.ALL'
       INCLUDE 'MENU.ALL'
       INCLUDE 'PARAM.P2'
C Local parameters:
```

```
        CHARACTER*80 DESCRP          !program description
        DIMENSION DESCRP(22)
        INTEGER NHEAD,NTEXT,NGRF     !number of lines for each description
CCCCCCCCCCCCCCCCCCCCCCCCCCCCCCCCCCCCCCCCCCCCCCCCCCCCCCCCCCCCCCCCCCCCCCCC
        CALL SETUP                   !get environment parameters
C
C       display header screen
        DESCRP(1)= 'PROJECT 2'
        DESCRP(2)= 'The structure of white dwarf stars'
        NHEAD=2
C
C       text output description
        DESCRP(3)= 'mass, radius, and energy of each star'
        NTEXT=1
C
C       graphics output description
        DESCRP(4)= 'total mass vs. final radius for all models'
        DESCRP(5)= 'density vs. radius for each model'
        NGRF=2
C
        CALL HEADER(DESCRP,NHEAD,NTEXT,NGRF)
C
C       calculate constants
        MEORMP = 1./1836.
C
C       setup menu
        CALL MENU                            !setup constant part of menu
C
        MPRMPT(4)='2) (not used)'
C
        MTYPE(13)=FLOAT
        MPRMPT(13)= 'Enter electron fraction, YE'
        MTAG(13)= 'Electron fraction, YE'
        MLOLIM(13)=.001
        MHILIM(13)=10.
        MREALS(13)=1.0
C
        MTYPE(14)=FLOAT
        MPRMPT(14)= 'Enter central density (gm/cm^3) for the first model'
        MTAG(14)= 'Initial central density (gm/cm^3)'
        MLOLIM(14)=1.E5
        MHILIM(14)=1.E15
        MREALS(14)=1.E5
C
        MTYPE(15)=FLOAT
        MPRMPT(15)= 'Enter central density (gm/cm^3) for the last model'
        MTAG(15)= 'Final central density (gm/cm^3)'
        MLOLIM(15)=1.E5
        MHILIM(15)=1.E15
        MREALS(15)=1.E11
C
        MTYPE(16)=NUM
```

```
      MPRMPT(16)= 'Enter the number of models to calculate'
      MTAG(16)= 'Number of models to calculate'
      MLOLIM(16)=1.
      MHILIM(16)=MAXMOD
      MINTS(16)=4.
C
      MTYPE(17)=SKIP
      MREALS(17)=35.
C
      MTYPE(36)=SKIP
      MREALS(36)=60.
C
      MSTRNG(MINTS(75))= 'proj2.txt'
C
      MTYPE(76)=SKIP
      MREALS(76)=80.
C
      MSTRNG(MINTS(86))= 'proj2.grf'
C
      MTYPE(87)=SKIP
      MREALS(87)=90.
C
      RETURN
      END
CCCCCCCCCCCCCCCCCCCCCCCCCCCCCCCCCCCCCCCCCCCCCCCCCCCCCCCCCCCCCCCCCCCCCCC
      SUBROUTINE PARAM
C gets parameters from screen
C ends program on request
C closes old files
C maps menu variables to program variables
C opens new files
C calculates all derivative parameters
CCCCCCCCCCCCCCCCCCCCCCCCCCCCCCCCCCCCCCCCCCCCCCCCCCCCCCCCCCCCCCCCCCCCCCC
C Global variables:
      INCLUDE 'MENU.ALL'
      INCLUDE 'IO.ALL'
      INCLUDE 'PARAM.P2'
      INCLUDE 'MAP.P2'
C Function:
      LOGICAL LOGCVT            !converts 1 and 0 to true and false
CCCCCCCCCCCCCCCCCCCCCCCCCCCCCCCCCCCCCCCCCCCCCCCCCCCCCCCCCCCCCCCCCCCCCCC
C     get input from terminal
      CALL CLEAR
      CALL ASK(1,ISTOP)
C
C     stop program if requested
      IF (MREALS(IMAIN) .EQ. STOP) CALL FINISH
C
C     close files if necessary
      IF (TNAME .NE. MSTRNG(MINTS(ITNAME)))
     +     CALL FLCLOS(TNAME,TUNIT)
      IF (GNAME .NE. MSTRNG(MINTS(IGNAME)))
```

```
      +     CALL FLCLOS(GNAME,GUNIT)
C
C     set new parameter values
C     physical and numerical:
      YE=MREALS(IYE)
      RHO1=MREALS(IRHO1)
      RHO2=MREALS(IRHO2)
      NMODEL=MINTS(INMODL)
C
C     text output
      TTERM=LOGCVT(MINTS(ITTERM))
      TFILE=LOGCVT(MINTS(ITFILE))
      TNAME=MSTRNG(MINTS(ITNAME))
C
C     graphics output
      GTERM=LOGCVT(MINTS(IGTERM))
      GHRDCP=LOGCVT(MINTS(IGHRD))
      GFILE=LOGCVT(MINTS(IGFILE))
      GNAME=MSTRNG(MINTS(IGNAME))
C
C     open files
      IF (TFILE) CALL FLOPEN(TNAME,TUNIT)
      IF (GFILE) CALL FLOPEN(GNAME,GUNIT)
C     files may have been renamed
      MSTRNG(MINTS(ITNAME))=TNAME
      MSTRNG(MINTS(IGNAME))=GNAME
C
      CALL PCHECK
      CALL CLEAR
C
C     calculate parameter dependent quantities
      R0=7.72E+08*YE              !scaling radius
      M0=5.67E+33*(YE**2)         !scaling mass
      RHO0=979000./YE             !scaling density
      E0=YE*MEORMP*9*(M0/1E+31)   !scaling energy
      RHOCUT=1000./RHO0           !nearly zero density
C
      RETURN
      END
CCCCCCCCCCCCCCCCCCCCCCCCCCCCCCCCCCCCCCCCCCCCCCCCCCCCCCCCCCCCCCCCCCCCCCCCC
      SUBROUTINE PCHECK
C ensure that RHO1 < RHO2
CCCCCCCCCCCCCCCCCCCCCCCCCCCCCCCCCCCCCCCCCCCCCCCCCCCCCCCCCCCCCCCCCCCCCCCCC
C Global parameters:
      INCLUDE 'PARAM.P2'
      INCLUDE 'MENU.ALL'
      INCLUDE 'IO.ALL'
      INCLUDE 'MAP.P2'
CCCCCCCCCCCCCCCCCCCCCCCCCCCCCCCCCCCCCCCCCCCCCCCCCCCCCCCCCCCCCCCCCCCCCCCCC
10    IF (RHO1 .GE. RHO2) THEN
          WRITE (OUNIT,15)
          CALL ASK(IRHO1,IRHO2)
```

```
       RHO1=MREALS(IRHO1)
       RHO2=MREALS(IRHO2)
     GOTO 10
     END IF
15   FORMAT
   + (' The initial density must be smaller than the final density')
     RETURN
     END
CCCCCCCCCCCCCCCCCCCCCCCCCCCCCCCCCCCCCCCCCCCCCCCCCCCCCCCCCCCCCCCCCCCCCCCCCCCC
     SUBROUTINE PRMOUT(MUNIT,NLINES)
C outputs parameters to the specified unit
CCCCCCCCCCCCCCCCCCCCCCCCCCCCCCCCCCCCCCCCCCCCCCCCCCCCCCCCCCCCCCCCCCCCCCCCCCCC
C Global variables
     INCLUDE 'PARAM.P2'
     INCLUDE 'IO.ALL'
C Passed variables
     INTEGER MUNIT              !unit number for output (input)
     INTEGER NLINES             !number of lines written so far (I/O)
CCCCCCCCCCCCCCCCCCCCCCCCCCCCCCCCCCCCCCCCCCCCCCCCCCCCCCCCCCCCCCCCCCCCCCCCCCCC
     IF (MUNIT .EQ. OUNIT) CALL CLEAR
C
     WRITE (MUNIT,19)
     WRITE (MUNIT,21)
     WRITE (MUNIT,25)  NMODEL
     WRITE (MUNIT,27)  YE
     WRITE (MUNIT,29)  RHO1,RHO2
     WRITE (MUNIT,19)
C
     IF (MUNIT .NE. GUNIT) THEN      !headers for text output only
        WRITE (MUNIT,31)
        WRITE (MUNIT,32)
        WRITE (MUNIT,33)
     END IF
C
     NLINES = 9
C
19   FORMAT (' ')
21   FORMAT (' Output from project 2: ',
   +         ' The structure of white dwarf stars ')
25   FORMAT (' Number of models = ',I3)
27   FORMAT (' Electron fraction YE= ',1PE9.3)
29   FORMAT (' First central density=',1PE9.3,
   +         5X,' Final central density=',1PE9.3)
31   FORMAT (3X,'Step',4X,'Cent Den',1X,'Nstep',3X,'Radius',
   +         5X,'Mass', 7X,'Kin E',6X,'Grav E',7X,'Tot E')
32   FORMAT (3X,' cm ',4X,'gm/cm^-3',1X,'      ',3X,'  cm  ',
   +         5X,' gm ', 7X,9X,'10^51 ergs ',4X,'      ')
33   FORMAT (3X,'----',4X,'--------',1X,'-----',3X,'------',
   +         5X,'----', 7X,'-----',6X,'------',7X,'-----')
C
     RETURN
     END
```

```
ccccccccccccccccccccccccccccccccccccccccccccccccccccccccccccccccccccc
      SUBROUTINE TXTOUT(MUNIT,NLINES,R,M,DR,RHOCEN,EKINT,EGRAV,IR)
C writes results for one model to the requested unit
ccccccccccccccccccccccccccccccccccccccccccccccccccccccccccccccccccccc
C Global variables:
      INCLUDE 'IO.ALL'
      INCLUDE 'PARAM.P2'
C Input variables:
      INTEGER MUNIT              !unit to which we are writing
      INTEGER NLINES             !number of lines printed out (I/O)
      REAL M,RHO                 !current mass and density
      REAL RHOCEN                !central density
      REAL R                     !current radius
      REAL DR                    !radial step
      REAL EGRAV                 !gravitational energy of star
      REAL EKINT                 !kinetic energy of star
      INTEGER IR                 !number of radial steps
ccccccccccccccccccccccccccccccccccccccccccccccccccccccccccccccccccccc
C     if screen is full, clear screen and retype headings
      IF ((MOD(NLINES,TRMLIN-4) .EQ. 0)
     +                    .AND. (MUNIT .EQ. OUNIT)) THEN
         CALL PAUSE('to continue',1)
         CALL CLEAR
         WRITE (MUNIT,31)
         WRITE (MUNIT,32)
         WRITE (MUNIT,33)
      END IF
C
C     write unscaled values to screen
      WRITE (MUNIT,35) DR*R0,RHOCEN*RHO0,IR,
     +                    R*R0,M*M0,EKINT,EGRAV,EKINT+EGRAV
C
C     keep track of printed lines only for terminal output
      IF (MUNIT .EQ. OUNIT) NLINES=NLINES+1
C
31    FORMAT (3X,'Step',4X,'Cent Den',1X,'Nstep',3X,'Radius',
     +        5X,'Mass', 7X,'Kin E',6X,'Grav E',7X,'Tot E')
32    FORMAT (3X,' cm ',4X,'gm/cm^-3',1X,'     ',3X,' cm ',
     +        5X,' gm ', 7X,9X,'10^51 ergs ',4X,'     ')
33    FORMAT (3X,'----',4X,'--------',1X,'-----',3X,'------',
     +        5X,'----', 7X,'-----',6X,'------',7X,'-----')
35    FORMAT (2(1X,1PE8.2),1X,I4,3(2X,1PE9.3),2(2X,1PE10.3))
C
      RETURN
      END
ccccccccccccccccccccccccccccccccccccccccccccccccccccccccccccccccccccc
      SUBROUTINE GRFOUT(DEVICE,MSTOR,RADIUS,STEP,RMIN,NSTEP,RHOSTR)
C outputs two plots:  1) total mass vs. final radius of star
C                     2) density vs. radius
ccccccccccccccccccccccccccccccccccccccccccccccccccccccccccccccccccccc
C Global variables
      INCLUDE 'IO.ALL'
```

```
      INCLUDE 'PARAM.P2'
      INCLUDE 'GRFDAT.ALL'
C Input variables:
      INTEGER DEVICE                    !which device is being used?
      REAL MSTOR(MAXMOD)                !mass of stars
      REAL RADIUS(MAXMOD)               !radius of stars
      REAL STEP(MAXMOD)                 !step size
      REAL RMIN(MAXMOD)                 !starting radius
      INTEGER NSTEP(MAXMOD)             !number of radial steps
      REAL RHOSTR(MAXMOD,0:MAXSTP)      !density(r) of star
C Local variables
      REAL R,DEN                        !radius and density for one model
      INTEGER IM,IR                     !current model and radius
      DIMENSION R(0:MAXSTP),DEN(0:MAXSTP) !radius and density
      CHARACTER*9 CYE                   !electron fraction as charac string
      INTEGER LEN                       !length of string
      INTEGER SCREEN                    !send to terminal
      INTEGER PAPER                     !make a hardcopy
      INTEGER FILE                      !send to a file
      DATA SCREEN,PAPER,FILE/1,2,3/
CCCCCCCCCCCCCCCCCCCCCCCCCCCCCCCCCCCCCCCCCCCCCCCCCCCCCCCCCCCCCCCCCCCCCCCCCCC
C     messages for the impatient
      IF (DEVICE .NE. SCREEN) WRITE (OUNIT,100)
C
C     define parameters which are the same for both plots
      IF (DEVICE .NE. FILE) THEN
          NPLOT=2                       !how many plots?
C
          XMIN=0.
          YOVAL=0.
C
          ILINE=1                       !line and symbol styles
          ISYM=1
          IFREQ=1
          NXTICK=5
          NYTICK=5
C
          CALL CONVRT(YE,CYE,LEN)
          TITLE = 'White Dwarf Stars with electron fraction='//CYE
C
          CALL GTDEV(DEVICE)                     !device nomination
          IF (DEVICE .EQ. SCREEN) CALL GMODE     !change to graphics mode
      END IF
C
      DO 20 IPLOT=1,NPLOT
C        define parameters which are different for each plot
         IF (DEVICE .NE. FILE) THEN
            IF (IPLOT .EQ. 1) THEN
                YMIN=MSTOR(1)
                XOVAL=MSTOR(1)
                XMAX=RADIUS(1)
                YMAX=MSTOR(NMODEL)
```

```
                    NPOINT=NMODEL
                    LABEL(1)= 'radius (10**8 cm)'
                    LABEL(2)= 'mass (10**33 gm)'
                    CALL LNLNAX
                ELSE
                    XMAX=RADIUS(1)
                    YMAX=RHO2
                    YMIN=RHOCUT*RHO0
                    X0VAL=RHOCUT*RHO0
                    IFREQ=0
                    LABEL(1)= 'radius (10**8 cm)'
                    LABEL(2)= 'density (gm*cm**(-3))'
                    CALL LGLNAX
                END IF
            END IF
C
C       output results
        IF (IPLOT .EQ. 1) THEN
C           total mass vs. final radius
            IF (DEVICE .EQ. FILE) THEN
                WRITE (GUNIT,80)
                WRITE (GUNIT,70) (RADIUS(IM),MSTOR(IM),IM=1,NMODEL)
            ELSE
                CALL XYPLOT (RADIUS,MSTOR)
            END IF
C
        ELSE
C           density(r) vs. r for each model
            DO 30 IM=1,NMODEL
                NPOINT=NSTEP(IM)+1
                DO 40 IR=0,NPOINT-1
                  R(IR)=RMIN(IM)+IR*STEP(IM)
                  DEN(IR)=RHOSTR(IM,IR)
40              CONTINUE
                IF (DEVICE .EQ. FILE) THEN
                    WRITE (GUNIT,85) IM
                    WRITE (GUNIT,70) (R(IR),DEN(IR),IR=0,NPOINT)
                ELSE
                    CALL XYPLOT(R,DEN)
                END IF
30          CONTINUE
        END IF
20      CONTINUE
C
        IPLOT=NPLOT
C       end graphing session
        IF (DEVICE .NE. FILE) CALL GPAGE(DEVICE)   !end graphing package
        IF (DEVICE .EQ. SCREEN) CALL TMODE         !switch to text mode
C
70      FORMAT (2(5X,E11.3))
80      FORMAT (/,10X,'radius',10X,'mass')
85      FORMAT (/,' Density vs. radius for model ',I2)
```

```
100    FORMAT (/,' Patience, please; output going to a file.')
C
       RETURN
       END
CCCCCCCCCCCCCCCCCCCCCCCCCCCCCCCCCCCCCCCCCCCCCCCCCCCCCCCCCCCCCCCCCCCCCC
CCCCCCCCCCCCCCCCCCCCCCCCCCCCCCCCCCCCCCCCCCCCCCCCCCCCCCCCCCCCCCCCCCCCCC
C param.p2
C
       REAL MEORMP             !mass of electron over mass of proton
       REAL YE                 !electron fraction
       REAL RHO1               !initial density
       REAL RHO2               !final density
       INTEGER NMODEL          !number of models to calculate
       REAL R0                 !scaling radius
       REAL M0                 !scaling mass
       REAL RHO0               !scaling density
       REAL E0                 !scaling energy
       REAL RHOCUT             !nearly zero density
       INTEGER MAXSTP           !array size for density
       INTEGER MAXMOD           !max number of models
C
       PARAMETER (MAXSTP=1000)
       PARAMETER (MAXMOD=25)
C
       COMMON / CONST /MEORMP
       COMMON / PPARAM / YE,RHO1,RHO2,NMODEL
       COMMON / PCALC / R0,M0,RHO0,E0,RHOCUT
CCCCCCCCCCCCCCCCCCCCCCCCCCCCCCCCCCCCCCCCCCCCCCCCCCCCCCCCCCCCCCCCCCCCCC
CCCCCCCCCCCCCCCCCCCCCCCCCCCCCCCCCCCCCCCCCCCCCCCCCCCCCCCCCCCCCCCCCCCCCC
C map.p2
       INTEGER IYE,IRHO1,IRHO2,INMODL
       PARAMETER (IYE =13)
       PARAMETER (IRHO1 =14)
       PARAMETER (IRHO2 =15)
       PARAMETER (INMODL =16)
```

C.3 Project 3

Algorithm   This program solves the Hartree-Fock equations in the filling approximation for atomic systems with electrons in the $1s$, $2s$, and $2p$ shells. To begin the iterative procedure, normalized hydrogenic wave functions (subroutine HYDRGN) and their energies (subroutine ENERGY) are calculated, the virial theorem is used to determine the variational parameter ZSTAR, which is in turn used to calculate the optimal scaled hydrogenic wave functions and their energies. These wave functions serve as our initial "guess".

Then for each iteration requested the single particle eigenvalue equations (III.28) are solved as inhomogeneous boundary value problems (subroutine SNGWFN) using the single particle energies, electrostatic potential, and Fock terms calculated from the previous wave functions. The homogeneous solutions are found using Numerov's method, and these solutions are then used to calculate the inhomogeneous solutions using the Green's function method. Care is taken to ensure that the $2s$ orbital is orthogonal to that of the $1s$. When the requested iterations are completed, you can request more or allow the calculation to terminate.

Much of the computational effort is in the calculation of the single particle energies (III.30) and total energy (III.26,27) for each set of single particle wave functions (subroutine ENERGY). In subroutine SOURCE the total electron density is calculated as a weighted average of the current and previous densities, and the Fock terms (III.28 b,c) are evaluated using the trapezoidal rule for the integrals and the three-j symbols from subroutine SQR3J. The electrostatic potential due to the electrons is calculated in subroutine POISSN where the Numerov method is used to integrate Poisson's equation (III.18), and the unwanted linear behavior is subtracted off. The remaining energy integrals are calculated in subroutine ENERGY using the trapezoidal rule.

Input   The physical parameters are the nuclear charge [6], and the occupations of the $1s$ [2], $2s$ [2], and $2p$ [2] states. The numerical parameters are the radial step size (Angstroms) [.01], the outer radius (Angstroms) [2], and the initial number of iterations. The program checks that the total number of radial steps does not exceed MAXSTP= 1000; if it does, you are prompted for a larger radial step.

Output   After each iteration all of the single particle energies (kinetic, electron-nucleus, electron-electron, exchange, total potential, and total) are displayed for each occupied state, as well as the total for all states. All energies are in electron volts. When iterations are completed, the graphics displays the partial and total electron densities as a function of radius.

```
cccccccccccccccccccccccccccccccccccccccccccccccccccccccccccccccccccc
      PROGRAM PROJ3
C     Project 3:  Hartree-Fock solutions of small atomic systems in the
C                 filling approximation
C  COMPUTATIONAL PHYSICS (FORTRAN VERSION)
C  by Steven E. Koonin and Dawn C. Meredith
C  Copyright 1989, Addison-Wesley Publishing Company
cccccccccccccccccccccccccccccccccccccccccccccccccccccccccccccccccccc
      CALL INIT          !display header screen, setup parameters
5     CONTINUE           !main loop/ execute once for each set of param
      CALL PARAM         !get input from screen
      CALL ARCHON        !find the Hartree-Fock wave functions
      GOTO 5
      END
cccccccccccccccccccccccccccccccccccccccccccccccccccccccccccccccccccc
      SUBROUTINE ARCHON
C find the Hartree-Fock wave functions for the specified atom
cccccccccccccccccccccccccccccccccccccccccccccccccccccccccccccccccccc
C Global variables:
      INCLUDE 'IO.ALL'
      INCLUDE 'PARAM.P3'
C Local variables:
      REAL E(MAXSTT+1,8)             !all energies of all states
      REAL FOCK(0:MAXSTP,MAXSTT)     !Fock terms
      REAL RHO(0:MAXSTP)             !density
      REAL PSTOR(0:MAXSTP,MAXSTT)    !radial wave function
      REAL PHI(0:MAXSTP)             !electron potential
      REAL ESP                       !single particle energy of state
      INTEGER ITER                   !iteration index
      INTEGER STATE                  !single particle state index
      REAL ZSTAR                     !optimal effective nuclear charge
      INTEGER DEVICE                 !current graphing device
      INTEGER ISTOP,ISTART           !current limits on iteration
      INTEGER NLINES                 !number of lines written to screen
      INTEGER SCREEN                 !send to terminal
      INTEGER PAPER                  !make a hardcopy
      INTEGER FILE                   !send to a file
C Functions
      INTEGER GETINT                 !integer screen input
      DATA SCREEN,PAPER,FILE/1,2,3/
cccccccccccccccccccccccccccccccccccccccccccccccccccccccccccccccccccc
C     begin iterations with a good guess
      MIX=1.                         !no old density to mix with new
      ZSTAR=Z
      CALL HYDRGN(ZSTAR,PSTOR)       !find hydrogenic wave functions
      CALL ENERGY(E,FOCK,RHO,PHI,PSTOR)    !and energy
C     optimal ZSTAR using Virial theorem
      ZSTAR=-Z*(E(NSTATE+1,IVTOT)/(2*E(NSTATE+1,IKTOT)) )
      CALL HYDRGN(ZSTAR,PSTOR)       !find new hydrogenic wave functions
```

```
            CALL ENERGY(E,FOCK,RHO,PHI,PSTOR)  ! and energies
      C

            !output summary of parameters
            IF (TTERM) CALL PRMOUT(OUNIT,ZSTAR,NLINES)
            IF (TFILE) CALL PRMOUT(TUNIT,ZSTAR,NLINES)
            IF (GFILE) CALL PRMOUT(GUNIT,ZSTAR,NLINES)
      C
            MIX=.5                          !mix old and new density for stability
            ITER=0                          !zero iteration counters
            ISTART=0
            ISTOP=0
      C
      C     output initial energies
            IF (TTERM) CALL TXTOUT(OUNIT,ITER,E,NLINES)
            IF (TFILE) CALL TXTOUT(TUNIT,ITER,E,NLINES)
      C
      10    CONTINUE                        !loop over iterations
              ISTART=ISTOP+1                !update iteration counters
              ISTOP=ISTOP+NITER
              DO 100 ITER=ISTART,ISTOP
                DO 50 STATE=1,NSTATE
                  IF (NOCC(STATE) .NE. 0) THEN  !loop over all occpd states
                    ESP=E(STATE,ITOT)           !single particle energy
                    IF (ESP .GT. 0) ESP=-10     !keep particle bound
      C             find single part wave function
                    CALL SNGWFN(STATE,ESP,PSTOR,FOCK,PHI)
                  END IF
      50        CONTINUE
      C         calculate new energies and output
                CALL ENERGY(E,FOCK,RHO,PHI,PSTOR)
                IF (TTERM) CALL TXTOUT(OUNIT,ITER,E,NLINES)
                IF (TFILE) CALL TXTOUT(TUNIT,ITER,E,NLINES)
      100     CONTINUE
      C
      C     allow for more iterations
            NITER=GETINT(0,0,20,'How many more iterations?')
            IF (NITER .NE. 0 ) GOTO 10
      C
            IF (TTERM) CALL CLEAR
      C
      C     graphics output
            IF (GTERM) CALL GRFOUT(SCREEN,PSTOR,RHO,E)
            IF (GFILE) CALL GRFOUT(FILE,PSTOR,RHO,E)
            IF (GHRDCP) CALL GRFOUT(PAPER,PSTOR,RHO,E)
      C
            RETURN
            END
      CCCCCCCCCCCCCCCCCCCCCCCCCCCCCCCCCCCCCCCCCCCCCCCCCCCCCCCCCCCCCCCCCCCCCCCCC
            SUBROUTINE HYDRGN(ZSTAR,PSTOR)
      C creates hydrogenic wave functions PSTOR with nuclear charge=ZSTAR
      CCCCCCCCCCCCCCCCCCCCCCCCCCCCCCCCCCCCCCCCCCCCCCCCCCCCCCCCCCCCCCCCCCCCCCCCC
      C Global variables:
```

```
      INCLUDE 'PARAM.P3'
C Input variables:
      REAL ZSTAR                              !effective nuclear charge
      REAL PSTOR(0:MAXSTP,MAXSTT)             !radial wave function
C Local variables:
      INTEGER IR                              !radial index
      REAL RSTAR                              !scaled radius
      REAL ERSTAR                             !useful exponential
      INTEGER STATE                           !state index
      REAL NORM                               !norm of wave function
CCCCCCCCCCCCCCCCCCCCCCCCCCCCCCCCCCCCCCCCCCCCCCCCCCCCCCCCCCCCCCCCCCCCCCCCC
C     store radial parts of hydrogenic wave functions
      DO 20 IR=0,NR
         RSTAR=IR*DR*ZSTAR/ABOHR              !scaled radius
         ERSTAR=EXP(-RSTAR/2.)                !useful exponential
         IF (NOCC(1) .NE. 0) PSTOR(IR,1)=RSTAR*ERSTAR**2
         IF (NOCC(2) .NE. 0) PSTOR(IR,2)=(2-RSTAR)*RSTAR*ERSTAR
         IF (NOCC(3) .NE. 0) PSTOR(IR,3)=(RSTAR**2)*ERSTAR
20    CONTINUE
C
C     normalize wave functions
      DO 40 STATE=1,NSTATE
         IF (NOCC(STATE) .NE. 0) THEN
            NORM=0.
            DO 30 IR=0,NR
               NORM=NORM+PSTOR(IR,STATE)**2
30          CONTINUE
            NORM=1./(SQRT(NORM*DR))
            DO 35 IR=0,NR
               PSTOR(IR,STATE)=PSTOR(IR,STATE)*NORM
35          CONTINUE
         END IF
40    CONTINUE
C
      RETURN
      END
CCCCCCCCCCCCCCCCCCCCCCCCCCCCCCCCCCCCCCCCCCCCCCCCCCCCCCCCCCCCCCCCCCCCCCCCC
      SUBROUTINE ENERGY(E,FOCK,RHO,PHI,PSTOR)
C subroutine to calculate the energies of a normalized
C set of single-particle wave functions (PSTOR);
C also calculates Fock terms, density, and electron potential
CCCCCCCCCCCCCCCCCCCCCCCCCCCCCCCCCCCCCCCCCCCCCCCCCCCCCCCCCCCCCCCCCCCCCCCCC
C Global variables:
      INCLUDE 'PARAM.P3'
C Passed variables:
      REAL PSTOR(0:MAXSTP,MAXSTT)        !radial wave function (input)
      REAL FOCK(0:MAXSTP,MAXSTT)         !Fock terms (output)
      REAL PHI(0:MAXSTP)                 !electron potential (output)
      REAL RHO(0:MAXSTP)                 !density (output)
      REAL E(MAXSTT+1,8)                 !all energies of all states (output)
C Local variables:
      INTEGER STATE                      !state index
```

```
            INTEGER IR                  !radial index
            INTEGER IE                  !energy index
            REAL R                      !current radius
            REAL LL1                    !square of angular momentum
            REAL PM,PZ,PZ2              !values to calc d(rho)/dr
cccccccccccccccccccccccccccccccccccccccccccccccccccccccccccccccccccccccccc
            CALL SOURCE(PSTOR,FOCK,RHO)  !calculate Fock terms and density
            CALL POISSN(PHI,RHO)         !calc potntl due to electron charge
C
            DO 20 IE=IKEN,ITOT           !zero total energies
                E(NSTATE+1,IE)=0
20          CONTINUE
C
            DO 38 STATE=1,NSTATE
                IF (NOCC(STATE) .NE. 0) THEN
C
                DO 10 IE=IKEN,ITOT       !zero energy for this state
                    E(STATE,IE)=0
10              CONTINUE
C
                LL1=ANGMOM(STATE)*(ANGMOM(STATE)+1)
                PM=0
                DO 48 IR=1,NR            !integrate the energy densities
                    R=IR*DR
                    PZ=PSTOR(IR,STATE)
                    PZ2=PZ**2
C
                    E(STATE,IKEN)=E(STATE,IKEN)+(PZ-PM)**2        !kinetic
                    E(STATE,ICEN)=E(STATE,ICEN)+PZ2*LL1/R**2      !centrifugal
                    E(STATE,IVEN)=E(STATE,IVEN)-PZ2/R             !elec-nucl
                    E(STATE,IVEE)=E(STATE,IVEE)+PHI(IR)*PZ2       !elec-elec
                    E(STATE,IVEX)=E(STATE,IVEX)+FOCK(IR,STATE)*PZ !exchange
C
                    PM=PZ       !roll values for derivative
48              CONTINUE
C
C               put in constant factors
                E(STATE,IKEN)=E(STATE,IKEN)*HBARM/(2*DR)
                E(STATE,ICEN)=E(STATE,ICEN)*DR*HBARM/2
                E(STATE,IVEN)=E(STATE,IVEN)*ZCHARG*DR
                E(STATE,IVEE)=E(STATE,IVEE)*DR
                E(STATE,IVEX)=E(STATE,IVEX)*DR
C
C               calculate totals for this state
                E(STATE,IKTOT)=E(STATE,IKEN)+E(STATE,ICEN)
                E(STATE,IVTOT)=E(STATE,IVEN)+E(STATE,IVEE)+E(STATE,IVEX)
                E(STATE,ITOT)=E(STATE,IVTOT)+E(STATE,IKTOT)
C
C               add this state's contribution to total energy
                DO 30 IE=IKEN,IVEX
                    E(NSTATE+1,IE)=E(NSTATE+1,IE)+E(STATE,IE)*NOCC(STATE)
30              CONTINUE
```

```
C
         END IF
38       CONTINUE
C
         STATE=NSTATE+1               !calculate total energies
C        don't double count electron-electron and exchange energies
         E(STATE,IVEE)=E(STATE,IVEE)/2
         E(STATE,IVEX)=E(STATE,IVEX)/2
C
C        find total kinetic, potential and total energies
         E(STATE,IKTOT)=E(STATE,IKEN)+E(STATE,ICEN)
         E(STATE,IVTOT)=E(STATE,IVEN)+E(STATE,IVEE)+E(STATE,IVEX)
         E(STATE,ITOT)=E(STATE,IVTOT)+E(STATE,IKTOT)
C
         RETURN
         END
CCCCCCCCCCCCCCCCCCCCCCCCCCCCCCCCCCCCCCCCCCCCCCCCCCCCCCCCCCCCCCCCCCCCCC
         SUBROUTINE SOURCE(PSTOR,FOCK,RHO)
C subroutine to compute the density and the Fock terms given PSTOR
CCCCCCCCCCCCCCCCCCCCCCCCCCCCCCCCCCCCCCCCCCCCCCCCCCCCCCCCCCCCCCCCCCCCCC
C Global variables:
         INCLUDE 'PARAM.P3'
C Passed variables:
         REAL PSTOR(0:MAXSTP,MAXSTT) !radial wave function (input)
         REAL RHO(0:MAXSTP)          !density (output)
         REAL FOCK(0:MAXSTP,MAXSTT)  !Fock terms (output)
C Local variables:
         REAL DF                     !increment in Fock term
         REAL FAC                    !constant factor in Fock term
         INTEGER IR                  !radial index
         REAL R                      !current radius
         REAL RLAM                   !r**lambda
         REAL RLAM1                  !r**(lambda+1)
         INTEGER STATE               !indexes state
         INTEGER STATE2              !indexes second state in Fock term
         REAL SUM                    !sum for Fock integrals
         REAL TERM                   !temporary term for sum
         INTEGER L1,L2               !angular momentum for two states
         INTEGER LSTART,LSTOP        !limits on sum of L1+L2
         INTEGER LAM                 !current value of ang mom sum
         REAL THREEJ                 !three j coefficient
CCCCCCCCCCCCCCCCCCCCCCCCCCCCCCCCCCCCCCCCCCCCCCCCCCCCCCCCCCCCCCCCCCCCCC
         DO 20 IR=1,NR               !include fraction of old density
            RHO(IR)=(1.-MIX)*RHO(IR)
20       CONTINUE
C
         DO 30 STATE=1,NSTATE
            IF (NOCC(STATE) .NE. 0) THEN   !loop over occupied states
C
               DO 40 IR=1,NR        !contribution of this state to density
                  RHO(IR)=RHO(IR)+MIX*NOCC(STATE)*(PSTOR(IR,STATE)**2)
                  FOCK(IR,STATE)=0        !zero FOCK term
```

```
40              CONTINUE
C
C               begin calculation of Fock term
                DO 50 STATE2=1,NSTATE
                   IF (NOCC(STATE2) .NE. 0) THEN    !loop over occupied states
C
                      L1=ANGMOM(STATE)
                      L2=ANGMOM(STATE2)
                      LSTART=IABS(L1-L2)            !limits on ang mom sum
                      LSTOP=L1+L2
C
                      DO 60 LAM=LSTART,LSTOP,2      !loop over ang mom values
C
                         CALL SQR3J(L1,L2,LAM,THREEJ)
                         FAC=-CHARGE/2*NOCC(STATE2)*THREEJ
C
                         SUM=0
                         DO 80 IR=1,NR              !outward integral for Fock
                            R=IR*DR
                            RLAM=R**LAM
                            TERM=PSTOR(IR,STATE2)*PSTOR(IR,STATE)*RLAM/2
                            SUM=SUM+TERM
                            DF=PSTOR(IR,STATE2)*FAC*SUM*DR/(RLAM*R)
                            FOCK(IR,STATE)=FOCK(IR,STATE)+DF
                            SUM=SUM+TERM
80                       CONTINUE
C
                         SUM=0
                         DO 90 IR=NR,1,-1          !inward integral for Fock
                            R=IR*DR
                            RLAM1=R**(LAM+1)
                            TERM=PSTOR(IR,STATE2)*PSTOR(IR,STATE)/RLAM1/2
                            SUM=SUM+TERM
                            DF=PSTOR(IR,STATE2)*FAC*SUM*DR*RLAM1/R
                            FOCK(IR,STATE)=FOCK(IR,STATE)+DF
                            SUM=SUM+TERM
90                       CONTINUE
C
60                    CONTINUE    !end loop over lam
                   END IF
50              CONTINUE          !end loop over state2
             END IF
30        CONTINUE                !end loop over state
C
       RETURN
       END
CCCCCCCCCCCCCCCCCCCCCCCCCCCCCCCCCCCCCCCCCCCCCCCCCCCCCCCCCCCCCCCCCCCCCCCCCC
       SUBROUTINE POISSN(PHI,RHO)
C subroutine to solve Poisson's equation for the direct potential
C given the electron density RHO
CCCCCCCCCCCCCCCCCCCCCCCCCCCCCCCCCCCCCCCCCCCCCCCCCCCCCCCCCCCCCCCCCCCCCCCCCC
C Global variables:
```

```
      INCLUDE 'PARAM.P3'
C Passed variables:
      REAL PHI(0:MAXSTP)              !electron potential (output)
      REAL RHO(0:MAXSTP)             !density (input)
C Local variables:
      REAL CON                       !useful constant
      INTEGER IR                     !radial index
      REAL M                         !linear behavior to subtract
      REAL R                         !radius
      REAL SM,SP,SZ                  !source terms
      REAL SUM                       !total charge
CCCCCCCCCCCCCCCCCCCCCCCCCCCCCCCCCCCCCCCCCCCCCCCCCCCCCCCCCCCCCCCCCCCCCCC
      SUM = 0.                       !quadrature of density to get
      DO 19 IR = 1,NR                ! initial value for PHI(1)
         SUM = SUM+RHO(IR)/REAL(IR)
19    CONTINUE
C
      CON = DR**2/12                 !initial values for outward integ
      SM = 0
      SZ = -CHARGE*RHO(1)/DR
      PHI(0) = 0
      PHI(1) = SUM*CHARGE*DR
C
      DO 29 IR = 1,NR-1              !Numerov algorithm
         SP=-CHARGE*RHO(IR+1)/((IR+1)*DR)
         PHI(IR+1)=2*PHI(IR)-PHI(IR-1)+CON*(10*SZ+SP+SM)
         SM=SZ
         SZ=SP
29    CONTINUE
C
      M=(PHI(NR)-PHI(NR-10))/(10*DR)  !subtract off linear behavior
      DO 39 IR=1,NR
         R=IR*DR
         PHI(IR)=PHI(IR)/R-M          !factor of 1/r for true potl
39    CONTINUE
C
      RETURN
      END
CCCCCCCCCCCCCCCCCCCCCCCCCCCCCCCCCCCCCCCCCCCCCCCCCCCCCCCCCCCCCCCCCCCCCCC
      SUBROUTINE SNGWFN(STATE,E,PSTOR,FOCK,PHI)
C subroutine to solve the single-particle wave function as an
C inhomogeneous boundary-value problem for a given state, energy E,
C source term (FOCK), and potential (PHI)
CCCCCCCCCCCCCCCCCCCCCCCCCCCCCCCCCCCCCCCCCCCCCCCCCCCCCCCCCCCCCCCCCCCCCCC
C Global variables:
      INCLUDE 'PARAM.P3'
C Passed variables:
      INTEGER STATE                  !which single particle state (input)
      REAL E                         !single particle energy (input)
      REAL FOCK(0:MAXSTP,MAXSTT)     !Fock terms (input)
      REAL PHI(0:MAXSTP)             !electron potential (input)
      REAL PSTOR(0:MAXSTP,MAXSTT)    !radial wave function (output)
```

```
C Local variables:
      REAL PSIIN(0:MAXSTP),PSIOUT(0:MAXSTP)   !homogeneous solutions
      INTEGER NR2                             !midpoint for the lattice
      REAL LL1                                !angular momentum
      REAL DRHBM                              !useful constant
      REAL K2M,K2Z,K2P                        !local wave numbers
      REAL NORM                               !normalization
      REAL R                                  !current radius
      REAL SUM                                !sum for integration
      REAL TERM                               !temp term in integration
      REAL WRON                               !Wronskian at middle of lattice
      INTEGER IR                              !radial index
CCCCCCCCCCCCCCCCCCCCCCCCCCCCCCCCCCCCCCCCCCCCCCCCCCCCCCCCCCCCCCCCCCCCCCCCCCCCCCCC
      DRHBM=DR**2/HBARM/6                     !useful constants
      LL1=ANGMOM(STATE)*(ANGMOM(STATE)+1)*HBARM/2
C
      K2M=0                                   !integrate outward homogeneous soln
      K2Z=DRHBM*(E-PHI(1)+(ZCHARG-LL1/DR)/DR)
      PSIOUT(0)=0
      PSIOUT(1)=1.0E-10
      DO 49 IR=2,NR
            R=DR*IR                           !Numerov algorithm
            K2P=DRHBM*(E-PHI(IR)+(ZCHARG-LL1/R)/R)
            PSIOUT(IR)=(PSIOUT(IR-1)*(2-10*K2Z)-
     +                 PSIOUT(IR-2)*(1+K2M))/(1+K2P)
            K2M=K2Z                           !roll values
            K2Z=K2P
49    CONTINUE
C
      K2P=0                                   !integrate inward homogeneous soln
      R=(NR-1)*DR
      K2Z=DRHBM*(E-PHI(NR-1)+(ZCHARG-LL1/R)/R)
      PSIIN(NR)=0
      PSIIN(NR-1)=1.0E-10
      DO 59 IR=NR-2,1,-1
            R=DR*IR                           !Numerov algorithm
            K2M=DRHBM*(E-PHI(IR)+(ZCHARG-LL1/R)/R)
            PSIIN(IR)=(PSIIN(IR+1)*(2-10*K2Z)-
     +                PSIIN(IR+2)*(1+K2P))/(1+K2M)
            K2P=K2Z                           !roll values
            K2Z=K2M
59    CONTINUE
C
      NR2=NR/2                                !Wronskian at middle of mesh
      WRON=(PSIIN(NR2+1)-PSIIN(NR2-1))/(2*DR)*PSIOUT(NR2)
      WRON=WRON-(PSIOUT(NR2+1)-PSIOUT(NR2-1))/(2*DR)*PSIIN(NR2)
C
      SUM=0                                   !outward integral in Green's soln
      DO 69 IR=1,NR
            TERM=-PSIOUT(IR)*FOCK(IR,STATE)/2
            SUM=SUM+TERM
            PSTOR(IR,STATE)=PSIIN(IR)*SUM*DR
```

```
                  SUM=SUM+TERM
69      CONTINUE
C
        SUM=0                              !inward integral in Green's soln
        DO 79 IR=NR,1,-1
              TERM=-PSIIN(IR)*FOCK(IR,STATE)/2
              SUM=SUM+TERM
              PSTOR(IR,STATE)=(PSTOR(IR,STATE)+PSIOUT(IR)*SUM*DR)/WRON
              SUM=SUM+TERM
79      CONTINUE
C
        IF (STATE .EQ. 2) THEN             !keep 1s and 2s states orthogonal
              SUM=0
              DO 89 IR=1,NR
                    SUM=SUM+PSTOR(IR,1)*PSTOR(IR,2)
89            CONTINUE
              SUM=SUM*DR
              DO 99 IR=1,NR
                    PSTOR(IR,2)=PSTOR(IR,2)-SUM*PSTOR(IR,1)
99            CONTINUE
        END IF
C
        NORM=0                             !normalize the soln
        DO 18 IR=1,NR
              NORM=NORM+PSTOR(IR,STATE)**2
18      CONTINUE
        NORM=1/SQRT(NORM*DR)
        DO 28 IR=1,NR
              PSTOR(IR,STATE)=PSTOR(IR,STATE)*NORM
28      CONTINUE
C
        RETURN
        END
CCCCCCCCCCCCCCCCCCCCCCCCCCCCCCCCCCCCCCCCCCCCCCCCCCCCCCCCCCCCCCCCCCCCCCCCC
        SUBROUTINE SQR3J(L1,L2,LAM,THREEJ)
C subroutine to calculate square of the 3-j coefficient appearing
C                  in the exchange energy
CCCCCCCCCCCCCCCCCCCCCCCCCCCCCCCCCCCCCCCCCCCCCCCCCCCCCCCCCCCCCCCCCCCCCCCCC
C Global variables:
        INCLUDE 'PARAM.P3'
C Passed variables:
        INTEGER L1,L2,LAM                  !angular momentum (input)
        REAL THREEJ                        !three j coefficient (output)
C Local variables:
        REAL DELTA                         !intermediate term for calc of 3J
        INTEGER IMAX,P                     !useful combinations of ang mom
CCCCCCCCCCCCCCCCCCCCCCCCCCCCCCCCCCCCCCCCCCCCCCCCCCCCCCCCCCCCCCCCCCCCCCCCC
        IMAX=L1+L2+LAM+1                   !useful combinations
        P=(L1+L2+LAM)/2
C
        DELTA=FACTRL(L1+L2-LAM)*FACTRL(-L1+L2+LAM)      !temp values
        DELTA=DELTA*FACTRL(L1-L2+LAM)/FACTRL(IMAX)
```

```
          THREEJ=DELTA*(FACTRL(P)**2)                      !calculate 3J squared
          THREEJ=THREEJ/(FACTRL(P-L1)**2)
          THREEJ=THREEJ/(FACTRL(P-L2)**2)
          THREEJ=THREEJ/(FACTRL(P-LAM)**2)
C
          RETURN
          END
CCCCCCCCCCCCCCCCCCCCCCCCCCCCCCCCCCCCCCCCCCCCCCCCCCCCCCCCCCCCCCCCCCCCCCC
          SUBROUTINE INIT
C initializes constants, displays header screen,
C initializes arrays for input parameters
CCCCCCCCCCCCCCCCCCCCCCCCCCCCCCCCCCCCCCCCCCCCCCCCCCCCCCCCCCCCCCCCCCCCCCC
C Global variables:
          INCLUDE 'IO.ALL'
          INCLUDE 'MENU.ALL'
          INCLUDE 'PARAM.P3'
C Local parameters:
          CHARACTER*80 DESCRP           !program description
          DIMENSION DESCRP(20)
          INTEGER NHEAD,NTEXT,NGRAPH    !number of lines for each description
          INTEGER I                     !index for factorial loop
CCCCCCCCCCCCCCCCCCCCCCCCCCCCCCCCCCCCCCCCCCCCCCCCCCCCCCCCCCCCCCCCCCCCCCC
          CALL SETUP                    !get environment parameters
C
C         display header screen
          DESCRP(1)= 'PROJECT 3'
          DESCRP(2)= 'Hartree-Fock solutions of small atomic systems'
          DESCRP(3)= 'in the filling approximation'
          NHEAD=3
C
C         text output description
          DESCRP(4)='kinetic, potential '
      +    //'(electron-nucleus, electron-electron,'
          DESCRP(5)= 'exchange and total) and total energy for each state'
          NTEXT=2
C
C         graphics output description
          DESCRP(6)= 'electron probability density for each state'
          NGRAPH=1
C
          CALL HEADER(DESCRP,NHEAD,NTEXT,NGRAPH)
C
C         calculate constants
          !atomic constants
          HBARM=7.6359               !hbar*2/(mass)
          CHARGE=14.409              !charge of the electron
          ABOHR=HBARM/CHARGE         !Bohr radius
C
          NSTATE=MAXSTT              !descriptions of states
          ANGMOM(1)=0
          ANGMOM(2)=0
          ANGMOM(3)=1
```

```
       ID(1)=' 1S '
       ID(2)=' 2S '
       ID(3)=' 2P '
       ID(4)='TOTAL'
C
       FACTRL(0)=1              !factorials
       DO 10 I=1,10
          FACTRL(I)=I*FACTRL(I-1)
10     CONTINUE
C
C      setup menu arrays
       CALL MENU               !setup constant part of menu
C
       MTYPE(13)=FLOAT
       MPRMPT(13)= 'Enter nuclear charge'
       MTAG(13)= 'Nuclear charge'
       MLOLIM(13)=1.
       MHILIM(13)=20.
       MREALS(13)=6.
C
       MTYPE(14)=NUM
       MPRMPT(14)= 'Enter number of electrons in the 1s state'
       MTAG(14)= 'Occupation of 1s state'
       MLOLIM(14)=0.
       MHILIM(14)=2.
       MINTS(14)=2.
C
       MTYPE(15)=NUM
       MPRMPT(15)= 'Enter number of electrons in the 2s state'
       MTAG(15)= 'Occupation of 2s state'
       MLOLIM(15)=0.
       MHILIM(15)=2.
       MINTS(15)=2.
C
       MTYPE(16)=NUM
       MPRMPT(16)= 'Enter number of electrons in the 2p state'
       MTAG(16)= 'Occupation of 2p state'
       MLOLIM(16)=0.
       MHILIM(16)=6.
       MINTS(16)=2.
C
       MTYPE(17)=SKIP
       MREALS(17)=35.
C
       MTYPE(38)=FLOAT
       MPRMPT(38)= 'Enter radial step size (Angstroms)'
       MTAG(38)= 'Radial step size (Angstroms)'
       MLOLIM(38)=.0001
       MHILIM(38)=.5
       MREALS(38)=.01
C
       MTYPE(39)=FLOAT
```

```
        MPRMPT(39)= 'Enter outer radius of the lattice (Angstroms)'
        MTAG(39)= 'Outer radius of lattice (Angstroms)'
        MLOLIM(39)=.01
        MHILIM(39)=10.
        MREALS(39)=2.
C
        MTYPE(40)=NUM
        MPRMPT(40)= 'Enter number of iterations'
        MTAG(40)= 'Number of iterations'
        MLOLIM(40)=1.
        MHILIM(40)=50.
        MINTS(40)=4.
C
        MTYPE(41)=SKIP
        MREALS(41)=60.
C
        MSTRNG(MINTS(75))= 'proj3.txt'
C
        MTYPE(76)=SKIP
        MREALS(76)=80.
C
        MSTRNG(MINTS(86))= 'proj3.grf'
C
        MTYPE(87)=SKIP
        MREALS(87)=90.
C
        RETURN
        END
CCCCCCCCCCCCCCCCCCCCCCCCCCCCCCCCCCCCCCCCCCCCCCCCCCCCCCCCCCCCCCCCCCCCCCCC
        SUBROUTINE PARAM
C gets parameters from screen
C ends program on request
C closes old files
C maps menu variables to program variables
C opens new files
C calculates all derivative parameters
CCCCCCCCCCCCCCCCCCCCCCCCCCCCCCCCCCCCCCCCCCCCCCCCCCCCCCCCCCCCCCCCCCCCCCCC
C Global variables:
        INCLUDE 'MENU.ALL'
        INCLUDE 'IO.ALL'
        INCLUDE 'PARAM.P3'
        INCLUDE 'MAP.P3'
C Local variables:
        INTEGER I                       !index of current state
C Function:
        LOGICAL LOGCVT                  !converts 1 to TRUE, others to FALSE
CCCCCCCCCCCCCCCCCCCCCCCCCCCCCCCCCCCCCCCCCCCCCCCCCCCCCCCCCCCCCCCCCCCCCCCC
C       get input from terminal
        CALL CLEAR
        CALL ASK(1,ISTOP)
C
C       stop program if requested
```

```
      IF (MREALS(IMAIN) .EQ. STOP) CALL FINISH
C
C     close files if necessary
      IF (TNAME .NE. MSTRNG(MINTS(ITNAME)))
     +    CALL FLCLOS(TNAME,TUNIT)
      IF (GNAME .NE. MSTRNG(MINTS(IGNAME)))
     +    CALL FLCLOS(GNAME,GUNIT)
C
C     set new parameter values
C     physical and numerical
      Z=MREALS(IZ)
      NOCC(1)=MINTS(IONE)
      NOCC(2)=MINTS(ITWO)
      NOCC(3)=MINTS(ITHREE)
      DR=MREALS(IDR)
      RMAX=MREALS(IRMAX)
      NITER=MINTS(INITER)
C
C     text output
      TTERM=LOGCVT(MINTS(ITTERM))
      TFILE=LOGCVT(MINTS(ITFILE))
      TNAME=MSTRNG(MINTS(ITNAME))
C
C     graphics output
      GTERM=LOGCVT(MINTS(IGTERM))
      GHRDCP=LOGCVT(MINTS(IGHRD))
      GFILE=LOGCVT(MINTS(IGFILE))
      GNAME=MSTRNG(MINTS(IGNAME))
C
C     open files
      IF (TFILE) CALL FLOPEN(TNAME,TUNIT)
      IF (GFILE) CALL FLOPEN(GNAME,GUNIT)
C     files may have been renamed
      MSTRNG(MINTS(ITNAME))=TNAME
      MSTRNG(MINTS(IGNAME))=GNAME
C
C     calculate derivative parameters:
      ZCHARG=Z*CHARGE        !nuclear charge
      NR=INT(RMAX/DR)        !number of radial steps
      NOCC(NSTATE+1)=0       !total number of electrons
      DO 10 I=1,NSTATE
         NOCC(NSTATE+1)=NOCC(NSTATE+1)+NOCC(I)
10    CONTINUE
C
      CALL PCHECK            !check input
      CALL CLEAR
C
      RETURN
      END
CCCCCCCCCCCCCCCCCCCCCCCCCCCCCCCCCCCCCCCCCCCCCCCCCCCCCCCCCCCCCCCCCCCCCCCCCC
      SUBROUTINE PCHECK
C ensure that the number of radial steps is not greater than the
```

```
C size of the arrays
cccccccccccccccccccccccccccccccccccccccccccccccccccccccccccccccccccccccc
C Global parameters:
      INCLUDE 'PARAM.P3'
      INCLUDE 'MENU.ALL'
      INCLUDE 'IO.ALL'
      INCLUDE 'MAP.P3'
C Functions:
      REAL GETFLT                         !get float from screen
cccccccccccccccccccccccccccccccccccccccccccccccccccccccccccccccccccccccc
10    IF ( NR .GT. MAXSTP) THEN
           WRITE (OUNIT,15) REAL(NR),MAXSTP
           MLOLIM(IDR)=RMAX/MAXSTP    !revise lower limit
           MREALS(IDR)=MLOLIM(IDR)
           MREALS(IDR) =  GETFLT(MREALS(IDR),MLOLIM(IDR),
     +                   MHILIM(IDR),'Enter a larger step')
           DR=MREALS(IDR)
           NR=INT(RMAX/DR)
        GOTO 10
        END IF
C
15    FORMAT (' Total number of radial steps (=',1PE9.2,
     +          ') is larger than maxstp (=',i5,')')
C
      RETURN
      END
cccccccccccccccccccccccccccccccccccccccccccccccccccccccccccccccccccccccc
      SUBROUTINE PRMOUT(MUNIT,ZSTAR,NLINES)
C outputs text results to the specified unit
cccccccccccccccccccccccccccccccccccccccccccccccccccccccccccccccccccccccc
C Global variables
      INCLUDE 'IO.ALL'
      INCLUDE 'PARAM.P3'
C Passed variables
      INTEGER MUNIT              !current unit number (input)
      REAL ZSTAR                 !optimal charge (input)
      INTEGER IS                 !indexes states (input)
      INTEGER NLINES             !number of lines written to screen (I/O)
cccccccccccccccccccccccccccccccccccccccccccccccccccccccccccccccccccccccc
      IF (MUNIT .EQ. OUNIT) CALL CLEAR
C
      WRITE (MUNIT,19)
      WRITE (MUNIT,21)
      WRITE (MUNIT,27)   Z,ZSTAR
      WRITE (MUNIT,23)   RMAX,DR
      WRITE (MUNIT,30)  (NOCC(IS),IS=1,NSTATE)
      WRITE (MUNIT,33)
C
      IF (MUNIT .EQ. GUNIT) THEN    !special heading for graphics file
           WRITE (MUNIT,19)
           WRITE (MUNIT,35)
           WRITE (MUNIT,40)
```

```
          WRITE (MUNIT,19)
      ELSE
          WRITE (MUNIT,19)
      END IF
C
      NLINES=7
C
19    FORMAT  (' ')
21    FORMAT  (' Output from project 3:',
     +         ' Hartree-Fock solutions for small atomic systems')
27    FORMAT  (' nuclear charge=',F6.3,5X,' zstar =',F6.3)
23    FORMAT  (' Rmax (Angstroms)=',F6.3,5X,' radial step (Angstroms)=',
     +         1PE12.5)
30    FORMAT  (' Occupation of the states are:',3(2X,I2))
33    FORMAT  (' All energies are in eV')
35    FORMAT  (4X,'   radius  ',4X,'1s density',5X,'2s density',
     +         5X,'2p density',4X,'total density')
40    FORMAT  (4X,'-----------',4X,'----------',5X,'----------',
     +         5X,'----------',4X,'------------')
C
      RETURN
      END
CCCCCCCCCCCCCCCCCCCCCCCCCCCCCCCCCCCCCCCCCCCCCCCCCCCCCCCCCCCCCCCCCCCCCCCCCCC
      SUBROUTINE TXTOUT(MUNIT,ITER,E,NLINES)
C writes energies for one iteration to the requested unit
CCCCCCCCCCCCCCCCCCCCCCCCCCCCCCCCCCCCCCCCCCCCCCCCCCCCCCCCCCCCCCCCCCCCCCCCCCC
C Global variables:
      INCLUDE 'IO.ALL'
      INCLUDE 'PARAM.P3'
C Passed variables:
      REAL E(MAXSTT+1,8)       !all energies of all states (input)
      INTEGER MUNIT            !current unit (input)
      INTEGER ITER             !current iteration (input)
      INTEGER NLINES           !number of lines written to screen (I/O)
C Local variables:
      INTEGER IS               !state index
      INTEGER DELINE           !number of lines written per call
CCCCCCCCCCCCCCCCCCCCCCCCCCCCCCCCCCCCCCCCCCCCCCCCCCCCCCCCCCCCCCCCCCCCCCCCCCC
      DELINE=3+NSTATE          !number of lines written per call
C     clear the screen at a convenient place
      IF ((NLINES+DELINE .GT. TRMLIN-6) .AND. (MUNIT .EQ. OUNIT)) THEN
          CALL PAUSE('to continue...',1)
          CALL CLEAR
          NLINES=0
      END IF
C
      WRITE (MUNIT,20) ITER
      WRITE (MUNIT,30)
C
      DO 50 IS=1,NSTATE+1
        IF (NOCC(IS) .NE. 0) THEN    !occupied states only
          WRITE (MUNIT,35) ID(IS),NOCC(IS),E(IS,IKTOT),E(IS,IVEN),
```

```
      +     E(IS,IVEE),E(IS,IVEX),E(IS,IVTOT),E(IS,ITOT)
            END IF
50          CONTINUE
C
20          FORMAT    (26X,'------- Iteration ',I2,' -------')
30          FORMAT    (2X,'State',2X,'Nocc',4X,'Ktot',7X,'Ven',7X,'Vee',7X,
      +               'Vex',6X,'Vtot',6X,'Etot')
35          FORMAT    (2X,A5,3X,I2,1X,6(1X,F9.3))
C
            IF (MUNIT .EQ. OUNIT) NLINES=NLINES+DELINE
C
            RETURN
            END
CCCCCCCCCCCCCCCCCCCCCCCCCCCCCCCCCCCCCCCCCCCCCCCCCCCCCCCCCCCCCCCCCCCCCCCCCC
            SUBROUTINE GRFOUT(DEVICE,PSTOR,RHO,E)
C graph the densities of each state and the total density
CCCCCCCCCCCCCCCCCCCCCCCCCCCCCCCCCCCCCCCCCCCCCCCCCCCCCCCCCCCCCCCCCCCCCCCCCC
C Global variables
            INCLUDE 'IO.ALL'
            INCLUDE 'PARAM.P3'
            INCLUDE 'GRFDAT.ALL'
C Input variables:
            INTEGER DEVICE                  !which device is being used?
            REAL PSTOR(0:MAXSTP,MAXSTT)     !radial wave function
            REAL RHO(0:MAXSTP)              !density
            REAL E(MAXSTT+1,8)              !all energies of all states
C Local variables
            REAL X,Y                        !radius and density
            REAL R                          !radius for GUNIT
            REAL DEN                        !density for gunit
            INTEGER IS                      !array of occupied states
            INTEGER OCC                     !number of occupied states
            INTEGER IR                      !indexes radius
            INTEGER STATE                   !current state
            CHARACTER*9 CZ,CE,COCC          !Z,E,NOCC  as a character strings
            INTEGER ZLEN,ELEN,OLEN          !character length
            DIMENSION X(0:MAXSTP),Y(0:MAXSTP)
            DIMENSION IS(MAXSTT+1),DEN(MAXSTT+1)
            INTEGER SCREEN                  !send to terminal
            INTEGER PAPER                   !make a hardcopy
            INTEGER FILE                    !send to a file
            DATA SCREEN,PAPER,FILE/1,2,3/
CCCCCCCCCCCCCCCCCCCCCCCCCCCCCCCCCCCCCCCCCCCCCCCCCCCCCCCCCCCCCCCCCCCCCCCCCC
C     messages for the impatient
            IF (DEVICE .NE. SCREEN) WRITE (OUNIT,100)
C
C     how many occupied states are there, and which are they?
            OCC=0
            DO 60 STATE=1,NSTATE
               IF (NOCC(STATE) .NE. 0) THEN
                  OCC=OCC+1
                  IS(OCC)=STATE
```

```
         END IF
60       CONTINUE
         IF (OCC .GT. 1) THEN               !include total density
            OCC=OCC+1                       ! if more than one sp state
            IS(OCC)=NSTATE+1
         END IF
C
C        define parameters which are the same for all plots
         IF (DEVICE .NE. FILE) THEN
            NPLOT=OCC                       !how many plots?
C
            XMIN=0.                         !x value is radius
            XMAX=RMAX
            YOVAL=0.
C
            YMAX=0.                         !y values are density
            DO 50 IR=1,NR
               IF (RHO(IR) .GT. YMAX) YMAX=RHO(IR)
50          CONTINUE
            YMIN=0.
            XOVAL=0.
C
            NPOINT=NR+1
            ILINE=1                         !line and symbol styles
            ISYM=1
            IFREQ=1
            NXTICK=5
            NYTICK=5
C
            CALL CONVRT(Z,CZ,ZLEN)          !titles and labels
            CALL CONVRT(E(OCC,8),CE,ELEN)
            TITLE = 'Atomic Hartree Fock, z='//CZ(1:ZLEN)//
     +              ', Total Energy='//CE
            LABEL(1)= 'radius (Angstroms)'
C
            CALL GTDEV(DEVICE)              !device nomination
            IF (DEVICE .EQ. SCREEN) CALL GMODE  !change to graphics mode
         END IF
C
         IF (DEVICE .NE. FILE) THEN
C        for graphics pack, do one plot for each occupied state
         DO 20 IPLOT=1,NPLOT
C
            STATE=IS(IPLOT)
            CALL CONVRT(E(IPLOT,8),CE,ELEN)
            CALL ICNVRT(NOCC(STATE),COCC,OLEN)
            INFO='Energy = '//CE(1:ELEN)//', occp='//COCC
            LABEL(2)=ID(STATE)//' probability density'
            CALL LNLNAX
C
            DO 30 IR=0,NR
               X(IR)=IR*DR
```

```
                 IF (STATE .EQ. NSTATE+1) THEN
                     Y(IR)=RHO(IR)
                 ELSE
                     Y(IR)=NOCC(STATE)*PSTOR(IR,STATE)**2
                 END IF
30           CONTINUE
             CALL XYPLOT(X,Y)
20       CONTINUE
C
         ELSE
C        for gunit, write one only line for each radius
         DO 120 IR=0,NR
           R=IR*DR
           DO 110 STATE=1,NSTATE+1
               IF (STATE .EQ. NSTATE+1) THEN
                   DEN(STATE)=RHO(IR)
               ELSE
                   DEN(STATE)=NOCC(STATE)*PSTOR(IR,STATE)**2
               END IF
110        CONTINUE
           WRITE (GUNIT,70) R,(DEN(STATE),STATE=1,NSTATE+1)
120      CONTINUE
C
         END IF
C
C        end graphing session
         IPLOT=NPLOT                                    !reset index
         IF (DEVICE .NE. FILE) CALL GPAGE(DEVICE)       !end graphics package
         IF (DEVICE .EQ. SCREEN) CALL TMODE             !switch to text mode
C
70       FORMAT (5(4X,1PE11.3))
100      FORMAT (/,' Patience, please; output going to a file.')
C
         RETURN
         END
CCCCCCCCCCCCCCCCCCCCCCCCCCCCCCCCCCCCCCCCCCCCCCCCCCCCCCCCCCCCCCCCCCCCCCCCC
CCCCCCCCCCCCCCCCCCCCCCCCCCCCCCCCCCCCCCCCCCCCCCCCCCCCCCCCCCCCCCCCCCCCCCCCC
C param.p3
C
         REAL MIX              !fraction of old density to mix in
         INTEGER NSTATE        !total number of states in the calculation
         REAL HBARM            !hbar**2/mass of an electron
         REAL CHARGE           !square of the electron charge
         REAL ABOHR            !Bohr radius
         INTEGER FACTRL        !array of factorials
         INTEGER ANGMOM        !angular momentum of each state
         CHARACTER*5 ID        !state identification
         REAL Z                !number of protons in nucleus
         INTEGER NOCC          !occupation of each state
         REAL DR               !radial step size
         REAL RMAX             !outer radius of lattice
         INTEGER NITER         !number of iterations
```

```
      REAL ZCHARG        !nuclear charge
      INTEGER NR         !number of radial steps
      INTEGER MAXSTT     !maximum number of states allowed
      INTEGER IKEN,ICEN,IVEN,IVEE,IVEX   !index energy types
      INTEGER IKTOT,IVTOT,ITOT !index energy totals
      INTEGER MAXSTP     !maximum size of data arrays
C
      PARAMETER (IKEN=1)
      PARAMETER (ICEN=2)
      PARAMETER (IVEN=3)
      PARAMETER (IVEE=4)
      PARAMETER (IVEX=5)
      PARAMETER (IKTOT=6)
      PARAMETER (IVTOT=7)
      PARAMETER (ITOT=8)
      PARAMETER (MAXSTT=3)
      PARAMETER (MAXSTP=1000)
C
      DIMENSION FACTRL(0:10),ANGMOM(MAXSTT),NOCC(MAXSTT+1),ID(MAXSTT+1)
C
      COMMON / CONST / MIX,NSTATE,HBARM,CHARGE,ABOHR,FACTRL,ANGMOM
      COMMON / PPARAM / Z,NOCC
      COMMON / NPARAM / DR,RMAX,NITER
      COMMON / PCALC / ZCHARG,NR
      COMMON / NAMES / ID
ccccccccccccccccccccccccccccccccccccccccccccccccccccccccccccccccccccccc
ccccccccccccccccccccccccccccccccccccccccccccccccccccccccccccccccccccccc
C map.p3
      INTEGER IZ,IONE,ITWO,ITHREE,IDR,IRMAX,INITER
      PARAMETER (IZ     =13 )
      PARAMETER (IONE   =14 )
      PARAMETER (ITWO   =15 )
      PARAMETER (ITHREE =16 )
      PARAMETER (IDR    =38 )
      PARAMETER (IRMAX  =39 )
      PARAMETER (INITER =40 )
```

C.4 Project 4

Algorithm This program finds the partial wave scattering solution for electrons incident on a square-well (radius 1.5 Å), a Gaussian well, or the Lenz-Jensen potential (subroutine POTNTL). Note that all of the potentials vanish for $r > r_{max}$. The scattering amplitude is calculated from Eq. (IV.4), with the Legendre polynomials evaluated by forward recursion in subroutines LPINIT and LEGPOL. To find the phase shift (Eq. (IV.8)) for each partial wave requested from LSTART to LSTOP, the radial Schroedinger equation (IV.2) is integrated outward for both the free and scattering wave functions (subroutine NUMERV); the required spherical Bessel functions are calculated in subroutine SBSSEL. Note that the starting value for the wave function at the second lattice point is determined from the approximation to the second derivative (1.7) at the first lattice point. Multiple-of-π ambiguities in the phase shifts are resolved in subroutine FIXDEL by comparing the number of nodes and antinodes in the free and scattered waves.

Input The physical parameters are the potential [Lenz-Jensen], the nuclear charge [4] (Lenz-Jensen only), or the well depth [50.] (square or Gaussian well), and the incident energy [20.]. All energies are in eV, and all lengths are in Angstroms. The numerical parameters are the number of integration points to $r^{(1)}$ [400], the number of extra integration points between $r^{(1)}$ and $r^{(2)}$ [60], and the number of angles at which to calculate the differential cross section [36]. The maximum number of integration points, extra integration points, and angles are fixed by MAXN= 1000, MAXE= 100, and MAXANG= 100, respectively.

At the beginning of the calculation, the program determines a reasonable value for LMAX based on the discussion in the book; this value is displayed and you are prompted for values for LSTART [0] and LSTOP [LMAX]. The maximum partial wave allowed is fixed by MAXL= 100.

Output The text output displays, for each value of L requested, the phase shift (in degrees) and the total cross section for that partial wave (in square Angstroms). When all partial waves are calculated, the program displays the total cross section followed by the real and imaginary scattering amplitudes and differential cross section at the requested number of angles (equally spaced from 0 to 180 degrees).

If you request graphics to the screen, the program will display the effective potential, free wave, and scattered wave *vs.* radius for each partial wave. (In order to avoid excessive output, an exception is made to the rule: the program does *not* send these graphs to the hardcopy device or a file.) Note that the energy scale on the graph goes from $-2E$ to $4\dot{E}$, where E is the incident energy, and the wave functions are centered about E. The phase shift for the partial wave is displayed in the legend. If you request both text and graphics to the screen, the text output appears after the graphics output. At the end of the calculation, the graphics displays the differential cross section *vs.* angle on a semi-log scale (this graph *can* be sent to hardcopy and file on request).

```
CCCCCCCCCCCCCCCCCCCCCCCCCCCCCCCCCCCCCCCCCCCCCCCCCCCCCCCCCCCCCCCCCCCCCCCCCC
      PROGRAM PROJ4
C     PROJECT 4: Partial-wave solution of quantum scattering
C COMPUTATIONAL PHYSICS (FORTRAN VERSION)
C by Steven E. Koonin and Dawn C. Meredith
C Copyright 1989, Addison-Wesley Publishing Company
CCCCCCCCCCCCCCCCCCCCCCCCCCCCCCCCCCCCCCCCCCCCCCCCCCCCCCCCCCCCCCCCCCCCCCCCCCC
      CALL INIT          !display header screen, setup parameters
5     CONTINUE           !main loop/ execute once for each set of param
      CALL PARAM         !get input from screen
      CALL ARCHON        !calculate total and differential cross section
      GOTO 5
      END
CCCCCCCCCCCCCCCCCCCCCCCCCCCCCCCCCCCCCCCCCCCCCCCCCCCCCCCCCCCCCCCCCCCCCCCCCCC
      SUBROUTINE ARCHON
C calculates differential and total cross section using
C partial-wave expansion; scattering is from a central potential
CCCCCCCCCCCCCCCCCCCCCCCCCCCCCCCCCCCCCCCCCCCCCCCCCCCCCCCCCCCCCCCCCCCCCCCCCCC
C Global variables:
      INCLUDE 'PARAM.P4'
      INCLUDE 'IO.ALL'
C Local variables:
      INTEGER  L              !which partial wave
      REAL DELTA(0:MAXL)      !phase shift
      REAL DELDEG(0:MAXL)     !phase shift in degrees
      REAL SIGMA(0:MAXL),SIGTOT !total cross section
      REAL DSIGMA(0:MAXANG)   !differential cross section
      COMPLEX F(0:MAXANG)     !scattering amplitude
      REAL CD,SD              !sins and cosines
      INTEGER ITHETA          !index the angle
      INTEGER NLINES          !number of lines written to terminal
      INTEGER DEVICE          !current graphics device
      INTEGER SCREEN          !send to terminal
      INTEGER PAPER           !make a hardcopy
      INTEGER FILE            !send to a file
      DATA SCREEN,PAPER,FILE/1,2,3/
CCCCCCCCCCCCCCCCCCCCCCCCCCCCCCCCCCCCCCCCCCCCCCCCCCCCCCCCCCCCCCCCCCCCCCCCCCC
C     output summary of parameters
      IF ((TTERM) .AND. (.NOT. GTERM)) CALL PRMOUT(OUNIT,NLINES)
      IF (TFILE) CALL PRMOUT(TUNIT,NLINES)
      IF (GFILE) CALL PRMOUT(GUNIT,NLINES)
```

```
C
C       find the phase shifts by integrating the time-indp Schrodinger eq.
        SIGTOT=0.
        DO 100 L=LSTART,LSTOP
          CALL NUMERV(L,DELTA)
          DELDEG(L)=180.0*DELTA(L)/PI
          SIGMA(L)=4*PI/K2*(2*L+1)*SIN(DELTA(L))**2
          SIGTOT=SIGTOT+SIGMA(L)
C
          IF (GTERM) THEN                    !output results for this L value
             CALL GRFOUT(SCREEN,L,DELDEG(L),SIGMA(L))
          ELSE IF (TTERM) THEN
             CALL TXTOUT(OUNIT,NLINES,L,DELDEG(L),SIGMA(L))
          END IF
          IF (TFILE) CALL TXTOUT(TUNIT,NLINES,L,DELDEG(L),SIGMA(L))
100     CONTINUE
C
C       output data for all L
        IF ((TTERM) .AND. (GTERM)) CALL DELOUT(DELDEG,SIGMA)
C
C       calculate differential cross section by summing over partial waves
        DO 300 ITHETA=0,NANG
          F(ITHETA)=(0.,0.)
          DO 200 L=LSTART,LSTOP
             SD=SIN(DELTA(L))
             CD=COS(DELTA(L))
             F(ITHETA)=F(ITHETA)+(2*L+1)*SD*PL(L,ITHETA)*CMPLX(CD,SD)
200       CONTINUE
          F(ITHETA)=F(ITHETA)/K
          DSIGMA(ITHETA)=(ABS(F(ITHETA)))**2
300     CONTINUE
C
        IF (TTERM) CALL PAUSE('to see total cross sections...',1)
        IF (TTERM) CALL CLEAR
C
        IF (TTERM) CALL SIGOUT(OUNIT,F,DSIGMA,SIGTOT)
        IF (TFILE) CALL SIGOUT(TUNIT,F,DSIGMA,SIGTOT)
C
        IF (TTERM) CALL PAUSE('to continue...',1)
        IF (TTERM) CALL CLEAR
C
C       graphics output
        IF (GTERM) CALL GRFSIG(SCREEN,DSIGMA,SIGTOT)
        IF (GFILE) CALL GRFSIG(FILE,DSIGMA,SIGTOT)
        IF (GHRDCP) CALL GRFSIG(PAPER,DSIGMA,SIGTOT)
C
        RETURN
        END
CCCCCCCCCCCCCCCCCCCCCCCCCCCCCCCCCCCCCCCCCCCCCCCCCCCCCCCCCCCCCCCCCCCCCCCCC
        Subroutine NUMERV(L,DELTA)
C find the phase shifts DELTA for fixed L by integrating
C the radial wave equation (for both the scattered and free wave)
```

```
C using the Numerov algorithm
ccccccccccccccccccccccccccccccccccccccccccccccccccccccccccccccccccc
C Global variables:
      INCLUDE 'PARAM.P4'
C Input/Output Variables:
      INTEGER  L              !which partial wave (input)
      REAL DELTA(0:MAXL)      !phase shift (output)
C Local Variables:
      REAL PSI(MAXN+MAXE),PSIF(MAXN+MAXE)   !wave functions
      REAL PSIMAX,PSFMAX      !maxima of wave functions
      REAL VEFF(MAXN+MAXE)    !effective pot for fixed L
      REAL CONST              !factor in Numerov algorithm
      REAL VFREE(MAXN+MAXE)   !centrifugal potential
      REAL KI,KIM1,KIP1       !Numerov coefficients for scatt wave
      REAL KIF,KIM1F,KIP1F    !Numerov coefficients for free wave
      REAL G,NUMER,DENOM      !factors to calculate phase shift
      REAL LL                 !useful constant for centrifugal pot
      INTEGER IR,MR           !index for radius
      COMMON/RESULT/PSI,PSIF,PSIMAX,PSFMAX,VEFF  !pass results to GRFOUT
ccccccccccccccccccccccccccccccccccccccccccccccccccccccccccccccccccc
      CONST=DR*DR/HBARM/6           !useful constant
C
C     calculate the centrifugal barrier and effective potential
      LL=L*(L+1)*HBARM/2.0
      DO 200 IR=1,NPTS+NXTRA
         VFREE(IR)=LL/(R(IR)*R(IR))
         VEFF(IR)=V(IR)+VFREE(IR)
200   CONTINUE
C
C     beginning values for Numerov algorithm
C     we are finding both scattered and free wave functions
      PSI(1)=1E-25                  !scattered wave initial conditions
      KIM1=CONST*(E-VEFF(1))        !k of Numerov algorithm
      PSI(2)=(2.-12.*KIM1)*PSI(1)   !use information from the second
      KI=CONST*(E-VEFF(2))          ! derivative to get started
C
      PSIF(1)=1E-25                 !free wave initial conditions
      KIM1F=CONST*(E-VFREE(1))
      PSIF(2)=(2.-12.*KIM1F)*PSIF(1)
      KIF=CONST*(E-VFREE(2))
C
      DO 300 IR=2,NPTS+NXTRA-1
C
         KIP1=CONST*(E-VEFF(IR+1))       !scattered wave K
         PSI(IR+1)=((2-10*KI)*PSI(IR)-(1+KIM1)*PSI(IR-1))/(1+KIP1)
         IF (PSI(IR+1).GT.1E+15) THEN    !prevent overflows
            DO 400 MR=1,IR+1
400            PSI(MR)=PSI(MR)*0.00001
         ENDIF
         KIM1=KI                         !roll values
         KI=KIP1
C
```

```
            KIP1F=CONST*(E-VFREE(IR+1))            !free wave  K
            PSIF(IR+1)=((2-10*KIF)*PSIF(IR)-(1+KIM1F)*PSIF(IR-1))/(1+KIP1F)
            IF (PSIF(IR+1).GT.1E+15) THEN          !prevent overflows
               DO 500 MR=1,IR+1
500               PSIF(MR)=PSIF(MR)*0.00001
            ENDIF
            KIM1F=KIF                              !roll values
            KIF=KIP1F
C
300     CONTINUE
C
C       calculate delta (see Eq. IV.8)
        G=RMAX*PSI(NPTS+NXTRA)/((RMAX+RXTRA)*PSI(NPTS))
        NUMER=G*JL(1,L)-JL(2,L)
        DENOM=G*NL(1,L)-NL(2,L)
        DELTA(L)=ATAN(NUMER/DENOM)
        CALL FIXDEL(DELTA(L),PSI,PSIF,PSIMAX,PSFMAX) !fix delta absolutely
C
        RETURN
        END
CCCCCCCCCCCCCCCCCCCCCCCCCCCCCCCCCCCCCCCCCCCCCCCCCCCCCCCCCCCCCCCCCCCCCCCC
        SUBROUTINE FIXDEL(DELTA,PSI,PSIF,PSIMAX,PSFMAX)
C resolves the ambiguity in DELTA using info from the wave functions
C PSI and PSIF; finds maxima in wave functions PSIMAX, PSIFMAX
CCCCCCCCCCCCCCCCCCCCCCCCCCCCCCCCCCCCCCCCCCCCCCCCCCCCCCCCCCCCCCCCCCCCCCCC
C Global variables:
        INCLUDE 'PARAM.P4'
C Passed Variables:
        REAL DELTA                              !phase shift (I/O)
        REAL PSI(MAXN+MAXE),PSIF(MAXN+MAXE)     !wave functions (input)
        REAL PSIMAX,PSFMAX                       !max of wave functions(output)
C Local Variables:
        INTEGER BUMP,NODES,BUMPF,NODESF         !number of nodes and antinodes
        INTEGER I                               !radial index
        LOGICAL DPOS                            !is the phase shift positive?
        INTEGER N                               !how many factors of pi to add
C Function:
        REAL SGN                                !returns sign of argument
CCCCCCCCCCCCCCCCCCCCCCCCCCCCCCCCCCCCCCCCCCCCCCCCCCCCCCCCCCCCCCCCCCCCCCCC
C       find number of maxima, nodes, and psimax for the scattered wave
        BUMP=0
        NODES=0
        PSIMAX=0.0
        DO 100 I=2,NPTS+NXTRA-1
            IF (SGN(PSI(I)) .NE. SGN(PSI(I-1))) NODES=NODES+1
            IF (ABS(PSI(I)).GT. PSIMAX) PSIMAX=ABS(PSI(I))
            IF ( SGN(PSI(I+1)-PSI(I)) .NE. SGN(PSI(I)-PSI(I-1)) )
     +          BUMP=BUMP+1
100     CONTINUE
C
C       find number of maxima, nodes, and psimax for the free wave
        BUMPF=0
```

```
      NODESF=0
      PSFMAX=0.0
      DO 200 I=2,NPTS+NXTRA-1
         IF (SGN(PSIF(I)) .NE. SGN(PSIF(I-1))) NODESF=NODESF+1
         IF (ABS(PSIF(I)).GT.PSFMAX) PSFMAX=ABS(PSIF(I))
         IF ( SGN(PSIF(I+1)-PSIF(I)) .NE. SGN(PSIF(I)-PSIF(I-1)) )
     +              BUMPF=BUMPF+1
200   CONTINUE
C
      DPOS=DELTA .GT. 0.0            !is the phase shift positive?
C
C     make sign of phase shift agree with convention
C     viz., an attractive potential has positive delta
      N=(NODES+BUMP-NODESF-BUMPF)
      IF (ATTRCT .AND. .NOT. DPOS) THEN
         N=(N+1)/2
      ELSE IF (.NOT. ATTRCT .AND. DPOS) THEN
         N=(N-1)/2
      ELSE
         N=N/2
      END IF
      DELTA=DELTA+N*PI
C
      RETURN
      END
CCCCCCCCCCCCCCCCCCCCCCCCCCCCCCCCCCCCCCCCCCCCCCCCCCCCCCCCCCCCCCCCCCCCCCCC
      REAL FUNCTION SGN(X)
C returns the sign of x
      REAL X
CCCCCCCCCCCCCCCCCCCCCCCCCCCCCCCCCCCCCCCCCCCCCCCCCCCCCCCCCCCCCCCCCCCCCCCC
      IF (X.EQ.0) THEN
         SGN=0.0
      ELSE IF (X.LT.0) THEN
         SGN=-1.0
      ELSE IF (X.GT.0) THEN
         SGN=1.0
      ENDIF
      RETURN
      END
CCCCCCCCCCCCCCCCCCCCCCCCCCCCCCCCCCCCCCCCCCCCCCCCCCCCCCCCCCCCCCCCCCCCCCCC
      SUBROUTINE POTNTL
C fill the array V for choice of potential
CCCCCCCCCCCCCCCCCCCCCCCCCCCCCCCCCCCCCCCCCCCCCCCCCCCCCCCCCCCCCCCCCCCCCCCC
C Global variables:
      INCLUDE 'PARAM.P4'
C Local variables:
      INTEGER IR                    !index for radius
      REAL U                        !temp value for Lenz-Jensen
CCCCCCCCCCCCCCCCCCCCCCCCCCCCCCCCCCCCCCCCCCCCCCCCCCCCCCCCCCCCCCCCCCCCCCCC
C     save values of the radius
      DO 100 IR=1,NPTS+NXTRA
         R(IR)=IR*DR
```

```
100   CONTINUE
C
      IF (POT .EQ. LENZ) THEN           !Lenz-Jensen
         DO 200 IR=1,NPTS
            U=4.5397*Z6*SQRT(R(IR))
            V(IR)=-((Z*E2)/R(IR))*EXP(-U)*
     +           (1+U+U*U*(0.3344+U*(0.0485+0.002647*U)))
200      CONTINUE
         ATTRCT=.TRUE.
C
      ELSE IF (POT .EQ. SQUARE) THEN       !square well
         DO 300 IR=1,NPTS*3/4
            V(IR)=-VZERO
300      CONTINUE
         DO 400 IR=NPTS*3/4,NPTS
            V(IR)=0.0
400      CONTINUE
         IF (VZERO .LE. 0) ATTRCT=.FALSE.
         IF (VZERO .GT. 0) ATTRCT=.TRUE.
C
      ELSE IF (POT .EQ. GAUSS) THEN        !Gaussian well
         DO 500 IR=1,NPTS
            V(IR)=-VZERO*EXP(-R(IR)*R(IR))
500      CONTINUE
         IF (VZERO .LE. 0) ATTRCT=.FALSE.
         IF (VZERO .GT. 0) ATTRCT=.TRUE.
C
      ENDIF
C
C     for all potentials, V is zero outside RMAX
      DO 600 IR=NPTS+1,NPTS+NXTRA
         V(IR)=0.
600   CONTINUE
C
      RETURN
      END
CCCCCCCCCCCCCCCCCCCCCCCCCCCCCCCCCCCCCCCCCCCCCCCCCCCCCCCCCCCCCCCCCCCCCCC
      SUBROUTINE LPINIT
C calculates the Legendre Polynomials P0 and P1
CCCCCCCCCCCCCCCCCCCCCCCCCCCCCCCCCCCCCCCCCCCCCCCCCCCCCCCCCCCCCCCCCCCCCCC
C Global Variables:
      INCLUDE 'PARAM.P4'
C Local Variables:
      INTEGER    ITHETA                !indexes the angle
CCCCCCCCCCCCCCCCCCCCCCCCCCCCCCCCCCCCCCCCCCCCCCCCCCCCCCCCCCCCCCCCCCCCCCC
      DO 100 ITHETA=0,NANG
         THETA(ITHETA)=ITHETA*PI/NANG
         DEGREE(ITHETA)=ITHETA*180/NANG
         CTHETA(ITHETA)=COS(THETA(ITHETA))
         PL(0,ITHETA)=1
         PL(1,ITHETA)=CTHETA(ITHETA)
100   CONTINUE
```

```
      LTABLE=1
C
      RETURN
      END
CCCCCCCCCCCCCCCCCCCCCCCCCCCCCCCCCCCCCCCCCCCCCCCCCCCCCCCCCCCCCCCCCCCCCCCC
      SUBROUTINE LEGPOL
C calculates the Legendre Polynomials from LTABLE to LSTOP
C using recursion relation
CCCCCCCCCCCCCCCCCCCCCCCCCCCCCCCCCCCCCCCCCCCCCCCCCCCCCCCCCCCCCCCCCCCCCCCC
C Global variables:
      INCLUDE    'PARAM.P4'
C Local Variables:
      INTEGER   ITHETA,L               !index angle and L
      REAL X                           !cos(theta)
CCCCCCCCCCCCCCCCCCCCCCCCCCCCCCCCCCCCCCCCCCCCCCCCCCCCCCCCCCCCCCCCCCCCCCCC
      DO 100 ITHETA=0,NANG
         X=CTHETA(ITHETA)
         DO 200 L=LTABLE,LSTOP-1
            PL(L+1,ITHETA)=((2*L+1)*X*PL(L,ITHETA)-L*PL(L-1,ITHETA))/(L+1)
200      CONTINUE
100   CONTINUE
      LTABLE=LSTOP
C
      RETURN
      END
CCCCCCCCCCCCCCCCCCCCCCCCCCCCCCCCCCCCCCCCCCCCCCCCCCCCCCCCCCCCCCCCCCCCCCCC
      SUBROUTINE SBSSEL
C calculates the Spherical Bessel functions at RMAX and RXTRA+RMAX
C for L values from 0 to LSTOP
CCCCCCCCCCCCCCCCCCCCCCCCCCCCCCCCCCCCCCCCCCCCCCCCCCCCCCCCCCCCCCCCCCCCCCCC
C Global variables:
      INCLUDE    'PARAM.P4'
C Local Variables:
      REAL       J,JM1,JP1             !temp values for JL backwd recursn
      REAL       NORM                  !normalizing factor for JL's
      INTEGER    L,LUPPER              !index of JL and NL
      INTEGER    IR                    !indexes RMAX or RMAX+RXTRA
      REAL       X                     !function argument
CCCCCCCCCCCCCCCCCCCCCCCCCCCCCCCCCCCCCCCCCCCCCCCCCCCCCCCCCCCCCCCCCCCCCCCC
      DO 1000 IR=1,2                   !needed at RMAX and RMAX+RXTRA
C        X values for which we are calculating spherical Bessel functions
         IF (IR .EQ. 1) X=K*RMAX
         IF (IR .EQ. 2) X=K*(RMAX+RXTRA)
C
C        obtain NL's by stable forward recursion
         NL(IR,0)=-COS(X)/X
         NL(IR,1)=-COS(X)/(X*X)-SIN(X)/X
         DO 100 L=1,LSTOP-1
            NL(IR,L+1)=(2*L+1)*NL(IR,L)/X-NL(IR,L-1)
100      CONTINUE
C
C        obtain JL's by stable backward recursion
```

```
          LUPPER=X+10.                    !start at L such that JL is neglgbl
          JP1=0
          J=9.9999999E-21                 !arbitrary beginning value
          DO 200 L=LUPPER,1,-1
              JM1=(2*L+1)*J/X-JP1
              IF ((L-1).LE. LSTOP) JL(IR,L-1)=JM1   !save needed L values
              JP1=J                                 !roll values
              J=JM1
200       CONTINUE
          NORM=SIN(X)/X/JL(IR,0)    !normalize using analytic form of JO
          DO 300 L=0,LSTOP
              JL(IR,L)=NORM*JL(IR,L)
300       CONTINUE
1000  CONTINUE
C
      RETURN
      END
CCCCCCCCCCCCCCCCCCCCCCCCCCCCCCCCCCCCCCCCCCCCCCCCCCCCCCCCCCCCCCCCCCCCCCCCCC
      SUBROUTINE INIT
C initializes constants, displays header screen,
C initializes arrays for input parameters
CCCCCCCCCCCCCCCCCCCCCCCCCCCCCCCCCCCCCCCCCCCCCCCCCCCCCCCCCCCCCCCCCCCCCCCCCC
C Global variables:
      INCLUDE 'IO.ALL'
      INCLUDE 'MENU.ALL'
      INCLUDE 'PARAM.P4'
C Local parameters:
      CHARACTER*80 DESCRP           !program description
      DIMENSION DESCRP(20)
      INTEGER NHEAD,NTEXT,NGRAPH    !number of lines for each description
CCCCCCCCCCCCCCCCCCCCCCCCCCCCCCCCCCCCCCCCCCCCCCCCCCCCCCCCCCCCCCCCCCCCCCCCCC
      CALL SETUP                    !get environment parameters
C
C     display header screen
      DESCRP(1)= 'PROJECT 4'
      DESCRP(2)= 'Partial-wave solution to quantum scattering'
      NHEAD=2
C
C     text output description
      DESCRP(3)= 'phase shift and total cross section for each L;'
      DESCRP(4)= 'differential cross section at several angles'
      NTEXT=2
C
C     graphics output description
      DESCRP(5)='Veffective, scattered wave, and free wave vs. radius;'
      DESCRP(6)='differential cross section vs. angle'
      NGRAPH=2
C
      CALL HEADER(DESCRP,NHEAD,NTEXT,NGRAPH)
C
C     set constant values
      PI=4.0*ATAN(1.)
```

```
      E2=14.409
      RMAX=2.0
      HBARM=7.6359
C
      CALL MENU                         !setup constant part of menu
C
      MTYPE(13)=MTITLE
      MPRMPT(13)='Potential Function Options:'
      MLOLIM(13)=2
      MHILIM(13)=1
C
      MTYPE(14)=MTITLE
      MPRMPT(14)='1) Lenz-Jensen Potential: electron & neutral atom'
      MLOLIM(14)=0
      MHILIM(14)=0
C
      MTYPE(15)=MTITLE
      MPRMPT(15)='2) Square Well'
      MLOLIM(15)=0
      MHILIM(15)=0
C
      MTYPE(16)=MTITLE
      MPRMPT(16)='3) Gaussian Well'
      MLOLIM(16)=0
      MHILIM(16)=1
C
      MTYPE(17)=MCHOIC
      MPRMPT(17)='Enter Choice'
      MTAG(17)='18 20 20'
      MLOLIM(17)=1
      MHILIM(17)=3
      MINTS(17)=1
      MREALS(17)=1.
C
      MTYPE(18)=NUM
      MPRMPT(18)='Enter charge of the atomic nucleus'
      MTAG(18)='Z'
      MLOLIM(18)=1
      MHILIM(18)=108
      MINTS(18)=4
C
      MTYPE(19)=SKIP
      MREALS(19)=21
C
      MTYPE(20)=FLOAT
      MPRMPT(20)='Enter depth of potential well (eV)'
      MTAG(20)='Vzero (eV)'
      MLOLIM(20)=-5000.
      MHILIM(20)=5000.
      MREALS(20)= 50.0
C
      MTYPE(21)=FLOAT
```

```
              MPRMPT(21)='Enter Energy (eV)'
              MTAG(21)='Energy (eV)'
              MLOLIM(21)=0.0001
              MHILIM(21)=1000.
              MREALS(21)=20.0
       C
              MTYPE(22)=SKIP
              MREALS(22)=35
       C
              MTYPE(38)=NUM
              MPRMPT(38)='Number of integration points to r1'
              MTAG(38)='number of integration points to r1'
              MLOLIM(38)=100
              MHILIM(38)=MAXN
              MINTS(38)=400
       C
              MTYPE(39)=NUM
              MPRMPT(39)='Number of integration points between r1 and r2'
              MTAG(39)='number of extra integration points'
              MLOLIM(39)=20
              MHILIM(39)=MAXE
              MINTS(39)=60
       C
              MTYPE(40)=NUM
              MPRMPT(40)='Number of angles'
              MTAG(40)='number of angles'
              MLOLIM(40)=1
              MHILIM(40)=MAXANG
              MINTS(40)=36
       C
              MTYPE(41)=SKIP
              MREALS(41)=60
       C
              MSTRNG(MINTS(75))= 'proj4.txt'
       C
              MTYPE(76)=SKIP
              MREALS(76)=80.
       C
              MSTRNG(MINTS(86))= 'proj4.grf'
       C
              MTYPE(87)=SKIP
              MREALS(87)=90.
       C
              RETURN
              END
       CCCCCCCCCCCCCCCCCCCCCCCCCCCCCCCCCCCCCCCCCCCCCCCCCCCCCCCCCCCCCCCCCCCCCCCCCCC
              SUBROUTINE PARAM
       C gets parameters from screen
       C ends program on request
       C closes old files
       C maps menu variables to program variables
       C opens new files
```

```
C calculates all derivative parameters
ccccccccccccccccccccccccccccccccccccccccccccccccccccccccccccccccccccccc
C Global Variables:
      INCLUDE 'MENU.ALL'
      INCLUDE 'IO.ALL'
      INCLUDE 'PARAM.P4'
C Local Variables:
      INTEGER GETINT            !get integer input from terminal
      INTEGER LDEF              !default value for L stop
C map between menu items and parameters
      INTEGER IPOT,IZ,IVZERO,IE,INPTS,INXTRA,INANG
      PARAMETER (IPOT   = 17 )
      PARAMETER (IZ     = 18 )
      PARAMETER (IVZERO = 20 )
      PARAMETER (IE     = 21 )
      PARAMETER (INPTS  = 38 )
      PARAMETER (INXTRA = 39 )
      PARAMETER (INANG  = 40 )
C Function:
      LOGICAL LOGCVT            !converts 1 and 0 to true and false
ccccccccccccccccccccccccccccccccccccccccccccccccccccccccccccccccccccccc
C     get input from terminal
      CALL CLEAR
      CALL ASK(1,ISTOP)
C
C     stop program if requested
      IF (MREALS(IMAIN) .EQ. STOP) CALL FINISH
C
C     close files if necessary
      IF (TNAME .NE. MSTRNG(MINTS(ITNAME)))
     +    CALL FLCLOS(TNAME,TUNIT)
      IF (GNAME .NE. MSTRNG(MINTS(IGNAME)))
     +    CALL FLCLOS(GNAME,GUNIT)
C
C     set new parameter values
C     physical and numerical
      POT=MINTS(IPOT)
      Z=MINTS(IZ)
      VZERO=MREALS(IVZERO)
      E=MREALS(IE)
      NPTS=MINTS(INPTS)
      NXTRA=MINTS(INXTRA)
C
C     calculate derivative parameters:
      Z6=Z**0.166667
      K2=2*E/HBARM
      K=SQRT(K2)
      LMAX=K*RMAX/2+4
      DR=RMAX/NPTS
      RXTRA=DR*NXTRA
C
      CALL POTNTL               !fill the potential array
```

```
C
C       prompt for range of partial waves after displaying LMAX
        WRITE (OUNIT,10) LMAX
10      FORMAT (' The value for LMAX at this energy is = ',I3)
        WRITE (OUNIT,*) ' '
        LSTART=GETINT(0,0,MAXL,' Enter a value for L start')
        LDEF=MAX(LSTART,LMAX)
        LSTOP=GETINT(LDEF,LSTART,MAXL,' Enter a value for L stop')
        CALL CLEAR
C
C       calculate Legendre Polynomials if not already done
        IF (NANG .NE. MINTS(INANG)) THEN            !angles have changed
            NANG=MINTS(INANG)
            CALL LPINIT
        END IF
C       need to have Legendre polynomials all the way to LSTOP
        IF (LTABLE .LT. LSTOP) CALL LEGPOL
C
C       calculate spherical bessel functions
        CALL SBSSEL
C
C       text output
        TTERM=LOGCVT(MINTS(ITTERM))
        TFILE=LOGCVT(MINTS(ITFILE))
        TNAME=MSTRNG(MINTS(ITNAME))
C
C       graphics output
        GTERM=LOGCVT(MINTS(IGTERM))
        GHRDCP=LOGCVT(MINTS(IGHRD))
        GFILE=LOGCVT(MINTS(IGFILE))
        GNAME=MSTRNG(MINTS(IGNAME))
C
C       open files
        IF (TFILE) CALL FLOPEN(TNAME,TUNIT)
        IF (GFILE) CALL FLOPEN(GNAME,GUNIT)
        !files may have been renamed
        MSTRNG(MINTS(ITNAME))=TNAME
        MSTRNG(MINTS(IGNAME))=GNAME
C
        RETURN
        END
CCCCCCCCCCCCCCCCCCCCCCCCCCCCCCCCCCCCCCCCCCCCCCCCCCCCCCCCCCCCCCCCCCCCCCC
        SUBROUTINE PRMOUT(MUNIT,NLINES)
C outputs parameters to MUNIT, keeping track of number of lines printed
CCCCCCCCCCCCCCCCCCCCCCCCCCCCCCCCCCCCCCCCCCCCCCCCCCCCCCCCCCCCCCCCCCCCCCC
C Global variables:
        INCLUDE 'IO.ALL'
        INCLUDE 'PARAM.P4'
C Passed variables:
        INTEGER MUNIT               !unit number for output (input)
        INTEGER NLINES              !number of lines written so far (I/O)
CCCCCCCCCCCCCCCCCCCCCCCCCCCCCCCCCCCCCCCCCCCCCCCCCCCCCCCCCCCCCCCCCCCCCCC
```

```
      IF (MUNIT .EQ. OUNIT) CALL CLEAR
C
      WRITE (MUNIT,2)
      WRITE (MUNIT,4)
      WRITE (MUNIT,6) E
      WRITE (MUNIT,8) LSTART,LSTOP
      WRITE (MUNIT,20) NPTS
      WRITE (MUNIT,22) NXTRA
      WRITE (MUNIT,24) NANG
C
      IF (POT .EQ. LENZ) THEN
          WRITE (MUNIT,10) Z
      ELSE IF (POT .EQ. SQUARE) THEN
          WRITE (MUNIT,12) VZERO
      ELSE IF (POT .EQ. GAUSS) THEN
          WRITE (MUNIT,14) VZERO
      END IF
C
C     different header for text and graphics files
      IF (MUNIT .EQ. GUNIT) THEN
        WRITE (MUNIT,2)
      ELSE
        WRITE (MUNIT,2)
        WRITE (MUNIT,16)
        WRITE (MUNIT,17)
        WRITE (MUNIT,18)
      END IF
C
      NLINES=11
C
2     FORMAT (' ')
4     FORMAT (' Output from project 4: Partial-wave solution'
     +         ' to quantum scattering')
6     FORMAT (' Energy (eV) =', 1PE10.3)
8     FORMAT (' L start = ' , I3, 10X,' L stop = ', I3)
10    FORMAT (' Lenz Jensen potential with Z = ', I3)
12    FORMAT (' Square well potential with Vzero = ' 1PE10.3)
14    FORMAT (' Gaussian potential with Vzero = ' 1PE10.3)
20    FORMAT (' Number of integration points = ', I4)
22    FORMAT (' Number of points between r1 and r2 = ', I4)
24    FORMAT (' Number of angles = ', I3)
16    FORMAT (15X,'L',15X,'Delta(L)',17X,'Sigma(L)')
17    FORMAT (15X,' ',15X,'degrees ',15X,'Angstroms**2')
18    FORMAT (15X,'-',15X,'--------',15X,'------------')
C
      RETURN
      END
CCCCCCCCCCCCCCCCCCCCCCCCCCCCCCCCCCCCCCCCCCCCCCCCCCCCCCCCCCCCCCCCCCCCCCC
      SUBROUTINE TXTOUT(MUNIT,NLINES,L,DELDEG,SIGMA)
C writes out phase shift and total cross section for each partial wave
C to MUNIT, keeping track of number of lines printed
CCCCCCCCCCCCCCCCCCCCCCCCCCCCCCCCCCCCCCCCCCCCCCCCCCCCCCCCCCCCCCCCCCCCCCC
```

```
C Global variables:
      INCLUDE 'IO.ALL'
      INCLUDE 'PARAM.P4'
C Passed variables:
      INTEGER MUNIT              !output unit specifier (input)
      INTEGER NLINES             !number of lines printed to screen(I/O)
      INTEGER L                  !partial wave number(input)
      REAL DELDEG,SIGMA          !phase shift and cross section(input)
CCCCCCCCCCCCCCCCCCCCCCCCCCCCCCCCCCCCCCCCCCCCCCCCCCCCCCCCCCCCCCCCCCCCCCCCCC
C     if screen is full, clear and type headings again
      IF ((MOD(NLINES,TRMLIN-4) .EQ. 0)
     +                         .AND. (MUNIT .EQ. OUNIT)) THEN
         CALL PAUSE('to continue...',1)
         CALL CLEAR
         WRITE (MUNIT,16)
         WRITE (MUNIT,17)
         WRITE (MUNIT,18)
         WRITE (MUNIT,2)
         NLINES=NLINES+4
      END IF
C
      WRITE (MUNIT,20) L,DELDEG,SIGMA
C     keep track of printed lines only for terminal output
      IF (MUNIT .EQ. OUNIT) NLINES=NLINES+1
C
20    FORMAT(13X,I3,13X,1PE10.3,13X,1PE15.8)
2     FORMAT (' ')
16    FORMAT (15X,'L',15X,'Delta(L)',17X,'Sigma(L)')
17    FORMAT (15X,' ',15X,'degrees ',15X,'Angstroms**2')
18    FORMAT (15X,'-',15X,'--------',15X,'------------')
C
      RETURN
      END
CCCCCCCCCCCCCCCCCCCCCCCCCCCCCCCCCCCCCCCCCCCCCCCCCCCCCCCCCCCCCCCCCCCCCCCCCC
      SUBROUTINE DELOUT(DELDEG,SIGMA)
C outputs delta and partial wave cross sections to screen
CCCCCCCCCCCCCCCCCCCCCCCCCCCCCCCCCCCCCCCCCCCCCCCCCCCCCCCCCCCCCCCCCCCCCCCCCC
C Global variables:
      INCLUDE 'IO.ALL'
      INCLUDE 'PARAM.P4'
C Passed variables:
      REAL DELDEG(0:MAXL)        !phase shift in degrees (input)
      REAL SIGMA(0:MAXL),SIGTOT  !total cross section (input)
      INTEGER L                  !partial wave index (input)
C Local variables:
      INTEGER NLINES             !number of lines written so far (I/O)
CCCCCCCCCCCCCCCCCCCCCCCCCCCCCCCCCCCCCCCCCCCCCCCCCCCCCCCCCCCCCCCCCCCCCCCCCC
      CALL PRMOUT(OUNIT,NLINES)
      DO 100 L=LSTART,LSTOP
         CALL TXTOUT(OUNIT,NLINES,L,DELDEG(L),SIGMA(L))
100   CONTINUE
      RETURN
```

```
      END
cccccccccccccccccccccccccccccccccccccccccccccccccccccccccccccccccccccc
      SUBROUTINE SIGOUT(MUNIT,F,DSIGMA,SIGTOT)
C writes out scattering angle, scattering amplitude,
C differential cross section and total cross section to MUNIT
cccccccccccccccccccccccccccccccccccccccccccccccccccccccccccccccccccccc
C Global variables:
      INCLUDE 'IO.ALL'
      INCLUDE 'PARAM.P4'
C Input variables:
      INTEGER MUNIT                  !output unit specifier
      REAL DSIGMA(0:MAXANG)          !differential cross section
      COMPLEX F(0:MAXANG)            !scattering amplitude
      INTEGER ITHETA                 !index of theta
      REAL SIGTOT                    !total cross section
C Local variables:
      INTEGER NLINES                 !keep track of lines printed out
cccccccccccccccccccccccccccccccccccccccccccccccccccccccccccccccccccccc
      WRITE (MUNIT,2)
      WRITE (MUNIT,2)
      WRITE (MUNIT,4) SIGTOT
      WRITE (MUNIT,2)
      WRITE (MUNIT,2)
      WRITE (MUNIT,10)
      WRITE (MUNIT,11)
      WRITE (MUNIT,12)
C
      NLINES=8
C
C     write out data, allowing time for user to examine full pages
      DO 100 ITHETA=0,NANG
         WRITE (MUNIT,14) DEGREE(ITHETA),F(ITHETA),DSIGMA(ITHETA)
         NLINES=NLINES+1
         IF ((MOD(NLINES,TRMLIN-4) .EQ. 0) .AND.
     +                 (MUNIT .EQ. OUNIT)) THEN
            CALL PAUSE('to continue...',1)
            CALL CLEAR
         END IF
100   CONTINUE
C
2     FORMAT (' ')
4     FORMAT (15X,'TOTAL CROSS SECTION = ',1PE15.8,' Angstroms**2')
10    FORMAT (10X,'Theta' ,14X,'Amplitude (Re,Im)',15X,'dSigma/dTheta')
11    FORMAT (9X,'degrees',13X,' Angstrons**2  ',15X,' Angstrcms**2')
12    FORMAT (9X,'-------',13X,'-----------------',15X,'-------------')
14    FORMAT (7X,F8.2,6X,'(',1PE15.8,',',1PE15.8,')',6X,1PE15.8)
C
      RETURN
      END
cccccccccccccccccccccccccccccccccccccccccccccccccccccccccccccccccccccc
      SUBROUTINE GRFOUT(DEVICE,L,DELDEG,SIGMA)
C outputs potential and both free and scattered wavefunctions for one L:
```

```
C does not allow for hardcopy or file output
C in order to avoid the creation of voluminous files
CCCCCCCCCCCCCCCCCCCCCCCCCCCCCCCCCCCCCCCCCCCCCCCCCCCCCCCCCCCCCCCCCCCCCC
C Global variables
      INCLUDE 'IO.ALL'
      INCLUDE 'PARAM.P4'
      INCLUDE 'GRFDAT.ALL'
C Input variables:
      INTEGER DEVICE                  !which device is being used?
      INTEGER L                       !partial wave
      REAL DELDEG,SIGMA               !phase shift and cross section
      REAL PSI(MAXN+MAXE),PSIF(MAXN+MAXE)   !wave functions
      REAL PSIMAX,PSFMAX              !maxima of wave functions
      REAL VEFF(MAXN+MAXE)           !effective pot for fixed L
      COMMON/RESULT/PSI,PSIF,PSIMAX,PSFMAX,VEFF!pass results from Numerv
C Local variables
      INTEGER IR                      !index of lattice
      REAL VGRF(MAXE+MAXN)            !V scaled for graphing
      REAL PSIG(MAXE+MAXN),PSIFG(MAXE+MAXN)   !rescaled wave functions
      REAL ENERGY(2),RADIUS(2)        !array for plotting energy
      CHARACTER*9 CL,CDEL,CSIG,CE     !character version of data
      INTEGER LEN,LCDEL,LCSIG         !length of strings
      INTEGER SCREEN                  !send to terminal
      INTEGER PAPER                   !make a hardcopy
      INTEGER FILE                    !send to a file
      DATA SCREEN,PAPER,FILE/1,2,3/
CCCCCCCCCCCCCCCCCCCCCCCCCCCCCCCCCCCCCCCCCCCCCCCCCCCCCCCCCCCCCCCCCCCCCC
C     messages for the impatient
      IF (DEVICE .NE. SCREEN) WRITE (OUNIT,100)
C
C     setup graphing parameters
      IF (DEVICE .NE. FILE) THEN
          NPLOT=1                       !how many plots?
          IPLOT=1
C
          XMIN=0.0                      !axis parameters
          XMAX=RMAX+RXTRA
          YOVAL=XMIN
          YMIN=-2*E                     !energy axis varies between
          YMAX=+4*E                     !-2E and 4E
          X0VAL=0.
C
          NPOINT=NPTS+NXTRA
C
          LABEL(1)= 'radius (Angstroms)' !titles and labels
          LABEL(2)= 'Veff(eV), scattered wave (solid),'
     +              //' and free (ooo) wave'
          CALL CONVRT(E,CE,LEN)
          IF (POT .EQ. LENZ) THEN
            TITLE = ' Lenz-Jensen Potential, Energy (eV)='//CE
          ELSE IF (POT .EQ. SQUARE) THEN
            TITLE = 'Square Well Potential, Energy (eV)='//CE
```

```
           ELSE IF (POT .EQ. GAUSS) THEN
             TITLE = 'Gaussian Potential, Energy (eV)='//CE
           END IF
C
C          arrays for plotting energy
           RADIUS(1)=0.
           RADIUS(2)=XMAX
           ENERGY(1)=E
           ENERGY(2)=E
C
C          plot Veffective between -2E and 4E
           DO 30 IR=1,NPTS+NXTRA              !normalize VEFF
             VGRF(IR)=VEFF(IR)
             IF (VGRF(IR) .GT. YMAX) VGRF(IR)=YMAX
             IF (VGRF(IR) .LT. YMIN) VGRF(IR)=YMIN
30         CONTINUE
C
C          rescale wave functions so that it fits on the energy scale
           DO 10 IR=1,NPTS+NXTRA
             PSIG(IR)=PSI(IR)*3*E/PSIMAX+E
             PSIFG(IR)=PSIF(IR)*3*E/PSFMAX+E
10         CONTINUE
C
           ILINE=1                            !line and symbol styles
           ISYM=1
           IFREQ=0
           NXTICK=5
           NYTICK=5
C
           CALL GTDEV(DEVICE)                 !device nomination
           IF (DEVICE .EQ. SCREEN) CALL GMODE !change to graphics mode
C
C          create the legend
           CALL ICNVRT(L,CL,LEN)
           CALL CONVRT(DELDEG,CDEL,LCDEL)
           CALL CONVRT(SIGMA,CSIG,LCSIG)
           INFO='Delta(L='//CL(1:LEN)//')='//CDEL(1:LCDEL)//
     +         ' deg'//
     +         ' Sig(L='//CL(1:LEN)//')='//CSIG(1:LCSIG)//' A**2'
C
           CALL LNLNAX
C
           CALL XYPLOT (R,VGRF)               !plot potential
           NPOINT=2
           CALL XYPLOT (RADIUS,ENERGY)        !plot energy
C
           NPOINT=NPTS+NXTRA
           CALL XYPLOT (R,PSIG)               !plot wave function
           IFREQ=4
           CALL XYPLOT (R,PSIFG)              !plot free wave function
C
         END IF
```

```
C
C       end graphing session
        IF (DEVICE .NE. FILE) CALL GPAGE(DEVICE)   !end graphics package
        IF (DEVICE .EQ. SCREEN) CALL TMODE         !switch to text mode
C
100     FORMAT (/,' Patience, please; output going to a file.')
C
        RETURN
        END
CCCCCCCCCCCCCCCCCCCCCCCCCCCCCCCCCCCCCCCCCCCCCCCCCCCCCCCCCCCCCCCCCCCCCCCCCC
        SUBROUTINE GRFSIG(DEVICE,DSIGMA,SIGTOT)
C graphs differential cross section vs. angle for all partial waves
C on a semi-log scale
CCCCCCCCCCCCCCCCCCCCCCCCCCCCCCCCCCCCCCCCCCCCCCCCCCCCCCCCCCCCCCCCCCCCCCCCCC
C Global variables
        INCLUDE 'IO.ALL'
        INCLUDE 'PARAM.P4'
        INCLUDE 'GRFDAT.ALL'
C Input variables:
        INTEGER DEVICE          !which device is being used?
        REAL SIGTOT             !total cross section
        REAL DSIGMA(0:MAXANG)   !differential cross section
C Local variables
        INTEGER ITHETA          !indexes angle
        INTEGER EXPMAX,EXPMIN   !min and max exp for diff cross section
        CHARACTER*9 CE,CSIG     !energy,sigma as a character string
        INTEGER LEN             !string length
        INTEGER SCREEN          !send to terminal
        INTEGER PAPER           !make a hardcopy
        INTEGER FILE            !send to a file
        DATA SCREEN,PAPER,FILE/1,2,3/
CCCCCCCCCCCCCCCCCCCCCCCCCCCCCCCCCCCCCCCCCCCCCCCCCCCCCCCCCCCCCCCCCCCCCCCCCC
C       messages for the impatient
        IF (DEVICE .NE. SCREEN) WRITE (OUNIT,100)
C
C       calculate parameters for graphing
        IF (DEVICE .NE. FILE) THEN
C
           NPLOT=1                      !how many plots?
           IPLOT=1
C
           YMAX=0.                      !find limits on data points
           YMIN=DSIGMA(1)
           DO 20 ITHETA=0,NANG
              IF (DSIGMA(ITHETA) .GT. YMAX) YMAX=DSIGMA(ITHETA)
              IF (DSIGMA(ITHETA) .LT. YMIN) YMIN=DSIGMA(ITHETA)
20         CONTINUE
           !find integer limits on exponent
           EXPMAX=INT(LOG10(YMAX))
           IF (YMAX .GT. 1.) EXPMAX =EXPMAX+1
           EXPMIN=INT(LOG10(YMIN))
           IF (YMIN .LT. 1.) EXPMIN=EXPMIN-1
```

```
      YMAX=10.**EXPMAX
      YMIN=10.**EXPMIN
C
      XMIN=DEGREE(0)
      XMAX=DEGREE(NANG)
      YOVAL=XMIN
      XOVAL=YMIN
C
      NPOINT=NANG
C
      ILINE=1                          !line and symbol styles
      ISYM=4
      IFREQ=1
      NXTICK=4
      NYTICK=EXPMAX-EXPMIN
      IF (NYTICK .GT. 8) THEN          !keep number of ticks small
         IF (MOD(NYTICK,2) .EQ. 0) THEN
            NYTICK=NYTICK/2
         ELSE
            NYTICK=8
         END IF
      END IF
C
      CALL CONVRT(E,CE,LEN)            !titles and labels
      IF (POT .EQ. LENZ) THEN
         TITLE = ' Lenz-Jensen Potential, Energy (eV)='//CE
      ELSE IF (POT .EQ. SQUARE) THEN
         TITLE = 'Square Well Potential, Energy (eV)='//CE
      ELSE IF (POT .EQ. GAUSS) THEN
         TITLE = 'Gaussian Potential, Energy (eV)='//CE
      END IF
      CALL CONVRT(SIGTOT,CSIG,LEN)
      INFO='Total cross section in A**2 = '//CSIG
      LABEL(1)= 'Angle (degrees)'
      LABEL(2)= 'Differential Cross Section (Angstroms**2)'
C
      CALL GTDEV(DEVICE)                     !device nomination
      IF (DEVICE .EQ. SCREEN) CALL GMODE     !change to graphics mode
      CALL LGLNAX                            !draw axes
      END IF
C
C     output results
      IF (DEVICE .EQ. FILE) THEN
         WRITE (GUNIT,10)
         WRITE (GUNIT,11)
         WRITE (GUNIT,12)
         WRITE (GUNIT,14) (DEGREE(ITHETA),DSIGMA(ITHETA),ITHETA=0,NANG)
      ELSE
         CALL XYPLOT (DEGREE,DSIGMA)
      END IF
C
C     end graphing session
```

```
          IF (DEVICE .NE. FILE) CALL GPAGE(DEVICE)  !end graphics package
          IF (DEVICE .EQ. SCREEN) CALL TMODE         !switch to text mode
C
10        FORMAT (21X,'Theta',   21X,'dSigma/dTheta')
11        FORMAT (20X,'degrees',20X,' Angstroms**2')
12        FORMAT (20X,'-------',20X,'-------------')
14        FORMAT (18X,1PE10.3,18X,1PE15.8)
100       FORMAT (/,' Patience, please; output going to a file.')
C

          RETURN
          END
CCCCCCCCCCCCCCCCCCCCCCCCCCCCCCCCCCCCCCCCCCCCCCCCCCCCCCCCCCCCCCCCCCCCCCCC
CCCCCCCCCCCCCCCCCCCCCCCCCCCCCCCCCCCCCCCCCCCCCCCCCCCCCCCCCCCCCCCCCCCCCCCC
C param.p4
C
          INTEGER MAXN,MAXE               !limits on integration steps
          INTEGER MAXL                    !limits on partial waves
          INTEGER MAXANG                  !limits on number of angles
C
          PARAMETER (MAXN=1000)
          PARAMETER (MAXE=100)
          PARAMETER (MAXL=100)
          PARAMETER (MAXANG=100)
C
          INTEGER Z                       !charge of nucleus
          REAL E                          !energy of particle
          REAL VZERO                      !depth of potential
          INTEGER POT                     !which potential
          INTEGER LENZ,SQUARE,GAUSS       !types of potentials
          REAL V(MAXN+MAXE)               !potential function
          LOGICAL ATTRCT                  !is V attractive?
C
          PARAMETER (LENZ=1)
          PARAMETER (SQUARE=2)
          PARAMETER (GAUSS=3)
C
          INTEGER NPTS,NXTRA              !number of integration points
          INTEGER LSTART,LSTOP            !range of partial waves
          INTEGER NANG                    !number of angles
C
          REAL E2                         !square of charge on electron
          REAL HBARM                      !Planck's constant**2/mass of elec
                                          !in eV * Angstroms**2
          REAL RMAX                       !maximum radius of potential
          REAL PI                         !constants
C
          REAL K,K2                       !wave number
          REAL Z6                         !sixth root of Z
          REAL DR                         !step size
          INTEGER LMAX                    !estimate of largest partial wave
          REAL RXTRA                      !value for r2-r1
          INTEGER LTABLE                  !how many Leg. pol. are in table
```

```
      REAL  THETA(0:MAXANG),CTHETA(0:MAXANG)  !theta and cos(theta)
      REAL  DEGREE(0:MAXANG)            !theta in degrees
      REAL  PL(0:MAXL,0:MAXANG)         !Legendre polynomials
      REAL  JL(2,0:MAXL),NL(2,0:MAXL)   !spherical Bessel func at r1,r2
      REAL  R(MAXN+MAXE)                !array of radii
C
      COMMON/PPARAM/Z,E,VZERO,POT
      COMMON/NPARAM/NPTS,NXTRA,LSTART,LSTOP,NANG
      COMMON/CONSTS/E2,HBARM,RMAX,PI
      COMMON/PCALC/Z6,K,K2,DR,LMAX,RXTRA,V,LTABLE,PL,JL,NL,THETA,
     +            CTHETA,R,DEGREE,ATTRCT
```

C.5 Project 5

Algorithm   This program solves the schematic shell model described in the text for a specified number of particles and coupling strength. The bases of states with even and odd m are treated separately. For the set of parameters input, the tri-diagonal Hamiltonian matrices of Eq. (V.9) are constructed (subroutine HAMLTN), and all of the negative eigenvalues are found by searching for zeros of the determinant of $(H - \lambda I)$. The Gerschgorin bounds on the eigenvalues, Eq. (5.11), are used to guide the search (subroutine DGNLZ), and the zeros are computed by the recursion formula (5.10) in subroutine PLYNML. After an eigenvalue has been found, the associated eigenvector is found by two inverse vector iterations (subroutine INVITR), if requested, where the matrix $(H - EI)$ is inverted in subroutine MATINV. In subroutine TXTOUT $\langle J_z \rangle$ is calculated for each eigenvector and the eigenvalues are sorted into ascending order.

Input   The physical parameters are the number of particles [14], the initial value of χ [1.], the increment for χ [.5], the number of values for χ [10], and whether or not the eigenvectors are to be calculated. The maximum number of χ values is fixed by MXNCHI= 500. Only even numbers of particles are allowed; this limitation is imposed to reduce the bookkeeping. The numerical parameters are the tolerance for the eigenvalue search [1.E-05] and the small factor used to keep $(H - EI)$ nonsingular [1.E-05].

Output   For each value of χ, the program displays χ, V, the negative eigenvalues, and $\langle J_z \rangle$ (if the eigenvectors have been requested). Once all eigenvalues for all χ values have been calculated, the graphics outputs the eigenvalues *vs.* χ (if more than one value of χ is requested) or simply the eigenvalues (if only one χ value is requested.)

```
CCCCCCCCCCCCCCCCCCCCCCCCCCCCCCCCCCCCCCCCCCCCCCCCCCCCCCCCCCCCCCCCCCCCCCCCCC
      PROGRAM PROJ5
C     PROJECT 5: Solution of a schematic shell model
C COMPUTATIONAL PHYSICS (FORTRAN VERSION)
C by Steven E. Koonin and Dawn C. Meredith
C Copyright 1989, Addison-Wesley Publishing Company
CCCCCCCCCCCCCCCCCCCCCCCCCCCCCCCCCCCCCCCCCCCCCCCCCCCCCCCCCCCCCCCCCCCCCCCCCC
      CALL INIT              !display header screen, setup parameters
5     CONTINUE               !main loop/ execute once for each set of param
      CALL PARAM             !get input from screen
      CALL ARCHON            !find eigenvalues and eigenvectors
      GOTO 5
      END
CCCCCCCCCCCCCCCCCCCCCCCCCCCCCCCCCCCCCCCCCCCCCCCCCCCCCCCCCCCCCCCCCCCCCCCCCC
      SUBROUTINE ARCHON
C calculates the eigenvalues and (optionally) the eigenvectors of a
```

```
C schematic shell model for several values of the coupling strength CHI
cccccccccccccccccccccccccccccccccccccccccccccccccccccccccccccccccccccccc
C Global variables:
      INCLUDE 'IO.ALL'
      INCLUDE 'PARAM.P5'
C Local variables
      REAL CEVEN(MAXDIM,MAXDIM),CODD(MAXDIM,MAXDIM)   !eigenvectors
      REAL EVEVEN(MXNCHI,MAXDIM),EVODD(MXNCHI,MAXDIM)!eigenvalues
      INTEGER JCHI                 !CHI index
      REAL V                       !coupling strength
      INTEGER NLINES               !number of lines printed out
      INTEGER SCREEN               !send to terminal
      INTEGER PAPER                !make a hardcopy
      INTEGER FILE                 !send to a file
      DATA SCREEN,PAPER,FILE/1,2,3/
cccccccccccccccccccccccccccccccccccccccccccccccccccccccccccccccccccccccc
C     output summary of parameters
      IF (TTERM) CALL PRMOUT(OUNIT,NLINES)
      IF (TFILE) CALL PRMOUT(TUNIT,NLINES)
      IF (GFILE) CALL PRMOUT(GUNIT,NLINES)
C
      DO 100 JCHI=1,NCHI      !for each value of CHI find the eigenvalues
        V=CHI(JCHI)/(NPART-1)    !and eigenvectors; output text results
        CALL HAMLTN(JCHI,V,CEVEN,CODD,EVEVEN,EVODD)
        CALL TXTOUT(JCHI,V,CEVEN,CODD,EVEVEN,EVODD,NLINES)
100   CONTINUE
      IF (TTERM) CALL PAUSE('to continue...',1)
      IF (TTERM) CALL CLEAR
C
C     graphics output; plot is different if there is only one CHI value
      IF (NCHI .NE. 1) THEN
        IF (GTERM) CALL GRFOUT(SCREEN,EVEVEN,EVODD)
        IF (GFILE) CALL GRFOUT(FILE,EVEVEN,EVODD)
        IF (GHRDCP) CALL GRFOUT(PAPER,EVEVEN,EVODD)
      ELSE
        IF (GTERM) CALL ONEOUT(SCREEN,EVEVEN,EVODD)
        IF (GFILE) CALL ONEOUT(FILE,EVEVEN,EVODD)
        IF (GHRDCP) CALL ONEOUT(PAPER,EVEVEN,EVODD)
      END IF
C
      RETURN
      END
cccccccccccccccccccccccccccccccccccccccccccccccccccccccccccccccccccccccc
      SUBROUTINE HAMLTN(JCHI,V,CEVEN,CODD,EVEVEN,EVODD)
C subroutine to set up and diagonalize the Hamiltonian for one
C value of the coupling, V
cccccccccccccccccccccccccccccccccccccccccccccccccccccccccccccccccccccccc
C Global variables:
      INCLUDE 'PARAM.P5'
C Input Variables
      INTEGER JCHI                 !index of the coupling
      REAL V                       !coupling strength
```

```
C Output variables:
      REAL CEVEN(MAXDIM,MAXDIM),CODD(MAXDIM,MAXDIM)   !eigenvectors
      REAL EVEVEN(MXNCHI,MAXDIM),EVODD(MXNCHI,MAXDIM)!eigenvalues
C Local Variables
      INTEGER IDIM,ICOMP                  !index on matrix elements
      INTEGER NDIM                        !size of matrix
      REAL DIAG(MAXDIM),LDIAG(MAXDIM)     !diagonal,off diagonal matrix elem
      INTEGER M                           !eigenvalue of Jz
      REAL TEMP                           !temp storage for LDIAG
      INTEGER NFIND                       !number of eigenvalues to find
      REAL EVAL(MAXDIM),TMPVEC(MAXDIM)    !temp storage for values and vect
CCCCCCCCCCCCCCCCCCCCCCCCCCCCCCCCCCCCCCCCCCCCCCCCCCCCCCCCCCCCCCCCCCCCCCCCCCC
C     find evectors and eigenvalues for even-m states
      NDIM=JJ+1                           !number of even-m states
      DO 100 IDIM=1,NDIM                  !set up the matrix elements
        M=-JJ+2*(IDIM-1)
        DIAG(IDIM)=M                        !diagonal matrix element
        IF (IDIM.NE.1) THEN                 !lower off-diag matrix elem
          TEMP=JJ1-M*(M-1)
          TEMP=TEMP*(JJ1-(M-1)*(M-2))
          LDIAG(IDIM)=-V/2*SQRT(ABS(TEMP))
        ENDIF
100   CONTINUE
C
      IF (MOD(JJ,2) .EQ. 0)THEN           !find only neg eigenvalues
        NFIND=1+JJ/2
      ELSE
        NFIND=(JJ+1)/2
      ENDIF
      CALL DGNLZ(DIAG,LDIAG,NDIM,NFIND,EVAL)   !find the eigenvalues
C
      DO 200 IDIM=1,NFIND
        EVEVEN(JCHI,IDIM)=EVAL(IDIM)          !save eigenvalues
        IF (VECTRS) THEN                      !sometimes find eigenvectors
          CALL INVITR(DIAG,LDIAG,EVAL(IDIM),NDIM,TMPVEC)
          DO 300 ICOMP=1,NDIM
            CEVEN(ICOMP,IDIM)=TMPVEC(ICOMP)
300       CONTINUE
        ENDIF
200   CONTINUE
C
C     find evectors and eigenvalues for odd-m states
      NDIM=JJ                             !number of odd-m states
      DO 400 IDIM=1,NDIM                  !set up the matrix elements
        M=-JJ-1+2*IDIM
        DIAG(IDIM)=M                        !diagonal matrix element
        IF (IDIM .NE. 1) THEN               !lower off-diag matrix elem
          TEMP=JJ1-M*(M-1)
          TEMP=TEMP*(JJ1-(M-1)*(M-2))
          LDIAG(IDIM)=-V/2*SQRT(ABS(TEMP))
        ENDIF
400   CONTINUE
```

```
C
      IF (MOD(JJ,2) .EQ. 0) THEN          !find only neg eigenvalues
         NFIND=JJ/2
      ELSE
         NFIND=(JJ+1)/2
      ENDIF
      CALL DGNLZ(DIAG,LDIAG,NDIM,NFIND,EVAL)   !find the eigenvalues
C
      DO 500 IDIM=1,NFIND
         EVODD(JCHI,IDIM)=EVAL(IDIM)       !save eigenvalues
         IF (VECTRS) THEN                  !sometimes find eigenvectors
            CALL INVITR(DIAG,LDIAG,EVAL(IDIM),NDIM,TMPVEC)
            DO 600 ICOMP=1,NDIM
               CODD(ICOMP,IDIM)=TMPVEC(ICOMP)
600         CONTINUE
         ENDIF
500   CONTINUE
C
      RETURN
      END
CCCCCCCCCCCCCCCCCCCCCCCCCCCCCCCCCCCCCCCCCCCCCCCCCCCCCCCCCCCCCCCCCCCCCCC
      SUBROUTINE DGNLZ(DIAG,LDIAG,NDIM,NFIND,EVAL)
C finds NFIND eigenvalues (EVAL) of a NDIM x NDIM symmetric tridiagonal
C matrix with DIAG and LDIAG matrix elements
CCCCCCCCCCCCCCCCCCCCCCCCCCCCCCCCCCCCCCCCCCCCCCCCCCCCCCCCCCCCCCCCCCCCCCC
C Global parameters:
      INCLUDE 'PARAM.P5'
C Input Variables
      REAL DIAG(MAXDIM),LDIAG(MAXDIM)      !diag,off-diag matrix elem
      INTEGER NDIM,NFIND     !matrix dim, number of values to find
C Output Variables:
      REAL EVAL(MAXDIM)        !eigenvalues
C Local Variables
      REAL UBOUND,LBOUND       !bounds on eigenvalues
      REAL RAD,GER             !temp storage for computing bounds
      REAL SPACNG              !estimate of eigenvalue spacing
      REAL DLAM                !step in eigenvalue search
      INTEGER IDIM             !index
      REAL LAMBDA              !current guess for the eigenvalue
      INTEGER COUNT,L          !how many egnvls are .lt. LAMBDA
CCCCCCCCCCCCCCCCCCCCCCCCCCCCCCCCCCCCCCCCCCCCCCCCCCCCCCCCCCCCCCCCCCCCCCC
C     find Gerschgorin bounds on eigenvalues (Eq. 5.11)
      LBOUND=DIAG(1)-ABS(LDIAG(2))
      UBOUND=DIAG(1)+ABS(LDIAG(2))
      DO 100 IDIM=2,NDIM-1
         RAD=ABS(LDIAG(IDIM+1))+ABS(LDIAG(IDIM))
         GER=DIAG(IDIM)-RAD
         LBOUND=AMIN1(GER,LBOUND)
         GER=DIAG(IDIM)+RAD
         UBOUND=AMAX1(GER,UBOUND)
100   CONTINUE
      GER=DIAG(NDIM)-ABS(LDIAG(NDIM))
```

```
        LBOUND=AMIN1(LBOUND,GER)
        GER=DIAG(NDIM)+ABS(LDIAG(NDIM))
        UBOUND=AMAX1(UBOUND,GER)
C
        LAMBDA=LBOUND                    !guess for first eigenvalue
        SPACNG=(UBOUND-LBOUND)/NDIM      !guess for eigenvalue spacing
C
        DO 200 L=1,NFIND                 !loop to find NFIND eigenvalues
           DLAM=SPACNG                   !initial guess for step
C
201        CONTINUE                      !coarse search to find upper
              LAMBDA=LAMBDA+DLAM         !bound for this eigenvalue
              CALL PLYNML(DIAG,LDIAG,LAMBDA,NDIM,COUNT)
           IF (COUNT .LT. L)  GOTO 201
C
           LAMBDA=LAMBDA-DLAM            !restart search
202        CONTINUE                      !search on a much finer scale
              CALL PLYNML(DIAG,LDIAG,LAMBDA,NDIM,COUNT)
              DLAM=DLAM/2                 !next step is half as big
              IF (COUNT .LE. L-1) THEN
                 LAMBDA=LAMBDA+DLAM      !not there yet, step forward
              ELSE
                 LAMBDA=LAMBDA-DLAM      !went too far, step backward
              ENDIF
           IF (DLAM .GT. LTOLE)  GOTO 202
           EVAL(L)=LAMBDA
200     CONTINUE
C
        RETURN
        END
CCCCCCCCCCCCCCCCCCCCCCCCCCCCCCCCCCCCCCCCCCCCCCCCCCCCCCCCCCCCCCCCCCCCCCCC
        SUBROUTINE PLYNML(DIAG,LDIAG,LAMBDA,NDIM,COUNT)
C counts the eigenvalues less than LAMBDA
C by evaluating the terms in the sequence of polynomials used to eval
C the determinant and looking for changes in sign of those terms
CCCCCCCCCCCCCCCCCCCCCCCCCCCCCCCCCCCCCCCCCCCCCCCCCCCCCCCCCCCCCCCCCCCCCCCC
C Global variables:
        INCLUDE 'PARAM.P5'
C Input Variables
        REAL   DIAG(MAXDIM),LDIAG(MAXDIM)   !diag,off-diag matrix elem
        INTEGER NDIM,NFIND                  !matrix dim, number of values to find
        REAL LAMBDA                         !guess for the eigenvalue
C Output Variables
        INTEGER COUNT                       !number sign changes in the sequence
C Local Variables
        REAL TEMP,TEMP1,DET                 !terms in the sequence
        INTEGER IDIM                        !index
CCCCCCCCCCCCCCCCCCCCCCCCCCCCCCCCCCCCCCCCCCCCCCCCCCCCCCCCCCCCCCCCCCCCCCCC
        TEMP1=1                             !first term in the sequence
        DET=DIAG(1)-LAMBDA                  !second term
        IF (DET .LT. 0) THEN               !initialize COUNT
           COUNT=1
```

```
        ELSE
          COUNT=0
        ENDIF
C
        DO 100 IDIM=2,NDIM            !recursion relation for determinant
        TEMP=(DIAG(IDIM)-LAMBDA)*DET-LDIAG(IDIM)**2*TEMP1
        TEMP1=DET                    !roll values
        DET=TEMP
        IF (DET*TEMP1 .LT. 0.) COUNT=COUNT+1    !sign change?
        IF (TEMP1 .EQ. 0.)     COUNT=COUNT+1    !count zeros once
        IF (ABS(DET) .GE. 100000) THEN          !keep things small
          DET=DET/100000
          TEMP1=TEMP1/100000
        ENDIF
100     CONTINUE
C
        RETURN
        END
CCCCCCCCCCCCCCCCCCCCCCCCCCCCCCCCCCCCCCCCCCCCCCCCCCCCCCCCCCCCCCCCCCCCCCCCC
        SUBROUTINE INVITR(DIAG,LDIAG,EVAL,NDIM,TMPVEC)
C finds the eigenvector TMPVEC corresponding to the eigenvalue EVAL
C of a symmetric, tri-diagonal NDIM x NDIM matrix with matrix elements
C DIAG and LDIAG
C eigenvector is found by two inverse iterations of (E-H)
CCCCCCCCCCCCCCCCCCCCCCCCCCCCCCCCCCCCCCCCCCCCCCCCCCCCCCCCCCCCCCCCCCCCCCCCC
C Global variables:
        INCLUDE 'PARAM.P5'
C Input Variables
        INTEGER NDIM                          !size of matrix
        REAL DIAG(MAXDIM),LDIAG(MAXDIM) !diagonal,off diagonal matrix elem
        REAL EVAL                             !eigenvalue
C Output Variables
        REAL TMPVEC(MAXDIM)                   !eigenvector
C Local Variables
        DOUBLE PRECISION A(MAXDIM,MAXDIM) !(E-H) and its inverse
        DOUBLE PRECISION NEWVEC(MAXDIM)   !temp storage for the evector
        INTEGER I,J                       !indices for matrix elements
        REAL NORM,SUM                     !variables for normalizing
        LOGICAL SINGLR                    !is the matrix singular?
        REAL TMPEPS                       !temp factor to avoid singularity
CCCCCCCCCCCCCCCCCCCCCCCCCCCCCCCCCCCCCCCCCCCCCCCCCCCCCCCCCCCCCCCCCCCCCCCCC
        TMPEPS=EPS                        !initial value is input value
C       find the inverse of (E-H)
10      DO 100 I=1,NDIM
          DO 200 J=1,NDIM
200       A(I,J)=0.0              .       !set A=(E-H)
          A(I,I)=(1.D+00+TMPEPS)*DBLE(EVAL)-DBLE(DIAG(I)) !diagonal elem
          IF (I .LT. NDIM) A(I,I+1)=-DBLE(LDIAG(I+1))  !off diagonal elem
          IF (I .GT. 1)    A(I,I-1)=-DBLE(LDIAG(I))
          TMPVEC(I)=1.                    !arbitrary initial vector
100     CONTINUE
        CALL MATINV(A,NDIM,SINGLR)        !find A inverse
```

```
            IF (SINGLR) THEN                   !if A is singular
                TMPEPS=TMPEPS*5.D+00           !try again with bigger EPS
                GOTO 10
            END IF
      C
            DO 300 I=1,NDIM                     !first inverse iteration
                SUM=0.0
                DO 400 J=1,NDIM
      400       SUM=SUM+A(I,J)*TMPVEC(J)
                NEWVEC(I)=SUM
      300   CONTINUE
      C
            NORM=0.0
           -DO_700 I=1,NDIM                     !second inverse iteration
                SUM=0.0                         !and normalization
                DO 800 J=1,NDIM
      800       SUM=SUM+A(I,J)*NEWVEC(J)
                TMPVEC(I)=SUM
                NORM=NORM+SUM*SUM
      700   CONTINUE
            NORM=1/SQRT(NORM)
            DO 900 I=1,NDIM
                TMPVEC(I)=TMPVEC(I)*NORM
      900   CONTINUE
      C
            RETURN
            END
cccccccccccccccccccccccccccccccccccccccccccccccccccccccccccccccccccccccccccc
            SUBROUTINE MATINV(A,NDIM,SINGLR)
C inverts the matrix A of dimension NDIM using Gauss-Jordan elimination
C A is replaced by its inverse
C SINGLR is true if A is singular
C on exit, A is replaced by its inverse
cccccccccccccccccccccccccccccccccccccccccccccccccccccccccccccccccccccccccccc
C Global variables:
            INCLUDE 'IO.ALL'
            INCLUDE 'PARAM.P5'
C Input/output variables:
            DOUBLE PRECISION A(MAXDIM,MAXDIM) !matrix for inverting
            INTEGER NDIM                       !dimension of matrix
            LOGICAL SINGLR                     !is the matrix singular?
C Local variables:
            DOUBLE PRECISION P(MAXDIM),Q(MAXDIM) !temporary storage
            LOGICAL ZEROED(MAXDIM)             !keeps track of who's zeroed
            INTEGER I,J,K                       !indices
            INTEGER KBIG                        !index of row being zeroed
            DOUBLE PRECISION PIVOT             !largest diag element in A
            REAL BIG                            !largest diag elem of non zeroed rows
cccccccccccccccccccccccccccccccccccccccccccccccccccccccccccccccccccccccccccc
            SINGLR=.FALSE.
      C     which rows are zeroed? none so far
            DO 10 I=1,NDIM
```

```
          ZEROED(I)=.FALSE.
10    CONTINUE
C
C     loop over all rows
      DO 100 I=1,NDIM
          KBIG=0                         !search for largest diag element
          BIG=0                          !in rows not yet zeroed
          DO 20 J=1,NDIM
             IF (.NOT. ZEROED(J)) THEN
                IF (ABS(A(J,J)) .GT. BIG) THEN
                   BIG=ABS(A(J,J))
                   KBIG=J
                END IF
             END IF
20        CONTINUE
C
C         store the largest diagonal element
          IF (I .EQ. 1) PIVOT=A(KBIG,KBIG)
C         if all diagonals are zero, then the matrix is singular
          IF (KBIG .EQ. 0) THEN
             WRITE (OUNIT,*) ' Matrix is singular'
             SINGLR=.TRUE.
             RETURN
          END IF
C         matrix is ill conditioned if the size of elements varies greatly
          IF (ABS(A(KBIG,KBIG)/PIVOT) .LT. 1.E-14) THEN
             WRITE (OUNIT,*) ' Matrix is ill conditioned'
             RETURN
          END IF
C
C         begin zeroing row KBIG
          ZEROED(KBIG)=.TRUE.
          Q(KBIG)=1/A(KBIG,KBIG)
          P(KBIG)=1.
          A(KBIG,KBIG)=0.
C
C         elements above the diagonal
          IF (KBIG .GT. 1) THEN
             DO 30 J=1,KBIG-1
                P(J)=A(J,KBIG)
                Q(J)=A(J,KBIG)*Q(KBIG)
                IF (.NOT. ZEROED(J)) Q(J)=-Q(J)
                A(J,KBIG)=0.
30           CONTINUE
          END IF
C
C         elements to the right of the diagonal
          IF (KBIG .LT. NDIM) THEN
             DO 40 J=KBIG+1,NDIM
                P(J)=A(KBIG,J)
                Q(J)=-A(KBIG,J)*Q(KBIG)
                IF (ZEROED(J)) P(J)=-P(J)
```

```
                        A(KBIG,J)=0.
       40            CONTINUE
                  END IF
       C
       C      transform all of A
              DO 60 J=1,NDIM
                 DO 50 K=J,NDIM
                    A(J,K)=A(J,K)+P(J)*Q(K)
       50        CONTINUE
       60     CONTINUE
       100    CONTINUE
       C
       C      symmetrize A**-1
              DO 110 J=2,NDIM
                 DO 120 K=1,J-1
                    A(J,K)=A(K,J)
       120       CONTINUE
       110    CONTINUE
       C
              RETURN
              END
CCCCCCCCCCCCCCCCCCCCCCCCCCCCCCCCCCCCCCCCCCCCCCCCCCCCCCCCCCCCCCCCCCCCCCCC
       SUBROUTINE TXTOUT(JCHI,V,CEVEN,CODD,EVEVEN,EVODD,NLINES)
C calculates <Jz> and outputs results
CCCCCCCCCCCCCCCCCCCCCCCCCCCCCCCCCCCCCCCCCCCCCCCCCCCCCCCCCCCCCCCCCCCCCCCC
C Global variables:
       INCLUDE    'IO.ALL'
       INCLUDE    'PARAM.P5'
C Input Variables
       REAL       CEVEN(MAXDIM,MAXDIM),CODD(MAXDIM,MAXDIM)   !eigenvectors
       REAL       EVEVEN(MXNCHI,MAXDIM),EVODD(MXNCHI,MAXDIM) !eigenvalues
       INTEGER    JCHI                                       !CHI index
       REAL       V                            !coupling strength
       INTEGER    NLINES                       !number of lines printed out
C Local Variables
       INTEGER    I,K,IDIM          !indices
       INTEGER    M                 !eigenvalue of Jz
       REAL       E,JZ              !eigenvalue and Jz
       INTEGER    DELINE            !number of lines this time round
CCCCCCCCCCCCCCCCCCCCCCCCCCCCCCCCCCCCCCCCCCCCCCCCCCCCCCCCCCCCCCCCCCCCCCCC
       DELINE=3+JJ+1               !number of lines written per call
C      clear the screen at a convenient place
       IF ((NLINES+DELINE  .GT. TRMLIN-6) .AND. (TTERM)) THEN
           CALL PAUSE('to continue...',1)
           CALL CLEAR
           NLINES=0
       END IF
C
       IF (TTERM) THEN                    !write out parameters
          WRITE (OUNIT,10) CHI(JCHI),V
          IF (VECTRS) THEN
             WRITE (OUNIT,20)
```

```
              WRITE (OUNIT,30)
         ELSE
              WRITE (OUNIT,50)
              WRITE (OUNIT,60)
         END IF
         NLINES=NLINES+3
      END IF
      IF (TFILE) THEN
         WRITE (TUNIT,10) CHI(JCHI),V
         IF (VECTRS) THEN
              WRITE (TUNIT,20)
              WRITE (TUNIT,30)
         ELSE
              WRITE (TUNIT,50)
              WRITE (TUNIT,60)
         END IF
      END IF
C
      DO 100 IDIM=1,JJ+1           !loop over all eigenvalues
         IF (MOD(IDIM,2) .EQ. 0) THEN      !putting them in order
              K=IDIM/2
              E=EVODD(JCHI,K)
              IF (VECTRS) THEN                   !find the expectation value
                 JZ=0                            ! of Jz if have eigenvectors
                 DO 300 I=1,JJ
                    M=-JJ-1+2*I
                    JZ=JZ+M*CODD(I,K)**2         !contribution to <Jz>
300              CONTINUE
              ENDIF
         ELSE                                    !even m states
              K=(IDIM+1)/2
              E=EVEVEN(JCHI,K)
              IF (VECTRS) THEN                   !their contribution to <Jz>
                 JZ=0
                 DO 200 I=1,JJ+1
                    M=-JJ+2*(I-1)
                    JZ=JZ+M*CEVEN(I,K)**2
200              CONTINUE
              ENDIF
         ENDIF
C        write it all out
         IF (VECTRS .AND. TTERM) WRITE (OUNIT,40) E,JZ
         IF (VECTRS .AND. TFILE) WRITE (TUNIT,40) E,JZ
         IF (.NOT. VECTRS .AND. TTERM) WRITE (OUNIT,70) E
         IF (.NOT. VECTRS .AND. TFILE) WRITE (TUNIT,70) E
         IF (TTERM) NLINES=NLINES+1
         IF ((TTERM) .AND. (NLINES .EQ. TRMLIN-6)) THEN
              CALL PAUSE('to continue...',1)
              CALL CLEAR
              NLINES=0
         END IF
100   CONTINUE
```

```
C
10     FORMAT (23X,'Chi=',1PE11.4,5X,'V=',1PE11.4)
20     FORMAT (20X,'Eigenvalue',25X,'<Jz>')
30     FORMAT (20X,'----------',25X,'----')
40     FORMAT (17X,1PE15.8,17X,1PE15.8)
50     FORMAT (35X,'Eigenvalue')
60     FORMAT (35X,'----------')
70     FORMAT (32X,1PE15.8)
C
       RETURN
       END
CCCCCCCCCCCCCCCCCCCCCCCCCCCCCCCCCCCCCCCCCCCCCCCCCCCCCCCCCCCCCCCCCCCCCCCC
       SUBROUTINE INIT
C initializes constants, displays header screen,
C initializes arrays for input parameters
CCCCCCCCCCCCCCCCCCCCCCCCCCCCCCCCCCCCCCCCCCCCCCCCCCCCCCCCCCCCCCCCCCCCCCCC
C Global variables:
       INCLUDE 'IO.ALL'
       INCLUDE 'MENU.ALL'
       INCLUDE 'PARAM.P5'
C Local parameters:
       CHARACTER*80 DESCRP          !program description
       DIMENSION DESCRP(20)
       INTEGER NHEAD,NTEXT,NGRAPH   !number of lines for each description
CCCCCCCCCCCCCCCCCCCCCCCCCCCCCCCCCCCCCCCCCCCCCCCCCCCCCCCCCCCCCCCCCCCCCCCC
       CALL SETUP                   !get environment parameters
C
C      display header screen
       DESCRP(1)= 'PROJECT 5'
       DESCRP(2)= 'Solution of a Schematic Shell Model'
       NHEAD=2
C
C      text output description
       DESCRP(3)= 'Negative eigenvalues and <Jz>'
       NTEXT=1
C
C      graphics output description
       DESCRP(4)= 'Eigenvalues vs. coupling strength'
       NGRAPH=1
C
       CALL HEADER(DESCRP,NHEAD,NTEXT,NGRAPH)
C
       CALL MENU                    !setup constant part of menu
C
       MTYPE(13)=NUM
       MPRMPT(13)='Enter the number of particles (must be even)'
       MTAG(13)='Number of particles'
       MLOLIM(13)=1
       MHILIM(13)=MAXN
       MINTS(13)=14
C
       MTYPE(14)=FLOAT
```

```
      MPRMPT(14)='Enter the intial value of coupling constant (Chi)'
      MTAG(14)= 'Initial coupling constant'
      MLOLIM(14)=0.
      MHILIM(14)=1000.
      MREALS(14)=1.0
C
      MTYPE(15)=FLOAT
      MPRMPT(15)='Enter increment in coupling constant (Chi)'
      MTAG(15)='Increment in coupling constant'
      MLOLIM(15)=0.
      MHILIM(15)=1000.
      MREALS(15)=.5
C
      MTYPE(16)=NUM
      MPRMPT(16)='Enter the number of values for coupling constant'
      MTAG(16)='Number of values of coupling constant'
      MLOLIM(16)=1
      MHILIM(16)=MXNCHI
      MINTS(16)=10
C
      MTYPE(17)=BOOLEN
      MPRMPT(17)='Do you want the eigenvectors <Jz> calculated?'
      MTAG(17)='Calculate eigenvectors'
      MINTS(17)=1
C
      MTYPE(18)=SKIP
      MREALS(18)=35
C
      MTYPE(38)=FLOAT
      MPRMPT(38)='Enter the tolerance for eigenvalue search'
      MTAG(38)='Tolerance for eigenvalue search'
      MLOLIM(38)=1.E-07
      MHILIM(38)=1.
      MREALS(38)=1.E-05
C
      MTYPE(39)=FLOAT
      MPRMPT(39)='Enter small factor to keep (E-H) nonsingular'
      MTAG(39)='Factor to keep (E-H) nonsingular'
      MLOLIM(39)=1.E-07
      MHILIM(39)=1.
      MREALS(39)=1.E-05
C
      MTYPE(40)=SKIP
      MREALS(40)=60
C
      MSTRNG(MINTS(75))= 'proj5.txt'
C
      MTYPE(76)=SKIP
      MREALS(76)=80.
C
      MSTRNG(MINTS(86))= 'proj5.grf'
C
```

```
              MTYPE(87)=SKIP
              MREALS(87)=90.
      C
              RETURN
              END
      CCCCCCCCCCCCCCCCCCCCCCCCCCCCCCCCCCCCCCCCCCCCCCCCCCCCCCCCCCCCCCCCCCCCCCCCCCC
              SUBROUTINE PARAM
      C gets parameters from screen
      C ends program on request
      C closes old files
      C maps menu variables to program variables
      C opens new files
      C calculates all derivative parameters
      CCCCCCCCCCCCCCCCCCCCCCCCCCCCCCCCCCCCCCCCCCCCCCCCCCCCCCCCCCCCCCCCCCCCCCCCCCC
      C Global variables:
              INCLUDE 'MENU.ALL'
              INCLUDE 'IO.ALL'
              INCLUDE 'PARAM.P5'
      C Local variables:
              INTEGER JCHI                    !CHI index
      C       map between menu indices and input parameters
              INTEGER INPART,ICHI,IDELCH,INCHI,IVEC,ILTOLE,IEPS,INGRF
              PARAMETER (INPART = 13)
              PARAMETER (ICHI   = 14)
              PARAMETER (IDELCH = 15)
              PARAMETER (INCHI  = 16)
              PARAMETER (IVEC   = 17)
              PARAMETER (ILTOLE = 38)
              PARAMETER (IEPS   = 39)
              PARAMETER (INGRF  = 87 )
      C Functions:
              LOGICAL LOGCVT                  !converts 1 and 0 to true and false
              INTEGER GETINT                  !get integer input from screen
      CCCCCCCCCCCCCCCCCCCCCCCCCCCCCCCCCCCCCCCCCCCCCCCCCCCCCCCCCCCCCCCCCCCCCCCCCCC
      C       get input from terminal
              CALL CLEAR
              CALL ASK(1,ISTOP)
      C
      C       stop program if requested
              IF (MREALS(IMAIN) .EQ. STOP) CALL FINISH
      C
      C       close files if necessary
              IF (TNAME .NE. MSTRNG(MINTS(ITNAME)))
             +      CALL FLCLOS(TNAME,TUNIT)
              IF (GNAME .NE. MSTRNG(MINTS(IGNAME)))
             +      CALL FLCLOS(GNAME,GUNIT)
      C
      C       set new parameter values
      C       physical and numerical
              NPART=MINTS(INPART)
              CHI(1)=MREALS(ICHI)
              NCHI=MINTS(INCHI)
```

```
        DELCHI=MREALS(IDELCH)
        VECTRS=LOGCVT(MINTS(IVEC))
        LTOLE=MREALS(ILTOLE)
        EPS=MREALS(IEPS)
C
C       text output
        TTERM=LOGCVT(MINTS(ITTERM))
        TFILE=LOGCVT(MINTS(ITFILE))
        TNAME=MSTRNG(MINTS(ITNAME))
C
C       graphics output
        GTERM=LOGCVT(MINTS(IGTERM))
        GHRDCP=LOGCVT(MINTS(IGHRD))
        GFILE=LOGCVT(MINTS(IGFILE))
        GNAME=MSTRNG(MINTS(IGNAMF):
C
C       open files
        IF (TFILE) CALL FLOPEN(TNAME,TUNIT)
        IF (GFILE) CALL FLOPEN(GNAME,GUNIT)
C       files may have been renamed
        MSTRNG(MINTS(ITNAME))=TNAME
        MSTRNG(MINTS(IGNAME))=GNAME
C
C       make sure that NPART is even (to avoid lots of bookkeeping)
10      IF (MOD(NPART,2) .NE. 0) THEN
            NPART=GETINT(NPART-1,2,MAXN,
     +              ' Number of particles must be even, try again')
            MINTS(INPART)=NPART
        GOTO 10
        END IF
        CALL CLEAR
C
C       calculated parameters
        JJ=NPART/2
        JJ1=JJ*(JJ+1)
        DO 100 JCHI=2,NCHI
            CHI(JCHI)=CHI(1)+(JCHI-1)*DELCHI
100     CONTINUE
C
        RETURN
        END
CCCCCCCCCCCCCCCCCCCCCCCCCCCCCCCCCCCCCCCCCCCCCCCCCCCCCCCCCCCCCCCCCCCCCCCC
        SUBROUTINE PRMOUT(MUNIT,NLINES)
C outputs parameter summary to the specified unit
CCCCCCCCCCCCCCCCCCCCCCCCCCCCCCCCCCCCCCCCCCCCCCCCCCCCCCCCCCCCCCCCCCCCCCCC
C Global variables:
        INCLUDE 'IO.ALL'
        INCLUDE 'PARAM.P5'
C Input variables:
        INTEGER MUNIT           !unit number for output
        INTEGER NLINES          !number of lines printed out (I/O)
CCCCCCCCCCCCCCCCCCCCCCCCCCCCCCCCCCCCCCCCCCCCCCCCCCCCCCCCCCCCCCCCCCCCCCCC
```

```
          IF (MUNIT .EQ. OUNIT) CALL CLEAR
C
          WRITE (MUNIT,2)
          WRITE (MUNIT,4)
          WRITE (MUNIT,6)NPART
          WRITE (MUNIT,10)JJ
C
          NLINES=4
C
C
2         FORMAT(1X,/)
4         FORMAT (' Output from project 5:',
        +    ' Solution of a schematic shell model')
6         FORMAT (1X,'Number of Particles = ',1I3)
10        FORMAT (1X,'Quasi Spin = ',1I3)
C
          RETURN
          END
CCCCCCCCCCCCCCCCCCCCCCCCCCCCCCCCCCCCCCCCCCCCCCCCCCCCCCCCCCCCCCCCCCCCCCCCCC
          SUBROUTINE GRFOUT(DEVICE,EVEVEN,EVODD)
C graphs eigenvalues vs. CHI
CCCCCCCCCCCCCCCCCCCCCCCCCCCCCCCCCCCCCCCCCCCCCCCCCCCCCCCCCCCCCCCCCCCCCCCCCC
C Global parameters:
          INCLUDE 'IO.ALL'
          INCLUDE 'PARAM.P5'
          INCLUDE 'GRFDAT.ALL'
C Input variables
          INTEGER DEVICE                   !which device are we calling
          REAL EVEVEN(MXNCHI,MAXDIM),EVODD(MXNCHI,MAXDIM)!eigenvalues
C Local parameters:
          INTEGER JCHI,IDIM                !CHI and DIM indices
          CHARACTER*9 CNPART               !NPART as a character string
          INTEGER LNPART                   !length of that string
          REAL EVAL(MAXDIM)                !one eigenvalue at several CHI
          INTEGER SCREEN                   !send to terminal
          INTEGER PAPER                    !make a hardcopy
          INTEGER FILE                     !send to a file
          DATA SCREEN,PAPER,FILE/1,2,3/
CCCCCCCCCCCCCCCCCCCCCCCCCCCCCCCCCCCCCCCCCCCCCCCCCCCCCCCCCCCCCCCCCCCCCCCCCC
C     messages for the impatient
          IF (DEVICE .NE. SCREEN) WRITE (OUNIT,100)
C
C     calculate parameters for graphing
          IF (DEVICE .NE. FILE) THEN
              NPLOT=1                      !how many plots
              IPLOT=1
C
              YMAX=0.                      !limits
              YMIN=EVEVEN(NCHI,1)
              DO 10 JCHI=1,NCHI-1
                  IF (EVEVEN(JCHI,1) .LT. YMIN) YMIN=EVEVEN(JCHI,1)
10            CONTINUE
              YMIN=YMIN+.01*(YMIN)
```

```
        XMIN=CHI(1)                         !leave a little extra room
        XMAX=CHI(NCHI)
        YOVAL=XMIN
        XOVAL=YMIN
C
        NPOINT=NCHI
C
        ILINE=1                             !line and symbol styles
        ISYM=1
        IFREQ=1
        NXTICK=5
        NYTICK=5
C
        CALL ICNVRT(NPART,CNPART,LNPART)    !titles
        TITLE='Schematic Shell Model with '
    +       //CNPART(1:LNPART)//' particles'
        LABEL(1)='CHI'
        LABEL(2)='eigenvalues'
C
        CALL GTDEV(DEVICE)                  !device nomination
        IF (DEVICE .EQ. SCREEN) CALL GMODE  !change to graphics mode
        CALL LNLNAX                         !draw axes
      END IF
C
C     output results
      DO 60 IDIM=1,JJ+1        !loop over all eigenvalues
        DO 50 JCHI=1,NCHI      !  and values of CHI
          IF (MOD(IDIM,2) .EQ. 0) EVAL(JCHI)=EVODD(JCHI,IDIM/2)
          IF (MOD(IDIM,2) .EQ. 1) EVAL(JCHI)=EVEVEN(JCHI,(IDIM+1)/2)
50      CONTINUE
        IF (DEVICE .EQ. FILE)THEN
          WRITE (GUNIT,70)
          WRITE (GUNIT,80) (CHI(JCHI),EVAL(JCHI),JCHI=1,NCHI)
        ELSE
          CALL XYPLOT(CHI,EVAL)
        END IF
60    CONTINUE
C
C     end graphing session
      IF (DEVICE .NE. FILE) CALL GPAGE(DEVICE)   !end graphics package
      IF (DEVICE .EQ. SCREEN) CALL TMODE         !switch to text mode
C
70    FORMAT (5X,'    CHI      ',5X,' Eigenvalue ')
80    FORMAT (2(5X,1PE15.8))
100   FORMAT (/,' Patience, please; output going to a file.')
C
      RETURN
      END
cccccccccccccccccccccccccccccccccccccccccccccccccccccccccccccccccccccccccccc
      SUBROUTINE ONEOUT(DEVICE,EVEVEN,EVODD)
C graphs eigenvalues for one value of CHI
C allows user to visually inspect eigenvalue spacings
```

```
CCCCCCCCCCCCCCCCCCCCCCCCCCCCCCCCCCCCCCCCCCCCCCCCCCCCCCCCCCCCCCCCCCCCCCCCC
C Global parameters:
      INCLUDE 'IO.ALL'
      INCLUDE 'PARAM.P5'
      INCLUDE 'GRFDAT.ALL'
C Input variables
      INTEGER DEVICE                     !which device are we calling
      REAL EVEVEN(MXNCHI,MAXDIM),EVODD(MXNCHI,MAXDIM) !eigenvalues
C Local parameters:
      INTEGER JCHI,IDIM                  !CHI and DIM indices
      CHARACTER*9 CNPART                 !NPART as a character string
      INTEGER LNPART                     !length of that string
      CHARACTER*9 CCHI                   !CHI as a character string
      INTEGER LCHI                       !length of that string
      REAL EVAL(2),X(2)                  !eigenvalue, dummy variable X
      INTEGER SCREEN                     !send to terminal
      INTEGER PAPER                      !make a hardcopy
      INTEGER FILE                       !send to a file
      DATA SCREEN,PAPER,FILE/1,2,3/
CCCCCCCCCCCCCCCCCCCCCCCCCCCCCCCCCCCCCCCCCCCCCCCCCCCCCCCCCCCCCCCCCCCCCCCCC
C     messages for the impatient
      IF (DEVICE .NE. SCREEN) WRITE (OUNIT,100)
C
C     calculate parameters for graphing
      IF (DEVICE .NE. FILE) THEN
          NPLOT=1                        !how many plots
          IPLOT=1
C
          YMAX=0.                        !limits
          YMIN=(1.01)*EVEVEN(1,1)
          XMIN=0.                        !x-axis is 'dummy' axis
          XMAX=10.
          YOVAL=XMIN
          X0VAL=YMIN
          X(1)=XMIN
          X(2)=1.
C
          NPOINT=2
C
          ILINE=1                        !line and symbol styles
          ISYM=1
          IFREQ=0
          NXTICK=5
          NYTICK=5
C
          CALL ICNVRT(NPART,CNPART,LNPART) !titles
          CALL CONVRT(CHI,CCHI,LCHI)
          TITLE='Schematic Shell Model with '
     +        //CNPART(1:LNPART)//' particles and CHI='//CCHI(1:LCHI)
          LABEL(1)=' '
          LABEL(2)='eigenvalues'
C
```

```
          CALL GTDEV(DEVICE)                    !device nomination
          IF (DEVICE .EQ. SCREEN) CALL GMODE    !change to graphics mode
          CALL LNLNAX                           !draw axes
       END IF
C
C      output results
       DO 60 IDIM=1,JJ+1          !loop over all eigenvalues
          DO 50 JCHI=1,2
             IF (MOD(IDIM,2) .EQ. 0) EVAL(JCHI)=EVODD(1,IDIM/2)
             IF (MOD(IDIM,2) .EQ. 1) EVAL(JCHI)=EVEVEN(1,(IDIM+1)/2)
50        CONTINUE
          IF (DEVICE .EQ. FILE) THEN
             WRITE (GUNIT,70)
             WRITE (GUNIT,80)    (X(JCHI),EVAL(JCHI),JCHI=1,2)
          ELSE
             CALL XYPLOT(X,EVAL)
          END IF
60     CONTINUE
C
C      end graphing session
       IF (DEVICE .NE. FILE) CALL GPAGE(DEVICE)   !end graphic package
       IF (DEVICE .EQ. SCREEN) CALL TMODE         !switch to text mode
C
70     FORMAT (5X,'      Dummy     ',5X,'  Eigenvalue   ')
80     FORMAT (2(5X,1PE15.8))
100    FORMAT (/,' Patience, please; output going to a file.')
C
       RETURN
       END
CCCCCCCCCCCCCCCCCCCCCCCCCCCCCCCCCCCCCCCCCCCCCCCCCCCCCCCCCCCCCCCCCCCCCCCCCC
CCCCCCCCCCCCCCCCCCCCCCCCCCCCCCCCCCCCCCCCCCCCCCCCCCCCCCCCCCCCCCCCCCCCCCCCCC
C param.p5
       INTEGER    MXNCHI             !maximum number of chi values
       INTEGER    MAXN               !maximum number of particles
       INTEGER    MAXDIM             !maximum dimension of matrices
       PARAMETER  (MAXN=100)
       PARAMETER  (MXNCHI=500)
       PARAMETER  (MAXDIM=MAXN/2+1)
C
       INTEGER    NPART              !number of particles
       REAL       CHI(MXNCHI),DELCHI !coupling constant and increment
       INTEGER    NCHI               !number of coupling constants
       LOGICAL    VECTRS             !do we want eigenvectors?
       REAL       LTOLE              !tolerance for eigenvalue search
       REAL       EPS                !factor to keep (E-H) nonsingular
C
       INTEGER JJ,JJ1                !J value, J*(J+1) value
C
       COMMON/PPARAM/NPART,CHI,NCHI,DELCHI,VECTRS,LTOLE,EPS
       COMMON/PCALC/JJ,JJ1
```

C.6 Project 6

Algorithm This program uses a relaxation method to solve for the stationary, incompressible flow around a plate in two dimensions; the geometry is as described in the text. The main relaxation loop begins at line 99 in subroutine ARCHON. Subroutines PRELAX and VRELAX relax the stream functions and vorticity, respectively, by solving Eqs. (VI.9) and (VI.10). Only those lines in the relaxation routines marked by the comment "interior point" are unaffected by the boundary conditions specified in the text. These sweeps can be restricted to a sublattice. For this problem, it is essential to under-relax the lattice in order to avoid crashing the program, especially at high Reynold's numbers. After each sweep the viscous and pressure forces are calculated in subroutine FORCES. Every NFREQ'th iteration you are given the option to continue iterating or to return to the main menu.

Input The input procedure is a bit different for this program, as it is for Example 6. For these two codes, you may find that during the iteration procedure you want to keep the field values and all parameters, except one or two, the same. For example, you may wish to alter just the stream relaxation parameter or change the frequency at which the fields are displayed. Such a procedure is straightforward, but only as long as the lattice dimensions remain the same. Therefore, the menu is broken into two pieces: the first level obtains the lattice dimensions; the second level (or main menu) obtains all other parameters. As long as you do not enter the first level menu, *all parameters and field values are unchanged unless you specifically request otherwise.*

The first level prompts you for the lattice parameters NX [20] and NY [20]. The maximum lattice sizes are fixed by two considerations: first, the parameters NXMAX= 100 and NYMAX= 100 which determine the array sizes; second, the size of your terminal as defined by TRMWID and TRMLIN in subroutine SETUP. This last restriction is imposed so that the fields can be displayed on the screen, using characters to indicate field strength (subroutine DISPLY). If you will be using a graphics package, you can remove the terminal size restrictions on NX and NY in subroutine LATTIC. If it is not your first time through this menu, you will also be given the option of ending program execution [no], and (if that answer is *no*) resetting input parameters to their default values. Note that all parameters that depend on lattice dimensions (i.e., the field values, the plate position, and sublattice parameters) are automatically reset every time you enter this first level menu.

The physical parameters are the half-width of the plate (in lattice spacings) [NY/2], the length of the plate [NX/4], the location of the front edge [NX/3], and the lattice Reynold's number [1.]. The numerical parameters are the relaxation parameters for the vorticity [.3] and stream function [.3], the option to relax a sublattice [no] specified by the lower left and upper right corners, and the initial field values. Initial field values at non-boundary points can: 1) be read in from a file [proj6.in] created earlier by this program; 2) be set to the free-streaming values (subroutine INTCND); or 3) remain unchanged. Option 2 is the default if the calculation is just beginning, otherwise option 3 is the default. The two graphics parameters are the number of contours for plotting [10] (used only if you have a graphics package), and the frequency at which the field is displayed [10]. This is also the frequency at which you are asked if you would like to continue iterations. The maximum number of contours is fixed by MAXLEV= 20.

Output After each relaxation sweep, the text output displays the iteration number (iteration count is zeroed on return from the main menu), pressure force, viscous force, and minimum and maximum values of the two fields. Every NFREQ'th iteration, the program will display the field. Note that the field will be displayed as contours (if you have requested graphics sent to the screen, see subroutine GRFOUT) or as characters (if you have requested text to the screen, see subroutine DISPLY). In the character display, small letters indicate negative values while numbers and capital letters indicate positive values; e.g., 'z' corresponds to the most negative field value; and 'Z', to the most positive. With this convention, the spacing in field values between the small letters will not be the same spacing as between the capital letters. In the character display the plate is indicated by blanks, while for the contour plots it is indicated by a dashed line. The appearance of the contour lines depends on your graphics package. Because display of the field can slow down calculations considerably, you may want to set NREQ to a large number.

Again, to avoid voluminous output, the field is not written out to a file or sent hardcopy device every NFREQ time. Instead, when you request to stop iterations, you will be asked if you want the field values to be written to requested output devices. Note that the field can be written out in three ways: as character data (if text is requested to a file), as numerical data (if graphics is requested to a file), or as a contour plot (if graphics is requested to a hardcopy device).

```
ccccccccccccccccccccccccccccccccccccccccccccccccccccccccccccccccccccc
         PROGRAM PROJ6
C    Project 6: 2-D viscous incompressible flow about a rectangular block
C    COMPUTATIONAL PHYSICS (FORTRAN VERSION)
C    by Steven E. Koonin and Dawn C. Meredith
C    Copyright 1989, Addison-Wesley Publishing Company
ccccccccccccccccccccccccccccccccccccccccccccccccccccccccccccccccccccc
         CALL INIT              !display header screen, setup parameters
5        CONTINUE               !main loop/ execute once for each set of param
         CALL LATTIC            !get lattice size, allow for ending
         CALL ARCHON            !get detailed bound cond and relax lattice
         GOTO 5
         END
ccccccccccccccccccccccccccccccccccccccccccccccccccccccccccccccccccccc
         SUBROUTINE ARCHON
C subroutine to get parameters, relax the lattice, and output results
ccccccccccccccccccccccccccccccccccccccccccccccccccccccccccccccccccccc
C Global variables:
         INCLUDE 'PARAM.P6'
         INCLUDE 'IO.ALL'
C Local variables:
         REAL P(NXMAX,NYMAX)               !stream function
         REAL XSI(NXMAX,NYMAX)             !vorticity
         REAL VFORCE,PFORCE                !viscous force and pressure force
         REAL XSIMAX,PMAX,XSIMIN,PMIN      !min,max values of XSI, P
         INTEGER XMIN,XMAX,YMIN,YMAX       !corners of relaxing lattice
         LOGICAL ITER                      !continue iterating?
         LOGICAL PRM                       !print out parameters?
         INTEGER CHOICE                    !write out field now?
         INTEGER NITER                     !number of iterations
         LOGICAL END                       !end this run?
         INTEGER IX,IY                     !lattice indices
         INTEGER NLINES                    !number of lines sent to terminal
         INTEGER SCREEN                    !send to terminal
         INTEGER PAPER                     !make a hardcopy
         INTEGER FILE                      !send to a file
         REAL FGRF(NXMAX,NYMAX)            !field for graphing
C Functions:
         REAL GETFLT
         INTEGER GETINT,YESNO
         LOGICAL LOGCVT
         DATA SCREEN,PAPER,FILE/1,2,3/
         DATA VRTCTY,STREAM /1,2/
ccccccccccccccccccccccccccccccccccccccccccccccccccccccccccccccccccccc
         END=.FALSE.                       !initialize values
         DO 5 IX=1,NX
           DO 5 IY=1,NY
             P(IX,IY)=0.
             XSI(IX,IY)=0.
5        CONTINUE
C
200      CONTINUE                          !allow for many runs with same lattice
```

```
C
      CALL PARAM(P,XSI,END)      !get new parameters
      IF (END) RETURN            !start fresh or end altogether
C
      IF (SUBLAT) THEN
         XMIN=NXLL               !set limits for relaxation
         XMAX=NXUR
         YMIN=NYLL
         YMAX=NYUR
      ELSE
         XMIN=1
         YMIN=1
         XMAX=NX
         YMAX=NY
      END IF
C
      IF (TFILE) CALL PRMOUT(TUNIT,NLINES)
      NITER=0
      PRM=.TRUE.
C
99    CONTINUE                   !begin iterations
      IF ((TTERM) .AND. (PRM)) CALL PRMOUT(OUNIT,NLINES)
      NITER=NITER+1
C
C     relax stream function, then vorticity; calculate forces
      CALL PRELAX(P,XSI,XMIN,XMAX,YMIN,YMAX,PMAX,PMIN)
      CALL VRELAX(P,XSI,XMIN,XMAX,YMIN,YMAX,XSIMAX,XSIMIN)
      CALL FORCES(XSI,VFORCE,PFORCE)
C
      IF (TFILE) CALL TXTOUT(TUNIT,NITER,
     +           VFORCE,PFORCE,XSIMAX,PMAX,XSIMIN,PMIN,NLINES)
      IF (TTERM) CALL TXTOUT(OUNIT,NITER,
     +           VFORCE,PFORCE,XSIMAX,PMAX,XSIMIN,PMIN,NLINES)
C
      ITER=.FALSE.
      PRM=.TRUE.
      IF (MOD(NITER,NFREQ) .NE. 0) THEN
         ITER=.TRUE.             ! continue iterating without asking
         PRM=.FALSE.            ! don't print out header
      ELSE                       ! otherwise display fields
         IF (GTERM) THEN
            CALL PAUSE('to see stream function and vorticity...',1)
            CALL CLEAR
            CALL GRFOUT(SCREEN,STREAM,P,PMIN,PMAX,FGRF,NX,NY)
            CALL GRFOUT(SCREEN,VRTCTY,XSI,XSIMIN,XSIMAX,FGRF,NX,NY)
         ELSE IF (TTERM) THEN
            CALL PAUSE('to see stream function...',1)
            CALL CLEAR
            CALL DISPLY(P,OUNIT,STREAM,PMAX,PMIN)
            CALL PAUSE('to see vorticity...',0)
            CALL CLEAR
            CALL DISPLY(XSI,OUNIT,VRTCTY,XSIMAX,XSIMIN)
```

```
                    END IF
                    IF (SUBLAT) THEN    !and provide options for continuing
                       SUBLAT=LOGCVT(YESNO(1,'Continue relaxing sublattice?'))
                       IF (SUBLAT) ITER=.TRUE.
                    ELSE
                       ITER=LOGCVT(YESNO(1,'Continue relaxing lattice?'))
                    END IF
                 END IF
C
              IF (ITER) THEN          !continue iterating
                 GOTO 99
C
              ELSE       `
C             prompt for writing out of field values
              IF ((GFILE) .OR. (GHRDCP) .OR. (TFILE)) THEN
                 CHOICE=YESNO(1,'Do you want to write out field now?')
                 IF (CHOICE .EQ. 1) THEN
                    IF (GFILE) CALL WRTOUT(P,XSI)    !write out field if requested
                    IF (GHRDCP) THEN
                       CALL GRFOUT(PAPER,STREAM,P,PMIN,PMAX,FGRF,NX,NY)
                       CALL GRFOUT(PAPER,VRTCTY,XSI,XSIMIN,XSIMAX,FGRF,NX,NY)
                    END IF
                    IF (TFILE) CALL DISPLY(P,TUNIT,STREAM,PMAX,PMIN)
                    IF (TFILE) CALL DISPLY(XSI,TUNIT,VRTCTY,XSIMAX,XSIMIN)
                 END IF
              END IF
              GOTO 200       !display menu
              END IF
C
       END
CCCCCCCCCCCCCCCCCCCCCCCCCCCCCCCCCCCCCCCCCCCCCCCCCCCCCCCCCCCCCCCCCCCCCCCCCCC
       SUBROUTINE FORCES(XSI,VFORCE,PFORCE)
C calculate pressure force (PFORCE) and viscous forces (VFORCE)
C from the vorticity (XSI)
CCCCCCCCCCCCCCCCCCCCCCCCCCCCCCCCCCCCCCCCCCCCCCCCCCCCCCCCCCCCCCCCCCCCCCCCCCC
C Global Variables:
       INCLUDE 'PARAM.P6'
C Passed Variables:
       REAL XSI(NXMAX,NYMAX)    !vorticity (input)
       REAL VFORCE,PFORCE       !viscous force and pressure force (output)
C Local variables:
       INTEGER IX,IY            !lattice indices
       REAL P                   !pressure
CCCCCCCCCCCCCCCCCCCCCCCCCCCCCCCCCCCCCCCCCCCCCCCCCCCCCCCCCCCCCCCCCCCCCCCCCCC
C      calculate viscous forces by integrating vorticity
C      along the top of the plate
       VFORCE=XSI(FRONT,HFWID)/2
       DO 10 IX=FRONT+1,BACK
          VFORCE=VFORCE+XSI(IX,HFWID)
10     CONTINUE
       VFORCE=VFORCE-XSI(BACK,HFWID)/2
       VFORCE=VFORCE/REYNLD/(HFWID-1)
```

```
C
C      calculate the pressure force by integrating vorticity along the
C      edges of the plate to obtain the pressure, then integrating
C      the pressure to obtain the pressure force
       P=0
       PFORCE=0
       DO 20 IY=2,HFWID             !up front
         P=P-(XSI(FRONT,IY)-XSI(FRONT-1,IY)
     +        +XSI(FRONT,IY-1)-XSI(FRONT-1,IY-1))/2/REYNLD
         PFORCE=PFORCE+P
20     CONTINUE
       PFORCE=PFORCE-P/2
C
       DO 30 IX=FRONT+1,BACK       !across top
         P=P+(XSI(IX,HFWID+1)-XSI(IX,HFWID)
     +        +XSI(IX-1,HFWID+1)-XSI(IX-1,HFWID))/2/REYNLD
30     CONTINUE
       PFORCE=PFORCE-P/2
C
       DO 40 IY=HFWID-1,1,-1       !down back
         P=P+(XSI(BACK+1,IY)-XSI(BACK,IY)
     +        +XSI(BACK+1,IY+1)-XSI(BACK,IY+1))/2/REYNLD
         PFORCE=PFORCE-P
40     CONTINUE
       PFORCE=PFORCE/(HFWID-1)
C
       RETURN
       END
CCCCCCCCCCCCCCCCCCCCCCCCCCCCCCCCCCCCCCCCCCCCCCCCCCCCCCCCCCCCCCCCCCCCCCC
       SUBROUTINE PRELAX(P,XSI,XMIN,XMAX,YMIN,YMAX,PMAX,PMIN)
C relaxes the stream function (P) given previous P and vorticity (XSI)
C on a lattice with corners at XMIN,XMAX,YMIN,YMAX;
C returns min, max value of P in PMIN, PMAX
CCCCCCCCCCCCCCCCCCCCCCCCCCCCCCCCCCCCCCCCCCCCCCCCCCCCCCCCCCCCCCCCCCCCCCC
C Global variables:
       INCLUDE 'PARAM.P6'
C Passed variables:
       REAL P(NXMAX,NYMAX)              !stream function (I/O)
       REAL XSI(NXMAX,NYMAX)           !vorticity (output)
       INTEGER XMIN,XMAX,YMIN,YMAX      !edges of relaxing lattice (input)
       REAL PMAX,PMIN                   !min,max value of P (output)
C Local variables:
       INTEGER IX,IY                    !lattice indices
       REAL TEMP                        !temporary storage
CCCCCCCCCCCCCCCCCCCCCCCCCCCCCCCCCCCCCCCCCCCCCCCCCCCCCCCCCCCCCCCCCCCCCCC
       PMAX=0.
       PMIN=0.
       DO 100 IX=XMIN,XMAX
         DO 200 IY=YMIN,YMAX
           IF (BNDCND(IX,IY) .EQ. 0) THEN   !relax only non-bound points
C
             IF (IY .EQ. NY-1) THEN         !just below top edge
```

```
C
            IF (IX .EQ. 2) THEN             !left corner
              TEMP=P(IX,IY-1)+P(IX+1,IY)-XSI(IX,IY)+1
              P(IX,IY)=SOMEGA/2*TEMP+MSOMEG*P(IX,IY)
              P(IX-1,IY)=P(IX,IY)
              P(IX-1,NY)=P(IX-1,IY)+1
C
            ELSE IF (IX .EQ. NX-1) THEN !right corner
              TEMP=P(IX,IY-1)+P(IX-1,IY)-XSI(IX,IY)+1
              P(IX,IY)=SOMEGA/2*TEMP+MSOMEG*P(IX,IY)
              P(NX,IY)=P(IX,IY)
              P(NX,NY)=P(NX,IY)+1
C
            ELSE                           !not a corner
              TEMP=P(IX,IY-1)+P(IX-1,IY)+P(IX+1,IY)-XSI(IX,IY)+1.
              P(IX,IY)=SOMEGA*TEMP/3+MSOMEG*P(IX,IY)
            END IF
C
            P(IX,NY)=P(IX,IY)+1            !top edge given by bound cond
C
        ELSE                              !not just below top edge
C
            IF (IX .EQ. 2) THEN            !front edge
              TEMP=P(IX,IY-1)+P(IX,IY+1)+P(IX+1,IY)-XSI(IX,IY)
              P(IX,IY)=SOMEGA*TEMP/3+MSOMEG*P(IX,IY)
              P(IX-1,IY)=P(IX,IY)
C
            ELSE IF (IX .EQ. NX-1) THEN !back edge
              TEMP=P(IX,IY-1)+P(IX-1,IY)+P(IX,IY+1)-XSI(IX,IY)
              P(IX,IY)=SOMEGA/3*TEMP+MSOMEG*P(IX,IY)
              P(NX,IY)=P(IX,IY)
C
            ELSE                          !interior point
              TEMP=P(IX,IY-1)+P(IX,IY+1)+P(IX-1,IY)+P(IX+1,IY)
     +            -XSI(IX,IY)
              P(IX,IY)=SOMEGA*TEMP/4+MSOMEG*P(IX,IY)
            END IF
C
          END IF
          IF (P(IX,IY) .GT. PMAX) PMAX=P(IX,IY)
          IF (P(IX,NY) .GT. PMAX) PMAX=P(IX,NY)
          IF (P(IX,IY) .LT. PMIN) PMIN=P(IX,IY)
          IF (P(IX,NY) .LT. PMIN) PMIN=P(IX,NY)
C
          END IF
200     CONTINUE
100   CONTINUE
      RETURN
      END
CCCCCCCCCCCCCCCCCCCCCCCCCCCCCCCCCCCCCCCCCCCCCCCCCCCCCCCCCCCCCCCCCCCCCC
      SUBROUTINE VRELAX(P,XSI,XMIN,XMAX,YMIN,YMAX,XSIMAX,XSIMIN)
C relaxes the vorticity (XSI) given previous XSI and stream function (P)
```

```
C on a lattice defined by XMIN,XMAX,YMIN,YMAX
C returns min, max value of XSI in XSIMIN, XSIMAX
CCCCCCCCCCCCCCCCCCCCCCCCCCCCCCCCCCCCCCCCCCCCCCCCCCCCCCCCCCCCCCCCCCCCCCCC
C Global variables:
      INCLUDE 'PARAM.P6'
C Passed variables:
      REAL P(NXMAX,NYMAX)            !stream function (input)
      REAL XSI(NXMAX,NYMAX)          !vorticity (I/O)
      INTEGER XMIN,XMAX,YMIN,YMAX    !edges of relaxing lattice (input)
      REAL XSIMAX,XSIMIN             !max and min value of XSI (output)
C Local variables:
      INTEGER IX,IY                  !lattice indices
      REAL TEMP,TEMP2,TEMP3          !temporary storage
CCCCCCCCCCCCCCCCCCCCCCCCCCCCCCCCCCCCCCCCCCCCCCCCCCCCCCCCCCCCCCCCCCCCCCCC
      XSIMAX=0.
      XSIMIN=0.
C     impose Dirichlet boundary conditions along the plate
      DO 10 IY=1,HFWID
         XSI(FRONT,IY)=2*P(FRONT-1,IY)
         XSI(BACK,IY)=2*P(BACK+1,IY)
         IF (XSI(FRONT,IY) .GT. XSIMAX) XSIMAX=XSI(FRONT,IY)
         IF (XSI(BACK,IY) .GT. XSIMAX) XSIMAX=XSI(BACK,IY)
         IF (XSI(FRONT,IY) .LT. XSIMIN) XSIMIN=XSI(FRONT,IY)
         IF (XSI(BACK,IY) .LT. XSIMIN) XSIMIN=XSI(BACK,IY)
10    CONTINUE
      DO 20 IX=FRONT+1,BACK-1
         XSI(IX,HFWID)=2*P(IX,HFWID+1)
         IF (XSI(IX,HFWID) .GT. XSIMAX) XSIMAX=XSI(IX,HFWID)
         IF (XSI(IX,HFWID) .LT. XSIMIN) XSIMIN=XSI(IX,HFWID)
20    CONTINUE
C
      DO 30 IX=XMIN,XMAX
         DO 40 IY=YMIN,YMAX
            IF (BNDCND(IX,IY) .EQ. 0) THEN   !don't relax bound points
C
            IF (IX .EQ. NX-1) THEN              !right edge, Neumann BC
               TEMP=XSI(IX,IY+1)+XSI(IX,IY-1)+XSI(IX-1,IY)
               TEMP2=REYND4*(P(IX,IY+1)-P(IX,IY-1))*XSI(IX-1,IY)
               TEMP3=REYND4*(P(IX+1,IY)-P(IX-1,IY))
     +               *(XSI(IX,IY+1)-XSI(IX,IY-1))
               XSI(IX,IY)=MVOMEG*XSI(IX,IY)+
     +          VOMEGA*(TEMP+TEMP2+TEMP3)/(3+REYND4*(P(IX,IY+1)-P(IX,IY-1)))
               XSI(NX,IY)=XSI(IX,IY)
C
            ELSE                              !interior point
               TEMP=XSI(IX,IY+1)+XSI(IX,IY-1)+XSI(IX+1,IY)+XSI(IX-1,IY)
               TEMP2=-REYND4*(P(IX,IY+1)-P(IX,IY-1))
     +               *(XSI(IX+1,IY)-XSI(IX-1,IY))
               TEMP3=REYND4*(P(IX+1,IY)-P(IX-1,IY))
     +               *(XSI(IX,IY+1)-XSI(IX,IY-1))
               XSI(IX,IY)=VOMEGA/4*(TEMP+TEMP2+TEMP3)+MVOMEG*XSI(IX,IY)
C
```

```
               END IF
               IF (XSI(IX,IY) .GT. XSIMAX) XSIMAX=XSI(IX,IY)
               IF (XSI(IX,IY) .LT. XSIMIN) XSIMIN=XSI(IX,IY)
C
               END IF
40       CONTINUE
30       CONTINUE
         RETURN
         END
CCCCCCCCCCCCCCCCCCCCCCCCCCCCCCCCCCCCCCCCCCCCCCCCCCCCCCCCCCCCCCCCCCCCCCCC
         SUBROUTINE INIT
C initializes constants, displays header screen,
C initializes menu arrays for input parameters
CCCCCCCCCCCCCCCCCCCCCCCCCCCCCCCCCCCCCCCCCCCCCCCCCCCCCCCCCCCCCCCCCCCCCCCC
C Global variables:
         INCLUDE 'IO.ALL'
         INCLUDE 'MENU.ALL'
         INCLUDE 'PARAM.P6'
C Local parameters:
         CHARACTER*80 DESCRP              !program description
         DIMENSION DESCRP(20)
         INTEGER NHEAD,NTEXT,NGRAPH       !number of lines for each description
CCCCCCCCCCCCCCCCCCCCCCCCCCCCCCCCCCCCCCCCCCCCCCCCCCCCCCCCCCCCCCCCCCCCCCCC
C        get environment parameters
         CALL SETUP
C
C        display header screen
         DESCRP(1)= 'PROJECT 6'
         DESCRP(2)= '2-D viscous incompressible flow about a '
     +               //'rectangular block'
         NHEAD=2
C
C        text output description
         DESCRP(3)= 'maximum vorticity and stream function,'
         DESCRP(4)= 'forces at each iteration'
         DESCRP(5)= '(all values are in scaled units)'
         NTEXT=3
C
C        graphics output description
         DESCRP(6)= 'stream function and vorticity'
         NGRAPH=1
C
         CALL HEADER(DESCRP,NHEAD,NTEXT,NGRAPH)
C
         FIRST=.TRUE.                     !is this the first time through the menu?
C
C        set up constant part of menu
         CALL MENU
C
C        item 9 has a different use in this program
         MTYPE(9)=MTITLE
         MPRMPT(9)='7) Change lattice size or end altogether'
```

```
        MLOLIM(9)=0
        MHILIM(9)=1
C
C       default values and limits of next 3 items are set in subroutine
C       LATTIC since they depend on the size of lattice chosen
        MTYPE(13)=NUM
        MPRMPT(13)= 'Enter half-width of plate in lattice spacings'
        MTAG(13)= 'Half-width of plate in lattice spacings'
C
        MTYPE(14)=NUM
        MPRMPT(14)= 'Enter length of plate in lattice spacings'
        MTAG(14)= 'Length of plate in lattice spacings'
C
        MTYPE(15)=NUM
        MPRMPT(15)=
       + 'Enter location of plate''s front edge in lattice spacings'
        MTAG(15)= 'Location of plate''s front edge in lattice spacings'
C
        MTYPE(16)=FLOAT
        MPRMPT(16)= 'Enter lattice Reynold''s number'
        MTAG(16)= 'Lattice Reynold''s number'
        MLOLIM(16)=0.01
        MHILIM(16)=20.
        MREALS(16)=1.
C
        MTYPE(17)=SKIP
        MREALS(17)=35.
C
        MTYPE(38)=FLOAT
        MPRMPT(38)= 'Enter vorticity relaxation parameter'
        MTAG(38)= 'Vorticity relaxation parameter'
        MLOLIM(38)=0.
        MHILIM(38)=1.9
        MREALS(38)=.3
C
        MTYPE(39)=FLOAT
        MPRMPT(39)= 'Enter stream function relaxation parameter'
        MTAG(39)= 'Stream function  relaxation parameter'
        MLOLIM(39)=0.
        MHILIM(39)=1.9
        MREALS(39)=.3
C
        MTYPE(40)=NOSKIP
        MPRMPT(40)='Do you want to specify a sublattice to relax first?'
        MTAG(40)='Relax a sublattice first?'
        MINTS(40)=0
        MREALS(40)=45.
C
        MTYPE(41)=NUM
        MPRMPT(41)=
       + 'Enter lower left X (in lattice spacings) for sublattice'
        MTAG(41)='Lower left X sublattice value'
```

```
            MLOLIM(41)=1
            MINTS(41)=1
      C
            MTYPE(42)=NUM
            MPRMPT(42)=
          + 'Enter lower left Y (in lattice spacings) for sublattice'
            MTAG(42)='Lower left Y sublattice value'
            MLOLIM(42)=1
            MINTS(42)=1
      C
            MTYPE(43)=NUM
            MPRMPT(43)='Enter upper right X value for sublattice'
            MTAG(43)='Upper right X sublattice value'
            MLOLIM(43)=1
            MINTS(43)=1
      C
            MTYPE(44)=NUM
            MPRMPT(44)='Enter upper right Y value for sublattice'
            MTAG(44)='Upper right Y sublattice value'
            MLOLIM(44)=1
            MINTS(44)=1
      C
            MTYPE(45)=MTITLE
            MPRMPT(45)='Starting Values (for stream function and Vorticity)'
            MLOLIM(45)=2
            MHILIM(45)=1
      C
            MTYPE(46)=MTITLE
            MPRMPT(46)='1) Read in starting values from a file'
            MLOLIM(46)=0
            MHILIM(46)=0
      C
            MTYPE(47)=MTITLE
            MPRMPT(47)='2) Set values to free flow values'
            MLOLIM(47)=0
            MHILIM(47)=0
      C
            MTYPE(48)=MTITLE
            MPRMPT(48)='3) Leave field values unchanged'
            MLOLIM(48)=0
            MHILIM(48)=0
      C
            MTYPE(49)=MCHOIC
            MPRMPT(49)='Enter Choice'
            MTAG(49)='50 51 51'
            MLOLIM(49)=1
            MHILIM(49)=3
            MINTS(49)=2
            MREALS(49)=2
      C
            MTYPE(50)=CHSTR
            MPRMPT(50)= 'Enter name of data file'
```

```
      MTAG(50)= 'File with initial values for the fields'
      MHILIM(50)=12.
      MINTS(50)=3.
      MSTRNG(MINTS(50))= 'proj6.in'
C
      MTYPE(51)=SKIP
      MREALS(51)=60.
C
      MSTRNG(MINTS(75))= 'proj6.txt'
C
      MTYPE(76)=SKIP
      MREALS(76)=80.
C
      MSTRNG(MINTS(86))= 'proj6.grf'
C
      MTYPE(87)=NUM
      MPRMPT(87)= 'Enter the display frequency for the fields'
      MTAG(87)= 'Field display frequency'
      MLOLIM(87)= 1.
      MHILIM(87)= 100.
      MINTS(87)= 10
C
      MTYPE(88)=NUM
      MPRMPT(88)= 'Enter number of contour levels'
      MTAG(88)= 'Number of contour levels'
      MLOLIM(88)= 1.
      MHILIM(88)= MAXLEV
      MINTS(88)= 10
C
      MTYPE(89)=SKIP
      MREALS(89)=90.
C
      RETURN
      END
CCCCCCCCCCCCCCCCCCCCCCCCCCCCCCCCCCCCCCCCCCCCCCCCCCCCCCCCCCCCCCCCCCCCCCC
      SUBROUTINE LATTIC
C gets lattice size from screen and calculates best way to display the
C field as ascii characters based on lattice size and terminal size;
C resets all boundary conditions and default menu values
CCCCCCCCCCCCCCCCCCCCCCCCCCCCCCCCCCCCCCCCCCCCCCCCCCCCCCCCCCCCCCCCCCCCCCC
C Global variables:
      INCLUDE 'PARAM.P6'
      INCLUDE 'IO.ALL'
      INCLUDE 'MENU.ALL'
      INCLUDE 'MAP.P6'
C Local variables:
      INTEGER END                      !end program
      INTEGER IX,IY,IBC                !lattice indices, BC index
      INTEGER NXHI,NYHI,NXLO,NYLO      !limits on lattice size
      INTEGER NXDEF,NYDEF              !default lattice sizes
      LOGICAL RESET                    !reset parameters?
C Functions:
```

```
        INTEGER YESNO,GETINT           !user input functions
        LOGICAL LOGCVT
CCCCCCCCCCCCCCCCCCCCCCCCCCCCCCCCCCCCCCCCCCCCCCCCCCCCCCCCCCCCCCCCCCCCCCCC
C       allow user to end the program
        CALL CLEAR
        IF (.NOT. FIRST) THEN
            END=YESNO(0,' Do you want to end the program?')
            IF (END .EQ. 1) CALL FINISH
        ELSE
C           the lattice size is determined by array size and terminal size;
C           if you're using graphics, terminal size won't matter
C           set NXHI=NXMAX and NYHI=NYMAX
            NXHI=MIN(TRMWID-2,NXMAX)
            NYHI=MIN(TRMLIN-3,NYMAX)
            NXDEF=MIN(NXHI,20)
            NYDEF=MIN(NYHI,20)
            NXLO=MIN(5,NXDEF)
            NYLO=MIN(5,NYDEF)
        END IF
C
C       get lattice parameters from the terminal
        NX=GETINT(NXDEF,NXLO,NXHI,' Enter number of X lattice points')
        NY=GETINT(NYDEF,NYLO,NYHI,' Enter number of Y lattice points')
        NXDEF=NX
        NYDEF=NY
C
C       calculate parameters for best looking display
        IF (2*NX .LE. TRMWID) THEN
            XSKIP=.TRUE.                !skip spaces in x
            XCNTR=(TRMWID-2*NX)/2       !how to center display
        ELSE
            XSKIP=.FALSE.
            XCNTR=(TRMWID-NX)/2
        END IF
        IF (XCNTR .LT. 1) XCNTR=1
C
        IF (2*NY .LE. TRMLIN-3) THEN
            YSKIP=.TRUE.                !skip lines in y
            YCNTR=(TRMLIN-2*NY)/2-2     !how to center display
        ELSE
            YSKIP=.FALSE.
            YCNTR=(TRMLIN-NY)/2-2
        END IF
        IF (YCNTR .LT. 0) YCNTR=0
C
C       set up default and limits for the plate location and geometry
        MHILIM(IWID)=NY
        MLOLIM(IWID)=2
        MINTS(IWID)=MAX(2,NY/3)
        MHILIM(ILEN)=NX
        MLOLIM(ILEN)=0
        MINTS(ILEN)=NX/4
```

```
      MHILIM(IFRNT)=NX
      MLOLIM(IFRNT)=1
      MINTS(IFRNT)=MAX(1,NX/3)
C
C     set up limits on sublattice
      MHILIM(INXLL)=NX
      MHILIM(INYLL)=NY
      MHILIM(INXUR)=NX
      MHILIM(INYUR)=NY
      MINTS(41)=1                       !sublattice  x, lower left
      MINTS(42)=1                       !sublattice  y, lower left
      MINTS(43)=1                       !sublattice  x, upper right
      MINTS(44)=1                       !sublattice  y, lower left
C
C     allow for resetting of defaults
      IF (FIRST) THEN
         FIRST=.FALSE.
      ELSE
         RESET=LOGCVT(YESNO(0,' Do you want to reset default values?'))
         IF (RESET) THEN
            MREALS(16)=1.                       !Reynolds number
            MREALS(38)=0.3                      !vorticity relaxation
            MREALS(39)=0.3                      !stream relaxation
            MINTS(40)=0                         !no sublattice
            MSTRNG(MINTS(50))= 'proj6.in'   !input file
            MSTRNG(MINTS(75))= 'proj6.txt'  !text file
            MSTRNG(MINTS(86))= 'proj6.grf'  !graphics file
            MINTS(87)= 10                       !graphing frequency
            MINTS(88)= 10                       !number of contours
            MINTS(49)=2                         !start from free
            MREALS(49)=2                        !  streaming values
            MINTS(73)=TXTTRM                    !default i/o
            MINTS(74)=TXTFIL
            MINTS(83)=GRFTRM
            MINTS(84)=GRFHRD
            MINTS(85)=GRFFIL
         END IF
      END IF
C
      RETURN
      END
CCCCCCCCCCCCCCCCCCCCCCCCCCCCCCCCCCCCCCCCCCCCCCCCCCCCCCCCCCCCCCCCCCCCCCCCCC
      SUBROUTINE PARAM(P,XSI,END)
C gets parameters from screen
C ends program on request
C closes old files
C maps menu variables to program variables
C opens new files
C calculates all derivative parameters
C performs checks on sublattice parameters
C set the field to its initial values (P,XSI)
C and controls ending of program (END)
```

```
cccccccccccccccccccccccccccccccccccccccccccccccccccccccccccccccccccccc
C Global variables:
      INCLUDE 'MENU.ALL'
      INCLUDE 'IO.ALL'
      INCLUDE 'PARAM.P6'
      INCLUDE 'MAP.P6'
C Input/output variables:
      REAL P(NXMAX,NYMAX)         !stream function
      REAL XSI(NXMAX,NYMAX)       !vorticity
      LOGICAL END                 !end program?
C Local variables:
      INTEGER IX,IY               !lattice indices
C Functions:
      LOGICAL LOGCVT              !converts 1 and 0 to true and false
      INTEGER GETINT              !get integer from screen
cccccccccccccccccccccccccccccccccccccccccccccccccccccccccccccccccccccc
C     get input from terminal
      CALL CLEAR
      CALL ASK(1,ISTOP)
C
C     start fresh or end altogether, if so requested
      IF (MREALS(IMAIN) .EQ. STOP)   THEN
          END=.TRUE.
          RETURN
      END IF
C
C     close files if necessary
      IF (TNAME .NE. MSTRNG(MINTS(ITNAME)))
     +      CALL FLCLOS(TNAME,TUNIT)
C
C     physical and numerical parameters
      HFWID=MINTS(IWID)
      LENGTH=MINTS(ILEN)
      FRONT=MINTS(IFRNT)
      REYNLD=MREALS(IRNLDS)
      VOMEGA=MREALS(IVOMEG)
      SOMEGA=MREALS(ISOMEG)
      SUBLAT=LOGCVT(MINTS(ISUB))
      NXLL=MINTS(INXLL)
      NYLL=MINTS(INYLL)
      NXUR=MINTS(INXUR)
      NYUR=MINTS(INYUR)
      PTYPE=ABS(MREALS(IPTYPE))
      PFILE=MSTRNG(MINTS(IPFILE))
C
C     text output
      TTERM=LOGCVT(MINTS(ITTERM))
      TFILE=LOGCVT(MINTS(ITFILE))
      TNAME=MSTRNG(MINTS(ITNAME))
C
C     graphics output
      GTERM=LOGCVT(MINTS(IGTERM))
```

```
         GHRDCP=LOGCVT(MINTS(IGHRD))
         GFILE=LOGCVT(MINTS(IGFILE))
         GNAME=MSTRNG(MINTS(IGNAME))
         NFREQ=MINTS(INFREQ)
         NLEV=MINTS(INLEV)
C
C        open files
         IF (TFILE) CALL FLOPEN(TNAME,TUNIT)
         !files may have been renamed
         MSTRNG(MINTS(ITNAME))=TNAME
C
C        check sublattice parameters
39       IF ((NXLL .GT. NXUR) .OR. (NYLL .GT. NYUR)) THEN
            WRITE (OUNIT,40)
            CALL ASK(41,44)
            NXLL=MINTS(INXLL)
            NYLL=MINTS(INYLL)
            NXUR=MINTS(INXUR)
            NYUR=MINTS(INYUR)
            GOTO 39
         END IF
40       FORMAT (' Sublattice parameters must have upper right values '
       +        'greater than lower left')
         CALL CLEAR
C
C        calculate derivative quantities
         BACK=FRONT+LENGTH-1
         IF (BACK .GT. NX) BACK=NX   !make sure block doesn't extend too far
         MVOMEG=1.-VOMEGA
         MSOMEG=1.-SOMEGA
         REYND4=REYNLD/4
C
         CALL INTCND(P,XSI)          !get starting values
C
C        reset PTYPE menu parameters so that P remains unchanged during
C        runs unless the user explicitly requests otherwise
         MINTS(IPTYPE)=3
         MREALS(IPTYPE)=3
C
         RETURN
         END
CCCCCCCCCCCCCCCCCCCCCCCCCCCCCCCCCCCCCCCCCCCCCCCCCCCCCCCCCCCCCCCCCCCCCCCC
         SUBROUTINE INTCND(P,XSI)
C get starting values for stream function (P) and vorticity (XSI)
C and set boundary conditions in BNDCND, depending on plate location
CCCCCCCCCCCCCCCCCCCCCCCCCCCCCCCCCCCCCCCCCCCCCCCCCCCCCCCCCCCCCCCCCCCCCCCC
C Global variables:
         INCLUDE 'PARAM.P6'
C Output variables:
         REAL P(NXMAX,NYMAX)          !stream function
         REAL XSI(NXMAX,NYMAX)        !vorticity
C Local variables:
```

```
          INTEGER IX,IY                    !lattice indices
CCCCCCCCCCCCCCCCCCCCCCCCCCCCCCCCCCCCCCCCCCCCCCCCCCCCCCCCCCCCCCCCCCCCCCC
C       set P to its initial value
        IF (PTYPE .EQ. 1) CALL READIN(P,XSI)        !read data in
C       sometimes in READIN, PTYPE is changed
        IF (PTYPE .EQ. 2) THEN
           DO 60 IX=1,NX
              DO 70 IY=1,NY                 !set to free stream values
                 P(IX,IY)=REAL(IY-1)
                 XSI(IX,IY)=0.
70            CONTINUE
60         CONTINUE
        ELSE IF (PTYPE .EQ. 3) THEN        !keep P and XSI the same
           CONTINUE
        END IF
C
C       release previous boundary conditions
        DO 61 IX=1,NX
           DO 71 IY=1,NY
              BNDCND(IX,IY)=0.
71         CONTINUE
61      CONTINUE
C
C       set up boundary conditions on plate and lattice edges
        DO 100 IY=1,NY                     !left and right edges are boundaries
           BNDCND(1,IY)=1
           BNDCND(NX,IY)=1
100     CONTINUE
        DO 110 IX=1,NX                     !upper and lower edges are boundaries
           BNDCND(IX,1)=1
           BNDCND(IX,NY)=1
110     CONTINUE
        DO 120 IX=FRONT,BACK               !top of plate is boundary
           P(IX,HFWID)=0
           BNDCND(IX,HFWID)=1
120     CONTINUE
        DO 130 IY=1,HFWID                  !front and back of plate are bound
           P(FRONT,IY)=0.
           P(BACK,IY)=0.
           BNDCND(FRONT,IY)=1
           BNDCND(BACK,IY)=1
130     CONTINUE
        DO 80 IX=FRONT+1,BACK-1            !plate interior is a boundary
           DO 90 IY=1,HFWID-1
              P(IX,IY)=0.
              XSI(IX,IY)=0.
              BNDCND(IX,IY)=2              !2 indicates actual plate
90         CONTINUE
80      CONTINUE
C
        RETURN
        END
```

```
ccccccccccccccccccccccccccccccccccccccccccccccccccccccccccccccccc
      SUBROUTINE READIN(P,XSI)
C read in stream function P and vorticity XSI from file written
C by subroutine WRTOUT; files must have same lattice parameters
ccccccccccccccccccccccccccccccccccccccccccccccccccccccccccccccccc
C Global variables:
      INCLUDE 'IO.ALL'
      INCLUDE 'PARAM.P6'
      INCLUDE 'MENU.ALL'
      INCLUDE 'MAP.P6'
C Output variables:
      REAL P(NXMAX,NYMAX)             !stream function
      REAL XSI(NXMAX,NYMAX)           !vorticity
C Local variables:
      INTEGER IX,IY                   !lattice indices
      CHARACTER*80 JUNK               !first line if of no interest
      LOGICAL SUCESS                  !did we find a file to open?
      INTEGER CHOICE,YESNO            !get yesno input from user
      INTEGER MX,MY,MOMEGA            !parameters from PFILE
C Function:
      CHARACTER*40 CHARAC             !returns character input
ccccccccccccccccccccccccccccccccccccccccccccccccccccccccccccccccc
10    CALL FLOPN2(PFILE,DUNIT,SUCESS)    !open the file for input
      MSTRNG(MINTS(IPFILE))=PFILE        !file may have been renamed
C
      IF (.NOT. SUCESS) THEN
         CALL REASK                      !prompt again for init P
         RETURN                          !if no file was found
      ELSE
C
         READ (DUNIT,5) JUNK             !skip over title
         READ (DUNIT,5) JUNK
5        FORMAT (A)
C
C        lattice sizes must match; if they don't, allow for other options
         READ (DUNIT,*) MX,MY
         IF ((MX .NE. NX) .OR. (MY .NE. NY)) THEN
            CALL FLCLOS(PFILE,DUNIT)         !close it up
            WRITE (OUNIT,15) MX,MY
15          FORMAT (' Input file has does not have the correct'
     +             ' lattice size, it is ',I2,' by ',I2)
            CHOICE=YESNO(1,' Do you want to try another file?')
            IF (CHOICE .EQ. 0) THEN
               CALL REASK                    !prompt again for init P
               RETURN
            ELSE
               PFILE=CHARAC(PFILE,12,'Enter another filename')
               MSTRNG(MINTS(IPFILE))=PFILE
               GOTO 10         !try to open this one
            END IF
         END IF
      END IF
C
```

```
C       if we've gotten this far, we've opened a file with data
C       from a lattice of the same size; finally read in field
        READ (DUNIT,5) JUNK                    !skip over parameter values
        READ (DUNIT,5) JUNK
        READ (DUNIT,5) JUNK
C
        DO 100 IX=1,NX
          READ (DUNIT,110) (P(IX,IY),IY=1,NY)
100     CONTINUE
        DO 200 IX=1,NX
          READ (DUNIT,110) (XSI(IX,IY),IY=1,NY)
200     CONTINUE
110     FORMAT (5(2X,1PE14.7))
        CALL FLCLOS(PFILE,DUNIT)               !close file
        END IF
C
        RETURN
        END
CCCCCCCCCCCCCCCCCCCCCCCCCCCCCCCCCCCCCCCCCCCCCCCCCCCCCCCCCCCCCCCCCCCCCCCCC
        SUBROUTINE REASK
C prompt again for initial P,XSI configuration;
C called only if reading in from a file is request and failed
CCCCCCCCCCCCCCCCCCCCCCCCCCCCCCCCCCCCCCCCCCCCCCCCCCCCCCCCCCCCCCCCCCCCCCCCC
C Global variables:
        INCLUDE 'MENU.ALL'
        INCLUDE 'MAP.P6'
        INCLUDE 'PARAM.P6'
CCCCCCCCCCCCCCCCCCCCCCCCCCCCCCCCCCCCCCCCCCCCCCCCCCCCCCCCCCCCCCCCCCCCCCCCC
C       redisplay initial Field Value Menu, but disallow choice 1
        MPRMPT(46)='1) (not allowed)'
        MLOLIM(49)=2
        MINTS(49)=2
        MREALS(49)=2
        CALL ASK(45,51)
C
        PTYPE=ABS(MREALS(IPTYPE))              !set parameter choices
C
        MPRMPT(46)='1) Read in starting values from a file'   !reset menu
        MLOLIM(49)=1
C
        RETURN
        END
CCCCCCCCCCCCCCCCCCCCCCCCCCCCCCCCCCCCCCCCCCCCCCCCCCCCCCCCCCCCCCCCCCCCCCCCC
        SUBROUTINE PRMOUT(MUNIT,NLINES)
C write out parameter summary of length NLINES to MUNIT
CCCCCCCCCCCCCCCCCCCCCCCCCCCCCCCCCCCCCCCCCCCCCCCCCCCCCCCCCCCCCCCCCCCCCCCCC
C Global variables:
        INCLUDE 'IO.ALL'
        INCLUDE 'PARAM.P6'
C Input variables:
        INTEGER MUNIT                          !fortran unit number
        INTEGER NLINES                         !number of lines sent to terminal (I/O)
```

```
ccccccccccccccccccccccccccccccccccccccccccccccccccccccccccccccccccc
      IF (MUNIT .EQ. OUNIT) CALL CLEAR
C
      WRITE (MUNIT,5)
      WRITE (MUNIT,6)
      IF (MUNIT .NE. GUNIT) THEN
          WRITE (MUNIT,10) NX,NY
      ELSE
          WRITE (MUNIT,*) NX,NY
      END IF
      WRITE (MUNIT,20) VOMEGA,SOMEGA
      WRITE (MUNIT,30) HFWID,LENGTH,FRONT
      WRITE (MUNIT,40) REYNLD
      IF (SUBLAT) WRITE (MUNIT,45) NXLL,NYLL,NXUR,NYUR
C
      IF (MUNIT .NE. GUNIT) THEN
          WRITE (MUNIT,*) ' '
          WRITE (MUNIT,50)
          WRITE (MUNIT,60)
          WRITE (MUNIT,70)
      END IF
C
      NLINES=8
      IF (SUBLAT) NLINES=NLINES+1
C
5     FORMAT (' Output from project 6:')
6     FORMAT (' 2-D viscous incompressible'
     +        ' flow around a rectangular block')
10    FORMAT (' NX =',I3,5X,' NY =',I3)
20    FORMAT (' Vorticity relaxation= ',F5.3,5X,'Stream relaxation= ',
     +        F5.3)
30    FORMAT (' Plate half width=',I3,5X,' length='I3,5X,
     +        'and front edge='I3)
40    FORMAT (' Lattice Reynold''s number =',F7.3)
45    FORMAT (' Sublattice defined by  (',I3,',',I3,') and (',
     +        I3,',',I3,')')
50    FORMAT (3X,'Iter',4X,'Pressure',6X,'Viscous',11X,'Vorticity',
     +   10X,'Stream Function')
60    FORMAT (12X,'Force',9X,'Force',13X,'min,max',15X,'min,max')
70    FORMAT (3X,'----',4X,'--------',6X,'-------',11X,'---------',
     +   10X,'---------------')
C
      RETURN
      END
cccccccccccccccccccccccccccccccccccccccccccccccccccccccccccccccccccc
      SUBROUTINE
     +  TXTOUT (MUNIT,NITER,VFORCE,PFORCE,XSIMAX,PMAX,XSIMIN,PMIN,NLINES)
C output forces and min, max field values to MUNIT
cccccccccccccccccccccccccccccccccccccccccccccccccccccccccccccccccccc
C Global variables:
      INCLUDE 'IO.ALL'
C Input variables:
```

```
      INTEGER NITER                  !number of iterations
      REAL VFORCE,PFORCE             !viscous force and pressure
      REAL XSIMAX,PMAX,XSIMIN,PMIN!min,max values of XSI, P
      INTEGER MUNIT                  !fortran unit number
      INTEGER NLINES                 !number of lines sent to terminal(I/O)
CCCCCCCCCCCCCCCCCCCCCCCCCCCCCCCCCCCCCCCCCCCCCCCCCCCCCCCCCCCCCCCCCCCCCCCCCCCCC
C     if screen is full, clear screen and retype headings
      IF ((MOD(NLINES,TRMLIN-6) .EQ. 0)
     +                            .AND. (MUNIT .EQ. OUNIT)) THEN
         CALL PAUSE('to continue...',1)
         CALL CLEAR
         WRITE (MUNIT,50)
         WRITE (MUNIT,60)
         WRITE (MUNIT,70)
      END IF
C
      IF (MUNIT .EQ. OUNIT) NLINES=NLINES+1
      WRITE (MUNIT,40) NITER,PFORCE,VFORCE,XSIMIN,XSIMAX,PMIN,PMAX
C
40    FORMAT (2X,I5,2(2X,1PE12.5),2X,4(1PE10.3,1X))
50    FORMAT (3X,'Iter',4X,'Pressure',6X,'Viscous',11X,'Vorticity',
     +   10X,'Stream Function')
60    FORMAT (12X,'Force',9X,'Force',13X,'min,max',15X,'min,max')
70    FORMAT (3X,'----',4X,'--------',6X,'-------',11X,'----------',
     +   10X,'---------------')
C
      RETURN
      END
CCCCCCCCCCCCCCCCCCCCCCCCCCCCCCCCCCCCCCCCCCCCCCCCCCCCCCCCCCCCCCCCCCCCCCCCCCCCC
      SUBROUTINE DISPLY(F,MUNIT,ITYPE,FMAX,FMIN)
C display stream function (P,ITYPE=STREAM) or vorticity
C (XSI,ITYPE=VRTCTY) as letters; positive values are capitals or
C numbers, negative values are small letters
CCCCCCCCCCCCCCCCCCCCCCCCCCCCCCCCCCCCCCCCCCCCCCCCCCCCCCCCCCCCCCCCCCCCCCCCCCCCC
C Global variables:
      INCLUDE 'PARAM.P6'
      INCLUDE 'IO.ALL'
C Input variables:
      REAL F(NXMAX,NYMAX)            !stream function or vorticity
      INTEGER MUNIT                  !unit we're writing to
      INTEGER ITYPE                  !which field are we displaying?
      REAL FMAX,FMIN                 !min and max field values
C Local variables:
      INTEGER IX,IY                  !lattice indices
      INTEGER TEMP                   !field at current lattice site
      CHARACTER*1 FIELD(NXMAX)       !field as character data
      CHARACTER*80 BLNK              !blanks for centering in X
      CHARACTER*1 ASKII(0:35),NEGASK(1:26)!charac data for display
      DATA BLNK /' '/
      DATA ASKII/'0','1','2','3','4','5','6','7','8','9',
     +           'A','B','C','D','E','F','G','H','I','J','K','L','M',
     +           'N','O','P','Q','R','S','T','U','V','W','X','Y','Z'/
```

```
      DATA NEGASK/'a','b','c','d','e','f','g','h','i','j','k','l','m',
     +         'n','o','p','q','r','s','t','u','v','w','x','y','z'/
      DATA VRTCTY,STREAM /1,2/
cccccccccccccccccccccccccccccccccccccccccccccccccccccccccccccccccccc
      IF (MUNIT .EQ. OUNIT) THEN
         DO 20 IY=1,YCNTR                  !center output
            WRITE (OUNIT,*) ' '
20       CONTINUE
      ELSE                                 !or display which field
         IF (ITYPE .EQ. STREAM) WRITE (MUNIT,*) ' stream function:'
         IF (ITYPE .EQ. VRTCTY) WRITE (MUNIT,*) ' vorticity:'
      END IF
C
      DO 100 IY=NY,1,-1
         DO 50 IX=1,NX
            IF (F(IX,IY) .GE. 0) THEN
               IF (FMAX .NE. 0) THEN
                  TEMP=NINT(F(IX,IY)*35/FMAX)   !scale field
               ELSE
                  TEMP=0.
               END IF
               FIELD(IX)=ASKII(TEMP)            !convert to ascii
            ELSE
               IF (FMIN .NE. 0) THEN
                  TEMP=NINT(F(IX,IY)*26/FMIN)   !scale field
               ELSE
                  TEMP=0
               END IF
               IF (TEMP .NE. 0) THEN
                  FIELD(IX)=NEGASK(TEMP)        !convert to ascii
               ELSE
                  FIELD(IX)=ASKII(TEMP)
               END IF
            END IF
C           leave blanks to indicate the plate
            IF (BNDCND(IX,IY) .EQ. 2) FIELD(IX)=BLNK(1:1)
50       CONTINUE
C
C        write out a line at a time (no centering done for TUNIT)
         IF (MUNIT .EQ. TUNIT) THEN
               WRITE (TUNIT,16) (FIELD(IX),IX=1,NX)
         ELSE
            IF (XSKIP) THEN
               WRITE (OUNIT,10) BLNK(1:XCNTR),(FIELD(IX),IX=1,NX)
            ELSE
               WRITE (OUNIT,15) BLNK(1:XCNTR),(FIELD(IX),IX=1,NX)
            END IF
            IF (YSKIP) WRITE (OUNIT,*) ' '
         END IF
10       FORMAT (1X,A,100(A1,1X))
15       FORMAT (1X,A,100(A1))
16       FORMAT (1X,100(A1))
```

```
      100   CONTINUE
      C
            RETURN
            END
      CCCCCCCCCCCCCCCCCCCCCCCCCCCCCCCCCCCCCCCCCCCCCCCCCCCCCCCCCCCCCCCCCCCCCCCCCC
            SUBROUTINE GRFOUT(DEVICE,FIELD,F,FMIN,FMAX,FGRF,MX,MY)
      C display contours of the scaled stream function P and vorticity PSI
      C to DEVICE
      C     the field values must be in an array that is exactly NX by NY;
      C     this can be accomplished with implicit dimensioning which
      C     requires that FGRF and its dimensions be passed to this routine
      CCCCCCCCCCCCCCCCCCCCCCCCCCCCCCCCCCCCCCCCCCCCCCCCCCCCCCCCCCCCCCCCCCCCCCCCCC
      C Global variables:
            INCLUDE 'IO.ALL'
            INCLUDE 'PARAM.P6'
            INCLUDE 'GRFDAT.ALL'
      C Input variables:
            INTEGER DEVICE              !which device
            REAL F(NXMAX,NYMAX)         !field
            INTEGER MX,MY               !NX and NY in disguise
            REAL FGRF(MX,MY)            !field for graphing
            REAL FMIN,FMAX              !min,max field values
            INTEGER FIELD               !which field
      C Local variables:
            INTEGER SCREEN              !send to terminal
            INTEGER PAPER               !make a hardcopy
            INTEGER FILE                !send to a file
            INTEGER IX,IY               !level index, lattice indices
            REAL PX(4),PY(4)            !edges of plate for graphing
            CHARACTER*9 CMIN,CMAX,CREY  !data as characters
            INTEGER LMIN,LMAX,RLEN      !length of string
            DATA SCREEN,PAPER,FILE/1,2,3/
            DATA VRTCTY,STREAM /1,2/
      CCCCCCCCCCCCCCCCCCCCCCCCCCCCCCCCCCCCCCCCCCCCCCCCCCCCCCCCCCCCCCCCCCCCCCCCCC
            DO 10 IX=1,MX
               DO 15 IY=1,MY
                  FGRF(IX,IY)=F(IX,IY)        !load field into FGRF
      15       CONTINUE
      10    CONTINUE
      C
      C     messages for the impatient
            IF ((DEVICE .NE. SCREEN) .AND. (FIELD .EQ. STREAM))
           +                    WRITE (OUNIT,100)
      C
      C     calculate parameters for graphing
            NPLOT=2                    !how many plots
      C
            YMAX=MY
            YMIN=1.
            XMIN=1.
            XMAX=MX
            YOVAL=XMIN
```

```
        X0VAL=YMIN
        NXTICK=5
        NYTICK=5
        NPOINT=5
        LABEL(1)='NX'
        LABEL(2)='NY'
        CALL CONVRT(FMIN,CMIN,LMIN)
        CALL CONVRT(FMAX,CMAX,LMAX)
        CALL CONVRT(REYNLD,CREY,RLEN)
C
        IF (FIELD .EQ. STREAM) THEN
          IPLOT=1
          INFO='Pmin='//CMIN(1:LMIN)//' Pmax='//CMAX(1:LMAX)
        ELSE IF (FIELD .EQ. VRTCTY) THEN
          IPLOT=2
          INFO='XSImin='//CMIN(1:LMIN)//' XSImax='//CMAX(1:LMAX)
        END IF
        TITLE='Stream Function and Vorticity, Reynold''s number ='
       +             //CREY(1:RLEN)
C
        IF (FIELD .EQ. STREAM) THEN  ·
          CALL GTDEV(DEVICE)                 !device nomination
          IF (DEVICE .EQ. SCREEN) CALL GMODE  !change to graphics mode
        END IF
        CALL LNLNAX                          !draw axes
C
C       draw in plate if it's big enough
        IF (LENGTH .GT. 2) THEN
          PX(1)=REAL(FRONT+1)
          PY(1)=1.
          PX(2)=PX(1)
          PY(2)=REAL(HFWID-1)
          PX(3)=REAL(BACK-1)
          PY(3)=PY(2)
          PX(4)=PX(3)
          PY(4)=1.
          NPOINT=4
          ILINE=2
          CALL XYPLOT(PX,PY)
        END IF
C
        CALL CONTOR(FGRF,MX,MY,FMIN,FMAX,NLEV)
C
C       end graphing session
        IF (FIELD .EQ. VRTCTY) THEN
          IF (DEVICE .NE. FILE) CALL GPAGE(DEVICE)  !end graphics package
          IF (DEVICE .EQ. SCREEN) CALL TMODE        !switch to text mode
        END IF
C
100     FORMAT (/,' Patience, please; output going to a file.')
C
        RETURN
```

```
       END
CCCCCCCCCCCCCCCCCCCCCCCCCCCCCCCCCCCCCCCCCCCCCCCCCCCCCCCCCCCCCCCCCCCCCCCC
       SUBROUTINE WRTOUT(P,XSI)
C write out stream function (P) and vorticity (XSI) to GUNIT for reading
C back in as initial conditions or for graphing with an external package
CCCCCCCCCCCCCCCCCCCCCCCCCCCCCCCCCCCCCCCCCCCCCCCCCCCCCCCCCCCCCCCCCCCCCCCC
C Global variables:
       INCLUDE 'MENU.ALL'
       INCLUDE 'IO.ALL'
       INCLUDE 'PARAM.P6'
C Input variables:
       REAL P(NXMAX,NYMAX)               !stream function
       REAL XSI(NXMAX,NYMAX)             !vorticity
C Local variables:
       INTEGER IX,IY                     !lattice indices
       INTEGER NLINES                    !number of lines written to file
CCCCCCCCCCCCCCCCCCCCCCCCCCCCCCCCCCCCCCCCCCCCCCCCCCCCCCCCCCCCCCCCCCCCCCCC
       CALL FLOPEN(GNAME,GUNIT)              !open file
       MSTRNG(MINTS(IGNAME))=GNAME           !name may have changed
       CALL PRMOUT(GUNIT,NLINES)             !write out header
C
       DO 100 IX=1,NX
          WRITE (GUNIT,10) (P(IX,IY),IY=1,NY)
100    CONTINUE
       DO 200 IX=1,NX
          WRITE (GUNIT,10) (XSI(IX,IY),IY=1,NY)
200    CONTINUE
10     FORMAT (5(2X,1PE14.7))
       CALL FLCLOS(GNAME,GUNIT)              !close it up
C
       RETURN
       END
CCCCCCCCCCCCCCCCCCCCCCCCCCCCCCCCCCCCCCCCCCCCCCCCCCCCCCCCCCCCCCCCCCCCCCCC
CCCCCCCCCCCCCCCCCCCCCCCCCCCCCCCCCCCCCCCCCCCCCCCCCCCCCCCCCCCCCCCCCCCCCCCC
C param.p6
C
       INTEGER NXMAX,NYMAX               !maximum lattice size
       PARAMETER (NXMAX=100,NYMAX=100)
       INTEGER MAXLEV                    !maximum number of contour levels
       PARAMETER (MAXLEV=20)
       INTEGER VRTCTY,STREAM             !display flags
C
       LOGICAL FIRST                     !first time through menu?
       INTEGER NX,NY                     !lattice size
       REAL VOMEGA                       !vorticity relaxation parameter
       REAL SOMEGA                       !stream relaxation parameter
       REAL MVOMEG                       !one minus vort omega
       REAL MSOMEG                       !one minus stream omega
       INTEGER HFWID                     !half width of plate
       INTEGER LENGTH                    !length of plate
       INTEGER FRONT                     !front edge of plate
       INTEGER BACK                      !back edge of plate
```

```
      REAL REYNLD                      !lattice Reynolds number
      REAL REYND4                      !Reynolds number / 4
C
      INTEGER PTYPE                    !choice for initial field
      CHARACTER*12 PFILE               !file to input init data
      LOGICAL SUBLAT                   !is there a sublattice?
      INTEGER NXLL,NYLL,NXUR,NYUR      !sublattice parameters
C
      INTEGER NFREQ                    !frequency of display
      INTEGER NLEV                     !number of contour levels
C
      INTEGER BNDCND(NXMAX,NYMAX)      !flag to indicate bound cond
      !0=not a boundary; 1=boundary; 2=plate interior
C
      LOGICAL XSKIP,YSKIP              !skip spaces or lines in display
      INTEGER XCNTR,YCNTR              !how to center display
C
      COMMON / FLAG   / FIRST
      COMMON / PPARAM / NX,NY,REYNLD
      COMMON / NPARAM / VOMEGA,SOMEGA,SUBLAT,NXLL,NYLL,NXUR,NYUR,PTYPE
      COMMON / BCPRM  / BNDCND,HFWID,LENGTH,FRONT
      COMMON / CPARAM / BACK,MVOMEG,MSOMEG,REYND4
      COMMON / GPARAM / NLEV,NFREQ,XSKIP,YSKIP,XCNTR,YCNTR
      COMMON / ASCII  / PFILE
CCCCCCCCCCCCCCCCCCCCCCCCCCCCCCCCCCCCCCCCCCCCCCCCCCCCCCCCCCCCCCCCCCCCCCC
CCCCCCCCCCCCCCCCCCCCCCCCCCCCCCCCCCCCCCCCCCCCCCCCCCCCCCCCCCCCCCCCCCCCCCC
C map.p6
      INTEGER IWID,ILEN,IFRNT,IRNLDS,IVOMEG,ISOMEG
      INTEGER ISUB,INXLL,INYLL,INXUR,INYUR,IPTYPE,IPFILE
      INTEGER INFREQ,INLEV
      PARAMETER (IWID   = 13 )
      PARAMETER (ILEN   = 14 )
      PARAMETER (IFRNT  = 15 )
      PARAMETER (IRNLDS = 16 )
      PARAMETER (IVOMEG = 38 )
      PARAMETER (ISOMEG = 39 )
      PARAMETER (ISUB   = 40 )
      PARAMETER (INXLL  = 41 )
      PARAMETER (INYLL  = 42 )
      PARAMETER (INXUR  = 43 )
      PARAMETER (INYUR  = 44 )
      PARAMETER (IPTYPE = 49 )
      PARAMETER (IPFILE = 50 )
      PARAMETER (INFREQ = 87 )
      PARAMETER (INLEV  = 88 )
```

C.7 Project 7

Algorithm   This program solves the non-linear reaction-diffusion equations of the Brusselator [Eqs. (VII.3)] in two dimensions ($0 \leq x \leq 1$ and $0 \leq y \leq 1$) with no-flux boundary conditions. You can make the problem one-dimensional by setting NY= 1 (if you set NX= 1 instead, the reaction terms are lost, and it becomes a simple diffusion problem). Initial conditions for the X and Y concentrations are taken to be random fluctuations about their equilibrium values (DO loop 5 in subroutine ARCHON). Time evolution (DO loop 20 in ARCHON) is then carried out by the alternating-direction method (7.19 along with 7.11–7.16) in subroutine EVOLVE. Note that some of the coefficients required for the inversions of the tri-diagonal matrices are calculated only once for each value of DT (subroutine TRDIAG and between lines 10 and 15 in subroutine ARCHON). Every NSTEP'th iteration, you have the option to continue iterations, change DT or NSTEP, or return to the main menu.

Note that X and Y refer both to species and direction (horizontal or vertical); if they appear twice (e.g., BETAXX) the first reference is to the species, the second to the direction; if they appear once, the meaning should be clear from the context.

Input   The physical parameters are the concentrations of A [2.] and B [4.], diffusion constants for X species [.001] and Y species [.003], and the size of the initial fluctuations for X species [.01] and Y species [.01]. All values are scaled as described in the text. After leaving the main menu, subroutine SUGGST calculates and displays limiting values of B based on the linear stability analysis presented in the text (e.g., what values of B give stable or unstable oscillations). You are then given an opportunity to revise your value of B based on this information. The program will also calculate and display the frequencies (VII.10) for the largest and smallest m values (these limits are based on boundary conditions and lattice size). The numerical parameters are the number of lattice points in x [20] and y [20] directions and the random number seed used for the initial fluctuations. The maximum number of x points is fixed by MAXX= 79 and y points by MAXY= 20. These limits are to ensure that the character display of the concentrations will fit on the typical screen, which is 24 lines long and 80 characters wide. If you are using graphics, or your terminal is not 24 × 80, you can adjust these parameters. The graphics parameter is the number of contours for plotting the concentrations and is used only if you have graphics. After you have left the main menu, you will be prompted for the time step [.07], the number of steps (NSTEP) until you can end the iterations, and the frequency for display of the concentrations (NFREQ). You should choose the time step so that the highest mode described by the lattice diffuses little during one time step.

Note that the default random number seed is a large integer [54765]. Compilers differ in how many bytes they use to store integers. If your compiler allows two or four bytes for integers, choose the four byte option. If your compiler only allows two bytes, then you must change the default, high, and low limits for the random number seed so that they do not exceed $2^{15} = 32768$. You will have to do the same for Example 8 and Project 8 as well.

Output At each time step the text output displays the time and the lowest, highest and average value for both X and Y concentrations. Every NFREQ'th step the concentrations (relative to their equilibrium values) are displayed as characters (if you request text to the screen) or as a contour plot (if you request graphics to the screen). For the character display (subroutine DISPLY) small letters indicate species values smaller than equilibrium, while capital letters, values larger than equilibrium. The appearance of the contour plot depends on your graphics package. If the problem is one-dimensional, the graphics displays a one-dimensional plot. Every NSTEP'th iteration, graphics are output to a file or hardcopy device, if requested. Note that the concentrations can be written out in three ways: as character data (if text is requested to a file), as numerical data (if graphics is requested a file), or as a contour plot (if graphics is requested to a hardcopy device).

Subroutine SUGGST fixes the graphing scales based on the stability analysis: if the system is unstable, the scale is equal to the maximum real frequency times the equilibrium value; if the system is stable the scale is equal to twice the initial fluctuations. Since the scale is fixed for the run, it may happen that values at the beginning (for an unstable system) or the end (for an exponentially decaying system) will not show up on the plots. However, a fixed scale is necessary to make clear the relative sizes of the fields at different times.

```
CCCCCCCCCCCCCCCCCCCCCCCCCCCCCCCCCCCCCCCCCCCCCCCCCCCCCCCCCCCCCCCCCCCCCCCCCC
      PROGRAM PROJ7
C     Project 7: The Brusselator in two dimensions
C  COMPUTATIONAL PHYSICS (FORTRAN VERSION)
C  by Steven E. Koonin and Dawn C. Meredith
C  Copyright 1989, Addison-Wesley Publishing Company
CCCCCCCCCCCCCCCCCCCCCCCCCCCCCCCCCCCCCCCCCCCCCCCCCCCCCCCCCCCCCCCCCCCCCCCCCC
      CALL INIT           !display header screen, setup parameters
5     CONTINUE            !main loop/ execute once for each set of param
```

```
          CALL PARAM       !get input from screen
          CALL ARCHON      !calculate time evolution of the chem reactions
          GOTO 5
          END
CCCCCCCCCCCCCCCCCCCCCCCCCCCCCCCCCCCCCCCCCCCCCCCCCCCCCCCCCCCCCCCCCCCCCCCCCC
          SUBROUTINE ARCHON
C calculates the time evolution of X and Y concentrations
C according to diffusion-reaction equations
CCCCCCCCCCCCCCCCCCCCCCCCCCCCCCCCCCCCCCCCCCCCCCCCCCCCCCCCCCCCCCCCCCCCCCCCCC
C Global variables:
          INCLUDE 'PARAM.P7'
          INCLUDE 'IO.ALL'
C Local variables:
          REAL X(MAXX,MAXY)          !X species concentration
          REAL Y(MAXX,MAXY)          !Y species concentration
          REAL TIME                  !time
          REAL DT                    !time step
          REAL DTMIN,DTMAX           !limits on time step
          INTEGER IT                 !time index
          INTEGER IX,IY              !horiz and vert indices
          INTEGER NLINES             !number of lines printed to terminal
          INTEGER SCREEN             !send to terminal
          INTEGER PAPER              !make a hardcopy
          INTEGER FILE               !send to a file
          INTEGER NSTEP              !number of time steps to take
          LOGICAL MORE,NEWDT         !options for continuing
          INTEGER NFREQ              !graphing frequency (movies only)
          REAL XLO,XHI,XTOT          !min,max, and total conc of species X
          REAL YLO,YHI,YTOT          !min,max, and total conc of species Y
          REAL SGRF(MAXX,MAXY)       !species values for graphing
          INTEGER XSPEC,YSPEC        !flag to indicate species
C Functions:
          REAL GETFLT                !get floating point number from screen
          INTEGER YESNO              !get.yes/no answer from screen
          LOGICAL LOGCVT             !change from 1,0 to true and false
          INTEGER GETINT             !get integer data from screen
          REAL RANNOS                !returns uniform random number
          DATA SCREEN,PAPER,FILE/1,2,3/
          DATA XSPEC,YSPEC/1,2/
CCCCCCCCCCCCCCCCCCCCCCCCCCCCCCCCCCCCCCCCCCCCCCCCCCCCCCCCCCCCCCCCCCCCCCCCCC
C     output summary of parameters
      IF (TTERM) CALL PRMOUT(OUNIT,NLINES)
      IF (TFILE) CALL PRMOUT(TUNIT,NLINES)
      IF (GFILE) CALL PRMOUT(GUNIT,NLINES)
C
C     initialize X and Y concentrations
      DO 5 IX=1,NX               !species concentrations fluctuate
        DO 6 IY=1,NY             !about equil values
          X(IX,IY)=A*(1.+XFLUCT*(RANNOS(DSEED)-.5))
          Y(IX,IY)=B/A*(1.+YFLUCT*(RANNOS(DSEED)-.5))
6       CONTINUE
5     CONTINUE
```

```
C
      TIME=0.                    !initialize time
      NSTEP=50                   !default num of time steps until next prompt
      NFREQ=10                   !default graphing freq
      DT=.07                     !default, min, and max time steps
      DTMIN=1.E-05
      DTMAX=1.
C
10    CONTINUE                   !loop over different DT and/or NSTEP
      DT=GETFLT(DT,-DTMAX,DTMAX,'Enter time step')
      IF (ABS(DT) .LT. DTMIN) DT=SIGN(DTMIN,DT)
      NSTEP=GETINT(NSTEP,1,1000,'Enter number of time steps')
      IF ((TTERM) .OR. (GTERM))
     +    NFREQ=GETINT(NFREQ,1,1000,
     +    'Enter display frequency for the concentration')
      NLINES=NLINES+6
C
C     calculate time step dependent constants
      IF (NX .GT. 1) THEN
         XXPLUS=-XDIFF*DT/HX/HX
         YXPLUS=-YDIFF*DT/HX/HX
      ELSE
         XXPLUS=0.
         YXPLUS=0.
      END IF
      IF (NY .GT. 1) THEN
         XYPLUS=-XDIFF*DT/HY/HY
         YYPLUS=-YDIFF*DT/HY/HY
      ELSE
         XYPLUS=0.
         YYPLUS=0.
      END IF
      XXZERO=1-2*XXPLUS
      YXZERO=1-2*YXPLUS
      XYZERO=1-2*XYPLUS
      YYZERO=1-2*YYPLUS
      ADT=A*DT
      BDT=B*DT
      BP1DT=1.-(B+1.)*DT
      CALL TRDIAG                !calculate ALPHA and GAMMA for this DT
C
      IF (TFILE) CALL TITLES(TUNIT,NLINES,DT)
C
15    CONTINUE                   !loop over sets of NSTEP time steps
      IF (TTERM) CALL TITLES(OUNIT,NLINES,DT)
C
      DO 20 IT=1,NSTEP           !time evolution
         TIME=TIME+DT
         CALL EVOLVE(X,Y,DT,XLO,XHI,XTOT,YLO,YHI,YTOT)
C
         IF (MOD(IT,NFREQ) .EQ. 0) THEN !graphics output
            IF (GTERM) THEN
```

```
             IF (TTERM) CALL PAUSE('to see concentrations...',1)
             CALL GRFOUT(SCREEN,XSPEC,X,XLO,XHI,XTOT,SGRF,NX,NY)
             CALL GRFOUT(SCREEN,YSPEC,Y,YLO,YHI,YTOT,SGRF,NX,NY)
             IF ((TTERM) .AND. (IT .NE. NSTEP))
   +               CALL TITLES(OUNIT,NLINES,DT)
           ELSE IF (TTERM) THEN
             CALL PAUSE('to see X concentration...',1)
             CALL DISPLY(X,OUNIT,XLO,XHI,XTOT,XSPEC)
             CALL DISPLY(Y,OUNIT,YLO,YHI,YTOT,YSPEC)
             CALL CLEAR
             NLINES=0
             IF (IT .NE. NSTEP) CALL TITLES(OUNIT,NLINES,DT)
           END IF
           ELSE                           !text output
             IF (TTERM)
   +         CALL TXTOUT(OUNIT,NLINES,TIME,XLO,XHI,XTOT,YLO,YHI,YTOT)
             IF (TFILE)
   +         CALL TXTOUT(TUNIT,NLINES,TIME,XLO,XHI,XTOT,YLO,YHI,YTOT)
           END IF
20         CONTINUE
C
           IF (TFILE) THEN                  !graphics output
             CALL DISPLY(X,TUNIT,XLO,XHI,XTOT,XSPEC)
             CALL DISPLY(Y,TUNIT,YLO,YHI,YTOT,YSPEC)
             CALL TITLES(TUNIT,NLINES,DT)
           END IF
           IF (GHRDCP) THEN
             CALL GRFOUT(PAPER,XSPEC,X,XLO,XHI,XTOT,SGRF,NX,NY)
             CALL GRFOUT(PAPER,YSPEC,Y,YLO,YHI,YTOT,SGRF,NX,NY)
           END IF
           IF (GFILE) CALL WRTOUT(X,Y,TIME,DT)
C
           MORE=LOGCVT(YESNO(1,'Continue iterating?'))
           IF (MORE) THEN
             NLINES=0
             IF (TTERM) CALL CLEAR
             GOTO 15
           END IF
C
           NEWDT=
   +       LOGCVT(YESNO(1,'Change time step or frequency and continue?'))
           IF (NEWDT) THEN
             NLINES=0
             IF (TTERM) CALL CLEAR
             GOTO 10
           END IF
C
           RETURN
           END
CCCCCCCCCCCCCCCCCCCCCCCCCCCCCCCCCCCCCCCCCCCCCCCCCCCCCCCCCCCCCCCCCCCCCCCCC
           SUBROUTINE EVOLVE(X,Y,DT,XLO,XHI,XTOT,YLO,YHI,YTOT)
C evolves the species X and Y according to the diffusion-reaction
```

```
C equation and calculates minima, maxima, and average species values;
C all the reaction terms are in the horiz sweep of the lattice;
C see Eq 7.11-7.16 and VII.3a,b
CCCCCCCCCCCCCCCCCCCCCCCCCCCCCCCCCCCCCCCCCCCCCCCCCCCCCCCCCCCCCCCCCCCCCCCC
C Global variables:
      INCLUDE 'PARAM.P7'
C Input/Output variables:
      REAL X(MAXX,MAXY)      !X species concentration (I/O)
      REAL Y(MAXX,MAXY)      !Y species concentration (I/O)
      REAL DT                !time step (I)
      REAL XLO,XHI,XTOT       !min,max, and total conc of species X (out)
      REAL YLO,YHI,YTOT       !min,max, and total conc of species Y (out)
C Local variables:
      INTEGER IX,IY          !horiz and vert indices
                             !variables for matrix inversion
      REAL BETAXX(0:MAXX),BETAYX(0:MAXX),BETAXY(0:MAXY),BETAYY(0:MAXY)
      REAL X2YDT,XRHS,YRHS    !temp variables
CCCCCCCCCCCCCCCCCCCCCCCCCCCCCCCCCCCCCCCCCCCCCCCCCCCCCCCCCCCCCCCCCCCCCCCC
      IF (NX .GT. 1) THEN
         DO 10 IY=1,NY             !for each value of y, do an x sweep
            BETAXX(NX)=0.          !no flux bound. cond
            BETAYX(NX)=0.
            DO 20 IX=NX-1,0,-1     !backward recursion
               X2YDT=Y(IX+1,IY)*DT*X(IX+1,IY)*X(IX+1,IY)
               XRHS=ADT+BP1DT*X(IX+1,IY)+X2YDT
               YRHS=-X2YDT+BDT*X(IX+1,IY)+Y(IX+1,IY)
               BETAXX(IX)=GAMXX(IX+1)*(XXPLUS*BETAXX(IX+1)-XRHS)
               BETAYX(IX)=GAMYX(IX+1)*(YXPLUS*BETAYX(IX+1)-YRHS)
20          CONTINUE
C
            X(1,IY)=BETAXX(0)/(1.-ALPHXX(0))     !no-flux b.c.
            Y(1,IY)=BETAYX(0)/(1.-ALPHYX(0))
            DO 30 IX=2,NX                         !forward recursion
               X(IX,IY)=ALPHXX(IX-1)*X(IX-1,IY)+BETAXX(IX-1)
               Y(IX,IY)=ALPHYX(IX-1)*Y(IX-1,IY)+BETAYX(IX-1)
30          CONTINUE
10       CONTINUE
      END IF
C
      IF (NY .GT. 1) THEN
         DO 40 IX=1,NX             !for each value of x, do a y sweep
            BETAXY(NY)=0.          !no flux bound. cond
            BETAYY(NY)=0.
            DO 50 IY=NY-1,0,-1     !backward recursion
               BETAXY(IY)=GAMXY(IY+1)*(XYPLUS*BETAXY(IY+1)-X(IX,IY+1))
               BETAYY(IY)=GAMYY(IY+1)*(YYPLUS*BETAYY(IY+1)-Y(IX,IY+1))
50          CONTINUE
C
            X(IX,1)=BETAXY(0)/(1.-ALPHXY(0))     !no-flux b.c.
            Y(IX,1)=BETAYY(0)/(1.-ALPHYY(0))
            DO 60 IY=2,NY                         !forward recursion
               X(IX,IY)=ALPHXY(IY-1)*X(IX,IY-1)+BETAXY(IY-1)
```

```
                     Y(IX,IY)=ALPHYY(IY-1)*Y(IX,IY-1)+BETAYY(IY-1)
60              CONTINUE
40         CONTINUE
      END IF
C
      XTOT=0.                  !calculate min, max, and averages
      YTOT=0.
      XLO=X(1,1)
      YLO=Y(1,1)
      XHI=XLO
      YHI=YLO
      DO 70 IX=1,NX
          DO 80 IY=1,NY
              IF (X(IX,IY) .GT. XHI) XHI=X(IX,IY)
              IF (X(IX,IY) .LT. XLO) XLO=X(IX,IY)
              IF (Y(IX,IY) .GT. YHI) YHI=Y(IX,IY)
              IF (Y(IX,IY) .LT. YLO) YLO=Y(IX,IY)
              XTOT=XTOT+X(IX,IY)
              YTOT=YTOT+Y(IX,IY)
80         CONTINUE
70    CONTINUE
      XTOT=XTOT/NX/NY          !average value
      YTOT=YTOT/NX/NY
C
      RETURN
      END
CCCCCCCCCCCCCCCCCCCCCCCCCCCCCCCCCCCCCCCCCCCCCCCCCCCCCCCCCCCCCCCCCCCCCCCCCC
      SUBROUTINE TRDIAG
C calculate ALPHA and GAMMA for the inversion of the tridiagonal matrix
C see Eq. 7.11-7.16 and VII.3a,b
CCCCCCCCCCCCCCCCCCCCCCCCCCCCCCCCCCCCCCCCCCCCCCCCCCCCCCCCCCCCCCCCCCCCCCCCCC
C Global variables:
      INCLUDE 'PARAM.P7'
C Local variables:
      INTEGER IX,IY                      !lattice indices
CCCCCCCCCCCCCCCCCCCCCCCCCCCCCCCCCCCCCCCCCCCCCCCCCCCCCCCCCCCCCCCCCCCCCCCCCC
C     horizontal recursion
      IF (NX .GT. 1) THEN
          ALPHXX(NX)=1.              !no-flux boundary conditions
          ALPHYX(NX)=1.
          DO 10 IX=NX,1,-1              !backward recursion
              GAMXX(IX)=-1./(XXZERO+XXPLUS*ALPHXX(IX))
              ALPHXX(IX-1)=GAMXX(IX)*XXPLUS
              GAMYX(IX)=-1./(YXZERO+YXPLUS*ALPHYX(IX))
              ALPHYX(IX-1)=GAMYX(IX)*YXPLUS
10        CONTINUE
      END IF
C
C     vertical recursion
      IF (NY .GT. 1) THEN
          ALPHXY(NY)=1.                 !no-flux boundary conditions
          ALPHYY(NY)=1.
```

```
          DO 20 IY=NY,1,-1                !backward recursion
             GAMXY(IY)=-1./(XYZERO+XYPLUS*ALPHXY(IY))
             ALPHXY(IY-1)=GAMXY(IY)*XYPLUS
             GAMYY(IY)=-1./(YYZERO+YYPLUS*ALPHYY(IY))
             ALPHYY(IY-1)=GAMYY(IY)*YYPLUS
20        CONTINUE
          END IF
C
          RETURN
          END
CCCCCCCCCCCCCCCCCCCCCCCCCCCCCCCCCCCCCCCCCCCCCCCCCCCCCCCCCCCCCCCCCCCCC
          SUBROUTINE SUGGST
C suggests values for B based on linear stability analysis
C (see section VII.2),
C also calculates graphing scales based on this analysis
CCCCCCCCCCCCCCCCCCCCCCCCCCCCCCCCCCCCCCCCCCCCCCCCCCCCCCCCCCCCCCCCCCCCC
C Global variables:
          INCLUDE 'MENU.ALL'
          INCLUDE 'IO.ALL'
          INCLUDE 'PARAM.P7'
          INCLUDE 'MAP.P7'
C Local variables:
          REAL PI                        !3.14159
          REAL TEMP                      !temp variable to find oscillations
          REAL DELTM,MPISQ               !functions of the mode M
          COMPLEX DISC                   !discriminant
          REAL M,MOSC                    !wave modes
          REAL BLO,BHI,BCRIT             !critical B values
          INTEGER IM                     !mode index
          REAL AM,BM                     !temp values to calculate frequencies
          REAL MAXFRQ                    !maximum real part of the frequencies
          LOGICAL UNSTBL                 !is the system probably unstable?
C Function:
          REAL GETFLT                    !get floating point number from screen
CCCCCCCCCCCCCCCCCCCCCCCCCCCCCCCCCCCCCCCCCCCCCCCCCCCCCCCCCCCCCCCCCCCCC
C for 2-dimensions Meffective=sqrt(mx**2+my**2) and so needn't be
C an integer; if we replace M by Meffective in all the formulas
C we have the correct 2-dim analysis
C
          WRITE (OUNIT,5)
          WRITE (OUNIT,6)
          WRITE (OUNIT,*) ' '
5         FORMAT (' The following numbers from linear stability analysis',
     +             ' will guide you')
6         FORMAT (' in your choice of B for the chosen ',
     +             'values of A, XDIFF, and YDIFF:')
C
          MLO=0.    !lowest M value; =0 for no-flux bc; =SQRT(2) for fixed bc
C
C         what is the highest mode which gives rise to oscillations in time?
C         what values of B give oscillations? instability?
          PI=4.*ATAN(1.0)
```

```
          TEMP=1./(YDIFF-XDIFF)
C

          IF (TEMP .GT. 0.) THEN   !oscillations are possible
            MOSC=(SQRT(TEMP)/PI)
            WRITE (OUNIT,10) MOSC
10          FORMAT (' The highest mode to give oscillations is M = ',F7.3)
            DO 20 IM=1,2                   !loop over small and large M
              IF (IM .EQ. 1) M=MLO              !smallest M
              IF (IM .EQ. 2) M=MOSC            !largest M  to give osc
              MPISQ=(M*PI)**2
              DELTM=1+MPISQ*(XDIFF-YDIFF)
              BCRIT=1+A**2+MPISQ*(XDIFF+YDIFF)    !smallest B to give instb.
              DELTM=SQRT(DELTM)
              BLO=(A-DELTM)**2                !smallest B to give osc
              BHI=(A+DELTM)**2                !largest B to give osc
              WRITE (OUNIT,*) ' '
              WRITE (OUNIT,30) M,BCRIT
              WRITE (OUNIT,40) BLO,BHI
30            FORMAT (' for M = ',F7.3,
     +        ' the smallest B to give unstable oscillations is = ',F7.3)
40            FORMAT (' and oscillations arise only for B between ',F7.3,
     +        ' and ', F7.3)
20          CONTINUE
C

          ELSE IF (TEMP .LE. 0.) THEN   !oscillations are not possible
            MOSC=0.
            WRITE (OUNIT,50)
50          FORMAT (' There are no oscillations for these parameter values')
          END IF
C
C         what values of B and M give unstable, non-oscillatory behavior?
          M=SQRT(A/PI/PI/SQRT(XDIFF*YDIFF))
          MPISQ=M*M*PI*PI
          BCRIT=1+A**2*(XDIFF/YDIFF+1./YDIFF/MPISQ)+XDIFF*MPISQ
          WRITE (OUNIT,*) ' '
          WRITE (OUNIT,55)
          WRITE (OUNIT,60) M,BCRIT
55        FORMAT (' Instability w.r.t. exponentially growing behavior')
60        FORMAT (' first occurs for M = ',F7.3,' and B = ',F7.3)
C
C         allow user to adjust value of B using this information
          WRITE (OUNIT,*) ' '
          B=GETFLT(B,0.,20.,'Enter a value for B concentration')
          MREALS(IB)=B
C
C         calculate the frequencies for smallest M (depends on b.c.)
C         and largest M (which depends on the lattice spacing)
          MHI=SQRT(REAL(NX/2)**2+REAL(NY/2)**2)
          DO 70 IM=1,2
            IF (IM .EQ. 1) MPISQ=MLO*MLO*PI*PI
            IF (IM .EQ. 2) MPISQ=MHI*MHI*PI*PI
            AM=B-1-MPISQ*XDIFF
```

```
        BM=A**2+MPISQ*YDIFF
        DISC=(AM+BM)**2-4*A**2*B
        OMEGAP(IM)=(AM-BM+SQRT(DISC))/2
        OMEGAM(IM)=(AM-BM-SQRT(DISC))/2
70      CONTINUE
C
        UNSTBL=.FALSE.                  !is the system unstable?
        MAXFRQ=REAL(OMEGAP(1))          !find the max freq
        IF (REAL(OMEGAP(2)) .GT. MAXFRQ) MAXFRQ=REAL(OMEGAP(2))
        IF (REAL(OMEGAM(1)) .GT. MAXFRQ) MAXFRQ=REAL(OMEGAM(1))
        IF (REAL(OMEGAM(2)) .GT. MAXFRQ) MAXFRQ=REAL(OMEGAM(2))
        IF (MAXFRQ .GE. 0) UNSTBL=.TRUE.
        IF (MAXFRQ .LT. .1) MAXFRQ=2*MAXFRQ  !adjust very small scales
        IF (UNSTBL) THEN                !scale for graphing
           YSCALE=MAXFRQ*B/A            !this scale is set for all times
           XSCALE=MAXFRQ*A              !for these values of the parameters
        ELSE
           YSCALE=2*YFLUCT
           XSCALE=2*XFLUCT
        END IF
C
        RETURN
        END
CCCCCCCCCCCCCCCCCCCCCCCCCCCCCCCCCCCCCCCCCCCCCCCCCCCCCCCCCCCCCCCCCCCCCCC
        SUBROUTINE INIT
C initializes constants, displays header screen,
C initializes menu arrays for input parameters
CCCCCCCCCCCCCCCCCCCCCCCCCCCCCCCCCCCCCCCCCCCCCCCCCCCCCCCCCCCCCCCCCCCCCCC
C Global variables:
        INCLUDE 'IO.ALL'
        INCLUDE 'MENU.ALL'
        INCLUDE 'PARAM.P7'
C Local parameters:
        CHARACTER*80 DESCRP             !program description
        DIMENSION DESCRP(20)
        INTEGER NHEAD,NTEXT,NGRAPH      !number of lines for each description
CCCCCCCCCCCCCCCCCCCCCCCCCCCCCCCCCCCCCCCCCCCCCCCCCCCCCCCCCCCCCCCCCCCCCCC
C       get environment parameters
        CALL SETUP
C
C       display header screen
        DESCRP(1)= 'PROJECT 7'
        DESCRP(2)= 'The Brusselator in two dimensions'
        NHEAD=2
C
C       text output description
        DESCRP(3)= 'time, X and Y species minima, maxima and averages'
        NTEXT=1
C
C       graphics output description
        DESCRP(4)= 'X and Y species concentrations'
        NGRAPH=1
```

```
C
        CALL HEADER(DESCRP,NHEAD,NTEXT,NGRAPH)
C
C       setup menu arrays, beginning with constant part
        CALL MENU
C
        MTYPE(13)=FLOAT
        MPRMPT(13)='Enter value for A species concentration'
        MTAG(13)='A concentration'
        MLOLIM(13)=0.001
        MHILIM(13)=20.
        MREALS(13)=2.
C
        MTYPE(14)=FLOAT
        MPRMPT(14)='Enter value for B species concentration'
        MTAG(14)='B concentration'
        MLOLIM(14)=0.
        MHILIM(14)=20.
        MREALS(14)=4.
C
        MTYPE(15)=FLOAT
        MPRMPT(15)='Enter value for XDIFF (X species diffusion constant)'
        MTAG(15)='X species diffusion constant'
        MLOLIM(15)=0.
        MHILIM(15)=10.
        MREALS(15)=0.001
C
        MTYPE(16)=FLOAT
        MPRMPT(16)='Enter value for YDIFF (Y species diffusion constant)'
        MTAG(16)='Y species diffusion constant'
        MLOLIM(16)=0.
        MHILIM(16)=10.
        MREALS(16)=0.003
C
        MTYPE(17)=FLOAT
        MPRMPT(17)='Enter value for XFLUCT (X species fluctuations)'
        MTAG(17)='X species fluctuation'
        MLOLIM(17)=-1.
        MHILIM(17)=1.
        MREALS(17)=.01
C
        MTYPE(18)=FLOAT
        MPRMPT(18)='Enter value for YFLUCT (Y species fluctuations)'
        MTAG(18)='Y species fluctuation'
        MLOLIM(18)=-1.
        MHILIM(18)=1.
        MREALS(18)=.01
C
        MTYPE(19)=SKIP
        MREALS(19)=35.
C
        MTYPE(38)=NUM
```

```
      MPRMPT(38)= 'Enter number of X lattice points'
      MTAG(38)= 'Number of X lattice points'
      MLOLIM(38)=1.
      MHILIM(38)=MAXX
      MINTS(38)=20
C
      MTYPE(39)=NUM
      MPRMPT(39)= 'Enter number of Y lattice points'
      MTAG(39)= 'Number of Y lattice points'
      MLOLIM(39)=1.
      MHILIM(39)=MAXY
      MINTS(39)=20
C
      MTYPE(40)=NUM
      MPRMPT(40)= 'Integer random number seed for init fluctuations'
      MTAG(40)= 'Random number seed'
      MLOLIM(40)=1000.
      MHILIM(40)=99999.
      MINTS(40)=54765
C
      MTYPE(41)=SKIP
      MREALS(41)=60.
C
      MSTRNG(MINTS(75))= 'proj7.txt'
C
      MTYPE(76)=SKIP
      MREALS(76)=80.
C
      MSTRNG(MINTS(86))= 'proj7.grf'
C
      MTYPE(87)=NUM
      MPRMPT(87)= 'Enter number of contour levels'
      MTAG(87)= 'Number of contour levels'
      MLOLIM(87)=2.
      MHILIM(87)=50.
      MINTS(87)=6
C
      MTYPE(88)=SKIP
      MREALS(88)=90.
C
      RETURN
      END
ccccccccccccccccccccccccccccccccccccccccccccccccccccccccccccccccccccccc
      SUBROUTINE PARAM
C gets parameters from screen
C ends program on request
C closes old files
C maps menu variables to program variables
C opens new files
C calculates all derivative parameters
C performs checks on parameters
ccccccccccccccccccccccccccccccccccccccccccccccccccccccccccccccccccccccc
```

```
C Global variables:
      INCLUDE 'MENU.ALL'
      INCLUDE 'IO.ALL'
      INCLUDE 'PARAM.P7'
      INCLUDE 'MAP.P7'
C Functions:
      LOGICAL LOGCVT            !converts 1 and 0 to true and false
      REAL GETFLT               !get floating point number from screen
      INTEGER GETINT            !get integer from screen
CCCCCCCCCCCCCCCCCCCCCCCCCCCCCCCCCCCCCCCCCCCCCCCCCCCCCCCCCCCCCCCCCCCCCCCCCCC
C     get input from terminal
      CALL CLEAR
      CALL ASK(1,ISTOP)
C
C     stop program if requested
      IF (MREALS(IMAIN) .EQ. STOP) CALL FINISH
C
C     close files if necessary
      IF (TNAME .NE. MSTRNG(MINTS(ITNAME)))
     +     CALL FLCLOS(TNAME,TUNIT)
      IF (GNAME .NE. MSTRNG(MINTS(IGNAME)))
     +     CALL FLCLOS(GNAME,GUNIT)
C
C     set new parameter values
C     physical and numerical
      A=MREALS(IA)
      B=MREALS(IB)
      XDIFF=MREALS(IXDIFF)
      YDIFF=MREALS(IYDIFF)
      XFLUCT=MREALS(IXFL)
      YFLUCT=MREALS(IYFL)
      NX=MINTS(INX)
      NY=MINTS(INY)
      DSEED=DBLE(MINTS(IDSEED))
C
C     text output
      TTERM=LOGCVT(MINTS(ITTERM))
      TFILE=LOGCVT(MINTS(ITFILE))
      TNAME=MSTRNG(MINTS(ITNAME))
C
C     graphics output
      GTERM=LOGCVT(MINTS(IGTERM))
      GHRDCP=LOGCVT(MINTS(IGHRD))
      GFILE=LOGCVT(MINTS(IGFILE))
      GNAME=MSTRNG(MINTS(IGNAME))
      NLEV=MINTS(INLEV)
C
C     open files
      IF (TFILE) CALL FLOPEN(TNAME,TUNIT)
      IF (GFILE) CALL FLOPEN(GNAME,GUNIT)
      !files may have been renamed
      MSTRNG(MINTS(ITNAME))=TNAME
```

```
        MSTRNG(MINTS(IGNAME))=GNAME
C
C     make sure that problem is at least one-dimensional
60    IF ((NX .EQ. 1) .AND. (NY .EQ. 1)) THEN
         WRITE (OUNIT,50)
50       FORMAT (' Both NX and NY can''t be = 1')
         NX=GETINT(20,1,MAXX,'Enter new value for NX')
         NY=GETINT(20,1,MAXY,'Enter new value for NY')
         MINTS(INX)=NX
         MINTS(INY)=NY
         GOTO 60
      END IF
C
      CALL SUGGST                        !suggest values for B
      CALL CLEAR
C
C     derivative parameters
      HX=0.                              !allow for one-dimensional systems
      HY=0.
      IF (NX .GT. 1) HX=1./(NX-1)        !calculate space steps
      IF (NY .GT. 1) HY=1./(NY-1)
C     calculate parameters for best looking text display
      IF (2*NX .LE. TRMWID) THEN
          XSKIP=.TRUE.                   !skip spaces in x
          XCNTR=(TRMWID-2*NX)/2          !how to center display
      ELSE
          XSKIP=.FALSE.
          XCNTR=(TRMWID-NX)/2
      END IF
      IF (XCNTR .LT. 1) XCNTR=1
C
      IF (2*NY .LE. TRMLIN-4) THEN
          YSKIP=.TRUE.                   !skip lines in y
          YCNTR=(TRMLIN-2*NY)/2-2        !how to center display
      ELSE
          YSKIP=.FALSE.
          YCNTR=(TRMLIN-NY)/2-2
      END IF
      IF (YCNTR .LT. 0) YCNTR=0
C
      RETURN
      END
CCCCCCCCCCCCCCCCCCCCCCCCCCCCCCCCCCCCCCCCCCCCCCCCCCCCCCCCCCCCCCCCCCCCCCCC
      SUBROUTINE DISPLY(S,MUNIT,SLO,SHI,STOT,SPEC)
C display species (S) as ascii characters
C values above the equilibrium value are displayed as capital letters,
C those less than the equilibrium value are displayed as small letters
CCCCCCCCCCCCCCCCCCCCCCCCCCCCCCCCCCCCCCCCCCCCCCCCCCCCCCCCCCCCCCCCCCCCCCCC
C Global variables:
      INCLUDE 'PARAM.P7'
      INCLUDE 'IO.ALL'
C Input variables:
```

```
            REAL S(MAXX,MAXY)              !species values
            INTEGER MUNIT                  !unit we're writing to
            REAL SLO,SHI,STOT              !min,max, and total conc of species S
            INTEGER SPEC                   !which species
      C Local variables:
            INTEGER IX,IY                  !lattice indices
            INTEGER STEMP                  !S as an integer
            CHARACTER*1 SPECS(MAXX)        !species as character data
            REAL SZERO                     !'zero' value for species
            REAL SCALE                     !scale for species
            CHARACTER*80 BLNK              !blanks for centering in X
            INTEGER XSPEC,YSPEC            !flag to indicate species
            INTEGER CSCALE                 !scale of ascii data
            CHARACTER*1 ASKII(0:25),NEGASK(0:25)!charac data for display
            DATA BLNK /' '/
            DATA ASKII/'A','B','C','D','E','F','G','H','I','J','K','L','M',
           +          'N','O','P','Q','R','S','T','U','V','W','X','Y','Z'/
            DATA NEGASK/'a','b','c','d','e','f','g','h','i','j','k','l','m',
           +          'n','o','p','q','r','s','t','u','v','w','x','y','z'/
            DATA XSPEC,YSPEC/1,2/
            DATA CSCALE/25/                          !since there are 26 charac
      CCCCCCCCCCCCCCCCCCCCCCCCCCCCCCCCCCCCCCCCCCCCCCCCCCCCCCCCCCCCCCCCCCCCCCCCCCC
            IF (MUNIT .EQ. OUNIT) THEN
               CALL CLEAR
               DO 20 IY=1,YCNTR                      !center output
                  WRITE (OUNIT,*) ' '
      20       CONTINUE                              !print out data
               IF (SPEC .EQ. XSPEC) WRITE (OUNIT,25) SLO,SHI,STOT
               IF (SPEC .EQ. YSPEC) WRITE (OUNIT,26) SLO,SHI,STOT
      25       FORMAT(15X,'X min =',F7.3,'   X max =',F7.3,'   X average =',F7.3)
      26       FORMAT(15X,'Y min =',F7.3,'   Y max =',F7.3,'   Y average =',F7.3)
            ELSE
               WRITE (MUNIT,*) ' '
               IF (SPEC .EQ. XSPEC) WRITE(MUNIT,*)
           +                ' X species concentration - A:'
               IF (SPEC .EQ. YSPEC) WRITE(MUNIT,*)
           +                ' Y species concentration - A/B:'
            END IF
      C
            IF (SPEC .EQ. XSPEC) THEN                !set scales
               SZERO=A                               !X equil value
               SCALE=XSCALE/CSCALE
            ELSE IF (SPEC .EQ. YSPEC) THEN
               SZERO=B/A                             !Y equil value
               SCALE=YSCALE/CSCALE
            END IF
      C
            DO 100 IY=NY,1,-1
               DO 50 IX=1,NX      !STEMP is scaled deviation from SZERO
                  STEMP=NINT((S(IX,IY)-SZERO)/SCALE)
                  IF (STEMP .GT. CSCALE)  STEMP=CSCALE
                  IF (STEMP .LT. -CSCALE) STEMP=-CSCALE
```

```
          IF (STEMP .GT. 0) THEN           !convert species to ASCII
             SPECS(IX)=ASKII(STEMP)
          ELSE IF (STEMP .LT. 0) THEN
             SPECS(IX)=NEGASK(-STEMP)
          ELSE IF (STEMP .EQ. 0) THEN
             IF (S(IX,IY).GT. SZERO) THEN
                SPECS(IX)=ASKII(0)
             ELSE
                SPECS(IX)=ASKII(0)
             END IF
          END IF
50        CONTINUE
C
C         write out a line at a time (no centering done for TUNIT)
          IF (MUNIT .EQ. TUNIT) THEN
                WRITE (TUNIT,16) (SPECS(IX),IX=1,NX)
          ELSE
           IF (XSKIP) THEN
              WRITE (OUNIT,10) BLNK(1:XCNTR),(SPECS(IX),IX=1,NX)
           ELSE
              WRITE (OUNIT,15) BLNK(1:XCNTR),(SPECS(IX),IX=1,NX)
           END IF
           IF (YSKIP) WRITE (OUNIT,*) ' '
          END IF
10        FORMAT (1X,A,100(A1,1X))
15        FORMAT (1X,A,100(A1))
16        FORMAT (1X,100(A1))
100       CONTINUE
          IF (MUNIT .EQ. OUNIT) THEN
             IF (SPEC .EQ. XSPEC) THEN
                CALL PAUSE('to see Y concentration...',0)
             ELSE
                CALL PAUSE('to continue...',0)
             END IF
          END IF
C
       RETURN
       END
ccccccccccccccccccccccccccccccccccccccccccccccccccccccccccccccccccccccc
       SUBROUTINE PRMOUT(MUNIT,NLINES)
C outputs parameter summary to MUNIT
ccccccccccccccccccccccccccccccccccccccccccccccccccccccccccccccccccccccc
C Global variables:
       INCLUDE 'IO.ALL'
       INCLUDE 'PARAM.P7'
C Input/Output variables:
       INTEGER MUNIT           !unit number for output (input)
       INTEGER NLINES          !number of lines written so far (output)
ccccccccccccccccccccccccccccccccccccccccccccccccccccccccccccccccccccccc
       IF (MUNIT .EQ. OUNIT) CALL CLEAR
C
       WRITE (MUNIT,2)
```

```
          WRITE (MUNIT,4)
          WRITE (MUNIT,6) A,B
          WRITE (MUNIT,8) XDIFF,YDIFF
          WRITE (MUNIT,10) XFLUCT,YFLUCT
          WRITE (MUNIT,12) NX,NY
          WRITE (MUNIT,16) MLO,OMEGAP(1),OMEGAM(1)
          WRITE (MUNIT,16) MHI,OMEGAP(2),OMEGAM(2)
          WRITE (MUNIT,14)
          WRITE (MUNIT,2)
C
          NLINES=10
C
2         FORMAT (' ')
4         FORMAT (' Project 7: Brusselator in two dimensions')
6         FORMAT (' A concentration = ',F8.3,5X,'B concentration =',F8.3)
8         FORMAT (' X species diffusion = ',1PE12.5,5X,
     +            ' Y species diffusion = ',1PE12.5)
10        FORMAT (' X species init fluct = '1PE12.5,5X,
     +            ' Y species init fluct = '1PE12.5)
12        FORMAT (' NX = ',I4,5X,' NY = ',I4)
14        FORMAT (' all values are in scaled units')
16        FORMAT (' for M=',F7.3,' the frequencies are ',
     +            2( 2X,'(',F7.3,',',F7.3,')' ) )
C
          RETURN
          END
CCCCCCCCCCCCCCCCCCCCCCCCCCCCCCCCCCCCCCCCCCCCCCCCCCCCCCCCCCCCCCCCCCCCCCCCC
          SUBROUTINE TXTOUT(MUNIT,NLINES,TIME,XLO,XHI,XTOT,YLO,YHI,YTOT)
C writes results for one time step to MUNIT
CCCCCCCCCCCCCCCCCCCCCCCCCCCCCCCCCCCCCCCCCCCCCCCCCCCCCCCCCCCCCCCCCCCCCCCCC
C Global variables:
          INCLUDE 'IO.ALL'
C Input variables:
          INTEGER MUNIT              !output unit specifier
          REAL TIME                  !time
          INTEGER NLINES             !number of lines printed so far (I/O)
          REAL XLO,XHI,XTOT          !min,max, and total conc of species X
          REAL YLO,YHI,YTOT          !min,max, and total conc of species Y
CCCCCCCCCCCCCCCCCCCCCCCCCCCCCCCCCCCCCCCCCCCCCCCCCCCCCCCCCCCCCCCCCCCCCCCCC
C         if screen is full, clear screen and retype headings
          IF ((MOD(NLINES,TRMLIN-4) .EQ. 0)
     +        .AND. (MUNIT .EQ. OUNIT) .AND. (.NOT. GTERM)) THEN
             CALL PAUSE('to continue...',1)
             CALL CLEAR
             WRITE (MUNIT,14)
             WRITE (MUNIT,16)
             NLINES=NLINES+2
          END IF
C
          WRITE (MUNIT,40) TIME,XLO,XHI,XTOT,YLO,YHI,YTOT
C         keep track of printed lines only for terminal output
          IF (MUNIT .EQ. OUNIT) NLINES=NLINES+1
```

```
C
40    FORMAT (3X,1PE12.5,5X,6(0PF8.3,2X))
14    FORMAT (8X,'time',11X,'Xlo',7X,'Xhi',7X,'Xave',7X,'Ylo',
     +         7X,'Yhi',6X,'Yave')
16    FORMAT (8X,'----',11X,'---',7X,'---',7X,'----',7X,'---',
     +         7X,'---',6X,'----')
C
      RETURN
      END
CCCCCCCCCCCCCCCCCCCCCCCCCCCCCCCCCCCCCCCCCCCCCCCCCCCCCCCCCCCCCCCCCCCCC
      SUBROUTINE TITLES(MUNIT,NLINES,DT)
C write out time step and column titles to MUNIT
CCCCCCCCCCCCCCCCCCCCCCCCCCCCCCCCCCCCCCCCCCCCCCCCCCCCCCCCCCCCCCCCCCCCC
C Global variables:
      INCLUDE 'IO.ALL'
C Passed variables:
      INTEGER MUNIT          !unit number for output (input)
      INTEGER NLINES         !number of lines written so far (output)
      REAL DT                !time step (input)
CCCCCCCCCCCCCCCCCCCCCCCCCCCCCCCCCCCCCCCCCCCCCCCCCCCCCCCCCCCCCCCCCCCCC
      IF ((MOD(NLINES,TRMLIN-7) .EQ. 0)
     +    .AND. (MUNIT .EQ. OUNIT) .AND. (.NOT. GTERM))
     +    CALL CLEAR
C
      WRITE (MUNIT,*)  ' '
      WRITE (MUNIT,20) DT
      WRITE (MUNIT,14)
      WRITE (MUNIT,16)
C
      IF (MUNIT .EQ. OUNIT) NLINES=NLINES+4
C
20    FORMAT (' time step = ',1PE12.5)
14    FORMAT (8X,'time',11X,'Xlo',7X,'Xhi',7X,'Xave',7X,'Ylo',
     +         7X,'Yhi',6X,'Yave')
16    FORMAT (8X,'----',11X,'---',7X,'---',7X,'----',7X,'---',
     +         7X,'---',6X,'----')
C
      RETURN
      END
CCCCCCCCCCCCCCCCCCCCCCCCCCCCCCCCCCCCCCCCCCCCCCCCCCCCCCCCCCCCCCCCCCCCC
      SUBROUTINE GRFOUT(DEVICE,SPEC,S,SLO,SHI,STOT,SGRF,MX,MY)
C display contours of the species-equilibrium value
C this routine will also do 1-dim plots if either NX or NY =1
C    the field values must be in an array that is exactly NX by NY;
C    this can be accomplished with implicit dimensioning which
C    requires that SGRF and its dimensions be passed to this routine
CCCCCCCCCCCCCCCCCCCCCCCCCCCCCCCCCCCCCCCCCCCCCCCCCCCCCCCCCCCCCCCCCCCCC
C Global variables:
      INCLUDE 'IO.ALL'
      INCLUDE 'PARAM.P7'
      INCLUDE 'GRFDAT.ALL'
C Input variables:
```

```
      INTEGER DEVICE               !which device
      INTEGER SPEC                 !which species
      REAL S(MAXX,MAXY)            !species values
      INTEGER MX,MY                !NX and NY in disguise
      REAL SGRF(MX,MY)             !species values for graphing
      REAL SLO,SHI,STOT            !min,max, average species values
C Local variables:
      REAL SZERO                   !'zero' value of species
      INTEGER IX,IY                !lattice indices
      INTEGER SCREEN               !send to terminal
      INTEGER PAPER                !make a hardcopy
      INTEGER FILE                 !send to a file
      INTEGER NCONT                !number of contours
      REAL LATTIC(1:MAXX)          !array for 1-dim plots
      CHARACTER*9 CLO,CHI,CTOT     !data as characters
      CHARACTER*9 CA,CB            !data as characters
      INTEGER XSPEC,YSPEC          !flag to indicate species
      INTEGER LLO,LHI,LTOT,LA,LB   !length of char strings
      REAL SMIN,SMAX
      DATA XSPEC,YSPEC/1,2/
      DATA SCREEN,PAPER,FILE/1,2,3/
CCCCCCCCCCCCCCCCCCCCCCCCCCCCCCCCCCCCCCCCCCCCCCCCCCCCCCCCCCCCCCCCCCCCCCCC
      IF (SPEC .EQ. XSPEC) SZERO=A     !set equil values
      IF (SPEC .EQ. YSPEC) SZERO=B/A
      DO 10 IX=1,MX
         DO 11 IY=1,MY
            SGRF(IX,IY)=S(IX,IY)-SZERO    !load field into SGRF
11       CONTINUE                         !display deviations from Szero
10    CONTINUE
      IF (MX .EQ. 1) THEN
         DO 15 IY=1,MY
            LATTIC(IY)=REAL(IY)           !need to reload data for
            SGRF(IY,1)=S(1,IY)-SZERO      !one-dim plots
15       CONTINUE
      END IF
      IF (MY .EQ. 1) THEN
         DO 16 IX=1,MX
            LATTIC(IX)=REAL(IX)
16       CONTINUE
      END IF
C
C     messages for the impatient
      IF ((DEVICE .NE. SCREEN) .AND. (SPEC .EQ. XSPEC)) WRITE(OUNIT,100)
C
C     calculate parameters for graphing
      NPLOT=2                          !how many plots
C
      YMAX=MY                          !axis data
      YMIN=1.
      XMIN=1.
      XMAX=MX
      YOVAL=XMIN
```

```
      X0VAL=YMIN
      NXTICK=5
      NYTICK=5
      NPOINT=5
C
      LABEL(1)='NX'                    !graph description
      LABEL(2)='NY'
      CALL CONVRT(SLO,CLO,LLO)
      CALL CONVRT(SHI,CHI,LHI)
      CALL CONVRT(STOT,CTOT,LTOT)
      CALL CONVRT(A,CA,LA)
      CALL CONVRT(B,CB,LB)
      TITLE='Brusselator with A='//CA(1:LA)//' and B='//CB(1:LB)
C
      IF (MX .EQ. 1) THEN              !labels and axes are different for
         XMIN=1.                       !one-dim plots
         XMAX=MY
         Y0VAL=XMIN
         LABEL(1)='NY'
         IF (SPEC .EQ. XSPEC) THEN
            LABEL(2)='X Concentration - A'
            YMIN=MAX(-XSCALE,-A)           !scale is the same
            YMAX=XSCALE                    !for all times
         ELSE IF (SPEC .EQ. YSPEC) THEN
            LABEL(2)='Y Concentration - B/A'
            YMIN=MAX(-YSCALE,-B/A)
            YMAX=YSCALE
         END IF
         X0VAL=YMIN
         NPOINT=NY
      ELSE IF (MY .EQ. 1) THEN
         IF (SPEC .EQ. XSPEC) THEN
            LABEL(2)='X Concentration - A'
            YMIN=MAX(-XSCALE,-A)
            YMAX=XSCALE
         ELSE IF (SPEC .EQ. YSPEC) THEN
            LABEL(2)='Y Concentration - B/A'
            YMIN=MAX(-YSCALE,-B/A)
            YMAX=YSCALE
         END IF
         X0VAL=YMIN
         NPOINT=NX
      END IF
C
      IF (SPEC .EQ. XSPEC) THEN     !species-dependent parameters
         IPLOT=1
         SMIN=MAX(-XSCALE,-A)       !scale for species
         SMAX=XSCALE
         INFO='Xmin='//CLO(1:LLO)//' Xmax='
     +        //CHI(1:LHI)//' Xave='//CTOT(1:LTOT)
         CALL GTDEV(DEVICE)                     !device nomination
         IF (DEVICE .EQ. SCREEN) CALL GMODE     !change to graphics mode
```

```
          ELSE IF (SPEC .EQ. YSPEC) THEN
             IPLOT=2
             SMIN=MAX(-YSCALE,-B/A)
             SMAX=YSCALE
             INFO='Ymin='//CLO(1:LLO)//' Ymax='
     +            //CHI(1:LHI)//' Yave='//CTOT(1:LTOT)
          END IF
C
          CALL LNLNAX                           !draw axes
C
          IF ((MX .EQ. 1) .OR. (MY .EQ. 1)) THEN
             CALL XYPLOT(LATTIC,SGRF)
          ELSE
             CALL CONTOR(SGRF,MX,MY,SMIN,SMAX,NLEV)
          END IF
C
C      end graphing session
          IF (SPEC .EQ. YSPEC) THEN
           IF (DEVICE .NE. FILE) CALL GPAGE(DEVICE)  !close graphics package
           IF (DEVICE .EQ. SCREEN) CALL TMODE        !switch to text mode
          END IF
C
100    FORMAT (/,' Patience, please; output going to a file.')
C
       RETURN
       END
CCCCCCCCCCCCCCCCCCCCCCCCCCCCCCCCCCCCCCCCCCCCCCCCCCCCCCCCCCCCCCCCCCCCCCC
       SUBROUTINE WRTOUT(X,Y,TIME,DT)
C write out X and Y species concentrations to GUNIT
C for graphing with an external package
CCCCCCCCCCCCCCCCCCCCCCCCCCCCCCCCCCCCCCCCCCCCCCCCCCCCCCCCCCCCCCCCCCCCCCC
C Global variables:
       INCLUDE 'MENU.ALL'
       INCLUDE 'IO.ALL'
       INCLUDE 'PARAM.P7'
C Input variables:
       REAL TIME                       !time
       REAL DT                         !time step
       REAL X(MAXX,MAXY)               !X species concentration
       REAL Y(MAXX,MAXY)               !Y species concentration
C Local variables:
       INTEGER IX,IY                   !lattice indices
CCCCCCCCCCCCCCCCCCCCCCCCCCCCCCCCCCCCCCCCCCCCCCCCCCCCCCCCCCCCCCCCCCCCCCC
       WRITE (GUNIT,20) TIME,DT
20     FORMAT (' time = ',1PE12.5,' time step=',1PE12.5)
       WRITE (GUNIT,*) ' X concentration'
       DO 100 IX=1,NX
         WRITE (GUNIT,10) (X(IX,IY),IY=1,NY)
100    CONTINUE
       WRITE (GUNIT,*) ' Y concentration'
       DO 200 IX=1,NX
         WRITE (GUNIT,10) (Y(IX,IY),IY=1,NY)
```

```
200    CONTINUE
10     FORMAT (5(2X,1PE14.7))
C

       RETURN
       END
ccccccccccccccccccccccccccccccccccccccccccccccccccccccccccccccccccccccccc
ccccccccccccccccccccccccccccccccccccccccccccccccccccccccccccccccccccccccc
C param.p7
C
       REAL A,B                  !A and B species concentrations
       REAL XDIFF,YDIFF          !X and Y species diff constants
       REAL XFLUCT,YFLUCT        !X and Y species init fluctuations
       INTEGER NX,NY             !horiz and vert number of lattice points
       REAL HX,HY                !horiz and vert steps
       DOUBLE PRECISION DSEED    !random number seed for init cond
       INTEGER XCNTR,YCNTR       !data for centering display
       LOGICAL XSKIP,YSKIP       !data for centering display
       INTEGER NLEV              !number of contour levels
       COMPLEX OMEGAM(2),OMEGAP(2)   !freq from linear pert theory
       REAL MLO,MHI              !smallest, largest mode for this lattice
       REAL XSCALE,YSCALE        !graphing scales
C
       REAL XXPLUS,XYPLUS,YXPLUS,YYPLUS   !useful constants; first letter
       REAL XXZERO,XYZERO,YXZERO,YYZERO   !refers to species, second to
       REAL ADT,BDT,BP1DT                 !horiz or vert
C
       INTEGER MAXX,MAXY         !maximum horiz and vert dimensions
       PARAMETER (MAXX=79)       !these are set assuming that you
       PARAMETER (MAXY=20)       !have no graphics and the length of
                                 !your terminal=24
C
       !following are the terms in the recursion (see Eq. 7.12-7.16)
       REAL ALPHXX(0:MAXX),GAMXX(MAXX)   !horiz recur of X
       REAL ALPHYX(0:MAXX),GAMYX(MAXX)   !horiz recur of Y
       REAL ALPHXY(0:MAXY),GAMXY(MAXY)   !vert recur of X
       REAL ALPHYY(0:MAXY),GAMYY(MAXY)   !vert recur of Y
C
       COMMON / PPARAM / A,B,XDIFF,YDIFF,XFLUCT,YFLUCT
       COMMON / NPARAM / NX,NY,DSEED
       COMMON / GPARAM / NLEV,XCNTR,YCNTR,XSKIP,YSKIP,XSCALE,YSCALE
       COMMON / PCALC / HX,HY,ALPHXX,ALPHXY,ALPHYX,ALPHYY,
      +    GAMXX,GAMXY,GAMYX,GAMYY,ADT,BDT,BP1DT,
      +    XXPLUS,XYPLUS,YXPLUS,YYPLUS,XXZERO,XYZERO,YXZERO,YYZERO,
      +    OMEGAP,OMEGAM,MHI,MLO
ccccccccccccccccccccccccccccccccccccccccccccccccccccccccccccccccccccccccc
ccccccccccccccccccccccccccccccccccccccccccccccccccccccccccccccccccccccccc
C map.p7
       INTEGER IA,IB,IXDIFF,IYDIFF,IXFL,IYFL,INX,INY,IDSEED,INLEV
       PARAMETER (IA     = 13)
       PARAMETER (IB     = 14)
       PARAMETER (IXDIFF = 15)
       PARAMETER (IYDIFF = 16)
```

```
PARAMETER (IXFL   = 17)
PARAMETER (IYFL   = 18)
PARAMETER (INX    = 38)
PARAMETER (INY    = 39)
PARAMETER (IDSEED = 40)
PARAMETER (INLEV  = 87)
```

C.8 Project 8

Algorithm This program uses variational or Path Integral Monte Carlo methods to solve the two-center, two-electron problem of the H_2 molecule using the trial wave functions specified by Eqs. (VIII.6,8). The program calculates either the electronic eigenvalue or the correlations in the energy. Before calculations begin, Eq. (VII.8) is solved for a (subroutine PARAM), and the initial configuration (for the variational calculation, subroutine INTCFG) or ensemble (for PIMC, subroutine INTENS) are generated. The main calculations are done in subroutine ARCHON: thermalization in loop 10 and data-taking in loop 20. The Metropolis steps for the variational calculation are taken in subroutine METROP, while time steps for the PIMC calculations are taken in subroutine TSTEP. Both methods use functions ELOCAL (to find the local energy (VIII.5) of a given configuration) and PHI (to calculate the wave function for a given configuration); each of these in turn calls subroutine RADII to calculate relative distances. In addition, TSTEP also calls subroutine DRIFT to calculate the drift vector and uses function GAUSS to generate a random number from a Gaussian distribution with zero mean and unit variance. The electronic eigenvalue is found from either method, using observations taken every NFREQ'th step, and divided into groups to estimate the step-to-step correlations in the energy. The energy correlations are found by setting NFREQ=NSMPL=1, storing the energies found from each sweep, and calculating the averages in Eq. 8.17 (subroutine CRLTNS). When the requested number of groups have been calculated, you will be prompted for the number of additional groups to include [10.].

Input The physical parameters are the inter-proton separation [0.] (so that a neutral Helium atom is being described) and the variational parameter β [.25]. (All lengths are in Angstroms, all energies are in eV.) The numerical parameters include the method of calculation [variational] (rather than PIMC), ensemble size (PIMC only) [20], time step (PIMC only) [.01], sampling step in configuration space [.4], number of thermalization sweeps [20], quantity to calculate [Energy] (rather than correlations), sampling frequency (energy only) [6], group size (energy only) [10], maximum correlation length (correlations only) [40], number of groups [10], and random number seed [34767]. Note that the sampling step for PIMC is set equal to $1.5a$ in subroutine PARAM. For correlations, the frequency and group size are set equal to one. The maximum ensemble size is fixed by MAXENS= 100, and the maximum number of groups allowed for correlations is fixed by MAXCRR= 500. Note that for correlations the number of groups must be significantly larger than the maximum correlation length (so that the averages in Eq. 8.17 are statistically significant), but must also be smaller than MAXCRR. These checks are performed in subroutine PARAM. The text output parameter allows you to choose the short version of the output for which the sample sums are not displayed [long version enabled]. Please see the note in C.7 concerning the random number seed and compiler storage of integers.

Output   Every NFREQ'th step the text output displays the group index, sample index, group size, and energy (unless the short version of the output is requested). For the electronic eigenvalue calculation, when each group is finished (every NFREQ*NSMPL'th step), the text displays the group index, number of groups, average group energy, and group uncertainty as well as the total (including all groups calculated so far) average energy, the two estimates for the uncertainty in the energy, the total energy of the H_2 system, and the acceptance ratio (variational method only). For the correlation calculations, when all groups are complete, the program displays correlation *vs.* correlation length; the results are plotted if graphics are requested, or else are written as text.

```
CCCCCCCCCCCCCCCCCCCCCCCCCCCCCCCCCCCCCCCCCCCCCCCCCCCCCCCCCCCCCCCCCCCCCCC
      PROGRAM PROJ8
C     Project 8: Monte Carlo solution of the H2 molecule
C COMPUTATIONAL PHYSICS (FORTRAN VERSION)
C by Steven E. Koonin and Dawn C. Meredith
C Copyright 1989, Addison-Wesley Publishing Company
CCCCCCCCCCCCCCCCCCCCCCCCCCCCCCCCCCCCCCCCCCCCCCCCCCCCCCCCCCCCCCCCCCCCCCC
      CALL INIT            !display header screen, setup parameters
5     CONTINUE             !main loop/ execute once for each set of param
      CALL PARAM           !get input from screen
      CALL ARCHON          !calculate the eigenvalue for this value of S
      GOTO 5
      END
CCCCCCCCCCCCCCCCCCCCCCCCCCCCCCCCCCCCCCCCCCCCCCCCCCCCCCCCCCCCCCCCCCCCCCC
      SUBROUTINE ARCHON
C calculates the electronic eigenvalue or energy auto-correlation
C for a given separation of the protons
CCCCCCCCCCCCCCCCCCCCCCCCCCCCCCCCCCCCCCCCCCCCCCCCCCCCCCCCCCCCCCCCCCCCCCC
C Global variables:
      INCLUDE 'PARAM.P8'
      INCLUDE 'IO.ALL'
C Local variables:
C     energy has two indices
C     first index is the level: sweep, group, or total
C     second index is the value: quantity, quant**2, or sigma**2
      REAL ENERGY(3,3)         !energy
      REAL CONFIG(NCOORD)      !configuration
```

```
      REAL W                        !weight for single variational config
      REAL WEIGHT(MAXENS)           !weight of ensemble members
      REAL ENSMBL(NCOORD,MAXENS)    !ensemble of configurations
      REAL ESAVE(MAXCRR)            !array of local energies for corr
      REAL EPSILN                   !local energy of CONFIG
      REAL ACCPT                    !acceptance ratio
      INTEGER ITHERM                !thermalization index
      INTEGER ISWP,ISMPL            !sweep and sample index
      INTEGER IQUANT                !quantity index
      INTEGER IGRP                  !group index
      INTEGER NLINES                !number of lines printed to terminal
      INTEGER MORE                  !how many more groups
      INTEGER SWEEP,GROUP,TOTAL     !which level of calculation
      INTEGER VALUE,SQUARE,SIGSQ    !which quantity
      INTEGER CORR,EPS              !what is being calculated?
      INTEGER PIMC,VARY             !which method?
C Functions:
      REAL ELOCAL                   !local energy
      INTEGER GETINT                !get integer data from screen
      DATA SWEEP,GROUP,TOTAL/1,2,3/
      DATA VALUE,SQUARE,SIGSQ/1,2,3/
      DATA EPS,CORR /1,2/
      DATA VARY,PIMC /1,2/
CCCCCCCCCCCCCCCCCCCCCCCCCCCCCCCCCCCCCCCCCCCCCCCCCCCCCCCCCCCCCCCCCCCCCCCC
C     output summary of parameters
      IF (TTERM) CALL PRMOUT(OUNIT,NLINES)
      IF (TFILE) CALL PRMOUT(TUNIT,NLINES)
      IF (GFILE) CALL PRMOUT(GUNIT,NLINES)
C
C     generate initial configuration or ensemble of configurations
      IF (METHOD .EQ. VARY) THEN
         CALL INTCFG(CONFIG,W)
      ELSE IF (METHOD .EQ. PIMC) THEN
         CALL INTENS(ENSMBL,WEIGHT,CONFIG)
      END IF
C
C     take thermalization steps
      DO 10 ITHERM=1,NTHERM
         IF (ITHERM .EQ. 1) WRITE (OUNIT,*) ' Thermalizing...'
         IF (ITHERM .EQ. NTHERM) WRITE (OUNIT,*) ' '
         IF (METHOD .EQ. VARY) THEN
            CALL METROP(CONFIG,W,ACCPT)
         ELSE IF (METHOD .EQ. PIMC) THEN
            CALL TSTEP(ENSMBL,WEIGHT,EPSILN)
         END IF
10    CONTINUE
C
      DO 11 IQUANT=1,3                      !zero total sums
         ENERGY(TOTAL,IQUANT)=0.
11    CONTINUE
      ACCPT=0                              !zero acceptance
      MORE=NGROUP                          !initial number of groups
```

```
C
15    CONTINUE                                 !allow for more groups
          DO 20 IGRP=NGROUP-MORE+1,NGROUP      !loop over groups
C
          DO 21 IQUANT=1,3                      !zero group sums
              ENERGY(GROUP,IQUANT)=0.
21        CONTINUE
C
          DO 30 ISWP=1,NFREQ*NSMPL             !loop over sweeps
C
          IF (METHOD .EQ. VARY) THEN           !take a Metrop step
              CALL METROP(CONFIG,W,ACCPT)
          ELSE IF (METHOD .EQ. PIMC) THEN !or a time step
              CALL TSTEP(ENSMBL,WEIGHT,EPSILN)
          END IF
C
          IF (MOD(ISWP,NFREQ) .EQ. 0) THEN !sometimes save the energy
              ISMPL=ISWP/NFREQ
              IF (METHOD .EQ. VARY) EPSILN=ELOCAL(CONFIG)
              ENERGY(GROUP,VALUE)=ENERGY(GROUP,VALUE)+EPSILN
              ENERGY(GROUP,SQUARE)=ENERGY(GROUP,SQUARE)+EPSILN**2
              IF (.NOT. TERSE) THEN
                  IF (TTERM) WRITE (OUNIT,100) IGRP,ISMPL,NSMPL,EPSILN
                  IF (TFILE) WRITE (TUNIT,100) IGRP,ISMPL,NSMPL,EPSILN
100               FORMAT (5X,' Group ',I4, ',  sample ',I4,' of ',I4,5X,
     +                    'Energy =',F9.4)
              END IF
          END IF
C
30        CONTINUE                             !this group is done
          IF (CALC .EQ. CORR) THEN             !save energy for corr
              ESAVE(IGRP)=ENERGY(GROUP,VALUE)
          ELSE                                 !or calc averages
              CALL AVERAG(ENERGY,ACCPT,IGRP)
          END IF
C
20    CONTINUE
      IF (CALC .EQ. CORR) CALL CRLTNS(ESAVE)   !calc corr
.C
C     allow for more groups, taking care not to exceed array bounds
      MORE=GETINT(10,0,1000,'How many more groups?')
      IF ((CALC .EQ. CORR) .AND. (NGROUP+MORE .GT. MAXCRR)) THEN
          WRITE (OUNIT,200) MAXCRR-NGROUP
200       FORMAT(' You will run out of storage space for '
     +           'corr if you do more than ',I3,' more groups')
          MORE=GETINT(MAXCRR-NGROUP,0,MAXCRR-NGROUP,
     +                     'How many more groups?')
      END IF
      IF (MORE .GT. 0) THEN
          NGROUP=NGROUP+MORE
          NLINES=0
          IF (TTERM) CALL CLEAR
```

```
      GOTO 15
      END IF
C
      RETURN
      END
CCCCCCCCCCCCCCCCCCCCCCCCCCCCCCCCCCCCCCCCCCCCCCCCCCCCCCCCCCCCCCCCCCCCCCCCC
      SUBROUTINE TSTEP(ENSMBL,WEIGHT,EPSILN)
C take a time step using the Path Integral Monte Carlo method
CCCCCCCCCCCCCCCCCCCCCCCCCCCCCCCCCCCCCCCCCCCCCCCCCCCCCCCCCCCCCCCCCCCCCCCCC
C Global variables:
      INCLUDE 'PARAM.P8'
C Input/Output variables
      REAL WEIGHT(MAXENS)        !weight of ensemble members (I/O)
      REAL ENSMBL(NCOORD,MAXENS) !ensemble of configurations (I/O)
      REAL EPSILN                !local energy of CONFIG (output)
C Local variables:
      REAL CONFIG(NCOORD)        !configuration
      REAL W                     !weight for single config
      REAL EBAR,WBAR             !ensmble average local energy and weight
      INTEGER IENSEM             !ensemble index
      INTEGER ICOORD             !coordinate index
      REAL NORM                  !normalization of weights
      REAL SHIFT(NCOORD)         !array containing drift vector
C Functions:
      REAL GAUSS                 !Gaussian random number
      REAL ELOCAL                !local energy of the configuration
CCCCCCCCCCCCCCCCCCCCCCCCCCCCCCCCCCCCCCCCCCCCCCCCCCCCCCCCCCCCCCCCCCCCCCCCC
      EBAR=0.                    !zero sums
      WBAR=0.
      DO 10 IENSEM=1,NENSEM                    !loop over ensemble
         DO 20 ICOORD=1,NCOORD
            CONFIG(ICOORD)=ENSMBL(ICOORD,IENSEM)  !get a configuration
20       CONTINUE
         CALL DRIFT(CONFIG,SHIFT)              !calc shifts
         DO 30 ICOORD=1,NCOORD
            CONFIG(ICOORD)=CONFIG(ICOORD)+     !shift configuration
     +               GAUSS(DSEED)*SQHBDT+SHIFT(ICOORD)
30       CONTINUE
C
         EPSILN=ELOCAL(CONFIG)                 !calculate energy
         WEIGHT(IENSEM)=WEIGHT(IENSEM)*EXP(-EPSILN*DT)  !calc weight
         EBAR=EBAR+WEIGHT(IENSEM)*EPSILN       !update sums
         WBAR=WBAR+WEIGHT(IENSEM)
C
         DO 40 ICOORD=1,NCOORD
            ENSMBL(ICOORD,IENSEM)=CONFIG(ICOORD)  !save configuration
40       CONTINUE
10    CONTINUE
C
      EPSILN=EBAR/WBAR                         !weighted average energy
      NORM=NENSEM/WBAR
      DO 50 IENSEM=1,NENSEM                    !renormalize weights
```

```
                WEIGHT(IENSEM)=NORM*WEIGHT(IENSEM)
50      CONTINUE
C
        RETURN
        END
CCCCCCCCCCCCCCCCCCCCCCCCCCCCCCCCCCCCCCCCCCCCCCCCCCCCCCCCCCCCCCCCCCCCCCC
        SUBROUTINE METROP(CONFIG,W,ACCPT)
C take a Metropolis step
CCCCCCCCCCCCCCCCCCCCCCCCCCCCCCCCCCCCCCCCCCCCCCCCCCCCCCCCCCCCCCCCCCCCCCC
C Global variables:
        INCLUDE 'PARAM.P8'
C Input/Output variables:
        REAL CONFIG(NCOORD)             !configuration
        REAL W                          !weight for single config
        REAL ACCPT                      !acceptance ratio
C Local variables:
        INTEGER ICOORD                  !coordinate index
        REAL CSAVE(NCOORD)              !temp storage for last config
        REAL WTRY                       !weight for trial config
C Function:
        REAL PHI                        !total wave function
        REAL RANNOS                     !uniform random number
CCCCCCCCCCCCCCCCCCCCCCCCCCCCCCCCCCCCCCCCCCCCCCCCCCCCCCCCCCCCCCCCCCCCCCC
        DO 10 ICOORD=1,NCOORD
          CSAVE(ICOORD)=CONFIG(ICOORD)            !save previous values
          CONFIG(ICOORD)=CONFIG(ICOORD)+DELTA*(RANNOS(DSEED)-.5)!trial step
10      CONTINUE
        WTRY=PHI(CONFIG)**2                       !trial weight
C
        IF (WTRY/W .GT. RANNOS(DSEED)) THEN   !sometimes accept the step
          W=WTRY                              !save new weight
          ACCPT=ACCPT+1                       !update accept ratio
        ELSE
          DO 20 ICOORD=1,NCOORD
            CONFIG(ICOORD)=CSAVE(ICOORD)      !or else restore old config
20        CONTINUE
        END IF
        RETURN
        END
CCCCCCCCCCCCCCCCCCCCCCCCCCCCCCCCCCCCCCCCCCCCCCCCCCCCCCCCCCCCCCCCCCCCCCC
        REAL FUNCTION ELOCAL(CONFIG)
C calculate the local energy for CONFIG
CCCCCCCCCCCCCCCCCCCCCCCCCCCCCCCCCCCCCCCCCCCCCCCCCCCCCCCCCCCCCCCCCCCCCCC
C Global variable:
        INCLUDE 'PARAM.P8'
C Input variables:
        REAL CONFIG(NCOORD)             !configuration
C Local variables:
        REAL TPOP,VPOP                  !kinetic and potential contributions
        REAL EECORR                     !elec-elec correlation
        REAL CROSS,CROSS1,CROSS2        !cross terms
        REAL ONEE1,ONEE2                !one electron terms
```

```
      REAL X1,X2,Y1,Y2,Z1,Z2      !coordinates of 2 electrons
      REAL R1L,R1R,R2L,R2R,R12     !relative distances
      REAL CHI1,CHI2,F             !parts of the wave function
      REAL DOTR1L,DOTR2L,DOTR1R,DOTR2R !dot products with R12
      REAL R12DR1,SR12Z            !temp vars for dot products
      REAL CHI,FDCHI,SDCHI,LAPCHI  !atomic orbitals
      REAL FEE,FDFEE,SDFEE,LAPFEE  !elec-elec correlations
      REAL DIST                    !Euclidean distance
      REAL R,X,Y,Z                 !dummy variables
CCCCCCCCCCCCCCCCCCCCCCCCCCCCCCCCCCCCCCCCCCCCCCCCCCCCCCCCCCCCCCCCCCCCCCC
C     define functions
      CHI(R)=EXP(-R/A)                       !atomic orbital
      FDCHI(R)=-CHI(R)/A                     !its first derivative,
      SDCHI(R)=CHI(R)/A/A                    !second derivative,
      LAPCHI(R)=SDCHI(R)+2*FDCHI(R)/R        !and Laplacian
C
      FEE(R)=EXP(R/(ALPHA*(1+BETA*R)))       !elec-elec correlation
      FDFEE(R)=FEE(R)/(ALPHA*(1.+BETA*R)**2) !its first,second deriv,
      SDFEE(R)=FDFEE(R)**2/FEE(R)-2.*BETA*FEE(R)/ALPHA/(1+BETA*R)**3
      LAPFEE(R)=SDFEE(R)+2*FDFEE(R)/R        !and Laplacian
C
      DIST(X,Y,Z)=SQRT(X**2+Y**2+Z**2)       !Euclidean distance
CCCCCCCCCCCCCCCCCCCCCCCCCCCCCCCCCCCCCCCCCCCCCCCCCCCCCCCCCCCCCCCCCCCCCCC
C     get coordinates and radii
      CALL RADII(X1,X2,Y1,Y2,Z1,Z2,R1L,R2L,R1R,R2R,R12,CONFIG)
C
C     calculate dot products with R12
      R12DR1=X1*(X1-X2)+Y1*(Y1-Y2)+Z1*(Z1-Z2) !convenient starting place
      SR12Z=S*(Z1-Z2)/2            !useful constant
      DOTR1L=R12DR1+SR12Z          !dot products with R12
      DOTR1R=R12DR1-SR12Z
      DOTR2L=DOTR1L-R12**2
      DOTR2R=DOTR1R-R12**2
      DOTR1L=DOTR1L/R12/R1L        !dot products of unit vectors
      DOTR2L=DOTR2L/R12/R2L
      DOTR1R=DOTR1R/R12/R1R
      DOTR2R=DOTR2R/R12/R2R
C
      CHI1=CHI(R1R)+CHI(R1L)       !pieces of the total wave function
      CHI2=CHI(R2R)+CHI(R2L)
      F=FEE(R12)
C
      EECORR=2*LAPFEE(R12)/F       !correlation contribution
      ONEE1=(LAPCHI(R1L)+LAPCHI(R1R))/CHI1        !electron one
      ONEE2=(LAPCHI(R2L)+LAPCHI(R2R))/CHI2        !electron two
      CROSS1=(FDCHI(R1L)*DOTR1L+FDCHI(R1R)*DOTR1R)/CHI1 !cross terms
      CROSS2=(FDCHI(R2L)*DOTR2L+FDCHI(R2R)*DOTR2R)/CHI2
      CROSS=2*FDFEE(R12)*(CROSS1-CROSS2)/F
C
      TPOP=-HBM*(EECORR+ONEE1+ONEE2+CROSS)/2              !kinetic
      VPOP=-E2*(1./R1L + 1./R1R + 1./R2L + 1./R2R - 1./R12) !potential
      ELOCAL=TPOP+VPOP                                   !total
```

```
C
      RETURN
      END
CCCCCCCCCCCCCCCCCCCCCCCCCCCCCCCCCCCCCCCCCCCCCCCCCCCCCCCCCCCCCCCCCCCCC
      SUBROUTINE DRIFT(CONFIG,SHIFT)
C calculate the drift vector (SHIFT) for CONFIG
CCCCCCCCCCCCCCCCCCCCCCCCCCCCCCCCCCCCCCCCCCCCCCCCCCCCCCCCCCCCCCCCCCCCC
C Global variables:
      INCLUDE 'PARAM.P8'
C Input/Output variables:
      REAL CONFIG(NCOORD)           !configuration (input)
      REAL SHIFT(NCOORD)            !array containing drift vector (output)
C Local variables:
      INTEGER ICOORD                !coordinate index
      REAL X1,X2,Y1,Y2,Z1,Z2        !coordinates of 2 electrons
      REAL R1L,R1R,R2L,R2R,R12      !relative distances
      REAL CHI1,CHI2,F              !parts of the wave function
      REAL CHI,FDCHI,SDCHI,LAPCHI   !atomic orbital
      REAL FEE,FDFEE,SDFEE,LAPFEE   !elec-elec correlations
      REAL R                        !dummy variables
      REAL FACTA,FACTB,FACTE        !useful factors
CCCCCCCCCCCCCCCCCCCCCCCCCCCCCCCCCCCCCCCCCCCCCCCCCCCCCCCCCCCCCCCCCCCCC
C     define functions
      CHI(R)=EXP(-R/A)                           !atomic orbital
      FDCHI(R)=-CHI(R)/A                         !its first derivative,
      SDCHI(R)=CHI(R)/A/A                        !second derivative,
      LAPCHI(R)=SDCHI(R)+2*FDCHI(R)/R            !and Laplacian
C
      FEE(R)=EXP(R/(ALPHA*(1+BETA*R)))           !elec-elec correlation
      FDFEE(R)=FEE(R)/(ALPHA*(1.+BETA*R)**2)     !its first, second deriv,
      SDFEE(R)=FDFEE(R)**2/FEE(R)-2.*BETA*FEE(R)/ALPHA/(1+BETA*R)**3
      LAPFEE(R)=SDFEE(R)+2*FDFEE(R)/R            !and Laplacian
CCCCCCCCCCCCCCCCCCCCCCCCCCCCCCCCCCCCCCCCCCCCCCCCCCCCCCCCCCCCCCCCCCCCC
C     get coordinates and radii
      CALL RADII(X1,X2,Y1,Y2,Z1,Z2,R1L,R2L,R1R,R2R,R12,CONFIG)
C
      CHI1=CHI(R1R)+CHI(R1L)            !pieces of the total wave function
      CHI2=CHI(R2R)+CHI(R2L)
      F=FEE(R12)
C
      FACTA=HBMDT*(FDCHI(R1L)/R1L+FDCHI(R1R)/R1R)/CHI1   !useful factors
      FACTB=HBMDT*(FDCHI(R1L)/R1L-FDCHI(R1R)/R1R)/CHI1
      FACTE=HBMDT*FDFEE(R12)/F/R12
C
      SHIFT(1)=FACTA*X1+FACTE*(X1-X2)   !shift for electron one
      SHIFT(2)=FACTA*Y1+FACTE*(Y1-Y2)
      SHIFT(3)=FACTA*Z1+FACTE*(Z1-Z2)+FACTB*S2
C
      FACTA=HBMDT*(FDCHI(R2L)/R2L+FDCHI(R2R)/R2R)/CHI2
      FACTB=HBMDT*(FDCHI(R2L)/R2L-FDCHI(R2R)/R2R)/CHI2
C
      SHIFT(4)=FACTA*X2-FACTE*(X1-X2)   !shift for electron two
```

```
         SHIFT(5)=FACTA*Y2-FACTE*(Y1-Y2)
         SHIFT(6)=FACTA*Z2-FACTE*(Z1-Z2)+FACTB*S2
C
         RETURN
         END
CCCCCCCCCCCCCCCCCCCCCCCCCCCCCCCCCCCCCCCCCCCCCCCCCCCCCCCCCCCCCCCCCCCCCCC
         REAL FUNCTION GAUSS(DSEED)
C returns a Gaussian random number with zero mean and unit variance
CCCCCCCCCCCCCCCCCCCCCCCCCCCCCCCCCCCCCCCCCCCCCCCCCCCCCCCCCCCCCCCCCCCCCCC
         INTEGER IGAUSS              !sum index
         DOUBLE PRECISION DSEED      !random number seed
         REAL RANNOS                 !uniform random number
CCCCCCCCCCCCCCCCCCCCCCCCCCCCCCCCCCCCCCCCCCCCCCCCCCCCCCCCCCCCCCCCCCCCCCC
         GAUSS=0.                    !sum 12 uniform random numbers
         DO 10 IGAUSS=1,12
            GAUSS=GAUSS+RANNOS(DSEED)
10       CONTINUE
         GAUSS=GAUSS-6.              !subtract six so that mean=0
         RETURN
         END
CCCCCCCCCCCCCCCCCCCCCCCCCCCCCCCCCCCCCCCCCCCCCCCCCCCCCCCCCCCCCCCCCCCCCCC
         REAL FUNCTION PHI(CONFIG)
C calculates the total variational wave function for CONFIG
CCCCCCCCCCCCCCCCCCCCCCCCCCCCCCCCCCCCCCCCCCCCCCCCCCCCCCCCCCCCCCCCCCCCCCC
C Global variables:
         INCLUDE 'PARAM.P8'
C Input variables:
         REAL CONFIG(NCOORD)        !configuration
C Local variables:
         REAL X1,X2,Y1,Y2,Z1,Z2     !coordinates of 2 electrons
         REAL R1L,R1R,R2L,R2R,R12   !relative distances
         REAL CHI1R,CHI1L,CHI2R,CHI2L,F !parts of the wave function
         REAL CHI,FEE               !terms in the wave function
         REAL R                     !radius
CCCCCCCCCCCCCCCCCCCCCCCCCCCCCCCCCCCCCCCCCCCCCCCCCCCCCCCCCCCCCCCCCCCCCCC
         CHI(R)=EXP(-R/A)                   !atomic orbital
         FEE(R)=EXP(R/(ALPHA*(1+BETA*R)))   !electron-electron correlation
CCCCCCCCCCCCCCCCCCCCCCCCCCCCCCCCCCCCCCCCCCCCCCCCCCCCCCCCCCCCCCCCCCCCCCC
C        calculate the radii
         CALL RADII(X1,X2,Y1,Y2,Z1,Z2,R1L,R2L,R1R,R2R,R12,CONFIG)
C
         CHI1R=CHI(R1R)            !pieces of the total wave function
         CHI1L=CHI(R1L)
         CHI2R=CHI(R2R)
         CHI2L=CHI(R2L)
         F=FEE(R12)
         PHI=(CHI1L +CHI1R)*(CHI2L+CHI2R)*F  !the whole thing
C
         RETURN
         END
CCCCCCCCCCCCCCCCCCCCCCCCCCCCCCCCCCCCCCCCCCCCCCCCCCCCCCCCCCCCCCCCCCCCCCC
         SUBROUTINE INTENS(ENSMBL,WEIGHT,CONFIG)
```

```
C generate the ENSMBL at t=0 for PIMC
CCCCCCCCCCCCCCCCCCCCCCCCCCCCCCCCCCCCCCCCCCCCCCCCCCCCCCCCCCCCCCCCCCCCCCCCCCCC
C Global variables:
      INCLUDE 'PARAM.P8'
C Output variables:
      REAL CONFIG(NCOORD)             !configuration
      REAL WEIGHT(MAXENS)             !weight of ensemble members
      REAL ENSMBL(NCOORD,MAXENS)      !ensemble of configurations
C Local variables:
      INTEGER ISTEP                   !step index
      INTEGER ICOORD                  !coordinate index
      REAL W                          !weight for single config
      REAL ACCPT                      !acceptance ratio
      INTEGER IENSEM                  !ensemble index
CCCCCCCCCCCCCCCCCCCCCCCCCCCCCCCCCCCCCCCCCCCCCCCCCCCCCCCCCCCCCCCCCCCCCCCCCCCC
      CALL INTCFG(CONFIG,W)           !generate a single intial configuration
C
      DO 10 ISTEP=1,20                !do 20 thermalization steps
         CALL METROP(CONFIG,W,ACCPT)  !using Metropolis algorithm
10    CONTINUE
C
      DO 30 ISTEP=1,10*NENSEM         !generate the ensemble
         CALL METROP(CONFIG,W,ACCPT)  !take a Metrop step
         IF (MOD(ISTEP,10) .EQ. 0) THEN
            IENSEM=ISTEP/10           !save every 10th config
            DO 20 ICOORD=1,NCOORD
               ENSMBL(ICOORD,IENSEM)=CONFIG(ICOORD)
20          CONTINUE
            WEIGHT(IENSEM)=1.         !set all weights=1
         END IF
30    CONTINUE
      RETURN
      END
CCCCCCCCCCCCCCCCCCCCCCCCCCCCCCCCCCCCCCCCCCCCCCCCCCCCCCCCCCCCCCCCCCCCCCCCCCCC
      SUBROUTINE INTCFG(CONFIG,W)
C generate a configuration (CONFIG) and calculate its weight (W)
CCCCCCCCCCCCCCCCCCCCCCCCCCCCCCCCCCCCCCCCCCCCCCCCCCCCCCCCCCCCCCCCCCCCCCCCCCCC
C Global variable:
      INCLUDE 'PARAM.P8'
C Output variables:
      REAL CONFIG(NCOORD)             !configuration
      REAL W                          !weight for single config
C Local variables:
      INTEGER ICOORD                  !coordinate index
C Function:
      REAL PHI                        !total wave function
      REAL RANNOS                     !uniform random number
CCCCCCCCCCCCCCCCCCCCCCCCCCCCCCCCCCCCCCCCCCCCCCCCCCCCCCCCCCCCCCCCCCCCCCCCCCCC
      DO 10 ICOORD=1,NCOORD           !pick configuration at random
         CONFIG(ICOORD)=A*(RANNOS(DSEED)-.5)
10    CONTINUE
      CONFIG(3)=CONFIG(3)+S2          !center elec 1. at right
```

```
        CONFIG(6)=CONFIG(6)-S2      !center elec 2. at left
        W=PHI(CONFIG)**2            !weight=phi**2
        RETURN
        END
CCCCCCCCCCCCCCCCCCCCCCCCCCCCCCCCCCCCCCCCCCCCCCCCCCCCCCCCCCCCCCCCCCCCCC
        SUBROUTINE CRLTNS(ESAVE)
C calculate the energy auto-correlations
CCCCCCCCCCCCCCCCCCCCCCCCCCCCCCCCCCCCCCCCCCCCCCCCCCCCCCCCCCCCCCCCCCCCCC
C Global variables:
        INCLUDE 'PARAM .P8'
        INCLUDE 'IO.ALL'
C Input variables:
        REAL ESAVE(MAXCRR)          !array of local energies for corr.
C Local variables:
        REAL EI,EIK,ESQI,ESQIK,EIEK !sums
        INTEGER I,K                 !index of ESAVE
        REAL ECORR(0:MAXCRR)        !energy auto-correlations
        INTEGER NI                  !number of energies in sum
        INTEGER SCREEN              !send to terminal
        INTEGER PAPER               !make a hardcopy
        INTEGER FILE                !send to a file
        DATA SCREEN,PAPER,FILE/1,2,3/
CCCCCCCCCCCCCCCCCCCCCCCCCCCCCCCCCCCCCCCCCCCCCCCCCCCCCCCCCCCCCCCCCCCCCC
        DO 10 K=0,NCORR             !loop over correlation lengths
           EI=0.                    !zero sums
           EIK=0.
           ESQI=0.
           ESQIK=0.
           EIEK=0.
           NI=NGROUP-K
           DO 20 I=1,NI
              EI=EI+ESAVE(I)        !calculate sums
              EIK=EIK+ESAVE(I+K)
              ESQI=ESQI+ESAVE(I)**2
              ESQIK=ESQIK+ESAVE(I+K)**2
              EIEK=EIEK+ESAVE(I)*ESAVE(I+K)
20         CONTINUE
           EI=EI/NI                 !calculate averages
           EIK=EIK/NI
           ESQI=ESQI/NI
           ESQIK=ESQIK/NI
           EIEK=EIEK/NI
           ECORR(K)=(EIEK-EI*EIK)/(SQRT(ESQI-EI**2))/(SQRT(ESQIK-EIK**2))
10      CONTINUE
C
        IF (GTERM) THEN             !display results
           CALL PAUSE ('to see the energy auto-correlations...',1)
           CALL GRFOUT(SCREEN,ECORR)
        ELSE IF (TTERM) THEN
           CALL PAUSE ('to see the energy auto-correlations...',1)
           CALL CRROUT(OUNIT,ECORR)
        END IF
```

```
      IF (TFILE) CALL CRROUT(TUNIT,ECORR)
      IF (GHRDCP) CALL GRFOUT(PAPER,ECORR)
      IF (GFILE) CALL GRFOUT(FILE,ECORR)
C
      RETURN
      END
CCCCCCCCCCCCCCCCCCCCCCCCCCCCCCCCCCCCCCCCCCCCCCCCCCCCCCCCCCCCCCCCCCCCCCC
      SUBROUTINE AVERAG(ENERGY,ACCPT,IGRP)
C calculate group averages, add to totals, print out
CCCCCCCCCCCCCCCCCCCCCCCCCCCCCCCCCCCCCCCCCCCCCCCCCCCCCCCCCCCCCCCCCCCCCCC
C Global variables:
      INCLUDE 'PARAM.P8'
      INCLUDE 'IO.ALL'
C Input variables:
C     energy has two indices
C     first index is the level: sweep, group, or total
C     second index is the value: quantity, quant**2, or sigma**2
      REAL ENERGY(3,3)            !energy
      INTEGER IGRP                !group index
      REAL ACCPT                  !acceptance ratio
C Local variables:
      REAL EVALUE                 !current average energy
      REAL SIG1,SIG2              !uncertainties in energy
      REAL U                      !total pot energy of the system
      INTEGER NLINES              !number of lines printed to terminal
      INTEGER SWEEP,GROUP,TOTAL   !which level of calculation
      INTEGER VALUE,SQUARE,SIGSQ  !which quantity
      DATA SWEEP,GROUP,TOTAL/1,2,3/
      DATA VALUE,SQUARE,SIGSQ/1,2,3/
CCCCCCCCCCCCCCCCCCCCCCCCCCCCCCCCCCCCCCCCCCCCCCCCCCCCCCCCCCCCCCCCCCCCCCC
C     calculate group averages and uncertainties
      ENERGY(GROUP,VALUE)=ENERGY(GROUP,VALUE)/NSMPL
      ENERGY(GROUP,SQUARE)=ENERGY(GROUP,SQUARE)/NSMPL
      ENERGY(GROUP,SIGSQ)=
     +   (ENERGY(GROUP,SQUARE)-ENERGY(GROUP,VALUE)**2)/NSMPL
      IF (ENERGY(GROUP,SIGSQ) .LT. 0.) ENERGY(GROUP,SIGSQ)=0.
C
C     add to totals
      ENERGY(TOTAL,VALUE)=ENERGY(TOTAL,VALUE)+ENERGY(GROUP,VALUE)
      ENERGY(TOTAL,SQUARE)=ENERGY(TOTAL,SQUARE)+ENERGY(GROUP,SQUARE)
      ENERGY(TOTAL,SIGSQ)=ENERGY(TOTAL,SIGSQ)+ENERGY(GROUP,SIGSQ)
C
C     calculate current grand averages
      EVALUE=ENERGY(TOTAL,VALUE)/IGRP
      SIG1=(ENERGY(TOTAL,SQUARE)/IGRP-EVALUE**2)/IGRP/NSMPL
      IF (SIG1 .LT. 0.) SIG1=0.
      SIG1=SQRT(SIG1)
      SIG2=SQRT(ENERGY(TOTAL,SIGSQ))/IGRP
C
C     calculate total energy of the system
      IF (S .GT. .01) THEN
           U=EVALUE+E2/S+E2/ABOHR
```

```
      ELSE
         U=0.
      END IF
C
      IF (TTERM) CALL TXTOUT(IGRP,ENERGY,EVALUE,SIG1,SIG2,U,ACCPT,OUNIT)
      IF (TFILE) CALL TXTOUT(IGRP,ENERGY,EVALUE,SIG1,SIG2,U,ACCPT,TUNIT)
C
      RETURN
      END
CCCCCCCCCCCCCCCCCCCCCCCCCCCCCCCCCCCCCCCCCCCCCCCCCCCCCCCCCCCCCCCCCCCCCCCCC
      SUBROUTINE RADII(X1,X2,Y1,Y2,Z1,Z2,R1L,R2L,R1R,R2R,R12,CONFIG)
C calculates cartesian coordinates and radii given CONFIG
CCCCCCCCCCCCCCCCCCCCCCCCCCCCCCCCCCCCCCCCCCCCCCCCCCCCCCCCCCCCCCCCCCCCCCCCC
C Global variable:
      INCLUDE 'PARAM.P8'
C Input variables:
      REAL CONFIG(NCOORD)          !configuration
C Output variables:
      REAL X1,X2,Y1,Y2,Z1,Z2       !coordinates of 2 electrons
      REAL R1L,R1R,R2L,R2R,R12     !relative distances
      REAL DIST                    !Euclidean distance
CCCCCCCCCCCCCCCCCCCCCCCCCCCCCCCCCCCCCCCCCCCCCCCCCCCCCCCCCCCCCCCCCCCCCCCCC
      DIST(X1,Y1,Z1)=SQRT(X1**2+Y1**2+Z1**2)      !Euclidean distance
C
      X1=CONFIG(1)                 !give the CONFIG elements their real names
      X2=CONFIG(4)
      Y1=CONFIG(2)
      Y2=CONFIG(5)
      Z1=CONFIG(3)
      Z2=CONFIG(6)
C
      R1L=DIST(X1,Y1,Z1+S2)        !calculate separations
      R1R=DIST(X1,Y1,Z1-S2)
      R2L=DIST(X2,Y2,Z2+S2)
      R2R=DIST(X2,Y2,Z2-S2)
      R12=DIST(X1-X2,Y1-Y2,Z1-Z2)
C
      RETURN
      END
CCCCCCCCCCCCCCCCCCCCCCCCCCCCCCCCCCCCCCCCCCCCCCCCCCCCCCCCCCCCCCCCCCCCCCCCC
      SUBROUTINE INIT
C initializes constants, displays header screen,
C initializes menu arrays for input parameters
CCCCCCCCCCCCCCCCCCCCCCCCCCCCCCCCCCCCCCCCCCCCCCCCCCCCCCCCCCCCCCCCCCCCCCCCC
C Global variables:
      INCLUDE 'IO.ALL'
      INCLUDE 'MENU.ALL'
      INCLUDE 'PARAM.P8'
C Local parameters:
      CHARACTER*80 DESCRP          !program description
      DIMENSION DESCRP(20)
      INTEGER NHEAD,NTEXT,NGRAPH   !number of lines for each description
```

```
cccccccccccccccccccccccccccccccccccccccccccccccccccccccccccccccccccccccc
c      get environment parameters
       CALL SETUP
c
c      display header screen
       DESCRP(1)= 'PROJECT 8'
       DESCRP(2)= 'Monte Carlo solution of the H2 molecule'
       NHEAD=2
c
c      text output description
       DESCRP(3)= 'electronic eigenvalue and its uncertainty'
       DESCRP(4)= 'or energy auto-correlation'
       NTEXT=2
c
c      graphics output description
       DESCRP(5)= 'energy auto-correlation vs. correlation length'
       NGRAPH=1
c
       CALL HEADER(DESCRP,NHEAD,NTEXT,NGRAPH)
c
c      calculate constants
       HBM=7.6359              !hbar*2/(mass)
       E2=14.409               !charge of the electron
       ABOHR=HBM/E2
c
c      setup menu arrays, beginning with constant part
       CALL MENU
c
       MTYPE(12)=TITLE
       MPRMPT(12)= 'PHYSICAL PARAMETERS'
       MLOLIM(12)=0.
       MHILIM(12)=1.
c
       MTYPE(13)=FLOAT
       MPRMPT(13)='Enter the interproton separation S (Angstroms)'
       MTAG(13)='Inter proton separation (Angstroms)'
       MLOLIM(13)=0.
       MHILIM(13)=10.
       MREALS(13)=0.
c
       MTYPE(14)=FLOAT
       MPRMPT(14)=
      + 'Enter value for variational parameter Beta (Angstroms**-1)'
       MTAG(14)='variational parameter Beta (Angstroms**-1)'
       MLOLIM(14)=0.
       MHILIM(14)=10.
       MREALS(14)=.25
c
       MTYPE(15)=SKIP
       MREALS(15)=35.
c
       MTYPE(37)=TITLE
```

```
      MPRMPT(37)= 'NUMERICAL PARAMETERS'
      MLOLIM(37)=1.
      MHILIM(37)=1.
C
      MTYPE(38)=TITLE
      MPRMPT(38)='Methods of calculation:'
      MLOLIM(38)=0.
      MHILIM(38)=0.
C
      MTYPE(39)=MTITLE
      MPRMPT(39)='1) Variational'
      MLOLIM(39)=0.
      MHILIM(39)=0.
C
      MTYPE(40)=MTITLE
      MPRMPT(40)='2) Path Integral Monte Carlo'
      MLOLIM(40)=0.
      MHILIM(40)=1.
C
      MTYPE(41)=MCHOIC
      MPRMPT(41)='Make a menu choice and press return'
      MTAG(41)='44 42'
      MLOLIM(41)=1.
      MHILIM(41)=2.
      MINTS(41)=1
      MREALS(41)=1.
C
      MTYPE(42)=NUM
      MPRMPT(42)= 'Enter size of the ensemble'
      MTAG(42)= 'Ensemble size'
      MLOLIM(42)=1.
      MHILIM(42)=MAXENS
      MINTS(42)=20.
C
      MTYPE(43)=FLOAT
      MPRMPT(43)='Enter time step (units of 1E-16 sec/hbar)'
      MTAG(43)='Time step (units of 1E-16 sec/hbar)'
      MLOLIM(43)=0.
      MHILIM(43)=10.
      MREALS(43)=.01
C
      MTYPE(44)=FLOAT
      MPRMPT(44)= 'Enter step size for sampling PHI (Angstroms)'
      MTAG(44)= 'Sampling step size (Angstroms)'
      MLOLIM(44)=.01
      MHILIM(44)=10.
      MREALS(44)=.4
C
      MTYPE(45)=NUM
      MPRMPT(45)= 'Number of thermalization sweeps'
      MTAG(45)= 'Thermalization sweeps'
      MLOLIM(45)=0
```

```
        MHILIM(45)=1000
        MINTS(45)=20
C
        MTYPE(46)=TITLE
        MPRMPT(46)='Quantity to calculate:'
        MLOLIM(46)=1.
        MHILIM(46)=0.
C
        MTYPE(47)=MTITLE
        MPRMPT(47)='1) Energy'
        MLOLIM(47)=0.
        MHILIM(47)=0.
C
        MTYPE(48)=MTITLE
        MPRMPT(48)='2) Correlations'
        MLOLIM(48)=0.
        MHILIM(48)=1.
C
        MTYPE(49)=MCHOIC
        MPRMPT(49)='Make a menu choice and press return'
        MTAG(49)='50 53'
        MLOLIM(49)=1.
        MHILIM(49)=2.
        MINTS(49)=1
        MREALS(49)=1.
C
        MTYPE(50)=NUM
        MPRMPT(50)= 'Enter sampling frequency (to avoid correlations)'
        MTAG(50)= 'Sampling frequency'
        MLOLIM(50)=1
        MHILIM(50)=100
        MINTS(50)=6
C
        MTYPE(51)=NUM
        MPRMPT(51)= 'Enter number of samples in a group'
        MTAG(51)= 'Group sample size'
        MLOLIM(51)=1
        MHILIM(51)=1000
        MINTS(51)=10
C
        MTYPE(52)=SKIP
        MREALS(52)=54.
C
        MTYPE(53)=NUM
        MPRMPT(53)= 'Enter maximum correlation length'
        MTAG(53)= 'Maximum correlation length'
        MLOLIM(53)=1
        MHILIM(53)=100.
        MINTS(53)=40
C
        MTYPE(54)=NUM
        MPRMPT(54)= 'Enter number of groups'
```

```
      MTAG(54)= 'Number of groups'
      MLOLIM(54)=1
      MHILIM(54)=1000
      MINTS(54)=10
C
      MTYPE(55)=NUM
      MPRMPT(55)= 'Integer random number seed for init fluctuations'
      MTAG(55)= 'Random number seed'
      MLOLIM(55)=1000.
      MHILIM(55)=99999.
      MINTS(55)=34767
C
      MTYPE(56)=SKIP
      MREALS(56)=60.
C
      MSTRNG(MINTS(75))= 'proj8.txt'
C
      MTYPE(76)=BOOLEN
      MPRMPT(76)='Do you want the SHORT version of the output?'
      MTAG(76)='Short version of output'
      MINTS(76)=0
C
      MTYPE(77)=SKIP
      MREALS(77)=80.
C
      MSTRNG(MINTS(86))= 'proj8.grf'
C
      MTYPE(87)=SKIP
      MREALS(87)=90.
C
      RETURN
      END
CCCCCCCCCCCCCCCCCCCCCCCCCCCCCCCCCCCCCCCCCCCCCCCCCCCCCCCCCCCCCCCCCCCCCCC
      SUBROUTINE PARAM
C gets parameters from screen
C ends program on request
C closes old files
C maps menu variables to program variables
C opens new files
C calculates all derivative parameters
C performs checks on parameters
CCCCCCCCCCCCCCCCCCCCCCCCCCCCCCCCCCCCCCCCCCCCCCCCCCCCCCCCCCCCCCCCCCCCCCC
C Global variables:
      INCLUDE 'MENU.ALL'
      INCLUDE 'IO.ALL'
      INCLUDE 'PARAM.P8'
C Local variables:
      REAL AOLD           !temp variable to search for A
      INTEGER CORR,EPS    !what is being calculated?
      INTEGER PIMC,VARY   !which method?
C gives the map between menu indices and parameters
      INTEGER IS,IBETA,IMETHD,INENSM,IDT,IDELTA,ITHERM,ICALC,
```

```
     +           INFREQ,INSMPL,INCORR,IGROUP,IDSEED,ITERSE
      PARAMETER (IS     = 13)
      PARAMETER (IBETA  = 14)
      PARAMETER (IMETHD = 41)
      PARAMETER (INENSM = 42)
      PARAMETER (IDT    = 43)
      PARAMETER (IDELTA = 44)
      PARAMETER (ITHERM = 45)
      PARAMETER (ICALC  = 49)
      PARAMETER (INFREQ = 50)
      PARAMETER (INSMPL = 51)
      PARAMETER (INCORR = 53)
      PARAMETER (IGROUP = 54)
      PARAMETER (IDSEED = 55)
      PARAMETER (ITERSE = 76)
C Functions:
      LOGICAL LOGCVT       !converts 1 and 0 to true and false
      INTEGER GETINT       !get integer from screen
      DATA VARY,PIMC /1,2/
      DATA EPS,CORR /1,2/
ccccccccccccccccccccccccccccccccccccccccccccccccccccccccccccccccccccccccccc
C     get input from terminal
      CALL CLEAR
      CALL ASK(1,ISTOP)
C
C     stop program if requested
      IF (MREALS(IMAIN) .EQ. STOP) CALL FINISH
C
C     close files if necessary
      IF (TNAME .NE. MSTRNG(MINTS(ITNAME)))
     +    CALL FLCLOS(TNAME,TUNIT)
      IF (GNAME .NE. MSTRNG(MINTS(IGNAME)))
     +    CALL FLCLOS(GNAME,GUNIT)
C
C     set new parameter values
C     physical and numerical
      S=MREALS(IS)
      BETA=MREALS(IBETA)
      METHOD=MINTS(IMETHD)
      NENSEM=MINTS(INENSM)
      DT=MREALS(IDT)
      DELTA=MREALS(IDELTA)
      NTHERM=MINTS(ITHERM)
      CALC=MINTS(ICALC)
      NFREQ=MINTS(INFREQ)
      NSMPL=MINTS(INSMPL)
      NCORR=MINTS(INCORR)
      NGROUP=MINTS(IGROUP)
      DSEED=DBLE(MINTS(IDSEED))
C
C     text output
      TTERM=LOGCVT(MINTS(ITTERM))
```

```
       TFILE=LOGCVT(MINTS(ITFILE))
       TNAME=MSTRNG(MINTS(ITNAME))
       TERSE=LOGCVT(MINTS(ITERSE))
C
C      graphics output
       GTERM=LOGCVT(MINTS(IGTERM))
       GHRDCP=LOGCVT(MINTS(IGHRD))
       GFILE=LOGCVT(MINTS(IGFILE))
       GNAME=MSTRNG(MINTS(IGNAME))
C
C      open files
       IF (TFILE) CALL FLOPEN(TNAME,TUNIT)
       IF (GFILE) CALL FLOPEN(GNAME,GUNIT)
       !files may have been renamed
       MSTRNG(MINTS(ITNAME))=TNAME
       MSTRNG(MINTS(IGNAME))=GNAME
C
C      check parameters for correlations, fix NFREQ, NSMPL
       IF (CALC .EQ. CORR) THEN
          NFREQ=1              !fixed for correlations
          NSMPL=1
          IF ((NGROUP .GT. MAXCRR) .OR. ((NGROUP-NCORR) .LE. 20)) THEN
             WRITE (OUNIT,*) ' '
             WRITE (OUNIT,20)
             WRITE (OUNIT,30) NGROUP,NCORR+20,MAXCRR
20           FORMAT (5X,' For reasonable values of the correlations ')
30           FORMAT (5X,' NGROUP (',I4,') must be between NCORR+20 (',
     +               I4,') and MAXCRR (',I4,')')
             WRITE (OUNIT,*) ' '
             NCORR=GETINT(NCORR,1,100,'Reenter NCORR')
             NGROUP=GETINT(NCORR+100,NCORR+20,MAXCRR,'Re-enter NGROUP')
             MINTS(INCORR)=NCORR
             MINTS(IGROUP)=NGROUP
          END IF
       END IF
C
       CALL CLEAR
C
C      calculate derivative parameters
       A=ABOHR
       AOLD=0.
10     IF (ABS(A-AOLD) .GT. 1.E-6) THEN
          AOLD=A
          A=ABOHR/(1+EXP(-S/AOLD))
       GOTO 10
       END IF
       S2=S/2
       HBMDT=HBM*DT
       SQHBDT=SQRT(HBMDT)
       ALPHA=2*ABOHR
       IF (METHOD .EQ. PIMC) DELTA=1.5*A
C
```

```
            RETURN
            END
cccccccccccccccccccccccccccccccccccccccccccccccccccccccccccccccccccccc
            SUBROUTINE PRMOUT(MUNIT,NLINES)
C write out parameter summary to MUNIT
cccccccccccccccccccccccccccccccccccccccccccccccccccccccccccccccccccccc
C Global variables:
            INCLUDE 'IO.ALL'
            INCLUDE 'PARAM.P8'
C Input variables:
            INTEGER MUNIT                      !fortran unit number
            INTEGER NLINES                     !number of lines sent to terminal
C Local variables:
            INTEGER CORR,EPS                   !what is being calculated?
            INTEGER PIMC,VARY                  !which method?
            DATA EPS,CORR /1,2/
            DATA VARY,PIMC /1,2/
cccccccccccccccccccccccccccccccccccccccccccccccccccccccccccccccccccccc
            IF (MUNIT .EQ. OUNIT) THEN
                CALL CLEAR
            ELSE
                WRITE (MUNIT,*) ' '
                WRITE (MUNIT,*) ' '
            END IF
C
            WRITE (MUNIT,5)
            WRITE (MUNIT,7) S
            WRITE (MUNIT,8) BETA
            WRITE (MUNIT,9) A
            WRITE (MUNIT,*) ' '
            IF (METHOD .EQ. PIMC) THEN
                WRITE (MUNIT,10) NENSEM, DT
            ELSE
                WRITE (MUNIT,11)
            END IF
            IF (CALC .EQ. CORR) WRITE (MUNIT,12) NCORR
            WRITE (MUNIT,13)DELTA
            WRITE (MUNIT,15) NTHERM
            WRITE (MUNIT,20) NFREQ,NSMPL
            WRITE (MUNIT,*) ' '
C
            NLINES=11
C
5           FORMAT (' Output from project 8:',
         +           ' Monte Carlo solution of the H2 molecule')
7           FORMAT (' Proton separation (Angstroms) = ',F7.4)
8           FORMAT (' Variational parameter Beta (Angstroms**-1) = ',F7.4)
9           FORMAT (' Wave function parameter A (Angstroms) = ',F7.4)
10          FORMAT (' Path Integral Monte Carlo with ensemble size = ',I4,
         +           ' and time step = ',1PE12.5)
11          FORMAT (' Variational Monte Carlo method')
12          FORMAT (' correlations will be calculated up to K = ', I4)
```

```
13   FORMAT (' Metropolis step in coordinate space (Angstroms)=',F7.4)
15   FORMAT (' number of thermalization sweeps =',I4)
20   FORMAT (' sweep frequency = ',I4,' group size =',I4)
C
     RETURN
     END
CCCCCCCCCCCCCCCCCCCCCCCCCCCCCCCCCCCCCCCCCCCCCCCCCCCCCCCCCCCCCCCCCCCCCCCCCCCC
     SUBROUTINE TXTOUT(IGRP,ENERGY,EVALUE,SIG1,SIG2,U,ACCPT,MUNIT)
C write out results to MUNIT
CCCCCCCCCCCCCCCCCCCCCCCCCCCCCCCCCCCCCCCCCCCCCCCCCCCCCCCCCCCCCCCCCCCCCCCCCCCC
C Global variables:
     INCLUDE 'PARAM.P8'
     INCLUDE 'IO.ALL'
C Input variables:
C    energy has two indices
C    first index is the level: sweep, group, or total
C    second index is the value: quantity, quant**2, or sigma**2
     REAL ENERGY(3,3)           !energy
     INTEGER IGRP               !group index
     REAL EVALUE                !current average energy
     REAL SIG1,SIG2             !uncertainties in energy
     REAL U                     !total energy of the system at this S
     REAL ACCPT                 !acceptance ratio
     INTEGER MUNIT              !unit to write to
C Local variables:
     INTEGER SWEEP,GROUP,TOTAL  !which level of calculation
     INTEGER VALUE,SQUARE,SIGSQ !which quantity
     INTEGER PIMC,VARY          !which method?
     DATA SWEEP,GROUP,TOTAL/1,2,3/
     DATA VALUE,SQUARE,SIGSQ/1,2,3/
     DATA VARY,PIMC /1,2/
CCCCCCCCCCCCCCCCCCCCCCCCCCCCCCCCCCCCCCCCCCCCCCCCCCCCCCCCCCCCCCCCCCCCCCCCCCCC
     WRITE (MUNIT,10) IGRP,NGROUP,
    +       ENERGY(GROUP,VALUE),SQRT(ENERGY(GROUP,SIGSQ))
     IF (METHOD .EQ. VARY) THEN
        WRITE (MUNIT,20) EVALUE,SIG1,SIG2,U,ACCPT/IGRP/NFREQ/NSMPL
     ELSE
        WRITE (MUNIT,30) EVALUE,SIG1,SIG2,U
     END IF
     IF (MUNIT .EQ. TUNIT) WRITE (MUNIT,*) ' '
C
     IF ((MUNIT .EQ. OUNIT) .AND. (.NOT. TERSE))
    +    CALL PAUSE('to continue...',1)
10   FORMAT (2X,'Group ', I4,' of ', I4,5X,'Eigenvalue = ',F9.4,
    +        ' +- ',F8.4)
20   FORMAT (2X,'Grand average E =',F9.4,'+-',F8.4,'/',F8.4,
    +        ' U=',F9.4,' acceptance=',F6.4)
30   FORMAT (2X,'Grand average E =',F9.4,'+-',F8.4,'/',F8.4,
    +        ' U=',F9.4)
     RETURN
     END
CCCCCCCCCCCCCCCCCCCCCCCCCCCCCCCCCCCCCCCCCCCCCCCCCCCCCCCCCCCCCCCCCCCCCCCCCCCC
```

```
         SUBROUTINE GRFOUT(DEVICE,ECORR)
C outputs energy auto-correlation vs. correlation length
CCCCCCCCCCCCCCCCCCCCCCCCCCCCCCCCCCCCCCCCCCCCCCCCCCCCCCCCCCCCCCCCCCCC
C Global variables
         INCLUDE 'IO.ALL'
         INCLUDE 'PARAM.P8'
         INCLUDE 'GRFDAT.ALL'
C Input variables:
         REAL ECORR(0:MAXCRR)           !energy auto-correlations
         INTEGER DEVICE                 !which device is being used?
C Local variables
         REAL K(0:MAXCRR)               !correlation length
         INTEGER IK                     !correlation length index
         CHARACTER*9 CB,CS,CG           !Beta, S, NGROUP as character data
         INTEGER SCREEN                 !send to terminal
         INTEGER PAPER                  !make a hardcopy
         INTEGER FILE                   !send to a file
         INTEGER LB,LS,LG               !true length of character data
         DATA SCREEN,PAPER,FILE/1,2,3/
CCCCCCCCCCCCCCCCCCCCCCCCCCCCCCCCCCCCCCCCCCCCCCCCCCCCCCCCCCCCCCCCCCCC
C     messages for the impatient
         IF (DEVICE .NE. SCREEN) WRITE (OUNIT,100)
C
C     calculate parameters for graphing
         IF (DEVICE .NE. FILE) THEN
              NPLOT=1                    !how many plots?
              IPLOT=1
C
              YMIN=-1.                   !limits on plot
              YMAX=1.
              XMIN=0.
              XMAX=NCORR
              X0VAL=0.
              Y0VAL=XMIN
C
              NPOINT=NCORR+1
C
              ILINE=1                    !line and symbol styles
              ISYM=1
              IFREQ=1
              NXTICK=5
              NYTICK=5
C
              CALL CONVRT(BETA,CB,LB)              !titles and labels
              CALL CONVRT(S,CS,LS)
              CALL ICNVRT(NGROUP,CG,LG)
              INFO='NGROUP = '//CG(1:LG)
              TITLE = 'H2 molecule, S='//CS(1:LS)//',  Beta='//CB(1:LB)
              LABEL(1) = 'Correlation length'
              LABEL(2) = 'Energy auto-correlation'
C
              CALL GTDEV(DEVICE)                   !device nomination
```

```
         IF (DEVICE .EQ. SCREEN) CALL GMODE   !change to graphics mode
         CALL LNLNAX                          !draw axes
      END IF
C
      DO 10 IK=0,NCORR                         !fill array of corr length
         K(IK)=REAL(IK)
10    CONTINUE
C
C     output results
      IF (DEVICE .EQ. FILE) THEN
         WRITE (GUNIT,*) ' '
         WRITE (GUNIT,25) NGROUP
         WRITE (GUNIT,70) (K(IK),ECORR(IK),IK=0,NCORR)
      ELSE
         CALL XYPLOT (K,ECORR)
      END IF
C
C     end graphing session
      IF (DEVICE .NE. FILE) CALL GPAGE(DEVICE) !close graphics package
      IF (DEVICE .EQ. SCREEN) CALL TMODE       !switch to text mode
C
70    FORMAT (2(5X,E11.3))
25    FORMAT (6X,'corr length',5X,
     +    'energy auto-correlation for NGROUP=',I5)
100   FORMAT (/,' Patience, please; output going to a file.')
      RETURN
      END
CCCCCCCCCCCCCCCCCCCCCCCCCCCCCCCCCCCCCCCCCCCCCCCCCCCCCCCCCCCCCCCCCCCCCCCCC
      SUBROUTINE CRROUT(MUNIT,ECORR)
C write out correlations to MUNIT
CCCCCCCCCCCCCCCCCCCCCCCCCCCCCCCCCCCCCCCCCCCCCCCCCCCCCCCCCCCCCCCCCCCCCCCCC
C Global variables:
      INCLUDE 'PARAM .P8'
      INCLUDE 'IO.ALL'
C Input variables:
      REAL ECORR(0:MAXCRR)        !energy auto-correlations
      INTEGER MUNIT               !unit to write to
C Local variables:
      INTEGER K                   !correlation length
      INTEGER NLINES              !number of lines written to screen
CCCCCCCCCCCCCCCCCCCCCCCCCCCCCCCCCCCCCCCCCCCCCCCCCCCCCCCCCCCCCCCCCCCCCCCCC
      IF (MUNIT .EQ. OUNIT) THEN
         CALL CLEAR
      ELSE
         WRITE (MUNIT,*) ' '
      END IF
C
      NLINES=1
      WRITE (MUNIT,30) NGROUP
30    FORMAT(' Correlations with NGROUP = ',I5)
      DO 10 K=0,NCORR
         NLINES=NLINES+1
```

```
            WRITE (MUNIT,20) K,ECORR(K)
            IF ((MUNIT .EQ. OUNIT) .AND. (MOD(NLINES,TRMLIN-3).EQ. 0)) THEN
               CALL PAUSE('to continue...',0)
               NLINES=0
            END IF
   10    CONTINUE
            IF (MUNIT .NE. OUNIT) WRITE (MUNIT,*) ' '
   20    FORMAT (5X,' Correlation length = ', I3, 5X,
         +           'Energy auto-correlation = ', F12.5)
            RETURN
            END
CCCCCCCCCCCCCCCCCCCCCCCCCCCCCCCCCCCCCCCCCCCCCCCCCCCCCCCCCCCCCCCCCCCCCCCCCC
CCCCCCCCCCCCCCCCCCCCCCCCCCCCCCCCCCCCCCCCCCCCCCCCCCCCCCCCCCCCCCCCCCCCCCCCCC
C param.p8
C
            REAL S                      !inter-proton separation
            REAL S2                     !S/2
            REAL BETA                   !variational parameter
            INTEGER METHOD,CALC         !which method?  which quant to calculate?
            INTEGER NENSEM              !ensemble size
            REAL DT                     !PIMC step size
            REAL DELTA                  !step size in configuration space
            INTEGER NCORR               !max correlation length
            INTEGER NGROUP              !initial group size
            DOUBLE PRECISION DSEED      !random number seed
            INTEGER NTHERM              !number of thermalization sweeps
            INTEGER NFREQ               !freq of sweeps to avoid correlations
            INTEGER NSMPL               !size of groups
            LOGICAL TERSE               !terse output?
C
            REAL A,ALPHA                !constants in PHI
            REAL HBM                    !hbar**2 divided by electron mass
            REAL E2                     !electron charge squared
            REAL ABOHR                  !Bohr radius (Angstroms)
            REAL HBMDT                  !hbar**2*dt/m
            REAL SQHBDT                 !sqrt(hbar**2*dt/m)
C
            INTEGER MAXENS              !maximum ensemble size
            INTEGER MAXCRR              !max number of groups for correlation
            INTEGER NCOORD              !number of coordinates
            PARAMETER (MAXENS=100,MAXCRR=500,NCOORD=6)
C
            COMMON / PPARAM / S,BETA
            COMMON / NPARAM / DSEED,NTHERM,NFREQ,NSMPL,METHOD,CALC,NENSEM,
         +               DT,DELTA,NCORR,NGROUP,TERSE
            COMMON / PCALC / A,S2,HBMDT,SQHBDT
            COMMON / CONST / E2,ABOHR,ALPHA,HBM
```

Common Utility Codes

D.1 Standardization Code

This program strips the non-standard '!' comment delimiter out of other FORTRAN codes. It is completely standard, requires no editing to run (except you must set TERM equal to the unit number connected to your keyboard), and prompts for the names of input and output files.

```
C STRIP.FOR
C Reads in a FORTRAN code and deletes all comments
C starting with the non-fortran-77 standard comment delimiter: "!";
C also deletes all blank lines
C
      CHARACTER*80 LINE
      CHARACTER*80 INFILE,OUTFIL
      CHARACTER*80 ANSWER,FIRST
      LOGICAL EXST,OPN,FOUND
      INTEGER EXCLM,SPACE
      INTEGER TERM,OUNIT,IUNIT
      INTEGER I,ASCII,NONBLK
C
      EXCLM=ICHAR('!')
      SPACE=ICHAR(' ')
C
C units for keyboard, INFILE and OUTFIL (change these if necessary)
      TERM=5
      IUNIT=10
      OUNIT=20
cccccccccccccccccccccccccccccccccccccccccccccccccccccccccccccccccccccccc
10    CONTINUE
C
C get name of input file and open file
C
      PRINT *, 'ENTER NAME OF FORTRAN PROGRAM'
      READ (TERM,60)  INFILE
```

```
C
20        INQUIRE (FILE=INFILE,EXIST=EXST,OPENED=OPN)
          IF (EXST .EQV. .FALSE.) THEN
             PRINT *, 'FILE DOES NOT EXIST'
             PRINT *, 'ENTER ANOTHER NAME'
             READ (TERM,60)  INFILE
          ELSE IF (OPN .EQV. .FALSE.) THEN
             OPEN (UNIT=IUNIT,FILE=INFILE,STATUS='OLD')
             OPN=.TRUE.
          END IF
          IF (OPN .EQV. .FALSE.) GOTO 20
C
C get name of output file and open
C
          PRINT *, 'ENTER NAME OF NEW FILE'
          READ (TERM,60)     OUTFIL
C
C this line is for a VAX only
C        OPEN (UNIT=OUNIT,FILE=OUTFIL,STATUS='NEW',CARRIAGECONTROL='LIST')
C this line is for any other machine
          OPEN (UNIT=OUNIT,FILE=OUTFIL,STATUS='NEW')
C
C read each line of input file and search for "!"
35        READ (IUNIT,60,END=50) LINE
          FOUND=.FALSE.
          I=0
          NONBLK=0
45        IF ((FOUND .EQV. .FALSE.) .AND. (I .LT. 80)) THEN
             I=I+1
             ASCII=ICHAR (LINE (I:I))
             IF (ASCII .EQ. EXCLM) THEN
                FOUND=.TRUE.
             ELSE IF (ASCII .NE. SPACE) THEN
                NONBLK=I
             END IF
          GOTO 45
          END IF
C
C        print up to last nonblank character, exclude "!"
          IF (NONBLK .GT. 0) WRITE (OUNIT,60) LINE (1:NONBLK)
C
          GOTO 35
50        CONTINUE
C
          CLOSE (UNIT=OUNIT)
          CLOSE (UNIT=IUNIT)
C
C allow for another file
          PRINT *, 'DO YOU WISH TO STANDARDIZE ANOTHER FILE? [Y]'
          READ (TERM,60)  ANSWER
C
          FIRST=ANSWER (1:1)
```

```
      IF ((FIRST .EQ. 'N') .OR. (FIRST .EQ. 'n')) STOP
C
      GOTO 10
C
60    FORMAT (A)
      END
```

D.2 Hardware and Compiler Specific Code

SETUP.FOR includes all parameters that depend on your computing environment. It therefore will require editing. For details on how to make the appropriate changes, see Appendix A.3.

```
C file SETUP.FOR
CCCCCCCCCCCCCCCCCCCCCCCCCCCCCCCCCCCCCCCCCCCCCCCCCCCCCCCCCCCCCCCCCCCCC
      SUBROUTINE SETUP
C allows users to supply i/o parameters for their computing environment
CCCCCCCCCCCCCCCCCCCCCCCCCCCCCCCCCCCCCCCCCCCCCCCCCCCCCCCCCCCCCCCCCCCCC
C Global variables:
      INCLUDE 'IO.ALL'
CCCCCCCCCCCCCCCCCCCCCCCCCCCCCCCCCCCCCCCCCCCCCCCCCCCCCCCCCCCCCCCCCCCCC
C     fortran unit numbers for i/o
C     unit for text output to a file
      TUNIT=10
C     unit for graphics output to file
      GUNIT=20
C     unit for input from keyboard
      IUNIT=5
C     unit for output to screen
      OUNIT=6
C     unit for input of data
      DUNIT=11
C
C     how many lines and columns of text fit on your screen?
      TRMLIN=24
      TRMWID=80
C
C     default output parameters
C     There are five forms of output provided, here you are choosing
C     which forms of output you will want MOST of the time (any
C     combination is possible), you always have the option to change
C     your mind at run time.
C     0=no   1=yes
C     do you want text sent to the screen?
      TXTTRM=1
C     do you want text sent to a file?
      TXTFIL=0
C     do you want graphics sent to the screen?
      GRFTRM=0
C     do you want graphics sent to a hardcopy device?
      GRFHRD=0
C     do you want graphics data sent to a file?
      GRFFIL=0
C
      RETURN
      END
CCCCCCCCCCCCCCCCCCCCCCCCCCCCCCCCCCCCCCCCCCCCCCCCCCCCCCCCCCCCCCCCCCCCC
      SUBROUTINE CLEAR
C clears text screen by sending an escape sequence;
```

```
C check your terminal manual for the correct sequence
C THIS IS NOT AN ESSENTIAL ROUTINE - YOU CAN LEAVE IT BLANK
CCCCCCCCCCCCCCCCCCCCCCCCCCCCCCCCCCCCCCCCCCCCCCCCCCCCCCCCCCCCCCCCCCCC
C Global variables:
      INCLUDE 'IO.ALL'
C Local variables:
      CHARACTER*1 ESC1(4),ESC2(6)     !escape characters
      INTEGER I,I1,I2                 !index of escape sequence arrays
CCCCCCCCCCCCCCCCCCCCCCCCCCCCCCCCCCCCCCCCCCCCCCCCCCCCCCCCCCCCCCCCCCCC
C VT200 terminal; text mode
      ESC1(1)=CHAR(27)                !<ESC>[2J
      ESC1(2)=CHAR(91)
      ESC1(3)=CHAR(50)
      ESC1(4)=CHAR(74)
      I1=4
      ESC2(1)=CHAR(27)                !<ESC>11;1f
      ESC2(2)=CHAR(91)
      ESC2(3)=CHAR(49)
      ESC2(4)=CHAR(59)
      ESC2(5)=CHAR(49)
      ESC2(6)=CHAR(102)
      I2=6
C
C TEK4010
C     ESC1(1)=CHAR(27)                !<ESC><FF>
C     ESC1(2)=CHAR(12)
C     I1=2
C     I2=0
C
C PST (Prime)
C     ESC1(1)=CHAR(27)                !<ESC>?
C     ESC1(2)=CHAR(63)
C     I1=2
C     I2=0
CCCCCCCCCCCCCCCCCCCCCCCCCCCCCCCCCCCCCCCCCCCCCCCCCCCCCCCCCCCCCCCCCCCC
C     WRITE(OUNIT,10) (ESC1(I),I=1,I1)
C     WRITE(OUNIT,10) (ESC2(I),I=1,I2)
10    FORMAT (1X,6A1)
      RETURN
      END
CCCCCCCCCCCCCCCCCCCCCCCCCCCCCCCCCCCCCCCCCCCCCCCCCCCCCCCCCCCCCCCCCCCC
      SUBROUTINE GMODE
C switches terminal from text to graphics mode
C by writing hardware dependent escape sequences to the terminal
C This routine contains the escape sequence for a Graphon terminal
C to switch between VT200 and TEK4014 modes
CCCCCCCCCCCCCCCCCCCCCCCCCCCCCCCCCCCCCCCCCCCCCCCCCCCCCCCCCCCCCCCCCCCC
C Global variables:
      INCLUDE 'IO.ALL'
C Local variables:
      CHARACTER*1 ESC(2)
      ESC(1)=CHAR(27)                 !ascii codes for <ESC> 1
```

```
          ESC(2)=CHAR(49)
CCCCCCCCCCCCCCCCCCCCCCCCCCCCCCCCCCCCCCCCCCCCCCCCCCCCCCCCCCCCCCCCCCCCCCCCC
C      WRITE(OUNIT,10) ESC(1),ESC(2)
10     FORMAT (1X,2A1)
       RETURN
       END
CCCCCCCCCCCCCCCCCCCCCCCCCCCCCCCCCCCCCCCCCCCCCCCCCCCCCCCCCCCCCCCCCCCCCCCCC
       SUBROUTINE TMODE
C switches terminal from graphics to text mode
C by writing hardware dependent escape sequences to the terminal
C This routine contains the escape sequence for a Graphon terminal
C to switch between TEK4014 and VT200 modes
CCCCCCCCCCCCCCCCCCCCCCCCCCCCCCCCCCCCCCCCCCCCCCCCCCCCCCCCCCCCCCCCCCCCCCCCC
C Global variables:
       INCLUDE 'IO.ALL'
C Local variables:
          CHARACTER*1 ESC(2)
          ESC(1)=CHAR(27)                       !ascii codes for <ESC> 2
          ESC(2)=CHAR(50)
CCCCCCCCCCCCCCCCCCCCCCCCCCCCCCCCCCCCCCCCCCCCCCCCCCCCCCCCCCCCCCCCCCCCCCCCC
C      WRITE(OUNIT,10) ESC(1),ESC(2)
10     FORMAT (1X,2A1)
       RETURN
       END
CCCCCCCCCCCCCCCCCCCCCCCCCCCCCCCCCCCCCCCCCCCCCCCCCCCCCCCCCCCCCCCCCCCCCCCCC
CCCCCCCCCCCCCCCCCCCCCCCCCCCCCCCCCCCCCCCCCCCCCCCCCCCCCCCCCCCCCCCCCCCCCCCCC
C io.all
C
C      environment dependent parameters
       INTEGER IUNIT      !unit number for input from screen
       INTEGER OUNIT      !unit number for output to screen
       INTEGER TUNIT      !unit number for text output to file
       INTEGER GUNIT      !unit number for graphics output to file
       INTEGER DUNIT      !unit number for data input from file
       INTEGER TRMLIN     !number of lines on terminal screen
       INTEGER TRMWID     !width of terminal screen
C
C      the following are default answers to i/o choices
C      1==yes   0 == no
       INTEGER TXTTRM     !send text output to terminal?
       INTEGER TXTFIL     !send text output to a file?
       INTEGER GRFTRM     !send graphics to terminal?
       INTEGER GRFHRD     !send graphics to a hard copy device?
       INTEGER GRFFIL     !send graphics data to a file?
C
C      i/o input parameters for this run
       LOGICAL TTERM          !write text output to terminal?
       LOGICAL TFILE          !write text output to a file?
       CHARACTER*12 TNAME     !name of text file
       LOGICAL GTERM          !send graphics output to terminal?
       LOGICAL GHRDCP         !send graphics output to hardcopy device?
       LOGICAL GFILE          !send graphics data to a file?
```

```
      CHARACTER*12 GNAME        !name of graphics data file
c
      COMMON /IO/TUNIT,GUNIT,IUNIT,OUNIT,DUNIT,TRMLIN,TRMWID,TXTTRM,
     +            TXTFIL,GRFTRM,GRFHRD,GRFFIL,
     +            TTERM,TFILE,GTERM,GFILE,GHRDCP
      COMMON / CIO / TNAME,GNAME
```

D.3 General Input/Output Codes

All of the utility routines that do not need editing are located in the file UTIL.FOR. They are as follows:

menu	MENU	: defines menu array variables that are the same for all programs
routines	ASK	: executes the menu
	PRTAGS	: prints menu items and default values
	PRBLKS	: prints blank lines
	PRYORN	: prints a 'yes' or 'no'
	PARSE	: controls menu branching by returning index of next menu item
file	FLOPEN	: opens a new file
manipulation	FLOPN2	: opens an existing file
	FLCLOS	: closes a file
user	GETFLT	: returns a floating point number
interfaces	GETINT	: returns an integer
	YESNO	: returns a 1 (yes) or 0 (no)
	CHARAC	: returns a character string
	PAUSE	: suspends execution until user presses Return key
conversion	CONVRT	: converts a real number to a string
routines	ICNVRT	: converts an integer to a string
	LOGCVT	: converts 1 to .TRUE., everything else to .FALSE.
miscellaneous	HEADER	: prints the introductory text screen
	FINISH	: closes open files and stops the program
	FLTDEF	: prints default value for real numbers in an appropriate format
	INTDEF	: prints default value for integers in an appropriate format
	LENTRU	: returns the length of a character string exclusive of trailing blanks
	RANNOS	: returns a uniformly distributed random number between 0 and 1.

```
ccccccccccccccccccccccccccccccccccccccccccccccccccccccccccccccccccccccccc
C   file UTIL.for
C   COMPUTATIONAL PHYSICS (FORTRAN VERSION)
C   by Steven E. Koonin and Dawn C. Meredith
C   Copyright 1989, Addison-Wesley Publishing Company
ccccccccccccccccccccccccccccccccccccccccccccccccccccccccccccccccccccccccc
```

```
      SUBROUTINE HEADER(DESCRP,NHEAD,NTEXT,NGRAPH)
C displays header and description of output to screen
CCCCCCCCCCCCCCCCCCCCCCCCCCCCCCCCCCCCCCCCCCCCCCCCCCCCCCCCCCCCCCCCCCCC
C Global variables:
      INCLUDE 'IO.ALL'
C Input variables:
      CHARACTER*(*) DESCRP(20)     !description of program and output
      INTEGER NHEAD,NTEXT,NGRAPH   !number of lines for each description
C Local variables:
      INTEGER N                    !current line number
      INTEGER LENGTH               !true length of character strings
      INTEGER NBLNKS               !num of blanks needed to center string
      CHARACTER*80 BLANKS          !array of blanks for centering
C Function:
      INTEGER LENTRU               !true length of character string
      DATA BLANKS/' '/
CCCCCCCCCCCCCCCCCCCCCCCCCCCCCCCCCCCCCCCCCCCCCCCCCCCCCCCCCCCCCCCCCCCC
      CALL CLEAR                   !vertically center output
      DO 190 N=1,(TRMLIN-18-NHEAD-NGRAPH-NTEXT)/2
         WRITE (OUNIT,20)
190   CONTINUE
C
C     write out constant part of header
      WRITE (OUNIT,40)
      WRITE (OUNIT,50)
      WRITE (OUNIT,60)
      WRITE (OUNIT,80)
      WRITE (OUNIT,20)
      WRITE (OUNIT,20)
C
C     write out chapter dependent section of the header
      DO 140 N=1,NHEAD+NTEXT+NGRAPH
         IF (N .EQ. NHEAD+1) THEN !text output header
            WRITE (OUNIT,110)
         END IF
         IF (N .EQ. NHEAD+NTEXT+1) THEN
            WRITE (OUNIT,115)      !graphics output header
         END IF
         LENGTH=LENTRU(DESCRP(N)) !horizontally center output
         NBLNKS=(80-LENGTH)/2
         WRITE (OUNIT,120) BLANKS(1:NBLNKS),DESCRP(N)(1:LENGTH)
140   CONTINUE
C
      CALL PAUSE('to begin the program...',1)
      CALL CLEAR
C
20    FORMAT (' ')
40    FORMAT (/,30X,'COMPUTATIONAL PHYSICS')
50    FORMAT (/,32X,'(FORTRAN VERSION)')
60    FORMAT (/,20X,'by Steven E. Koonin and Dawn C. Meredith')
80    FORMAT (/,14X,
     +   'Copyright 1989, Benjamin/Cummings Publishing Company')
```

```
110    FORMAT (/,30X, 'Text output displays')
115    FORMAT (/,28X, 'Graphics output displays')
120    FORMAT (A,A)
C
       RETURN
       END
CCCCCCCCCCCCCCCCCCCCCCCCCCCCCCCCCCCCCCCCCCCCCCCCCCCCCCCCCCCCCCCCCCCCCCC
       SUBROUTINE MENU
C sets up the part of the menu that is the same for all programs
CCCCCCCCCCCCCCCCCCCCCCCCCCCCCCCCCCCCCCCCCCCCCCCCCCCCCCCCCCCCCCCCCCCCCCC
C Global variables:
       INCLUDE 'MENU.ALL'
       INCLUDE 'IO.ALL'
CCCCCCCCCCCCCCCCCCCCCCCCCCCCCCCCCCCCCCCCCCCCCCCCCCCCCCCCCCCCCCCCCCCCCCC
C      main menu
       MTYPE(1)=CLRTRM
C
       MTYPE(2)=MTITLE
       MPRMPT(2)='MAIN MENU'
       MLOLIM(2)=2
       MHILIM(2)=1
C
       MTYPE(3)=MTITLE
       MPRMPT(3)='1) Change physical parameters'
       MLOLIM(3)=0
       MHILIM(3)=0
C
       MTYPE(4)=MTITLE
       MPRMPT(4)='2) Change numerical parameters'
       MLOLIM(4)=0
       MHILIM(4)=0
C
       MTYPE(5)=MTITLE
       MPRMPT(5)='3) Change output parameters'
       MLOLIM(5)=0
       MHILIM(5)=0
C
       MTYPE(6)=MTITLE
       MPRMPT(6)='4) Display physical and numerical parameters'
       MLOLIM(6)=0.
       MHILIM(6)=0.
C
       MTYPE(7)=MTITLE
       MPRMPT(7)='5) Display output parameters'
       MLOLIM(7)=0.
       MHILIM(7)=0.
C
       MTYPE(8)=MTITLE
       MPRMPT(8)='6) Run program'
       MLOLIM(8)=0
       MHILIM(8)=0
C
```

```
       MTYPE(9)=MTITLE
       MPRMPT(9)='7) Stop program'
       MLOLIM(9)=0
       MHILIM(9)=1
C
       MTYPE(10)=MCHOIC
       MPRMPT(10)= 'Make a menu choice'
       MTAG(10)='11 36 61 91 94 99 99'
       MLOLIM(10)=1
       MHILIM(10)=7
       MINTS(10)=6
       MREALS(10)=-6
C
C      physical parameters
       MTYPE(11)=CLRTRM
C
       MTYPE(12)=TITLE
       MPRMPT(12)= 'PHYSICAL PARAMETERS'
       MLOLIM(12)=2.
       MHILIM(12)=1.
C
       MTYPE(35)=SKIP
       MREALS(35)=1.
C
C      numerical parameters
       MTYPE(36)=CLRTRM
C
       MTYPE(37)=TITLE
       MPRMPT(37)= 'NUMERICAL PARAMETERS'
       MLOLIM(37)=2.
       MHILIM(37)=1.
C
       MTYPE(60)=SKIP
       MREALS(60)=1.
C
C      output menu
       MTYPE(61)=CLRTRM
C
       MTYPE(62)=MTITLE
       MPRMPT(62)= 'OUTPUT MENU'
       MLOLIM(62)=0.
       MHILIM(62)=1.
C
       MTYPE(63)=MTITLE
       MPRMPT(63)='1) Change text output parameters'
       MLOLIM(63)=0.
       MHILIM(63)=0.
C
       MTYPE(64)=MTITLE
       MPRMPT(64)='2) Change graphics output parameters'
       MLOLIM(64)=0.
       MHILIM(64)=0.
```

```
c
      MTYPE(65)=MTITLE
      MPRMPT(65)='3) Return to main menu'
      MLOLIM(65)=0.
      MHILIM(65)=1.
c
      MTYPE(66)=MCHOIC
      MPRMPT(66)= 'Make menu choice and press Return'
      MTAG(66)='71 81 01'
      MLOLIM(66)=1.
      MHILIM(66)=3.
      MINTS(66)=3.
c
c     text output parameters
      MTYPE(71)=CLRTRM
c
      MTYPE(72)=TITLE
      MPRMPT(72)= 'TEXT OUTPUT PARAMETERS'
      MLOLIM(72)=2.
      MHILIM(72)=1.
c
      MTYPE(73)=BOOLEN
      MPRMPT(73)= 'Do you want text output displayed on screen?'
      MTAG(73)= 'Text output to screen'
      MINTS(73)=TXTTRM
c
      MTYPE(74)=NOSKIP
      MPRMPT(74)= 'Do you want text output sent to a file?'
      MTAG(74)= 'Text output to file'
      MREALS(74)=76.
      MINTS(74)=TXTFIL
c
      MTYPE(75)=CHSTR
      MPRMPT(75)= 'Enter name of file for text output'
      MTAG(75)= 'File name for text output'
      MLOLIM(75)=1.
      MHILIM(75)=12.
      MINTS(75)=1
      MSTRNG(MINTS(75))= 'cmphys.txt'
c
      MTYPE(80)=SKIP
      MREALS(80)=61.
c
c     graphics output parameters
      MTYPE(81)=CLRTRM
c
      MTYPE(82)=TITLE
      MPRMPT(82)= 'GRAPHICS OUTPUT PARAMETERS'
      MLOLIM(82)=2.
      MHILIM(82)=1.
c
      MTYPE(83)=BOOLEN
```

```
      MPRMPT(83)= 'Do you want graphics sent to the terminal?'
      MTAG(83)= 'Graphics output to terminal'
      MINTS(83)=GRFTRM
C
      MTYPE(84)=BOOLEN
      MPRMPT(84)= 'Do you want graphics sent to the hardcopy device?'
      MTAG(84)= 'Graphics output to hardcopy device'
      MINTS(84)=GRFHRD
C
      MTYPE(85)=NOSKIP
      MPRMPT(85)= 'Do you want data for graphing sent to a file?'
      MTAG(85)= 'Data for graphing sent to file'
      MREALS(85)=87.
      MINTS(85)=GRFFIL
C
      MTYPE(86)=CHSTR
      MPRMPT(86)= 'Enter name of file for graphics data'
      MTAG(86)= 'File for graphics data'
      MLOLIM(86)=2.
      MHILIM(86)=12.
      MINTS(86)=2.
      MSTRNG(MINTS(86))= 'cmphys.grf'
C
      MTYPE(90)=SKIP
      MREALS(90)=61.
C
C     printing numerical and physical parameters
      MTYPE(91)=PPRINT
      MLOLIM(91)=11.
      MHILIM(91)=60.
C
      MTYPE(92)=SKIP
      MREALS(92)=1.
C
C     printing output parameters
      MTYPE(94)=PPRINT
      MLOLIM(94)=71.
      MHILIM(94)=90.
C
      MTYPE(95)=SKIP
      MREALS(95)=1.
C
      RETURN
      END
CCCCCCCCCCCCCCCCCCCCCCCCCCCCCCCCCCCCCCCCCCCCCCCCCCCCCCCCCCCCCCCCCCCCCCC
      SUBROUTINE ASK(START,END)
C executes menu items from START to END;
C see Appendix A for a description of the menu
CCCCCCCCCCCCCCCCCCCCCCCCCCCCCCCCCCCCCCCCCCCCCCCCCCCCCCCCCCCCCCCCCCCCCCC
C Global variables:
      INCLUDE 'MENU.ALL'
      INCLUDE 'IO.ALL'
```

```
C Input variables:
      INTEGER START,END          !starting/ending menu items to execute
C Local variables:
      INTEGER I                  !current menu item
      INTEGER ILOW,IHIGH         !integer limits for NUM type
      INTEGER NUMSKP             !number of blank lines to print
      INTEGER ICHOIC             !current menu choice
C Functions
      CHARACTER*40 CHARAC        !character input
      REAL GETFLT                !real input
      INTEGER GETINT             !integer input
      INTEGER PARSE              !determines menu branching
      INTEGER YESNO              !boolean input
CCCCCCCCCCCCCCCCCCCCCCCCCCCCCCCCCCCCCCCCCCCCCCCCCCCCCCCCCCCCCCCCCCCCCCC
      I=START
1000  CONTINUE
          IF (MTYPE(I) .EQ. FLOAT) THEN
              MREALS(I) =
     +          GETFLT(MREALS(I),MLOLIM(I),MHILIM(I),MPRMPT(I))
C
          ELSE IF ( MTYPE(I) .EQ. NUM) THEN
              ILOW = MLOLIM(I)
              IHIGH = MHILIM(I)
              MINTS(I) = GETINT(MINTS(I), ILOW, IHIGH, MPRMPT(I))
C
          ELSE IF (MTYPE(I) .EQ. BOOLEN) THEN
              MINTS(I) = YESNO(MINTS(I), MPRMPT(I))
C
          ELSE IF (MTYPE(I) .EQ. CHSTR) THEN
              MSTRNG(MINTS(I)) =
     +          CHARAC(MSTRNG(MINTS(I)),INT(MHILIM(I)),MPRMPT(I))
C
          ELSE IF ( MTYPE(I) .EQ. MCHOIC) THEN
              ILOW = MLOLIM(I)
              IHIGH = MHILIM(I)
              ICHOIC = GETINT(MINTS(I), ILOW, IHIGH, MPRMPT(I))
C             if MREALS is > 0, save ICHOIC and change default
              IF (MREALS(I) .GT. 0) THEN
                  MREALS(I)=REAL(ICHOIC)
                  MINTS(I)=ICHOIC
C             if MREALS is < 0, save ICHOIC but leave default the same
              ELSE IF (MREALS(I) .LT. 0) THEN
                  MREALS(I)=-REAL(ICHOIC)
              END IF
              I = PARSE (MTAG(I), ICHOIC) - 1
C
          ELSE IF (MTYPE(I) .EQ. TITLE .OR. MTYPE(I) .EQ. MTITLE) THEN
              NUMSKP = MLOLIM(I)
              CALL PRBLKS(NUMSKP)
              WRITE (OUNIT, 10) MPRMPT(I)
              NUMSKP = MHILIM(I)
              CALL PRBLKS(NUMSKP)
```

```
C
      ELSE IF (MTYPE(I) .EQ. YESKIP) THEN
         MINTS(I) = YESNO(MINTS(I), MPRMPT(I))
         IF (MINTS(I) .NE. 0) THEN
            I = MREALS(I) - 1
         END IF
C
      ELSE IF (MTYPE(I) .EQ. NOSKIP) THEN
         MINTS(I) = YESNO(MINTS(I), MPRMPT(I))
         IF (MINTS(I) .EQ. 0) THEN
            I = MREALS(I) - 1
         END IF
C
      ELSE IF (MTYPE(I) .EQ. SKIP) THEN
         I = MREALS(I) - 1
C
      ELSE IF (MTYPE(I) .EQ. WAIT) THEN
         WRITE(OUNIT, 10) MPRMPT(I)
         CALL PAUSE('to continue',1)
C
      ELSE IF (MTYPE(I) .EQ. CLRTRM) THEN
         CALL CLEAR
C
      ELSE IF (MTYPE(I) .EQ. QUIT) THEN
         I=END
C
      ELSE IF (MTYPE(I) .EQ. PPRINT) THEN
         ILOW = MLOLIM(I)
         IHIGH = MHILIM(I)
         CALL CLEAR
         CALL PRTAGS(ILOW,IHIGH)
         CALL PAUSE('to see the Main Menu...',1)
         CALL CLEAR
C
      END IF
C     display info about defaults
      IF (I .EQ. 1) THEN
         WRITE (OUNIT,*) ' '
         WRITE (OUNIT,100)
         WRITE (OUNIT,101)
      END IF
C
      I = I+1
      IF (I .LE. END) GO TO 1000
C
10    FORMAT( 1X, A )
11    FORMAT( 1X, A, 1PE11.3 )
12    FORMAT( 1X, A, I6 )
100   FORMAT (' To accept the default value [in brackets] for any item')
101   FORMAT (' just press Return at the prompt')
C
      RETURN
```

```
          END
CCCCCCCCCCCCCCCCCCCCCCCCCCCCCCCCCCCCCCCCCCCCCCCCCCCCCCCCCCCCCCCCCCCCCCCC
      SUBROUTINE PRTAGS(START,END)
C prints menu prompts and default values for items START to END
CCCCCCCCCCCCCCCCCCCCCCCCCCCCCCCCCCCCCCCCCCCCCCCCCCCCCCCCCCCCCCCCCCCCCCCC
C Global variables:
      INCLUDE 'MENU.ALL'
      INCLUDE 'IO.ALL'
C Input variables:
      INTEGER START,END          !limiting indices of printed menu items
C Local variables:
      INTEGER I                  !menu items index
      INTEGER NUMSKP             !number of lines to skip
      INTEGER INDEX              !subindex for menu items
      INTEGER PLEN               !length of prompt
      INTEGER ICHOIC             !menu/parameter choice
C Functions:
      INTEGER LENTRU             !true length of character string
      INTEGER PARSE              !menu choice
CCCCCCCCCCCCCCCCCCCCCCCCCCCCCCCCCCCCCCCCCCCCCCCCCCCCCCCCCCCCCCCCCCCCCCCC
      I=START
1000  CONTINUE
          IF (MTYPE(I) .EQ. FLOAT) THEN
              WRITE (OUNIT, 11) MTAG(I), MREALS(I)
C
          ELSE IF (MTYPE(I) .EQ. NUM) THEN
              WRITE (OUNIT, 12) MTAG(I), MINTS(I)
C
          ELSE IF (MTYPE(I) .EQ. BOOLEN) THEN
              CALL PRYORN(MTAG(I), MINTS(I))
C
          ELSE IF (MTYPE(I) .EQ. CHSTR) THEN
              WRITE( OUNIT, 13) MTAG(I), MSTRNG(MINTS(I))
C
          ELSE IF (MTYPE(I) .EQ. TITLE) THEN
              NUMSKP = MLOLIM(I)
              CALL PRBLKS(NUMSKP)
              WRITE (OUNIT, 10) MPRMPT(I)
              NUMSKP = MHILIM(I)
              CALL PRBLKS(NUMSKP)
C
          ELSE IF (MTYPE(I) .EQ. YESKIP) THEN
              CALL PRYORN(MTAG(I), MINTS(I))
              IF (MINTS(I) .NE. 0 .AND. MREALS(I) .GT. I) THEN
                  I = MREALS(I) - 1
              END IF
C
          ELSE IF (MTYPE(I) .EQ. NOSKIP) THEN
              CALL PRYORN(MTAG(I), MINTS(I))
              IF (MINTS(I) .EQ. 0 .AND. MREALS(I) .GT. I) THEN
                  I = MREALS(I) - 1
              END IF
```

```
C
        ELSE IF (MTYPE(I) .EQ. SKIP) THEN
           IF (MREALS(I) .GT. I) I=MREALS(I) - 1   !don't skip backwards
C
        ELSE IF (MTYPE(I) .EQ. MCHOIC)  THEN
           IF (MREALS(I) .GT. 0) THEN
C             for menu choices that are parameter choices, print out
C             choice, but first you must find it
              DO 20 INDEX=I-MHILIM(I),I-1
                 IF ((I+MREALS(I)-MHILIM(I)-1) .EQ. INDEX) THEN
                    PLEN=LENTRU(MPRMPT(INDEX))
                    WRITE (OUNIT,10) MPRMPT(INDEX)(4:PLEN)
                 END IF
20            CONTINUE
           END IF
           IF (MREALS(I) .NE. 0) THEN
              !branch to chosen parameter
              ICHOIC=ABS(INT(MREALS(I)))
              I=PARSE(MTAG(I),ICHOIC)-1
           END IF
C
        END IF
        I = I+1
        IF (I .LE. END) GO TO 1000
C
10      FORMAT( 1X, A )
11      FORMAT( 1X, A, 1PE11.3 )
12      FORMAT( 1X, A, I6 )
13      FORMAT( 1X, A, 5X, A)
        RETURN
        END
CCCCCCCCCCCCCCCCCCCCCCCCCCCCCCCCCCCCCCCCCCCCCCCCCCCCCCCCCCCCCCCCCCCCCC
        SUBROUTINE PRYORN(PMPT,YORN)
C print a 'yes' or 'no'
CCCCCCCCCCCCCCCCCCCCCCCCCCCCCCCCCCCCCCCCCCCCCCCCCCCCCCCCCCCCCCCCCCCCCC
C Global variables:
        INCLUDE 'IO.ALL'
C Input variables:
        INTEGER YORN                    !1 == 'YES', '0' == 'NO'
        CHARACTER*(*) PMPT              !string to print before y/n
C Functions:
        INTEGER LENTRU                  !actual length of prompt
CCCCCCCCCCCCCCCCCCCCCCCCCCCCCCCCCCCCCCCCCCCCCCCCCCCCCCCCCCCCCCCCCCCCCC
        IF (YORN .EQ. 0) THEN
           WRITE(OUNIT, 10) PMPT(1:LENTRU(PMPT))
        ELSE
           WRITE(OUNIT, 11) PMPT(1:LENTRU(PMPT))
        END IF
10      FORMAT( 1X, A, ': no')
11      FORMAT( 1X, A, ': yes')
        RETURN
        END
```

```
CCCCCCCCCCCCCCCCCCCCCCCCCCCCCCCCCCCCCCCCCCCCCCCCCCCCCCCCCCCCCCCCCCCCCCCCC
      SUBROUTINE PRBLKS(NUMLIN)
C prints NUMLIN blank lines on terminal
CCCCCCCCCCCCCCCCCCCCCCCCCCCCCCCCCCCCCCCCCCCCCCCCCCCCCCCCCCCCCCCCCCCCCCCCC
C Global variables:
      INCLUDE 'IO.ALL'
C Passed variables:
      INTEGER NUMLIN                    !number of blank lines to print
C Local variables:
      INTEGER I                         !dummy index
CCCCCCCCCCCCCCCCCCCCCCCCCCCCCCCCCCCCCCCCCCCCCCCCCCCCCCCCCCCCCCCCCCCCCCCCC
      DO 1000 I=1,NUMLIN
         WRITE( OUNIT,*) ' '
1000  CONTINUE
      RETURN
      END
CCCCCCCCCCCCCCCCCCCCCCCCCCCCCCCCCCCCCCCCCCCCCCCCCCCCCCCCCCCCCCCCCCCCCCCCC
      SUBROUTINE PAUSE(PHRASE,NSKIP)
C gives user time to read screen by waiting for dummy input;
C allows for printing of PHRASE to screen;
C skips NSKIP lines before printing PHRASE
CCCCCCCCCCCCCCCCCCCCCCCCCCCCCCCCCCCCCCCCCCCCCCCCCCCCCCCCCCCCCCCCCCCCCCCCC
C Global variables:
      INCLUDE 'IO.ALL'
C Passed variables:
      CHARACTER*(*) PHRASE              !phrase to be printed
      INTEGER NSKIP                     !number of lines to skip
C Local variables:
      CHARACTER*1 DUMMY                 !dummy variable
      INTEGER ISKIP                     !NSKIP index
CCCCCCCCCCCCCCCCCCCCCCCCCCCCCCCCCCCCCCCCCCCCCCCCCCCCCCCCCCCCCCCCCCCCCCCCC
      DO 10 ISKIP=1,NSKIP              !skip lines
         WRITE (OUNIT,5)
10    CONTINUE
5     FORMAT (' ')
      WRITE (OUNIT,15) PHRASE          !write phrase
      READ (IUNIT,20) DUMMY            !wait for dummy input
15    FORMAT (' Press return ',A)
20    FORMAT (A1)
      RETURN
      END
CCCCCCCCCCCCCCCCCCCCCCCCCCCCCCCCCCCCCCCCCCCCCCCCCCCCCCCCCCCCCCCCCCCCCCCCC
      SUBROUTINE FLOPEN(FNAME,FUNIT)
C opens a new file, unless one by the same name already exists
CCCCCCCCCCCCCCCCCCCCCCCCCCCCCCCCCCCCCCCCCCCCCCCCCCCCCCCCCCCCCCCCCCCCCCCCC
C Global variables:
      INCLUDE 'IO.ALL'
C Input variables:
      CHARACTER*(*) FNAME   !file name
      INTEGER FUNIT         !unit number
C Local variables:
      LOGICAL OPN           !is the file open?
```

```
      LOGICAL EXST        !does it exist?
      CHARACTER*40 CHARAC  !function that return character input
      INTEGER LENTRU
CCCCCCCCCCCCCCCCCCCCCCCCCCCCCCCCCCCCCCCCCCCCCCCCCCCCCCCCCCCCCCCCCCCCCCC
10    INQUIRE(FILE=FNAME,EXIST=EXST,OPENED=OPN)
C
      IF (OPN) RETURN
C
      IF (EXST) THEN
         WRITE (OUNIT,20) FNAME(1:LENTRU(FNAME))
20       FORMAT (' Output file ',A,' already exists')
         FNAME=CHARAC(FNAME,12, 'Enter another filename')
      ELSE
         OPEN(UNIT=FUNIT,FILE=FNAME,STATUS='NEW')
         RETURN
      END IF
      GOTO 10
      END
CCCCCCCCCCCCCCCCCCCCCCCCCCCCCCCCCCCCCCCCCCCCCCCCCCCCCCCCCCCCCCCCCCCCCCC
      SUBROUTINE FLOPN2(FNAME,FUNIT,SUCESS)
C opens an existing file for input data
CCCCCCCCCCCCCCCCCCCCCCCCCCCCCCCCCCCCCCCCCCCCCCCCCCCCCCCCCCCCCCCCCCCCCCC
C Global variables:
      INCLUDE 'IO.ALL'
C Input variables:
      CHARACTER*(*) FNAME  !file name
      INTEGER FUNIT        !unit number
      LOGICAL SUCESS       !did we find an existing file to open?
C Local variables:
      LOGICAL OPN          !is the file open?
      LOGICAL EXST         !does it exist?
      CHARACTER*40 CHARAC  !function that return character input
      INTEGER CHOICE       !choice for continuing
C Functions:
      INTEGER YESNO        !get yes or no input
      INTEGER LENTRU
CCCCCCCCCCCCCCCCCCCCCCCCCCCCCCCCCCCCCCCCCCCCCCCCCCCCCCCCCCCCCCCCCCCCCCC
10    INQUIRE(FILE=FNAME,EXIST=EXST,OPENED=OPN)
C
      IF ((.NOT. EXST) .OR. (OPN)) THEN
         WRITE (OUNIT,20) FNAME(1:LENTRU(FNAME))
20       FORMAT(' Input file ',A,' does not exist or is already open')
         CHOICE=YESNO(1,' Would you like to try another file name?')
C
         IF (CHOICE .EQ. 0) THEN
            SUCESS=.FALSE.
            RETURN      !leave without opening file for reading
         ELSE
            FNAME=CHARAC(FNAME,12, 'Enter another filename')
         END IF
      ELSE
         OPEN(UNIT=FUNIT,FILE=FNAME,STATUS='OLD')
```

```
               SUCESS=.TRUE.
               RETURN
          END IF
       GOTO 10
       END
CCCCCCCCCCCCCCCCCCCCCCCCCCCCCCCCCCCCCCCCCCCCCCCCCCCCCCCCCCCCCCCCCCCCCC
       SUBROUTINE FLCLOS(FNAME,FUNIT)
C checks on file status of file, and closes if open
CCCCCCCCCCCCCCCCCCCCCCCCCCCCCCCCCCCCCCCCCCCCCCCCCCCCCCCCCCCCCCCCCCCCCC
C Global variables:
       INCLUDE 'IO.ALL'
C Input variables:
       CHARACTER*(*) FNAME    !file name
       INTEGER FUNIT          !unit number
C Local variables:
       LOGICAL OPN            !is the file open
CCCCCCCCCCCCCCCCCCCCCCCCCCCCCCCCCCCCCCCCCCCCCCCCCCCCCCCCCCCCCCCCCCCCCC
       INQUIRE(FILE=FNAME,OPENED=OPN)
       IF (OPN) CLOSE(UNIT=FUNIT)
       RETURN
       END
CCCCCCCCCCCCCCCCCCCCCCCCCCCCCCCCCCCCCCCCCCCCCCCCCCCCCCCCCCCCCCCCCCCCCC
       SUBROUTINE FINISH
C closes files and stops execution
CCCCCCCCCCCCCCCCCCCCCCCCCCCCCCCCCCCCCCCCCCCCCCCCCCCCCCCCCCCCCCCCCCCCCC
C Global variables:
       INCLUDE 'IO.ALL'
CCCCCCCCCCCCCCCCCCCCCCCCCCCCCCCCCCCCCCCCCCCCCCCCCCCCCCCCCCCCCCCCCCCCCC
       CALL FLCLOS(TNAME,TUNIT)
       CALL FLCLOS(GNAME,GUNIT)
       STOP
       END
CCCCCCCCCCCCCCCCCCCCCCCCCCCCCCCCCCCCCCCCCCCCCCCCCCCCCCCCCCCCCCCCCCCCCC
       SUBROUTINE FLTDEF(XPRMPT,X)
C prints prompt for floating number
C and displays default X in a format dictated by size of X
CCCCCCCCCCCCCCCCCCCCCCCCCCCCCCCCCCCCCCCCCCCCCCCCCCCCCCCCCCCCCCCCCCCCCC
C Global variables:
       INCLUDE 'IO.ALL'
C Input variables:
       CHARACTER*(*) XPRMPT                 !prompt string
       REAL X                               !default value
C Function:
       INTEGER LENTRU                       !true length of string
CCCCCCCCCCCCCCCCCCCCCCCCCCCCCCCCCCCCCCCCCCCCCCCCCCCCCCCCCCCCCCCCCCCCCC
C      positive numbers (leave no room for a sign)
       IF (X .GT. 0) THEN
          IF ((ABS(X) .LT. 999.49) .AND. (ABS(X) .GE. 99.949)) THEN
             WRITE (OUNIT,5) XPRMPT(1:LENTRU(XPRMPT)),X
          ELSE IF ((ABS(X) .LT. 99.949) .AND. (ABS(X) .GE. 9.9949)) THEN
             WRITE (OUNIT,10) XPRMPT(1:LENTRU(XPRMPT)),X
          ELSE IF ((ABS(X) .LT. 9.9949) .AND. (ABS(X) .GE. .99949)) THEN
```

```
      WRITE (OUNIT,15) XPRMPT(1:LENTRU(XPRMPT)),X
   ELSE IF ((ABS(X) .LT. .99949) .AND. (ABS(X) .GE. .099949)) THEN
      WRITE (OUNIT,20) XPRMPT(1:LENTRU(XPRMPT)),X
   ELSE
      WRITE (OUNIT,25) XPRMPT(1:LENTRU(XPRMPT)),X
   END IF
C
C  negative numbers (leave room for the sign)
   ELSE
      IF ((ABS(X) .LT. 999.49) .AND. (ABS(X) .GE. 99.949)) THEN
      WRITE (OUNIT,105) XPRMPT(1:LENTRU(XPRMPT)),X
   ELSE IF ((ABS(X) .LT. 99.949) .AND. (ABS(X) .GE. 9.9949)) THEN
      WRITE (OUNIT,110) XPRMPT(1:LENTRU(XPRMPT)),X
   ELSE IF ((ABS(X) .LT. 9.9949) .AND. (ABS(X) .GE. .99949)) THEN
      WRITE (OUNIT,115) XPRMPT(1:LENTRU(XPRMPT)),X
   ELSE IF ((ABS(X) .LT. .99949) .AND. (ABS(X) .GE. .099949)) THEN
      WRITE (OUNIT,120) XPRMPT(1:LENTRU(XPRMPT)),X
   ELSE
      WRITE (OUNIT,125) XPRMPT(1:LENTRU(XPRMPT)),X
   END IF
   END IF
C
5     FORMAT (1X,A,1X,'[',F4.0,']')
10    FORMAT (1X,A,1X,'[',F4.1,']')
15    FORMAT (1X,A,1X,'[',F4.2,']')
20    FORMAT (1X,A,1X,'[',F4.3,']')
25    FORMAT (1X,A,1X,'[',1PE8.2,']')
105   FORMAT (1X,A,1X,'[',F5.0,']')
110   FORMAT (1X,A,1X,'[',F5.1,']')
115   FORMAT (1X,A,1X,'[',F5.2,']')
120   FORMAT (1X,A,1X,'[',F5.3,']')
125   FORMAT (1X,A,1X,'[',1PE9.2,']')
C
      RETURN
      END
CCCCCCCCCCCCCCCCCCCCCCCCCCCCCCCCCCCCCCCCCCCCCCCCCCCCCCCCCCCCCCCCCCCCCCC
      SUBROUTINE INTDEF(KPRMPT,K)
C prints prompt for integer input from screen
C and default value in appropriate format
CCCCCCCCCCCCCCCCCCCCCCCCCCCCCCCCCCCCCCCCCCCCCCCCCCCCCCCCCCCCCCCCCCCCCCC
C Global variables:
      INCLUDE 'IO.ALL'
C Input variables:
      CHARACTER *(*) KPRMPT       !prompt string
      INTEGER K                   !default values
C Function:
      INTEGER LENTRU              !true length of string
CCCCCCCCCCCCCCCCCCCCCCCCCCCCCCCCCCCCCCCCCCCCCCCCCCCCCCCCCCCCCCCCCCCCCCC
C  positive numbers (leave no room for a sign)
      IF (K .GE. 0 ) THEN
         IF ((IABS(K) .LE. 9999) .AND. (IABS(K) .GE. 1000)) THEN
            WRITE (OUNIT,10) KPRMPT(1:LENTRU(KPRMPT)),K
```

```
            ELSE IF ((IABS(K) .LE. 999) .AND. (IABS(K) .GE. 100)) THEN
          -    WRITE (OUNIT,20) KPRMPT(1:LENTRU(KPRMPT)),K
            ELSE IF ((IABS(K) .LE. 99) .AND. (IABS(K) .GE. 10)) THEN
                WRITE (OUNIT,30) KPRMPT(1:LENTRU(KPRMPT)),K
            ELSE IF ((IABS(K) .LE. 9) .AND. (IABS(K) .GE. 0)) THEN
                WRITE (OUNIT,40) KPRMPT(1:LENTRU(KPRMPT)),K
            ELSE
                WRITE (OUNIT,50) KPRMPT(1:LENTRU(KPRMPT)),K
            END IF
C
C     negative numbers (leave room for the sign)
      ELSE
          IF ((IABS(K) .LE. 9999) .AND. (IABS(K) .GE. 1000)) THEN
              WRITE (OUNIT,110) KPRMPT(1:LENTRU(KPRMPT)),K
          ELSE IF ((IABS(K) .LE. 999) .AND. (IABS(K) .GE. 100)) THEN
              WRITE (OUNIT,120) KPRMPT(1:LENTRU(KPRMPT)),K
          ELSE IF ((IABS(K) .LE. 99) .AND. (IABS(K) .GE. 10)) THEN
              WRITE (OUNIT,130) KPRMPT(1:LENTRU(KPRMPT)),K
          ELSE IF ((IABS(K) .LE. 9) .AND. (IABS(K) .GE. 1)) THEN
              WRITE (OUNIT,140) KPRMPT(1:LENTRU(KPRMPT)),K
          ELSE
              WRITE (OUNIT,150) KPRMPT(1:LENTRU(KPRMPT)),K
          END IF
      END IF
C
10    FORMAT (1X,A,1X,'[',I4,']')
20    FORMAT (1X,A,1X,'[',I3,']')
30    FORMAT (1X,A,1X,'[',I2,']')
40    FORMAT (1X,A,1X,'[',I1,']')
50    FORMAT (1X,A,1X,'[',I10,']')
110   FORMAT (1X,A,1X,'[',I5,']')
120   FORMAT (1X,A,1X,'[',I4,']')
130   FORMAT (1X,A,1X,'[',I3,']')
140   FORMAT (1X,A,1X,'[',I2,']')
150   FORMAT (1X,A,1X,'[',I10,']')
C
      RETURN
      END
CCCCCCCCCCCCCCCCCCCCCCCCCCCCCCCCCCCCCCCCCCCCCCCCCCCCCCCCCCCCCCCCCCCCC
      SUBROUTINE CONVRT(X,STRING,LEN)
C converts a real number x to a character variable string of length LEN
C for printing; the format is chosen according to the value of X,
C taking roundoff into account
CCCCCCCCCCCCCCCCCCCCCCCCCCCCCCCCCCCCCCCCCCCCCCCCCCCCCCCCCCCCCCCCCCCCC
C Passed variables:
      CHARACTER*9 STRING               !routine output
      REAL X                           !routine input
      INTEGER LEN                      !string length
C Function
      INTEGER LENTRU                   !gets string length
CCCCCCCCCCCCCCCCCCCCCCCCCCCCCCCCCCCCCCCCCCCCCCCCCCCCCCCCCCCCCCCCCCCCC
C     positive numbers (leave no room for a sign)
```

```
      IF (X .GT. 0) THEN
        IF ((ABS(X) .LT. 999.4) .AND. (ABS(X) .GE. 99.94)) THEN
          WRITE(STRING,5) X
        ELSE IF ((ABS(X) .LT. 99.94) .AND. (ABS(X) .GE. 9.994)) THEN
          WRITE(STRING,10) X
        ELSE IF ((ABS(X) .LT. 9.994) .AND. (ABS(X) .GE. .9994)) THEN
          WRITE (STRING,15) X
        ELSE IF ((ABS(X) .LT. .9994) .AND. (ABS(X) .GE. .09994)) THEN
          WRITE (STRING,20) X
        ELSE
          WRITE (STRING,25) X
        END IF
C
C     negative numbers (leave room for the sign)
      ELSE
        IF ((ABS(X) .LT. 999.4) .AND. (ABS(X) .GE. 99.94)) THEN
          WRITE(STRING,105) X
        ELSE IF ((ABS(X) .LT. 99.94) .AND. (ABS(X) .GE. 9.994)) THEN
          WRITE(STRING,110) X
        ELSE IF ((ABS(X) .LT. 9.994) .AND. (ABS(X) .GE. .9994)) THEN
          WRITE (STRING,115) X
        ELSE IF ((ABS(X) .LT. .9994) .AND. (ABS(X) .GE. .09994)) THEN
          WRITE (STRING,120) X
        ELSE
          WRITE (STRING,125) X
        END IF
      END IF
C
      LEN=LENTRU(STRING)
C
5     FORMAT (F4.0)
10    FORMAT (F4.1)
15    FORMAT (F4.2)
20    FORMAT (F4.3)
25    FORMAT (1PE8.2)
105   FORMAT (F5.0)
110   FORMAT (F5.1)
115   FORMAT (F5.2)
120   FORMAT (F5.3)
125   FORMAT (1PE9.2)
C
      RETURN
      END
cccccccccccccccccccccccccccccccccccccccccccccccccccccccccccccccccccccccc
      SUBROUTINE ICNVRT(I,STRING,LEN)
C converts an integer I to a character variable STRING for
C printing; the format is chosen according to the value of I
cccccccccccccccccccccccccccccccccccccccccccccccccccccccccccccccccccccccc
C Passed variables:
      CHARACTER*9 STRING              !routine output
      INTEGER I                      !routine input
      INTEGER LEN                    !length of string
```

```
cccccccccccccccccccccccccccccccccccccccccccccccccccccccccccccccccc
C    positive numbers (leave no room for a sign)
     IF (I .GE. 0) THEN
      IF ((ABS(I) .LE. 9) .AND. (ABS(I) .GE. 0)) THEN
         WRITE(STRING,5) I
         LEN=1
      ELSE IF ((ABS(I) .LE. 99) .AND. (ABS(I) .GE. 10)) THEN
         WRITE(STRING,10)  I
         LEN=2
      ELSE IF ((ABS(I) .LE. 999) .AND. (ABS(I) .GE. 100)) THEN
         WRITE (STRING,15) I
         LEN=3
      ELSE IF ((ABS(I) .LE. 9999) .AND. (ABS(I) .GE. 1000)) THEN
         WRITE (STRING,20) I
         LEN=4
      ELSE
         WRITE (STRING,25) REAL(I)
         LEN=8
      END IF
C
C    negative numbers (leave room for the sign)
     ELSE
      IF ((ABS(I) .LE. 9) .AND. (ABS(I) .GE. 1)) THEN
         WRITE(STRING,105) I
         LEN=2
      ELSE IF ((ABS(I) .LE. 99) .AND. (ABS(I) .GE. 10)) THEN
         WRITE(STRING,110)  I
         LEN=3
      ELSE IF ((ABS(I) .LE. 999) .AND. (ABS(I) .GE. 100)) THEN
         WRITE (STRING,115) I
         LEN=4
      ELSE IF ((ABS(I) .LE. 9999) .AND. (ABS(I) .GE. 1000)) THEN
         WRITE (STRING,120) I
         LEN=5
      ELSE
         WRITE (STRING,125) REAL(I)
         LEN=9
      END IF
     END IF
C
5     FORMAT (I1)
10    FORMAT (I2)
15    FORMAT (I3)
20    FORMAT (I4)
25    FORMAT (1PE8.2)
105   FORMAT (I2)
110   FORMAT (I3)
115   FORMAT (I4)
120   FORMAT (I5)
125   FORMAT (1PE9.2)
C
      RETURN
```

```
      END
CCCCCCCCCCCCCCCCCCCCCCCCCCCCCCCCCCCCCCCCCCCCCCCCCCCCCCCCCCCCCCCCCCCCCC
      INTEGER FUNCTION PARSE(STRING,CHOICE)
C determines branching in menu list
C
C breaks STRING (of the form 'nn nn nn nn nn nn ....') into pieces, and
C returns the integer value represented by the CHOICE group of digits
CCCCCCCCCCCCCCCCCCCCCCCCCCCCCCCCCCCCCCCCCCCCCCCCCCCCCCCCCCCCCCCCCCCCCC
C Input variables:
      CHARACTER*(*) STRING       !string to look at
      INTEGER CHOICE             !specific number to look at
C Local variables:
      INTEGER IPOS               !current character position in string
      INTEGER IGROUP             !current group of digits in string
CCCCCCCCCCCCCCCCCCCCCCCCCCCCCCCCCCCCCCCCCCCCCCCCCCCCCCCCCCCCCCCCCCCCCC
      IPOS=1
      DO 20 IGROUP = 1,CHOICE-1
40       IF ( STRING(IPOS:IPOS) .NE. ' ' ) THEN
            IPOS = IPOS+1
            GOTO 40
         END IF
         IPOS=IPOS+1
20    CONTINUE
      READ( STRING(IPOS:IPOS+2),10) PARSE
10    FORMAT( I2 )
      RETURN
      END
CCCCCCCCCCCCCCCCCCCCCCCCCCCCCCCCCCCCCCCCCCCCCCCCCCCCCCCCCCCCCCCCCCCCCC
      INTEGER FUNCTION LENTRU(CHARAC)
C finds the true length of a character string by searching
C backward for first nonblank character
CCCCCCCCCCCCCCCCCCCCCCCCCCCCCCCCCCCCCCCCCCCCCCCCCCCCCCCCCCCCCCCCCCCCCC
C Input variables:
      CHARACTER *(*) CHARAC      !string whose length we are finding
C Local variables:
      INTEGER ISPACE             !ascii value of a blank
      INTEGER I                  !index of elements in CHARAC
CCCCCCCCCCCCCCCCCCCCCCCCCCCCCCCCCCCCCCCCCCCCCCCCCCCCCCCCCCCCCCCCCCCCCC
      ISPACE=ICHAR(' ')
      I=LEN(CHARAC)
10    IF (ICHAR(CHARAC(I:I)) .EQ. ISPACE) THEN
         I=I-1
      ELSE
         LENTRU=I
         RETURN
      END IF
      IF (I .GT. 0) GOTO 10
      LENTRU=0
      RETURN
      END
CCCCCCCCCCCCCCCCCCCCCCCCCCCCCCCCCCCCCCCCCCCCCCCCCCCCCCCCCCCCCCCCCCCCCC
      REAL FUNCTION GETFLT(X,XMIN,XMAX,XPRMPT)
```

```
C get a floating point number GETFLT; make sure it is between XMIN
C and XMAX and prompt with XPRMPT
C
C If your compiler accepts (FMT=*) to an internal unit, comment out
C lines 3 and 5, and uncomment lines 2 and 4
CCCCCCCCCCCCCCCCCCCCCCCCCCCCCCCCCCCCCCCCCCCCCCCCCCCCCCCCCCCCCCCCCCCCCC
C Global variables:
         INCLUDE 'IO.ALL'
C Input variables:
         CHARACTER*(*) XPRMPT        !prompt
         REAL X                      !default value
         REAL XMIN,XMAX              !limits on input
C Local variables:
         CHARACTER*40 STRING         !internal unit
C Function
         INTEGER LENTRU              !returns true length of string
CCCCCCCCCCCCCCCCCCCCCCCCCCCCCCCCCCCCCCCCCCCCCCCCCCCCCCCCCCCCCCCCCCCCCC
C      prompt for float, display default value
10     CALL FLTDEF(XPRMPT,X)
       READ (IUNIT,35,ERR=10) STRING
C
C      accept default value X if STRING is empty
       IF (LENTRU(STRING) .EQ. 0) THEN
           GETFLT=X
       ELSE
C2           READ (UNIT=STRING,FMT=*,ERR=10) GETFLT
3            READ (UNIT=STRING,FMT=1,ERR=10) GETFLT
1            FORMAT (E9.2)
       END IF
C
C      make sure GETFLT is between XMIN and XMAX
40     IF ((GETFLT .LT. XMIN) .OR. (GETFLT .GT. XMAX)) THEN
50         WRITE (OUNIT,60) XMIN,XMAX
           READ (IUNIT,35,ERR=50) STRING
           IF (LENTRU(STRING) .EQ. 0) THEN
               GETFLT=X
           ELSE
C4               READ (UNIT=STRING,FMT=*,ERR=50) GETFLT
5                READ (UNIT=STRING,FMT=1,ERR=50) GETFLT
           END IF
       GOTO 40
       END IF
C
35     FORMAT (A40)
60     FORMAT (' Try again: input outside of range = [',1PE11.3,
      +        1PE11.3,']')
100    FORMAT (1PE9.2)
       RETURN
       END
CCCCCCCCCCCCCCCCCCCCCCCCCCCCCCCCCCCCCCCCCCCCCCCCCCCCCCCCCCCCCCCCCCCCCC
         INTEGER FUNCTION GETINT(K,KMIN,KMAX,KPRMPT)
C get an integer value GETINT;
```

```
C check that it lies between KMIN and KMAX and prompt with KPRMPT
C
C This function allows input of integers in a natural way (i.e., without
C preceding blanks or decimal points) even though we cannot use list
C directed READ (i.e., FMT=*) from internal units
CCCCCCCCCCCCCCCCCCCCCCCCCCCCCCCCCCCCCCCCCCCCCCCCCCCCCCCCCCCCCCCCCCCCCCCC
C Global variables:
      INCLUDE 'IO.ALL'
C Input variables:
      CHARACTER*(*) KPRMPT          !string prompt
      INTEGER K                     !default value
      INTEGER KMIN,KMAX             !upper and lower limits
C Local variables:
      CHARACTER*40 STRING           !internal unit
      REAL TEMP                     !temp var to allow for easier input
C Functions:
      INTEGER LENTRU                !returns true length of string
CCCCCCCCCCCCCCCCCCCCCCCCCCCCCCCCCCCCCCCCCCCCCCCCCCCCCCCCCCCCCCCCCCCCCCCC
C     prompt for input; display default
10    CALL INTDEF(KPRMPT,K)
      READ (IUNIT,35,ERR=10) STRING
C
C     accept default value K if STRING is empty
      IF (LENTRU(STRING) .EQ. 0) THEN
         GETINT=K
      ELSE
C        change the integer into a real number
         STRING=STRING(1:LENTRU(STRING))//'.'
C        read the real number from string
         READ (UNIT=STRING,FMT=1,ERR=10) TEMP
1        FORMAT(F7.0)
C        change it to an integer
         GETINT=INT(TEMP)
      END IF
C
C     check that GETINT lies between KMIN and KMAX
40    IF ((GETINT .LT. KMIN) .OR. (GETINT .GT. KMAX)) THEN
50       WRITE (OUNIT,60) KMIN,KMAX
         READ (IUNIT,35,ERR=50) STRING
         IF (LENTRU(STRING) .EQ. 0) THEN
            GETINT=K
         ELSE
            STRING=STRING(1:LENTRU(STRING))//'.'
            READ (UNIT=STRING,FMT=1,ERR=50) TEMP
            GETINT=INT(TEMP)
         END IF
      GOTO 40
      END IF
C
35    FORMAT (A40)
60    FORMAT (' Try again: input is outside of range = [',I6,I6,']')
100   FORMAT (I10)
```

```
C
      RETURN
      END
ccccccccccccccccccccccccccccccccccccccccccccccccccccccccccccccccccccc
      CHARACTER*40 FUNCTION CHARAC(C,CLNGTH,CPRMPT)
C gets character string CHARAC no longer than CLNGTH from the screen
ccccccccccccccccccccccccccccccccccccccccccccccccccccccccccccccccccccc
C Global variables:
      INCLUDE 'IO.ALL'
C Input variables:
      CHARACTER*(*) C                     !default value
      CHARACTER*(*) CPRMPT                 !prompt
      INTEGER CLNGTH                       !max length
C Local variables:
      CHARACTER*40 STRING                  !internal unit
C Functions:
      INTEGER LENTRU                       !returns true length of string
ccccccccccccccccccccccccccccccccccccccccccccccccccccccccccccccccccccc
C     data can't be longer than 40 characters due to fixed format
      IF (CLNGTH .GT. 40) CLNGTH=40
C
C     prompt for string; display default value C
10    WRITE (OUNIT,20) CPRMPT(1:LENTRU(CPRMPT)),C(1:LENTRU(C))
      READ (IUNIT,35,ERR=10) STRING
C
C     accept default value C if STRING is empty
      IF (LENTRU(STRING) .EQ. 0) THEN
         CHARAC=C
      ELSE
         READ (STRING,35,ERR=10) CHARAC
      END IF
C
C     find the true length of the input; verify that it is not too long
40    IF (LENTRU(CHARAC) .GT. CLNGTH) THEN
50       WRITE (OUNIT,60) CLNGTH
         READ (IUNIT,35,ERR=50) STRING
         IF (LENTRU(STRING) .EQ. 0) THEN
            CHARAC=C
         ELSE
            READ (STRING,35,ERR=50) CHARAC
         END IF
      GOTO 40
      END IF
C
20    FORMAT (1X,A,1X,'[',A,']')
35    FORMAT (A40)
60    FORMAT (' Try again: string is too long, maximum length = ',I2)
C
      RETURN
      END
ccccccccccccccccccccccccccccccccccccccccccccccccccccccccccccccccccccc
      INTEGER FUNCTION YESNO(BINARY,PROMPT)
```

```
C obtains YESNO from the screen; value is 0 for no, 1 for yes
CCCCCCCCCCCCCCCCCCCCCCCCCCCCCCCCCCCCCCCCCCCCCCCCCCCCCCCCCCCCCCCCCCCCC
C Global variables:
      INCLUDE 'IO.ALL'
C Input parameters:
      CHARACTER*(*) PROMPT        !prompt
      INTEGER BINARY              !default value
C Local variables:
      CHARACTER*3 STRING          !internal unit
C Functions:
      INTEGER LENTRU              !returns true length of string
CCCCCCCCCCCCCCCCCCCCCCCCCCCCCCCCCCCCCCCCCCCCCCCCCCCCCCCCCCCCCCCCCCCCC
1000  CONTINUE
C     write prompt and display default values
      IF (BINARY .EQ. 1) WRITE(OUNIT,10) PROMPT(1:LENTRU(PROMPT))
      IF (BINARY .EQ. 0) WRITE(OUNIT,11) PROMPT(1:LENTRU(PROMPT))
C
      READ (IUNIT,20,ERR=1000) STRING
C
C     accept default value; check that input is 'y' or 'n'
      IF (LENTRU(STRING) .EQ. 0) THEN
          YESNO = BINARY
      ELSE IF (STRING(1:1) .EQ. 'y' .OR. STRING(1:1) .EQ. 'Y') THEN
          YESNO = 1
      ELSE IF (STRING(1:1) .EQ. 'n' .OR. STRING(1:1) .EQ. 'N') THEN
          YESNO = 0
      ELSE
          WRITE (OUNIT,200)
          GOTO 1000
      END IF
C
10    FORMAT(1X,A,1X,'[yes]')
11    FORMAT(1X,A,1X,'[no]')
20    FORMAT(A)
200   FORMAT (' Try again, answer must be yes or no')
C
      RETURN
      END
CCCCCCCCCCCCCCCCCCCCCCCCCCCCCCCCCCCCCCCCCCCCCCCCCCCCCCCCCCCCCCCCCCCCC
      LOGICAL FUNCTION LOGCVT(IJK)
C converts 1 to true and anything else to false
CCCCCCCCCCCCCCCCCCCCCCCCCCCCCCCCCCCCCCCCCCCCCCCCCCCCCCCCCCCCCCCCCCCCC
      INTEGER IJK                 !input
      IF (IJK .EQ. 1) THEN
         LOGCVT=.TRUE.
      ELSE
         LOGCVT=.FALSE.
      END IF
      RETURN
      END
CCCCCCCCCCCCCCCCCCCCCCCCCCCCCCCCCCCCCCCCCCCCCCCCCCCCCCCCCCCCCCCCCCCCC
      REAL FUNCTION RANNOS(DSEED)
```

```
C returns a uniformly distributed random number between 0 and 1
CCCCCCCCCCCCCCCCCCCCCCCCCCCCCCCCCCCCCCCCCCCCCCCCCCCCCCCCCCCCCCCCCCCC
      DOUBLE PRECISION      DSEED
      DOUBLE PRECISION         D2P31M,D2P31
      DATA            D2P31M/2147483647.D0/
      DATA            D2P31 /2147483711.D0/
CCCCCCCCCCCCCCCCCCCCCCCCCCCCCCCCCCCCCCCCCCCCCCCCCCCCCCCCCCCCCCCCCCCC
      DSEED = MOD(16807.D0*DSEED,D2P31M)
      RANNOS = DSEED / D2P31
      RETURN
      END
CCCCCCCCCCCCCCCCCCCCCCCCCCCCCCCCCCCCCCCCCCCCCCCCCCCCCCCCCCCCCCCCCCCCCC
CCCCCCCCCCCCCCCCCCCCCCCCCCCCCCCCCCCCCCCCCCCCCCCCCCCCCCCCCCCCCCCCCCCCCC
C  menu.all
C
      INTEGER FLOAT,NUM,BOOLEN,YESKIP,NOSKIP,SKIP,QUIT
      INTEGER TITLE,WAIT,CHSTR,MTITLE,MCHOIC,PPRINT,CLRTRM
C
      INTEGER IMAIN,STOP,ITTERM,ITFILE,ITNAME,IGTERM,IGHRD
      INTEGER IGFILE,IGNAME,ISTOP
C
C     data arrays for menu
      CHARACTER*60 MPRMPT              !prompt string
      CHARACTER*60 MTAG                !terse description
      INTEGER MTYPE                    !data type
      INTEGER MINTS                    !default value for integer
      REAL MREALS                     !default value for real
      REAL MLOLIM                      !lower limit on input
      REAL MHILIM                      !high limit on input
      CHARACTER*40 MSTRNG              !default value for string
C
C     menu data types
      PARAMETER (FLOAT  = 0)          !floating point number
      PARAMETER (NUM    = 1)          !integer
      PARAMETER (BOOLEN = 2)          !yes or no user input
      PARAMETER (YESKIP = 3)          !yes or no, skip on YES
      PARAMETER (NOSKIP = 4)          !Yes or no, skip on NO
      PARAMETER (SKIP   = 5)          !Unconditional GOTO
      PARAMETER (QUIT   = 6)          !abort current ASK call
      PARAMETER (TITLE  = 7)          !print prompt (in ASK or PRTAGS)
      PARAMETER (WAIT   = 8)          !print prompt and invoke PAUSE
      PARAMETER (CHSTR  = 9)          !character string
      PARAMETER (MTITLE = 10)         !print MPRMPT() during ASK only
      PARAMETER (MCHOIC = 11)         !print prompt, get choice, branch
      PARAMETER (PPRINT = 12)         !print out parameters
      PARAMETER (CLRTRM = 13)         !clear screen
C
C     menu entries which are the same for all programs
      PARAMETER (IMAIN  = 10)         !main menu choice
      PARAMETER (STOP   = -7)         !flag to stop
      PARAMETER (ITTERM = 73 )        !text to terminal
      PARAMETER (ITFILE = 74 )        !text to file
```

```
       PARAMETER (ITNAME = 75 )        !text file name
       PARAMETER (IGTERM = 83 )        !graphics to terminal
       PARAMETER (IGHRD  = 84 )        !graphics hardcopy
       PARAMETER (IGFILE = 85 )        !graphics to file
       PARAMETER (IGNAME = 86 )        !graphics file name
       PARAMETER (ISTOP  = 98 )        !last entry
C
       COMMON/MVARS/MTYPE(100),MINTS(100),
     1           MREALS(100),MLOLIM(100),MHILIM(100)
       COMMON/CMVARS/MPRMPT(100),MTAG(100),MSTRNG(10)
```

D.4 Graphics Codes

The two FORTRAN files, GRAPHIT.HI and GRAPHIT.LO, are collections of subroutines that produce two-dimensional or contour plots. They allow one to four plots per graphics page. The two-dimensional axes can be either linear-linear, log-linear, or log-log, while both the axes for contour plots must be linear. To produce a plot, all of the variables listed in GRFDAT.ALL must be defined, as well as the data arrays (X and Y for two-dimensional plots; and Z, NCONT, NX, NY, ZMIN, and ZMAX for contour plots) and the device type (DEVICE). To create a graph the program must first call GTDEV to nominate a device and GMODE (in file SETUP.FOR) to change to graphics mode (if necessary). For each plot on the page, the program calls LNLNAX, LGLNAX, or LGLGAX to draw the appropriate axes; and for each data set to be plotted on the same set of axes, the program calls XYPLOT (two-dimensional plots) or CONTOR (contour plots). When the graphics page is finished, the program calls GPAGE to close the graphics package and TMODE (in file SETUP.FOR) to switch back to text mode (if necessary).

GRAPHIT.HI calls CA-DISSPLA (version 11.0) routines; GRAPHIT.LO calls UIS routines. For users who do not have access to either CA-DISSPLA or UIS graphics, these programs are intended as templates to facilitate the development of code for their graphics package. For users who do have one of these packages, please read the information below that is specific to your software.

CA-DISSPLA GRAPHIT.HI includes the following subroutines, which are not called directly by the physics programs: SUBPLT, which sets the area of the device to be used for plotting, BANNER, which prints the title, LABELS, which writes out axis labels, and LEGEND, which writes out an informational message. Device nomination is often *installation* dependent, so you might need to edit subroutine GTDEV before the code will work. Note that XOVAL and YOVAL are not used by CA-DISSPLA.

UIS Graphics GRAPHIT.LO calls to the UIS graphics routines, which are available on VAXstations. All non-standard FORTRAN-77 syntax is associated solely with UIS calls. The subroutines not explicitly called by the physics programs are SUBPLT, which sets the area for plotting, fixes character size, creates virtual displays along with windows and transformations, writes the title, and enables the display list (for hard copy); XAXIS and YAXIS, which draw and label the axes (either linear or log); LEGEND, which prints an informational message at the bottom of the page; GSYM, which draws symbols on the plot; and CONT2, which does the calculation and drawing of contours. Throughout the code, three different attribute blocks are used: block 1 writes text in a size appropriate to the plot size; block 3 writes the same size as block 1, but rotated 90 degrees; and block 2 contains the correct line type (e.g., solid, dashed, etc.)

Hardcopy is obtained by enabling the display list (subroutine SUBPLT), then writing out the display list using an HCUIS routine (subroutine GPAGE). Because the HCUIS software is optional it may not be currently available on your machine, but is easily installed. The device independent output file created by HCUIS is then translated by the DCL command RENDER into a file with the format necessary for a specific hardcopy device.

For more help on hardcopy, see *Micro VMS Workstation Guide to Printing Graphics* (order number AA-HQ85B-TN). For general help on UIS routines, see *Micro VMS Workstation Graphics Programming Guide* (order number AI-GI10B-TN), available from Digital Equipment Corporation.

```
ccccccccccccccccccccccccccccccccccccccccccccccccccccccccccccccccccccc
C file GRAPHIT.HI
ccccccccccccccccccccccccccccccccccccccccccccccccccccccccccccccccccccc
      SUBROUTINE LNLNAX
C draws linear-linear axes
ccccccccccccccccccccccccccccccccccccccccccccccccccccccccccccccccccccc
C Global variables
      INCLUDE 'GRFDAT.ALL'
      REAL XSIZE,YSIZE              !size of axes in inches
C Local variables
      REAL XSTP,YSTP               !length between ticks in user units
      REAL VSIZE,HSIZE             !XSIZE and YSIZE
      COMMON/SIZE/VSIZE,HSIZE
ccccccccccccccccccccccccccccccccccccccccccccccccccccccccccccccccccccc
C     write title
      IF (IPLOT .EQ. 1) CALL BANNER(TITLE)
C     set subplot area
      CALL SUBPLT(IPLOT,NPLOT,XSIZE,YSIZE)
C     label axes
      CALL LABELS(LABEL)
C     write informational message
      HSIZE=XSIZE
      VSIZE=YSIZE
      CALL LEGEND
C     draw linear-linear axes
      XSTP=(XMAX-XMIN)/NXTICK
      YSTP=(YMAX-YMIN)/NYTICK
      CALL GRAF(XMIN,XSTP,XMAX,YMIN,YSTP,YMAX)
C
```

```
      RETURN
      END
ccccccccccccccccccccccccccccccccccccccccccccccccccccccccccccccccccc
      SUBROUTINE LGLNAX
C draws log-linear axes
ccccccccccccccccccccccccccccccccccccccccccccccccccccccccccccccccccc
C Global variables
      INCLUDE 'GRFDAT.ALL'
      REAL XSIZE,YSIZE          !size of axes in inches
C Local variables
      REAL YCYCLE               !inches per cycle on y axis
      REAL XSTEP                !length in user units/length in inches
      REAL VSIZE,HSIZE           !XSIZE and YSIZE
      COMMON/SIZE/VSIZE,HSIZE
ccccccccccccccccccccccccccccccccccccccccccccccccccccccccccccccccccc
C     write title
      IF (IPLOT .EQ. 1) CALL BANNER(TITLE)
C     set subplot area
      CALL SUBPLT(IPLOT,NPLOT,XSIZE,YSIZE)
C     label axes
      CALL LABELS(LABEL)
C     write informational message
      HSIZE=XSIZE
      VSIZE=YSIZE
      CALL LEGEND
C     draw log-linear axes
      YCYCLE=YSIZE/(ALOG10(YMAX)-ALOG10(YMIN))
      XSTEP=(XMAX-XMIN)/XSIZE
      CALL YLOG(XMIN,XSTEP,YMIN,YCYCLE)
C
      RETURN
      END
ccccccccccccccccccccccccccccccccccccccccccccccccccccccccccccccccccc
      SUBROUTINE LGLGAX
C draws log-linear axes
ccccccccccccccccccccccccccccccccccccccccccccccccccccccccccccccccccc
C Global variables
      INCLUDE 'GRFDAT.ALL'
C Local variables
      REAL XSIZE,YSIZE          !size of axes in inches
      REAL XCYCLE,YCYCLE        !inches per cycle
      REAL VSIZE,HSIZE          !XSIZE and YSIZE
      COMMON/SIZE/VSIZE,HSIZE
ccccccccccccccccccccccccccccccccccccccccccccccccccccccccccccccccccc
C     write title
      IF (IPLOT .EQ. 1) CALL BANNER(TITLE)
C     set subplot area
      CALL SUBPLT(IPLOT,NPLOT,XSIZE,YSIZE)
C     label axes
      CALL LABELS(LABEL)
C     write informational message
      HSIZE=XSIZE
```

```
          VSIZE=YSIZE
          CALL LEGEND
C         draw log-log axes
          XCYCLE=XSIZE/(ALOG10(XMAX)-ALOG10(XMIN))
          YCYCLE=YSIZE/(ALOG10(YMAX)-ALOG10(YMIN))
          CALL LOGLOG(XMIN,XCYCLE,YMIN,YCYCLE)
C
          RETURN
          END
CCCCCCCCCCCCCCCCCCCCCCCCCCCCCCCCCCCCCCCCCCCCCCCCCCCCCCCCCCCCCCCCCCCCCC
          SUBROUTINE LEGEND
C write information at the top left of the plot
CCCCCCCCCCCCCCCCCCCCCCCCCCCCCCCCCCCCCCCCCCCCCCCCCCCCCCCCCCCCCCCCCCCCCC
C Global variables:
          INCLUDE 'GRFDAT.ALL'
C Passed variables:
          REAL XSIZE,YSIZE
C Local variables:
          INTEGER LEN,LENTRU            !length of char string
          REAL VSIZE,HSIZE              !XSIZE and YSIZE
          COMMON/SIZE/VSIZE,HSIZE
          DATA INFO/' '/
CCCCCCCCCCCCCCCCCCCCCCCCCCCCCCCCCCCCCCCCCCCCCCCCCCCCCCCCCCCCCCCCCCCCCC
C         prints message at top of plotting area
          LEN=LENTRU(INFO)
          IF (LEN .GT. 0) THEN
              CALL MESSAG(INFO(1:LEN),LEN,HSIZE*.05,VSIZE*1.03)
          END IF
          RETURN
          END
CCCCCCCCCCCCCCCCCCCCCCCCCCCCCCCCCCCCCCCCCCCCCCCCCCCCCCCCCCCCCCCCCCCCCC
          SUBROUTINE XYPLOT(X,Y)
C plots xy data
CCCCCCCCCCCCCCCCCCCCCCCCCCCCCCCCCCCCCCCCCCCCCCCCCCCCCCCCCCCCCCCCCCCCCC
C Global variables:
          INCLUDE 'GRFDAT.ALL'
          REAL X                        !independent variable data array
          REAL Y                        !dependent variable data array
          DIMENSION X(NPOINT),Y(NPOINT)
CCCCCCCCCCCCCCCCCCCCCCCCCCCCCCCCCCCCCCCCCCCCCCCCCCCCCCCCCCCCCCCCCCCCCC
C         set line type
          IF (ILINE .EQ. 2) CALL DOT
          IF (ILINE .EQ. 3) CALL DASH
          IF (ILINE .EQ. 4) CALL CHNDSH
          IF (ILINE .EQ. 5) IFREQ=-IFREQ    !no line at all
C
C         set symbol type
          IF (ISYM .EQ. 1) CALL MARKER(16)  !circle
          IF (ISYM .EQ. 2) CALL MARKER(2)   !triangle
          IF (ISYM .EQ. 3) CALL MARKER(0)   !square
          IF (ISYM .EQ. 4) CALL MARKER(4)   !cross
          IF (ISYM .EQ. 5) THEN             !point
```

```
                CALL SCLPIC(.3)
                CALL MARKER(15)
            END IF
C
C       plot
            CALL CURVE(X,Y,NPOINT,IFREQ)
C
C       reset line type and frequency
            IF (ILINE .EQ. 2) CALL RESET('DOT')
            IF (ILINE .EQ. 3) CALL RESET('DASH')
            IF (ILINE .EQ. 4) CALL RESET('CHNDSH')
            IF (ILINE .EQ. 5) IFREQ=-IFREQ
C
            RETURN
            END
CCCCCCCCCCCCCCCCCCCCCCCCCCCCCCCCCCCCCCCCCCCCCCCCCCCCCCCCCCCCCCCCCCCCCCCC
        SUBROUTINE CONTOR(Z,MX,MY,ZMIN,ZMAX,NCONT)
C drawn NCONT lines for data contained in Z
C lines alternate between solid and dashed;
C if possible, solid lines are labeled
CCCCCCCCCCCCCCCCCCCCCCCCCCCCCCCCCCCCCCCCCCCCCCCCCCCCCCCCCCCCCCCCCCCCCCCC
C Global variables:
            INCLUDE 'GRFDAT.ALL'
C Passed variables:
            INTEGER MX,MY               !dimensions of Z
            REAL Z(MX,MY)               !data
            REAL ZMIN,ZMAX              !limits on data
            INTEGER NCONT               !number of contours to draw
C Local variables:
            REAL ZINCR                  !increment between contours
            REAL WORK(4000)             !work space for DISSPLA
            COMMON WORK
CCCCCCCCCCCCCCCCCCCCCCCCCCCCCCCCCCCCCCCCCCCCCCCCCCCCCCCCCCCCCCCCCCCCCCCC
            CALL FRAME                  !draw a box around area
            CALL BCOMON(4000)           !pass information about workspace
            ZINCR=(ZMAX-ZMIN)/NCONT     !increments between contours
            CALL CONMAK(Z,MX,MY,ZINCR)  !make the contour lines
C
C       provide two line types
            CALL CONLIN(0,'SOLID','LABELS',1,10)
            CALL CONLIN(1,'DOT','NOLABELS',1,10)
            CALL CONTUR(2,'LABELS','DRAW')              !draw contour lines
C
            RETURN
            END
CCCCCCCCCCCCCCCCCCCCCCCCCCCCCCCCCCCCCCCCCCCCCCCCCCCCCCCCCCCCCCCCCCCCCCCC
        SUBROUTINE GTDEV(DEVICE)
C sets device for graphics output to DEVICE
CCCCCCCCCCCCCCCCCCCCCCCCCCCCCCCCCCCCCCCCCCCCCCCCCCCCCCCCCCCCCCCCCCCCCCCC
C Global variables:
            INCLUDE 'IO.ALL'
C Passed variables:
```

```
      INTEGER DEVICE               !device flag
      INTEGER SCREEN               !send to terminal
      INTEGER PAPER                !make a hardcopy
      INTEGER FILE                 !send to a file
      DATA SCREEN,PAPER,FILE/1,2,3/
cccccccccccccccccccccccccccccccccccccccccccccccccccccccccccccccccccccccc
C     reset output device
      CALL IOMGR(0,-102)
C     4014 tektronix screen at 9600 baud
      IF (DEVICE .EQ. SCREEN) CALL TEKALL(4014,960,0,0,0)
C     postscript Printer with default values
      IF (DEVICE .EQ. PAPER) CALL PSCRPT(0,0,0)
      RETURN
      END
cccccccccccccccccccccccccccccccccccccccccccccccccccccccccccccccccccccccc
      SUBROUTINE GPAGE(DEVICE)
C end graphics page
cccccccccccccccccccccccccccccccccccccccccccccccccccccccccccccccccccccccc
      INTEGER DEVICE               !which device is it?
cccccccccccccccccccccccccccccccccccccccccccccccccccccccccccccccccccccccc
      CALL ENDPL(0)
      CALL CLEAR
      CALL DONEPL
      RETURN
      END
cccccccccccccccccccccccccccccccccccccccccccccccccccccccccccccccccccccccc
C The following subroutines are not called directly by the user
cccccccccccccccccccccccccccccccccccccccccccccccccccccccccccccccccccccccc
      SUBROUTINE BANNER(TITLE)
C prints title to top of graphics page
cccccccccccccccccccccccccccccccccccccccccccccccccccccccccccccccccccccccc
C Global variables:
      CHARACTER*60 TITLE           !title to be printed
C Functions:
      INTEGER LENTRU               !returns string length
C Local variables:
      INTEGER LENTTL               !length of title
cccccccccccccccccccccccccccccccccccccccccccccccccccccccccccccccccccccccc
C     treat title as a graph by itself so that it is centered
      CALL AREA2D(7.5,9.)
      CALL HEIGHT(.14)
      LENTTL=LENTRU(TITLE)
      CALL HEADIN(TITLE(1:LENTTL),LENTTL,1.,1)
      CALL ENDGR(0)
      RETURN
      END
cccccccccccccccccccccccccccccccccccccccccccccccccccccccccccccccccccccccc
      SUBROUTINE LABELS(LABEL)
C labels both x and y axes
cccccccccccccccccccccccccccccccccccccccccccccccccccccccccccccccccccccccc
C Global variables:
      CHARACTER*60 LABEL(2)        !x and y labels
```

```
C Functions:
      INTEGER LENTRU                    !returns string length
C Local variables:
      INTEGER LENX,LENY                 !length of labels
CCCCCCCCCCCCCCCCCCCCCCCCCCCCCCCCCCCCCCCCCCCCCCCCCCCCCCCCCCCCCCCCCCCCCCCCC
      LENX=LENTRU(LABEL(1))
      LENY=LENTRU(LABEL(2))
      CALL XNAME(LABEL(1)(1:LENX),LENX)
      CALL YNAME(LABEL(2)(1:LENY),LENY)
      RETURN
      END
CCCCCCCCCCCCCCCCCCCCCCCCCCCCCCCCCCCCCCCCCCCCCCCCCCCCCCCCCCCCCCCCCCCCCCCCC
      SUBROUTINE SUBPLT(IPLOT,NPLOT,XSIZE,YSIZE)
C defines subplotting area
CCCCCCCCCCCCCCCCCCCCCCCCCCCCCCCCCCCCCCCCCCCCCCCCCCCCCCCCCCCCCCCCCCCCCCCCC
C The choice of subplot size and placement is dependent on the
C device and graphics package; those given here are for an 8.5 X 11
C inch page using DISSPLA
CCCCCCCCCCCCCCCCCCCCCCCCCCCCCCCCCCCCCCCCCCCCCCCCCCCCCCCCCCCCCCCCCCCCCCCCC
C Passed variables:
      INTEGER NPLOT                     !total number of plots on page
      INTEGER IPLOT                     !number of current plot
      REAL XSIZE,YSIZE                  !size of plotting area in inches
C Local variables:
      REAL XORIG,YORIG                  !location of lower left corner (inches)
CCCCCCCCCCCCCCCCCCCCCCCCCCCCCCCCCCCCCCCCCCCCCCCCCCCCCCCCCCCCCCCCCCCCCCCCC
C     finish last plot (sets DISSPLA back to level 1)
      IF (IPLOT .NE. 1) CALL ENDGR(0)
C
C     define subplotting area (in inches); whole page is 8.5 x 11
      IF (NPLOT .EQ. 1) THEN
         XSIZE=7.5
         YSIZE=9.
         CALL HEIGHT(.14)
      ELSE IF (NPLOT .EQ. 2) THEN
         YSIZE=4.25
         XSIZE=6.5
         XORIG=1.
         IF (IPLOT .EQ. 1) YORIG=5.75
         IF (IPLOT .EQ. 2) YORIG=.75
         CALL HEIGHT(.125)
      ELSE IF (NPLOT .EQ. 3) THEN
         YSIZE=2.80
         XSIZE=6.5
         XORIG=1.
         IF (IPLOT .EQ. 1) YORIG=7.20
         IF (IPLOT .EQ. 2) YORIG=3.9
         IF (IPLOT .EQ. 3) YORIG=.6
         CALL HEIGHT(.125)
      ELSE IF (NPLOT .EQ. 4)THEN
         XSIZE=3.25
         YSIZE=4.25
```

```
      CALL HEIGHT(.11)
      IF (IPLOT .EQ. 1) THEN
         XORIG=.5
         YORIG=5.75
      ELSE IF (IPLOT .EQ. 2) THEN
         XORIG=4.75
         YORIG=5.75
      ELSE IF (IPLOT .EQ. 3) THEN
         XORIG=.5
         YORIG=.5
      ELSE IF (IPLOT .EQ. 4) THEN
         XORIG=4.75
         YORIG=.5
      END IF
      END IF
C
C     use default origin if there is only one plot
      IF (NPLOT .NE. 1) CALL PHYSOR(XORIG,YORIG)
C
      CALL AREA2D(XSIZE,YSIZE)
C
      RETURN
      END
CCCCCCCCCCCCCCCCCCCCCCCCCCCCCCCCCCCCCCCCCCCCCCCCCCCCCCCCCCCCCCCCCCCCCC
CCCCCCCCCCCCCCCCCCCCCCCCCCCCCCCCCCCCCCCCCCCCCCCCCCCCCCCCCCCCCCCCCCCCCC
C file GRAPHIT.LO
CCCCCCCCCCCCCCCCCCCCCCCCCCCCCCCCCCCCCCCCCCCCCCCCCCCCCCCCCCCCCCCCCCCCCC
      SUBROUTINE LNLNAX
C draws linear xy axes
CCCCCCCCCCCCCCCCCCCCCCCCCCCCCCCCCCCCCCCCCCCCCCCCCCCCCCCCCCCCCCCCCCCCCC
C Global variables:
      INCLUDE 'GRFDAT.ALL'
      LOGICAL YLOG,XLOG              !flags to signal log plotting
      REAL X0,X1,X2                  !temp values for x0val,xmin,xmax
      REAL Y0,Y1,Y2                  !temp values for y0val,ymin,ymax
      COMMON/LOGLMT/X0,X1,X2,Y0,Y1,Y2
      COMMON/LOGFLG/YLOG,XLOG
CCCCCCCCCCCCCCCCCCCCCCCCCCCCCCCCCCCCCCCCCCCCCCCCCCCCCCCCCCCCCCCCCCCCCC
C These temporary values are essential for log plotting, but also must
C be set for linear plotting
      YLOG=.FALSE.
      XLOG=.FALSE.
      X1=XMIN
      X2=XMAX
      Y0=Y0VAL
      Y1=YMIN
      Y2=YMAX
      X0=X0VAL
C
      CALL SUBPLT                    !define subplot area
      CALL XAXIS                     !draw and label x axis
      CALL YAXIS                     !drawn and label y axis
```

```
            CALL LEGEND                        !write out information
      C
            RETURN
            END
      CCCCCCCCCCCCCCCCCCCCCCCCCCCCCCCCCCCCCCCCCCCCCCCCCCCCCCCCCCCCCCCCCCCCCCCC
            SUBROUTINE LGLNAX
      C draws log-linear xy axes
      CCCCCCCCCCCCCCCCCCCCCCCCCCCCCCCCCCCCCCCCCCCCCCCCCCCCCCCCCCCCCCCCCCCCCCCC
      C Global variables:
            INCLUDE 'GRFDAT.ALL'
            LOGICAL YLOG,XLOG                   !flags to signal log plotting
            REAL X0,X1,X2                       !temp values for x0val,xmin,xmax
            REAL Y0,Y1,Y2                       !temp values for y0val,ymin,ymax
            COMMON/LOGLMT/X0,X1,X2,Y0,Y1,Y2
            COMMON/LOGFLG/YLOG,XLOG
      CCCCCCCCCCCCCCCCCCCCCCCCCCCCCCCCCCCCCCCCCCCCCCCCCCCCCCCCCCCCCCCCCCCCCCCC
            YLOG=.TRUE.                         !convert to log plotting
            XLOG=.FALSE.
      C
      C     convert to log values if necessary, without changing global vars
            X1=XMIN
            X2=XMAX
            Y0=Y0VAL
            Y1=ALOG10(YMIN)
            Y2=ALOG10(YMAX)
            X0=ALOG10(X0VAL)
      C
            CALL SUBPLT                         !define subplot area
            CALL XAXIS                          !draw and label x axis
            CALL YAXIS                          !draw and label y axis
            CALL LEGEND                         !write out information
      C
            RETURN
            END
      CCCCCCCCCCCCCCCCCCCCCCCCCCCCCCCCCCCCCCCCCCCCCCCCCCCCCCCCCCCCCCCCCCCCCCCC
            SUBROUTINE LGLGAX
      C draws log-log xy axes
      CCCCCCCCCCCCCCCCCCCCCCCCCCCCCCCCCCCCCCCCCCCCCCCCCCCCCCCCCCCCCCCCCCCCCCCC
      C Global variables:
            INCLUDE 'GRFDAT.ALL'
            LOGICAL YLOG,XLOG                   !flags to signal log plotting
            REAL X0,X1,X2                       !temp values for x0val,xmin,xmax
            REAL Y0,Y1,Y2                       !temp values for y0val,ymin,ymax
            COMMON/LOGLMT/X0,X1,X2,Y0,Y1,Y2
            COMMON/LOGFLG/YLOG,XLOG
      CCCCCCCCCCCCCCCCCCCCCCCCCCCCCCCCCCCCCCCCCCCCCCCCCCCCCCCCCCCCCCCCCCCCCCCC
      C     convert to log plotting
            YLOG=.TRUE.
            XLOG=.TRUE.
      C
      C     convert to log values if necessary, without changing global vars
            X1=ALOG10(XMIN)
```

```
      X2=ALOG10(XMAX)
      Y0=ALOG10(YOVAL)
      Y1=ALOG10(YMIN)
      Y2=ALOG10(YMAX)
      X0=ALOG10(XOVAL)
C
      CALL SUBPLT                    !define subplot area
      CALL XAXIS                     !draw and label x axis
      CALL YAXIS                     !drawn and label y axis
      CALL LEGEND                    !write out information
C
      RETURN
      END
cccccccccccccccccccccccccccccccccccccccccccccccccccccccccccccccccccccc
      SUBROUTINE LEGEND
C write information at the bottom left of the plot
cccccccccccccccccccccccccccccccccccccccccccccccccccccccccccccccccccccc
C Global variables:
      INCLUDE 'UISGRF.ALL'
      INCLUDE 'GRFDAT.ALL'
      REAL X0,X1,X2                  !temp values for x0val,xmin,xmax
      REAL Y0,Y1,Y2                  !temp values for y0val,ymin,ymax
      COMMON/LOGLMT/X0,X1,X2,Y0,Y1,Y2
C Local variables:
      REAL XTXT,YTXT                 !location of text
      REAL SMALL                     !small increments along either axis
      INTEGER LEN                    !length of text
C Function:
      INTEGER LENTRU                 !returns length of text
      DATA SMALL/.01/
      DATA INFO/' '/
cccccccccccccccccccccccccccccccccccccccccccccccccccccccccccccccccccccc
      LEN=LENTRU(INFO)
      IF (LEN .GT. 0) THEN
         YTXT=Y1-5*SMALL*YLNGTH
         XTXT=X1+4*SMALL*XLNGTH
         CALL UIS$TEXT(TRID(IPLOT),1,INFO(1:LEN),XTXT,YTXT)
      END IF
      RETURN
      END
cccccccccccccccccccccccccccccccccccccccccccccccccccccccccccccccccccccc
      SUBROUTINE XYPLOT(X,Y)
C plots x,y data
cccccccccccccccccccccccccccccccccccccccccccccccccccccccccccccccccccccc
C Global variables:
      INCLUDE 'UISGRF.ALL'
      INCLUDE 'GRFDAT.ALL'
      LOGICAL YLOG,XLOG              !flags to signal log plotting
      COMMON/LOGFLG/YLOG,XLOG
C Passed variables:
      REAL X(NPOINT),Y(NPOINT)       !data arrays
C Local variables:
```

```
        INTEGER I                       !x and y indices
        REAL X1,X2,Y1,Y2                !points to be plotted
CCCCCCCCCCCCCCCCCCCCCCCCCCCCCCCCCCCCCCCCCCCCCCCCCCCCCCCCCCCCCCCCCCCCCCCCCC
C       set line type (type 5 is no line at all)
        IF (ILINE .EQ. 1) CALL
     +      UIS$SET_LINE_STYLE(TRID(IPLOT),0,2,'FFFFFFFF'X)
        IF (ILINE .EQ. 2) CALL
     +      UIS$SET_LINE_STYLE(TRID(IPLOT),0,2,'FF00FF00'X)
        IF (ILINE .EQ. 3) CALL
     +      UIS$SET_LINE_STYLE(TRID(IPLOT),0,2,'FFF0FFF0'X)
        IF (ILINE .EQ. 4) CALL
     +      UIS$SET_LINE_STYLE(TRID(IPLOT),0,2,'F000F000'X)
C
C       draw xy plot
        X1=X(1)
        Y1=Y(1)
C       convert data for log plots
        IF (XLOG) X1=ALOG10(X1)
        IF (YLOG) Y1=ALOG10(Y1)
        DO 100 I=1,NPOINT-1
           X2=X(I+1)
           Y2=Y(I+1)
           IF (XLOG) X2=ALOG10(X2)
           IF (YLOG) Y2=ALOG10(Y2)
           IF (IFREQ .NE. 0) THEN
C              plot symbols at ifreq frequency
               IF (MOD(I,IFREQ) .EQ. 0) CALL GSYM(X1,Y1)
           END IF
           IF (ILINE .NE. 5) CALL UIS$PLOT(TRID(IPLOT),2,X1,Y1,X2,Y2)
           X1=X2                !roll values
           Y1=Y2
100     CONTINUE
C
C       plot symbol at the last point
        IF (IFREQ .NE. 0) THEN
            IF (MOD(I,IFREQ) .EQ. 0) CALL GSYM(X1,Y1)
        END IF
C
        RETURN
        END
CCCCCCCCCCCCCCCCCCCCCCCCCCCCCCCCCCCCCCCCCCCCCCCCCCCCCCCCCCCCCCCCCCCCCCCCCC
        SUBROUTINE CONTOR(Z,MX,MY,ZMIN,ZMAX,NCONT)
C draws positive contours solid and negative contours dots
CCCCCCCCCCCCCCCCCCCCCCCCCCCCCCCCCCCCCCCCCCCCCCCCCCCCCCCCCCCCCCCCCCCCCCCCCC
C Global variables:
        INCLUDE 'IO.ALL'
        INCLUDE 'GRFDAT.ALL'
C Passed variables:
        INTEGER MX,MY           !dimensions of Z
        REAL Z(MX,MY)           !data
        REAL ZMIN,ZMAX          !limits on data
        INTEGER NCONT           !number of contour lines
```

```
C Local variables:
      REAL DELZ                !Z interval
      REAL ZERO                !ZMIN
      INTEGER NNEG             !number of negative contours
CCCCCCCCCCCCCCCCCCCCCCCCCCCCCCCCCCCCCCCCCCCCCCCCCCCCCCCCCCCCCCCCCCCCCCCCCC
C     all Z values are positive
      IF ((ZMAX .GT. 0.) .AND. (ZMIN .GE. 0)) THEN
         ILINE=1
         CALL CONT2(Z,MX,MY,ZMIN,ZMAX,NCONT)
C
C     Z values are both positive and negative
      ELSE IF ((ZMAX .GT. 0.) .AND. (ZMIN .LT. 0.)) THEN
         DELZ=(ZMAX-ZMIN)/(NCONT+1)
         NNEG=INT(-ZMIN/DELZ)
         ZERO=ZMIN+DELZ*(NNEG)         !min value for positive cont
         ILINE=1
         CALL CONT2(Z,MX,MY,ZERO,ZMAX,NCONT-NNEG)
         ILINE=4
         ZERO=ZMIN+DELZ*(NNEG+1)       !max value for neg cont
         CALL CONT2(Z,MX,MY,ZMIN,ZERO,NNEG)
C
C     Z values are all negative
      ELSE IF ((ZMAX .LE. 0.) .AND. (ZMIN .LT. 0)) THEN
         ILINE=4
         CALL CONT2(Z,MX,MY,ZMIN,ZMAX,NCONT)
      END IF
C
      RETURN
      END
CCCCCCCCCCCCCCCCCCCCCCCCCCCCCCCCCCCCCCCCCCCCCCCCCCCCCCCCCCCCCCCCCCCCCCCCCC
      SUBROUTINE CONT2(Z,MX,MY,ZMIN,ZMAX,NCONT)
C drawn NCONT contours between ZMIN and ZMAX
CCCCCCCCCCCCCCCCCCCCCCCCCCCCCCCCCCCCCCCCCCCCCCCCCCCCCCCCCCCCCCCCCCCCCCCCCC
C Global variables:
      INCLUDE 'IO.ALL'
      INCLUDE 'UISGRF.ALL'
      INCLUDE 'GRFDAT.ALL'
C Passed variables:
      INTEGER MX,MY            !dimensions of Z
      REAL Z(MX,MY)            !data
      REAL ZMIN,ZMAX           !limits on data
      INTEGER NCONT            !number of contour lines
C Local variables:
      REAL DELX,DELY,DELZ      !X,Y and Z intervals
      INTEGER IX,IY            !X and Y indices
      REAL ZLL,ZUR,ZUL,ZLR     !Z values on corners of a square
      REAL LL,UR,UL,LR         !differences: ZLVEV-ZLL, etc.
      REAL X,Y                 !X,Y values
      REAL ZLEV                !value of Z on this contour
      INTEGER ICONT            !index of contour
      INTEGER NCUT             !number of cuts of ZLEV on square
      REAL XX(4),YY(4)         !location of cuts on square
```

```
CCCCCCCCCCCCCCCCCCCCCCCCCCCCCCCCCCCCCCCCCCCCCCCCCCCCCCCCCCCCCCCCCCCCCCCC
C     draw box
      CALL UIS$SET_LINE_STYLE(TRID(IPLOT),0,2,'FFFFFFFF'X)
      CALL UIS$PLOT(TRID(IPLOT),2,XMIN,YMAX,XMAX,YMAX)
      CALL UIS$PLOT(TRID(IPLOT),2,XMAX,YMAX,XMAX,YMIN)
C     set line type
      IF (ILINE .EQ. 1) CALL
     +    UIS$SET_LINE_STYLE(TRID(IPLOT),0,2,'FFFFFFFF'X)
      IF (ILINE .EQ. 2) CALL
     +    UIS$SET_LINE_STYLE(TRID(IPLOT),0,2,'FF00FF00'X)
      IF (ILINE .EQ. 3) CALL
     +    UIS$SET_LINE_STYLE(TRID(IPLOT),0,2,'FFF0FFF0'X)
      IF (ILINE .EQ. 4) CALL
     +    UIS$SET_LINE_STYLE(TRID(IPLOT),0,2,'F0F0F0F0'X)
C
      DELX=(XMAX-XMIN)/(MX-1)          !step sizes
      DELY=(YMAX-YMIN)/(MY-1)
      DELZ=(ZMAX-ZMIN)/(NCONT+1)
C
      DO 100 IX=1,MX-1                 !loop over X values
         X=XMIN+DELX*IX
         DO 200 IY=1,MY-1             !loop over Y values
            Y=YMIN+DELY*IY
C
            ZLL=Z(IX,IY)             !value of Z on lower left
            ZUL=Z(IX,IY+1)          !value of Z on upper left
            ZUR=Z(IX+1,IY+1)        !on upper right
            ZLR=Z(IX+1,IY)          !on lower right
C
            DO 300 ICONT=1,NCONT     !loop over contour levels
               ZLEV=ZMIN+DELZ*(ICONT) !level value
               LL=ZLEV-ZLL            !differences between
               UL=ZLEV-ZUL           !corner values and ZLEV
               UR=ZLEV-ZUR
               LR=ZLEV-ZLR
C
C For each of the four sides, determine if the contour line cuts that
C side; if so find the location of the cut by linear interpolation.
C Note that the contour line can cut only 0, 2, or 4 of the 4 sides
               NCUT=0
               IF (LL*UL .LE. 0) THEN
                  NCUT=NCUT+1
                  XX(NCUT)=X-DELX
                  YY(NCUT)=Y-DELY*UL/(UL-LL)
               END IF
               IF (UL*UR .LE. 0) THEN
                  NCUT=NCUT+1
                  XX(NCUT)=X-DELX*UR/(UR-UL)
                  YY(NCUT)=Y
               END IF
               IF (UR*LR .LE. 0) THEN
                  NCUT=NCUT+1
```

```
                XX(NCUT)=X
                YY(NCUT)=Y-DELY*UR/(UR-LR)
             END IF
             IF (LR*LL .LE. 0) THEN
                NCUT=NCUT+1
                XX(NCUT)=X-DELX*LR/(LR-LL)
                YY(NCUT)=Y-DELY
             END IF
C
             IF (NCUT .EQ. 0) THEN
                CONTINUE                      !do nothing
             ELSE IF (NCUT .EQ. 2) THEN
                !connect the cut
                CALL UIS$PLOT(TRID(IPLOT),2,XX(1),YY(1),XX(2),YY(2))
             ELSE IF (NCUT .EQ. 4) THEN
                !connect cuts; there is arbitrariness here
                CALL UIS$PLOT(TRID(IPLOT),2,XX(1),YY(1),XX(2),YY(2))
                CALL UIS$PLOT(TRID(IPLOT),2,XX(3),YY(3),XX(4),YY(4))
             END IF
300          CONTINUE    !end loop over contours
200       CONTINUE       !end loop over Y
100    CONTINUE          !end loop over Z
C
       RETURN
       END
CCCCCCCCCCCCCCCCCCCCCCCCCCCCCCCCCCCCCCCCCCCCCCCCCCCCCCCCCCCCCCCCCCCCCCCCC
       SUBROUTINE GPAGE(DEVICE)
C close graphics packages
CCCCCCCCCCCCCCCCCCCCCCCCCCCCCCCCCCCCCCCCCCCCCCCCCCCCCCCCCCCCCCCCCCCCCCCCC
C Global variables:
       INCLUDE 'IO.ALL'
       INCLUDE 'UISGRF.ALL'
       INCLUDE 'GRFDAT.ALL'
       REAL X0,X1,X2              !temp values for x0val,xmin,xmax
       REAL Y0,Y1,Y2              !temp values for y0val,ymin,ymax
       COMMON/LOGLMT/X0,X1,X2,Y0,Y1,Y2
C Passed variables:
       INTEGER DEVICE             !which device is it?
C Local variables:
       REAL XTXT,YTXT             !location of text
       REAL SMALL                 !small increments along either axis
       CHARACTER*1 DUMMY          !dummy input
       INTEGER I                  !plot index
       INTEGER SCREEN             !send to terminal
       INTEGER PAPER              !make a hardcopy
       INTEGER FILE               !send to a file
       CHARACTER*8 CTIME          !time as character data
       CHARACTER*12 FNAME         !file name
       DATA SMALL/.01/
       DATA SCREEN,PAPER,FILE/1,2,3/
CCCCCCCCCCCCCCCCCCCCCCCCCCCCCCCCCCCCCCCCCCCCCCCCCCCCCCCCCCCCCCCCCCCCCCCCC
       IF (DEVICE .EQ. SCREEN) THEN
```

```
C       allow for inspection of graph before deleting
C       type informational message in lower right
        YTXT=Y1-5*SMALL*YLNGTH
        XTXT=X1+2*XLNGTH/3
        CALL UIS$TEXT(TRID(IPLOT),1,'Press return to continue',
     +                XTXT,YTXT)
        READ (IUNIT,20) DUMMY
20      FORMAT (A1)
C
        ELSE IF (DEVICE .EQ. PAPER) THEN
C       get file name, of the form CPtime.UIS, where time=hhmmss
C       this convention allows many plots, each with a unique name
        CALL TIME(CTIME)
        FNAME='CP'//CTIME(1:2)//CTIME(4:5)//CTIME(7:8)//'.UIS'
C       write data to file, disable display list
        CALL HCUIS$WRITE_DISPLAY(VDID,FNAME)
        CALL UIS$DISABLE_DISPLAY_LIST(VDID)
        END IF
C
C       delete the display
        CALL UIS$DELETE_DISPLAY(VDID)
C
        RETURN
        END
CCCCCCCCCCCCCCCCCCCCCCCCCCCCCCCCCCCCCCCCCCCCCCCCCCCCCCCCCCCCCCCCCCCC
        SUBROUTINE GTDEV(DEVICE)
C sets device for graphics output to DEVICE;
C for this graphics package, this work is actually done in SUBPLOT
CCCCCCCCCCCCCCCCCCCCCCCCCCCCCCCCCCCCCCCCCCCCCCCCCCCCCCCCCCCCCCCCCCCC
C Global variables:
        INCLUDE 'IO.ALL'
C passed variables:
        INTEGER DEVICE     !graphics to be sent to term or hard copy device
C local variables:
        INTEGER DEV                !DEVICE in disguise
        COMMON / DEVINFO/ DEV      !pass variable to SUBPLT
CCCCCCCCCCCCCCCCCCCCCCCCCCCCCCCCCCCCCCCCCCCCCCCCCCCCCCCCCCCCCCCCCCCC
        DEV=DEVICE
        RETURN
        END
CCCCCCCCCCCCCCCCCCCCCCCCCCCCCCCCCCCCCCCCCCCCCCCCCCCCCCCCCCCCCCCCCCCC
C The following routines are not called directly by the physics programs
CCCCCCCCCCCCCCCCCCCCCCCCCCCCCCCCCCCCCCCCCCCCCCCCCCCCCCCCCCCCCCCCCCCC
        SUBROUTINE SUBPLT
C defines subplotting area
C does all jobs which depend on the size of the plot
CCCCCCCCCCCCCCCCCCCCCCCCCCCCCCCCCCCCCCCCCCCCCCCCCCCCCCCCCCCCCCCCCCCC
C The choice of subplt size and placement is dependent on the
C device and graphics package; those given here are for an 30 x 20 cm
C screen using UIS routines
CCCCCCCCCCCCCCCCCCCCCCCCCCCCCCCCCCCCCCCCCCCCCCCCCCCCCCCCCCCCCCCCCCCC
C Global variables:
```

```
      INCLUDE 'GRFDAT.ALL'
      INCLUDE 'UISGRF.ALL'
      LOGICAL YLOG,XLOG           !flags to signal log plotting
      REAL X0,X1,X2               !temp values for x0val,xmin,xmax
      REAL Y0,Y1,Y2               !temp values for y0val,ymin,ymax
      INTEGER DEV    !graphics to be sent to term or hard copy device
      COMMON/LOGLMT/X0,X1,X2,Y0,Y1,Y2
      COMMON/LOGFLG/YLOG,XLOG
      COMMON / DEVINFO / DEV      !passed from GETDEV
C local variables:
      INTEGER LENTTL,LENTRU       !string lengths without blanks
      REAL DSXMIN,DSXMAX          !total x display size in user units
      REAL DSYMIN,DSYMAX          !total y display size in user units
      REAL XSMAL,YSMAL            !percentage of window in margin
      REAL XMARG,YMARG            !length of margin
      REAL WIDTH,HEIGHT           !character size
      INTEGER SCREEN              !send to terminal
      INTEGER PAPER               !make a hardcopy
      INTEGER FILE                !send to a file
      CHARACTER*60 FONTID         !type of font
      REAL WID_TO_LEN             !width to length ratio for title
      REAL XTXT                   !x location of title in centimeters
      INTEGER FIRST,YES,NO        !first time through?
      DATA SCREEN,PAPER,FILE/1,2,3/
      DATA FIRST,YES,NO/1,1,2/
CCCCCCCCCCCCCCCCCCCCCCCCCCCCCCCCCCCCCCCCCCCCCCCCCCCCCCCCCCCCCCCCCCCCC
C     length of axes
      XLNGTH=(X2-X1)
      YLNGTH=(Y2-Y1)
C     a reasonable margin
      IF ((NPLOT .EQ. 3) .OR. (NPLOT .EQ. 4)) THEN
         XSMAL=.11
         YSMAL=.11
      ELSE
         XSMAL=.08
         YSMAL=.08
      END IF
      XMARG=XSMAL*XLNGTH
      YMARG=YSMAL*YLNGTH
C     limits of virtual display in user coordinates
      DSXMIN=X1-XMARG
      DSXMAX=X2+XMARG
      DSYMIN=Y1-YMARG
      DSYMAX=Y2+2*YMARG
C
C     set size of window in centimeters
      XSIZE=30.
      YSIZE=25.
C
C     make corrections to some of the above specifications,
C     as a function of NPLOT and IPLOT
      IF (NPLOT .EQ. 1) THEN
```

```
              XSIZE=22.                  !make this window a little smaller
              YSIZE=27.
         ELSE IF (NPLOT .EQ. 2) THEN
              IF (IPLOT .EQ. 1) DSXMAX=X2+3*XMARG+XLNGTH
              IF (IPLOT .EQ. 2) DSXMIN=X1-3*XMARG-XLNGTH
         ELSE IF (NPLOT .EQ. 3) THEN
              IF (IPLOT .EQ. 1) DSYMIN=Y1-5*YMARG-2*YLNGTH
              IF (IPLOT .EQ. 2) THEN
                   DSYMIN=Y1-3*YMARG-YLNGTH
                   DSYMAX=Y2+4*YMARG+YLNGTH
              END IF
              IF (IPLOT .EQ. 3)   DSYMAX=Y2+6*YMARG+2*YLNGTH
         ELSE IF (NPLOT .EQ. 4)THEN
              IF (IPLOT .EQ. 1) THEN
                   DSXMAX=X2+3*XMARG+XLNGTH
                   DSYMIN=Y1-3*YMARG-YLNGTH
              ELSE IF (IPLOT .EQ. 2) THEN
                   DSXMIN=X1-3*XMARG-XLNGTH
                   DSYMIN=Y1-3*YMARG-YLNGTH
              ELSE IF (IPLOT .EQ. 3) THEN
                   DSXMAX=X2+3*XMARG+XLNGTH
                   DSYMAX=Y2+4*YMARG+YLNGTH
              ELSE IF (IPLOT .EQ. 4) THEN
                   DSXMIN=X1-3*XMARG-XLNGTH
                   DSYMAX=Y2+4*YMARG+YLNGTH
              END IF
         END IF
C
         IF (IPLOT .EQ. 1) THEN
C         create a virtual display
           VDID=UIS$CREATE_DISPLAY(DSXMIN,DSYMIN,DSXMAX,DSYMAX,
      +               XSIZE,YSIZE)
C
C         choose a device
           IF (DEV .EQ. SCREEN) THEN
             WDID=UIS$CREATE_WINDOW(VDID,
      +         'SYS$WORKSTATION','Computational Physics')
             CALL UIS$DISABLE_DISPLAY_LIST(VDID)
           ELSE IF (DEV .EQ. PAPER) THEN
             CALL UIS$ENABLE_DISPLAY_LIST(VDID)
           END IF
C
C         set character size according to window size
           IF (NPLOT .EQ. 3) THEN
             HEIGHT=.05*YLNGTH
             CALL UIS$SET_CHAR_SIZE(VDID,0,1,,,HEIGHT)
           ELSE IF (NPLOT .EQ. 4) THEN
             WIDTH=.017*XLNGTH
             CALL UIS$SET_CHAR_SIZE(VDID,0,1,,WIDTH)
           ELSE IF (NPLOT .EQ. 2) THEN
             WIDTH=.015*XLNGTH
             CALL UIS$SET_CHAR_SIZE(VDID,0,1,,WIDTH)
```

```
          ELSE
             CALL UIS$SET_CHAR_SIZE(VDID,0,1)
          END IF
C       allow for text which is rotated 90 degrees
          CALL UIS$SET_TEXT_SLOPE(VDID,1,3,90.0)
C
C       write out title
C       x scale is centimeters; yscale is user coordinates
          TRID(0)=UIS$CREATE_TRANSFORMATION(VDID,0.,DSYMIN,
      +               XSIZE,DSYMAX)
C       get information to center the title
          LENTTL=LENTRU(TITLE)
          IF (FIRST .EQ. YES) THEN
             CALL UIS$GET_FONT(VDID,0,FONTID)
             CALL UIS$GET_FONT_SIZE(FONTID,TITLE(1:LENTTL),WIDTH,HEIGHT)
             WID_TO_LEN=WIDTH/REAL(LENTTL)  !save width/length for other calls
             FIRST=NO
          END IF
          WIDTH=LENTTL*WID_TO_LEN
          XTXT=(XSIZE-WIDTH)/2          !width, height are in centimeters
C       write out title
          CALL UIS$TEXT(TRID(0),0,TITLE(1:LENTTL),XTXT,Y2+1.6*YMARG)
C
       END IF
C     create the transformation for each plot
       TRID(IPLOT)=UIS$CREATE_TRANSFORMATION(VDID,DSXMIN,DSYMIN,
      +               DSXMAX,DSYMAX)
C
       RETURN
       END
CCCCCCCCCCCCCCCCCCCCCCCCCCCCCCCCCCCCCCCCCCCCCCCCCCCCCCCCCCCCCCCCCCCCCCCC
       SUBROUTINE XAXIS
C draws and labels x (horizontal) axis
CCCCCCCCCCCCCCCCCCCCCCCCCCCCCCCCCCCCCCCCCCCCCCCCCCCCCCCCCCCCCCCCCCCCCCCC
C Global variables:
       INCLUDE 'UISGRF.ALL'
       INCLUDE 'GRFDAT.ALL'
       LOGICAL YLOG,XLOG               !flags to signal log plotting
       REAL X0,X1,X2                   !temp values for x0val,xmin,xmax
       REAL Y0,Y1,Y2                   !temp values for y0val,ymin,ymax
       COMMON/LOGLMT/X0,X1,X2,Y0,Y1,Y2
       COMMON/LOGFLG/YLOG,XLOG
C Local variables:
       REAL XTXT,YTXT                  !location of text
       INTEGER LENX,LENSTR,LENTRU      !string lengths without blanks
       REAL TCKMIN,TCKMAX              !y values at ends of tick marks
       REAL XTICK                      !x value at tick mark
       CHARACTER*9 STRING              !string value of xtick
       REAL SMALL                      !small increments along either axis
       INTEGER I                       !indexes tick marks
       DATA SMALL/.01/
CCCCCCCCCCCCCCCCCCCCCCCCCCCCCCCCCCCCCCCCCCCCCCCCCCCCCCCCCCCCCCCCCCCCCCCC
```

```
C         draw x axis
          CALL UIS$PLOT(TRID(IPLOT),0,X1,X0,X2,X0)
C
C         label x axis
          LENX=LENTRU(LABEL(1))
          XTXT=X2+5*SMALL*XLNGTH
          YTXT=X0+5*SMALL*YLNGTH
          CALL UIS$TEXT(TRID(IPLOT),3,LABEL(1)(1:LENX),XTXT,YTXT)
C
C         tick marks
C         set length of ticks
          TCKMIN=X0-SMALL*YLNGTH
          TCKMAX=X0
C
          DO 10 I=0,NXTICK
C            draw tick marks
             XTICK=X1+I*XLNGTH/NXTICK
             CALL UIS$PLOT(TRID(IPLOT),0,XTICK,TCKMIN,XTICK,TCKMAX)
C
C            label tick marks
             XTXT=XTICK
             IF (XLOG) XTICK=10**XTICK
             CALL CONVRT(XTICK,STRING,LENSTR)
             CALL UIS$TEXT(TRID(IPLOT),1,STRING(1:LENSTR),XTXT,TCKMIN)
10        CONTINUE
C
          RETURN
          END
CCCCCCCCCCCCCCCCCCCCCCCCCCCCCCCCCCCCCCCCCCCCCCCCCCCCCCCCCCCCCCCCCCCCCCC
          SUBROUTINE YAXIS
C draws and labels y (vertical) axis
CCCCCCCCCCCCCCCCCCCCCCCCCCCCCCCCCCCCCCCCCCCCCCCCCCCCCCCCCCCCCCCCCCCCCCC
C Global variables:
          INCLUDE 'UISGRF.ALL'
          INCLUDE 'GRFDAT.ALL'
          LOGICAL YLOG,XLOG                !flags to signal log plotting
          REAL X0,X1,X2                    !temp values for x0val,xmin,xmax
          REAL Y0,Y1,Y2                    !temp values for y0val,ymin,ymax
          COMMON/LOGLMT/X0,X1,X2,Y0,Y1,Y2
          COMMON/LOGFLG/YLOG,XLOG
C Local variables:
          REAL XTXT,YTXT                   !location of text
          INTEGER LENY,LENTRU,LENSTR       !string lengths without blanks
          REAL TCKMIN,TCKMAX               !x values at ends of tick marks
          REAL YTICK                       !y value at tick mark
          CHARACTER*9 STRING               !string value of xtick
          REAL SMALL                       !small increments along either axis
          INTEGER I                        !indexes tick marks
          DATA SMALL/.01/
CCCCCCCCCCCCCCCCCCCCCCCCCCCCCCCCCCCCCCCCCCCCCCCCCCCCCCCCCCCCCCCCCCCCCCC
C         draw y axis
          CALL UIS$PLOT(TRID(IPLOT),0,Y0,Y1,Y0,Y2)
```

```
C
C         label y axis
          LENY=LENTRU(LABEL(2))
          YTXT=Y2+5*SMALL*YLNGTH
          XTXT=Y0+4*SMALL*XLNGTH
          CALL UIS$TEXT(TRID(IPLOT),1,LABEL(2)(1:LENY),XTXT,YTXT)
C
C         tick marks
C         set tick length
          TCKMIN=Y0-SMALL*XLNGTH
          TCKMAX=Y0
C
          DO 20 I=0,NYTICK
C            draw marks
             YTICK=Y1+I*YLNGTH/NYTICK
             CALL UIS$PLOT(TRID(IPLOT),0,TCKMIN,YTICK,TCKMAX,YTICK)
C
C            label marks
             YTXT=YTICK-.02*YLNGTH
             IF (YLOG) YTICK=10**YTICK
             CALL CONVRT(YTICK,STRING,LENSTR)
             XTXT=Y0-.04*XLNGTH
             CALL UIS$TEXT(TRID(IPLOT),3,STRING(1:LENSTR),XTXT,YTXT)
20        CONTINUE
          RETURN
          END
CCCCCCCCCCCCCCCCCCCCCCCCCCCCCCCCCCCCCCCCCCCCCCCCCCCCCCCCCCCCCCCCCCCCCCCC
          SUBROUTINE GSYM(XCNTR,YCNTR)
C draws a symbol at XCNTR,YCNTR
CCCCCCCCCCCCCCCCCCCCCCCCCCCCCCCCCCCCCCCCCCCCCCCCCCCCCCCCCCCCCCCCCCCCCCCC
C Global variables
          INCLUDE 'UISGRF.ALL'
          INCLUDE 'GRFDAT.ALL'
C Passed variables:
          REAL XCNTR,YCNTR           !x,y coordinates of symbol center
C Local variables:
          REAL XSYM,YSYM             !symbol sizes in world coordinates
          REAL X1,X2,Y1,Y2           !edges of square, cross and triangle
CCCCCCCCCCCCCCCCCCCCCCCCCCCCCCCCCCCCCCCCCCCCCCCCCCCCCCCCCCCCCCCCCCCCCCCC
C         set symbol size
          XSYM=.005*XLNGTH
          YSYM=.005*YLNGTH
C
C         circle
          IF (ISYM .EQ. 1) THEN
              CALL UIS$CIRCLE(TRID(IPLOT),0,XCNTR,YCNTR,XSYM)
C
C         triangle
          ELSE IF (ISYM .EQ. 2) THEN
              X1=XCNTR-XSYM
              X2=XCNTR+XSYM
              Y1=YCNTR-YSYM
```

```
                Y2=YCNTR+YSYM
                CALL UIS$PLOT(TRID(IPLOT),0,X1,Y1,X2,Y1,XCNTR,Y2,X1,Y1)
C
C       square
        ELSE IF (ISYM .EQ. 3) THEN
                X1=XCNTR-XSYM
                X2=XCNTR+XSYM
                Y1=YCNTR-YSYM
                Y2=YCNTR+YSYM
                CALL UIS$PLOT(TRID(IPLOT),0,X1,Y1,X2,Y1,X2,Y2,X1,Y2,X1,Y1)
C
C       cross
        ELSE IF (ISYM .EQ. 4) THEN
                X1=XCNTR-XSYM
                X2=XCNTR+XSYM
                Y1=YCNTR-YSYM
                Y2=YCNTR+YSYM
                CALL UIS$PLOT(TRID(IPLOT),0,X1,Y1,X2,Y2)
                CALL UIS$PLOT(TRID(IPLOT),0,X1,Y2,X2,Y1)
C
C       point
        ELSE IF (ISYM .EQ. 5) THEN
                CALL UIS$CIRCLE(TRID(IPLOT),0,XCNTR,YCNTR,XSYM/2.)
        END IF
C
        RETURN
        END
CCCCCCCCCCCCCCCCCCCCCCCCCCCCCCCCCCCCCCCCCCCCCCCCCCCCCCCCCCCCCCCCCCCCCCCC
CCCCCCCCCCCCCCCCCCCCCCCCCCCCCCCCCCCCCCCCCCCCCCCCCCCCCCCCCCCCCCCCCCCCCCCC
C grfdat.all
C
        INTEGER NPOINT              !number of points to graph
        INTEGER ILINE               !code for line type
        INTEGER ISYM                !code for symbol type
        INTEGER IFREQ               !symbol frequency
        INTEGER NPLOT               !total number of plots on page
        INTEGER IPLOT               !number of current plot
        INTEGER NXTICK,NYTICK       !number of tick marks
C
        CHARACTER*60 LABEL(2)       !xaxis and yaxis labels
        CHARACTER*60 TITLE          !title for graphics page
        CHARACTER*60 INFO           !informational message
C
        REAL XMIN,XMAX              !limits on independent variable
        REAL YMIN,YMAX              !limits on dependent variable
        REAL X0VAL                  !y value along x axis
        REAL Y0VAL                  !x value along y axis
C
        COMMON/GSTYLE/NPOINT,ILINE,ISYM,IFREQ,NPLOT,IPLOT,
     +                NXTICK,NYTICK
        COMMON/LABELS/TITLE,LABEL,INFO
        COMMON/XYAXES/XMIN,XMAX,YMIN,YMAX,X0VAL,Y0VAL
```

```
ccccccccccccccccccccccccccccccccccccccccccccccccccccccccccccccccccccc
ccccccccccccccccccccccccccccccccccccccccccccccccccccccccccccccccccccc
C uisgrf.all  (for GRAPHIT.LO only)
      INCLUDE 'SYS$LIBRARY:UISENTRY'      !these are part of VWS
      INCLUDE 'SYS$LIBRARY:UISUSRDEF'     !and only available on a VAX
      INTEGER MAXPLT                      !maximum number of graphs/page
      INTEGER VDID,WDID                   !id's for display and window
      INTEGER TRID                        !id's for transformations
      REAL XSIZE,YSIZE                    !size of plotting area in centimeters
      REAL XLNGTH,YLNGTH                  !axes length in user units
      PARAMETER (MAXPLT=4)
C
      COMMON/DSPLAY/XLNGTH,YLNGTH,XSIZE,YSIZE
      COMMON/ID/VDID,WDID,TRID(0:MAXPLT)
```

Appendix E

Network
File Transfer

The ubiquity of network connections has made network file transfer a practical option for code distribution. To transfer files you need ftp (file transfer protocol) software and a connection to the Internet network (but not necessarily a direct connection). Check with your system manager if you are unsure of the availablity of either.

The files are available on the Internet node physicsftp.unh.edu in the directory pub/compphys. There are three subdirectories: /fortran contains all of the FORTRAN codes listed in this book, /basic contains all of the IBM-BASIC codes, and /graphics contains subdirectories with additional FORTRAN interfaces to graphics packages (i.e., interfaces other than graphit.hi and graphit.lo). (At the time of writing, only a GKS interface is available.)

E.1 Sample Session

Below is a sample session that transfers all of the files ending with the .all extension. Phrases that you type are in boldface, our comments are in italics, and computer responses are in plain type.

The host computer is a UNIX environment. For those not familiar with UNIX, you need only know the following commands: ls -l to list files in a directory, cd to change directories, and quit to logoff. The ftp commands are as follows: prompt to toggle the prompting for file transfer (the default is that it will verify each file before sending), and mget to get files from the remote host. mget must be followed by the name of the file to be sent; wild cards (e.g., *.* or *.all) are allowed. UNIX is case sensitive, so names must match both letter and case.

Note that some features of the host computer may change (e.g., the name of the computer, other subdirectories), but every attempt will be made to keep the Internet address and directories associated with *Computational Physics* the same. Also, some of the details (e.g., prompts) depend on the specific implementation of ftp on your machine; consult a local expert if you run into difficulties.

```
>  ftp physicsftp.unh.edu        ftp from your machine to ours
Connected to mozz.unh.edu.
220 mozz.unh.edu FTP server (ULTRIX Version 4.1 Tue Mar 19 00:38:17 EST
1991) ready.
Name (ftphost:): anonymous        login to the anonymous account
                on other ftp implementations you must type 'login anonymous'
331 Guest login ok, send ident as password.
Password:                         type in anything for the password
230 Guest login ok, access restrictions apply.
ftp> ls -l                        look at contents of directory
<Opening data connection for /bin/ls (132.177.128.8,3763) (0 bytes).
total 3
dr-xr-xr-x   2 0          512 May 24 10:26 bin
dr-xr-xr-x   2 0          512 May 24 10:26 etc
dr-xr-xr-x   4 0          512 May 24 10:26 pub
<Transfer complete.
                The 'd' at the beginning of the line indicates a directory;
                the name of the directory or file is at the end of the line.
ftp> cd pub                       change to the pub directory
<CWD command successful.
ftp> ls -l
<Opening data connection for /bin/ls (132.177.128.8,3764) (0 bytes).
total 17
drwxr-xr-x   5 0          512 May 24 10:26 compphys
drwxr-xr-x   5 0          512 May 24 10:26 NeXT
-rw-r--r--   1 0        14754 May 24 10:26 ncp.defines
<Transfer complete.
ftp> cd compphys
<CWD command successful.
ftp> ls -l
<Opening data connection for /bin/ls (132.177.128.8,3767) (0 bytes).
total 4
drwxr-xr-x   2 0          512 May 24 10:26 basic
drwxr-xr-x   2 0         1536 Jun 12 11:33 fortran
drwxr-xr-x   2 0          512 May 24 10:26 graphics
<Transfer complete.
ftp> cd fortran                   change to the fortran directory
```

```
250 CWD command successful.
ftp>  ls *.all                    show all files ending in .all
200 PORT command successful.
150 Opening data connection for /bin/ls (132.177.128.23,1790) (0 bytes).
grfdat.all
io.all
menu.all
uisgrf.all
226 Transfer complete.
538 bytes received in 0.27 seconds (2 Kbytes/s)
ftp>  prompt                      turn off the prompt
Interactive mode off.
ftp>  mget *.all                  get everything with the .all extension
200 PORT command successful.
150 Opening data connection for grfdat.all (132.177.128.23,1796) (1050
bytes).
226 Transfer complete.
local: grfdat.all remote: grfdat.all
1073 bytes received in 0.0039 seconds (2.7e+02 Kbytes/s)
200 PORT command successful.
150 Opening data connection for io.all (132.177.128.23,1797) (1528
bytes).
226 Transfer complete.
local: io.all remote: io.all
1560 bytes received in 0.0078 seconds (2e+02 Kbytes/s)
200 PORT command successful.
150 Opening data connection for menu.all (132.177.128.23,1798) (2455
bytes).
226 Transfer complete.
local: menu.all remote: menu.all
2504 bytes received in 0.0078 seconds (3.1e+02 Kbytes/s)
200 PORT command successful.
150 Opening data connection for uisgrf.all (132.177.128.23,1799) (625
bytes).
226 Transfer complete.
local: uisgrf.all remote: uisgrf.all
637 bytes received in 0.0039 seconds (1.6e+02 Kbytes/s)
ftp>  quit                        logoff remote host
221 Goodbye.
>                                 you're back to your own machine
```

References

[Ab64] *Handbook of Mathematical Functions*, eds. M. Abramowitz and I. A. Stegun (Dover, New York, 1964).

[Ab78] R. Abraham and J. E. Marsden, *Foundations of Mechanics, Second Edition* (Benjamin-Cummings Publishing Corp., Reading, 1978) Chapter 8.

[Ac70] F. S. Acton, *Numerical Methods that Work* (Harper and Row, New York, 1970).

[Ad84] S. L. Adler and T. Piran, Rev. Mod. Phys. **56**, 1 (1984).

[Ar68] V. I. Arnold and A. Avez, *Ergodic Problems in Classical Mechanics* (Benjamin Publishing Corp., New York, 1968).

[Ba64] A. Baker, Phys. Rev. **134**, B240 (1964).

[Be68] H. A. Bethe and R. W. Jackiw, *Intermediate Quantum Mechanics* (Benjamin Publishing Corp., New York, 1968).

[Be69] P. R. Bevington, *Data Reduction and Error Analysis for the Physical Sciences* (McGraw-Hill, New York, 1969).

[Bo74] R. A. Bonham and M. Fink, *High Energy Electron Scattering* (Van Nostrand-Reinhold, New York, 1974).

[Bo76] J. A. Boa and D. S. Cohen, SIAM J. Appl. Math., **30**, 123 (1976).

[Bu81] R. L. Burden, J. D. Faires, and A. C. Reynolds, *Numerical Analysis, Second Edition* (Prindle, Weber, and Schmidt, Boston, 1981).

[Br68] D. M. Brink and G. R. Satchler, *Angular Momentum, Second edition* (Clarendon Press, Oxford, 19680.

[Ca80] J. M. Cavedon, Thesis, Université de Paris-Sud, 1980.

[Ce79]   D. M. Ceperley and M. H. Kalos in *Monte Carlo Methods in Statistical Physics*, ed. K. Binder (Springer-Verlag, New York/Berlin, 1979).

[Ce80]   D. M. Ceperley and B. J. Alder, Phys. Rev. Lett. **45**, 566 (1980).

[Ch57]   S. Chandrasekhar, *An Introduction to the Study of Stellar Structure* (Dover, New York, 1957).

[Ch84]   S. Chandrasekhar, Rev. Mod. Phys. **56**, 137 (1984).

[Ch84a] S. A. Chin, J. W. Negele, and S. E. Koonin, Ann. Phys, **157**, 140 (1984).

[Fl78]   H. Flocard, S. E. Koonin, and M. S. Weiss, Phys. Rev. C **17**, 1682 (1978).

[Fo63]   L. D. Fosdick in *Methods in Computational Physics, vol. 1*, ed. B. Alder *et al.*, p. 245 (Academic Press, New York, 1963).

[Fo66]   T. deForest, Jr. and J. D. Walecka, Adv. Phys. **15**, 1 (1966).

[Fr73]   J. L. Friar and J. W. Negele, Nucl. Phys. **A212**, 93 (1973).

[Fr75]   J. L. Friar and J. W. Negele, Adv. Nucl. Phys. **8**, 219 (1975).

[Fr77]   B. Frois *et al.*, Phys. Rev. Lett. **38**, 152 (1977).

[Go49]   P. Gombas, *Die Statische Theorie des Atoms* (Springer, Vienna, 1949).

[Go67]   A. Goldberg, H.M. Schey, and J. L. Schwartz, Am. J. Phys. **35**, 177 (1967).

[Go80]   H. Goldstein, *Classical Mechanics, Second Edition* (Addison-Wesley Publishing Company, Reading, 1980).

[Ha64]   J. M. Hammersley and D. C. Handscomb, *The Monte Carlo Method* (Methuen, London, 1964).

[He50]   G. Herzberg, *Spectra of Diatomic Molecules* (D. Van Nostrand Company, Inc., New York, 1950).

[He64]   M. Hénon and C. Heiles, Astron. J. **69**, 73 (1964).

[He80] R. H. G. Helleman in *Fundamental Problems in Statistical Mechanics, vol. 5*, ed. E. G. D. Cohen (North Holland Publishing, Amsterdam, 1980) pp. 165–233.

[He82] M. Hénon, Physica **5D**, 412 (1982).

[Ho57] R. Hofstadter, Ann. Rev. Nuc. Sci., **7**, 231 (1957).

[Hu63] K. Huang, *Statistical Mechanics* (John Wiley and Sons, New York, 1963).

[Ka79] K. K. Kan, J. J. Griffin, P. C. Lichtner, and M. Dworzecka, Nucl. Phys. **A332**, 109 (1979).

[Ka81] M. H. Kalos, M. A. Lee, P. A. Whitlock. and G. V. Chester, Phys. Rev. B **24**, 115 (1981).

[Ka85] M. H. Kalos and P. A. Whitlock, *The Basics of Monte Carlo Methods* (J. Wiley and Sons, New York, in press).

[Ke78] B. W. Kernighan and P. J. Plauger, *The Elements of Programming Style, Second Edition* (McGraw-Hill Book Company, New York, 1978).

[Ki71] L. J. Kieffer, At. Data, **2**, 293 (1971).

[Kn69] D. Knuth *The Art of Computer Programming, Volume 2: Seminumerical Algorithms* (Addison-Wesley Publishing Company, Reading, 1969).

[Ko87] E. B. Koffman and F. L. Friedman, *Problem Solving and Structured Programming in FORTRAN 77* (Addison-Wesley Publishing Company, Inc., Reading, 1987).

[La59] L. D. Landau and E. M. Lifshitz, *Course of Theoretical Physics, Volume 6, Fluid Mechanics* (Pergamon Press, Oxford, 1959).

[Li65] H. Lipkin, N. Meshkov, and A. J. Glick, Nucl. Phys. **62**, 188 (1965), and the two papers following.

[Mc73] B. McCoy and T. T. Wu, *The Two-Dimensional Ising Model* (Harvard University Press, Cambridge, 1973).

[Mc80] J. B. McGrory and B. H. Wildenthal, Ann. Rev. Nuc. Sci., **30**, 383 (1980).

[Me53] N. Metropolis, A. Rosenbluth, M. Rosenbluth, A. Teller, and E. Teller, J. Chem. Phys. **21**, 1087 (1953).

[Me68] A. Messiah, *Quantum Mechanics* (J. Wiley & Sons, Inc., New York, 1968)

[Ne66] R. G. Newton, *Scattering Theory of Waves and Particles* (McGraw-Hill Book Co., New York, 1966).

[Ni77] G. Nicolis and I. Prigogine, *Self-organization in Nonequilibrium Systems* (J. Wiley and Sons, New York, 1977), Chapter 7.

[Pr86] W. H. Press, B. P. Flannery, S. A. Teukolsky, and W. T. Vetterling, *Numerical Recipes* (Cambridge University Press, Cambridge, 1986).

[Re82] P. J. Reynolds, D. M. Ceperley, B. J. Alder, and W. A. Lester, Jr., J. Chem. Phys. **77**, 5593 (1982).

[Ri67] R. D. Richtmeyer and K. W. Morton, *Difference Methods for Initial-value Problems, Second Edition* (Interscience, New York, 1967).

[Ri80] S. A. Rice in *Quantum Dynamics of Molecules (1980)*, ed. R. G. Wolley (Plenum Publishing Corp., New York, 1980) pp. 257–356.

[Ro76] P. J. Roache, *Computational Fluid Dynamics* (Hermosa, Albuquerque, 1976).

[Ru63] H. Rutishauser, Comm. of the ACM, **6**, 67, Algorithm 150 (February, 1963).

[Sh80] R. Shankar, Phys. Rev. Lett. **45**, 1088 (1980).

[Sh83] S. L. Shapiro and S. A. Teukolsky, *Black Holes, White Dwarfs, and Neutron Stars* (J. Wiley & Sons, Inc., New York, 1983) Chapter 3.

[Sh84] T. E. Shoup, *Applied Numerical Methods for the Microcomputer* (Prentice Hall, Inc., Englewood Cliffs, 1984).

[Si74] I. Sick, Nucl. Phys. **A218**, 509 (1974).

[Va62] R. Varga, *Matrix Iterative Analysis* (Prentice-Hall, Englewood Cliffs, 1962).

[Wa66] E. L. Wachspress, *Iterative Solution of Elliptic Systems and Applications to the Neutron Diffusion Equations of Reactor Physics* (Prentice-Hall, Englewood Cliffs, 1966).

[Wa67] T. G. Waech and R. B. Bernstein, J. Chem. Phys. **46**, 4905 (1967).

[Wa73] S. J. Wallace, Ann. Phys. **78**, 190 (1973); Phys. Rev. D8, 1846 (1973); Phys. Rev. D9, 406 (1974).

[We71] R. C. Weast, *Handbook of Chemistry and Physics, 52nd edition*, (The Chemical Rubber Company, Cleveland, 1971).

[We80] M. Weissbluth, *Atoms and Molecules* (Academic Press, New York, 1980).

[Wh77] R. R. Whitehead, A. Watt, B. J. Cole, and I. Morrison, Adv. Nucl. Phys. **9**, 123 (1977).

[Wi74] A. T. Winfree, Sci. Amer. **220**, 82 (June, 1974).

[Wu62] T.-Y. Wu and T. Ohmura, *Quantum Theory of Scattering* (Prentice-Hall, Englewood Cliffs, 1962).

[Ya82] L.G. Yaffe, Reviews of Modern Physics, **54**, 407 (1982).

[Ye54] D. R. Yennie, D. G. Ravenhall, and D. N. Wilson, Phys. Rev. **95**, 500 (1954).

Index

Printed in the United States
by Baker & Taylor Publisher Services